Lecture Notes in Computer Science 2858

Edited by G. Goos, J. Hartmanis, and J. van Leeuwen

Springer

Berlin
Heidelberg
New York
Hong Kong
London
Milan
Paris
Tokyo

Alex Veidenbaum   Kazuki Joe
Hideharu Amano   Hideo Aiso (Eds.)

High Performance Computing

5th International Symposium, ISHPC 2003
Tokyo-Odaiba, Japan, October 20-22, 2003
Proceedings

Springer

Volume Editors

Alex Veidenbaum
University of California, Information and Computer Science
444 Computer Science, Building 302, Irvine, CA 92697-3425, USA
E-mail: alexv@ics.uci.edu

Kazuki Joe
Nara Women's University, Department of Information and Computer Sciences
Kita-Uoya Nishimachi, Nara 630-8506, Japan
E-mail: joe@ics.nara-wu.ac.jp

Hideharu Amano
Keio University, Department of Information and Computer Science
3-14-1 Hiyoshi, Yokohama 223-8522, Japan
E-mail: hunga@am.ics.keio.ac.jp

Hideo Aiso
Tokyo University of Technology
1404-1 Katakura, Hachioji, Tokyo 192-0982, Japan
E-mail: aiso@media.teu.ac.jp

Cataloging-in-Publication Data applied for

A catalog record for this book is available from the Library of Congress

Bibliographic information published by Die Deutsche Bibliothek
Die Deutsche Bibliothek lists this publication in the Deutsche Nationalbibliographie;
detailed bibliographic data is available in the Internet at <http://dnb.ddb.de>.

CR Subject Classification (1998): D.1, D.2, F.2, E.4, G.1-4, J.1-2, J.6, I.6

ISSN 0302-9743
ISBN 3-540-20359-1 Springer-Verlag Berlin Heidelberg New York

Springer-Verlag Berlin Heidelberg New York
a member of BertelsmannSpringer Science+Business Media GmbH

http://www.springer.de

© Springer-Verlag Berlin Heidelberg 2003
Printed in Germany

Typesetting: Camera-ready by author, data conversion by Steingräber Satztechnik GmbH, Heidelberg
Printed on acid-free paper   SPIN: 10963302   06/3142   5 4 3 2 1 0

Foreword

I wish to welcome all of you to the proceedings of International Symposium on High Performance Computing 2003 (ISHPC–V), an event we were delighted to hold in Odaiba, a new city area of Tokyo. ISHPC 2003 was the fifth in the ISHPC series, ISHPC 1997 (Fukuoka, November 1997), ISHPC 1999 (Kyoto, May 1999), ISHPC 2000 (Tokyo, October 2000) and ISHPC 2002 (Kyoto, May 2002). The success of these symposia proves the importance of this area and indicates the strong interest of the research community. I am very pleased to serve as General Chair at a time when high–performance computing (HPC) plays a crucial role in the era of the "IT (Information Technology)" revolution.

The objective of this symposium was to provide a forum for the discussion of all aspects of HPC (from system architecture to real applications) in a more informal and friendly atmosphere. This symposium included excellent invited talks and workshops as well as high–quality technical papers. We hope that all the participants enjoyed not only the symposium but also the stay in Odaiba.

This symposium would not have been possible without the significant help of many people who devoted tremendous time and efforts. I thank all of those who worked diligently to make ISHPC 2003 a great success. In particular I would like to thank the Organizing Chair, Hideharu Amano of Keio University, and the Organizing Committee members for their great contribution to the planning and organization of ISHPC 2003. I would also like to thank the Program Chair, Alex Veidenbaum of UCI, the Program Co-chairs, Nicholas Carter of UIUC (architecture track), Jesus Labarta of UPC (software track), Yutaka Akiyama of CBRC (applications track), and the program committee members for their contribution to a technically excellent symposium program. Thanks are due to the Workshop Chairs, Mitsuhisa Sato of the University of Tsukuba and Eduard Ayguadé of UPC for organizing the International Workshop on OpenMP: Experiences and Implementations (WOMPEI 2003) as usual.

The ITBL special session was proposed and organized by Masahiro Fukuda of NAL. His contribution was very timely, as everybody is now looking at Grid Computing, and this made the symposium very practical. A last note of thanks goes to the people who were involved with the ITBL project.

October 2003

Hideo Aiso
General Chair

Preface

The 5th International Symposium on High Performance Computing (ISHPC–V) was held in Odaiba, Tokyo, Japan, October 20–22, 2003. The symposium was thoughtfully planned, organized, and supported by the ISHPC Organizing Committee and its collaborating organizations.

The ISHPC-V program included two keynote speeches, several invited talks, two panel discussions, and technical sessions covering theoretical and applied research topics in high–performance computing and representing both academia and industry. One of the regular sessions highlighted the research results of the ITBL project (IT–based research laboratory, http://www.itbl.riken.go.jp/). ITBL is a Japanese national project started in 2001 with the objective of realizing a virtual joint research environment using information technology. ITBL aims to connect 100 supercomputers located in main Japanese scientific research laboratories via high–speed networks.

A total of 58 technical contributions from 11 countries were submitted to ISHPC-V. Each paper received at least three peer reviews. After a thorough evaluation process, the program committee selected 14 regular (12-page) papers for presentation at the symposium. In addition, several other papers with favorable reviews were recommended for a poster session presentation. They are also included in the proceedings as short (8-page) papers.

The program committee gave a distinguished paper award and a best student paper award to two of the regular papers. The distinguished paper award was given for "Code and Data Transformations for Improving Shared Cache Performance on SMT Processors" by Dimitrios S. Nikolopoulos. The best student paper award was given for "Improving Memory Latency Aware Fetch Policies for SMT Processors" by Francisco J. Cazorla.

The third International Workshop on OpenMP: Experiences and Implementations (WOMPEI 2003) was held in conjunction with ISHPC–V. It was organized by Mitsuhisa Sato of University of Tsukuba, and Eduard Ayguadé of UPC. The ISHPC–V program committee decided to include the WOMPEI 2003 papers in the symposium proceedings.

We hope that the final program was of significant interest to the participants and served as a launching pad for interaction and debate on technical issues among the attendees. Last but not least, we thank the members of the program committee and the referees for all the hard work that made this meeting possible.

October 2003   Alex Veidenbaum

Organization

ISHPC–V Executive Committee

General Chair
Program Chair
Program Co-Chair

Organizing Chair
Publication and Treasury Chair
Local Arrangements Chair
Social Arrengements Chair
Workshop Chair

Hideo Aiso (Tokyo U. of Technology)
Alex Veidenbaum (UCI)
Nicholas Carter (UIUC)
Jesus Labarta (UPC)
Yutaka Akiyama (CBRC)
Hideharu Amano (Keio U.)
Kazuki Joe (NWU)
Hironori Nakajo (TUAT)
Mitaro Namiki (TUAT)
Mitsuhisa Sato (U. Tsukuba)
Eduard Ayguadé (UPC)

ISHPC–V Program Committee

Hideharu Amano (Keio U.)
Taisuke Boku (U. Tsukuba)
John Granacki (ISI/USC)
Sam Midkiff (Purdue U.)
Hironori Nakajo (TUAT)
Hitoshi Oi (Temple U.)
Constantine Polychronopoulos (UIUC)
Hong Shen (JAIST)
Rudolf Eigenmann (Purdue U.)
Mario Furnari (CNR)
Gabrielle Jost (NASA)
Hironori Kasahara (Waseda U.)
Yoshitoshi Kunieda (Wakayama U.)
Steve Lumetta (UIUC)
Dimitris Nikolopoulos (CWM)
Thierry Priol (INRIA)
Hamid R. Arabnia (U. Georgia)
Stratis Gallopoulos (U. Patras)
Takashi Nakamura (NAL)
Naoyuki Nide (NWU)
Noriyuki Fujimoto (Osaka U.)
Mariko Sasakura (Okayama U.)
Tomo Hiroyasu (Doshisha U.)
Shoichi Saito (Wakayama U.)

Ramon Beivide (U. Cantabria)
Matt Frank (UIUC)
Alex Nicolau (UCI)
Trevor Mudge (U. Michigan)
Hiroshi Nakashima (TUT)
Alex Orailoglu (UCSD)
Scott Rixner (Rice U.)
Mateo Valero (UPC)
Jose Castaños (IBM)
Dennis Gannon (U. Indiana)
Manolis Katevenis (FORTH/U. Crete)
Uli Kremer (Rutgers U.)
Ulrich Lang (U. Stuttgart)
Allen Malony (U. Oregon)
Leftaris Polychronopoulos (U. Patras)
Mitsuhisa Sato (U. Tsukuba)
Mitsuo Yokokawa (GTRC)
Kyle Gallivan (FSU)
Ophir Frieder (IIT)
Umpei Nagashima (AIST)
Hayaru Shouno (Yamaguchi U.)
Makoto Ando (JFE)
Atsushi Kubota (Hiroshima C.U.)
Kento Aida (TIT)

ISHPC–V Organizing Committee

Kazuki Joe (NWU)
Hironori Nakajo (TUAT)
Shin-ya Watanabe (CBRC)
Mitsuhisa Sato (U. Tsukuba)
Masahiro Fukuda (NAL)
Hitoshi Oi (Temple U.)

Chiemi Watanabe (NWU)
Mitaro Namiki (TUAT)
Chika Tanaka (CBRC)
Eduard Ayguadé (UPC)
Naoki Hirose (NAL)

Referees

Hiroyuki Abe
Manuel E. Acacio
Makoto Ando
Gabriel Antoniu
Ayon Basumallik
Christos D. Antonopoulos
Jose Angel Gregorio
Kazuki Joe
Yvon Jou
Keiji Kimura

Carmen Martinez
Hiroshi Nakamura
Zhelong Pan
Cristian Petrescu-Prahova
Oliverio J. Santana
Yasuhiro Suzuki
Jordi Torres
Ioannis E. Venetis

WOMPEI 2003 Organization

Workshop Chair   Mitsuhisa Sato (U. Tsukuba)
Workshop Chair   Eduard Ayguade (UPC)
Program Committee
    Barbara Chapman (U. Houston)
    Hironori Kasahara (Waseda U.)
    Hideki Saito (Intel)
    Larry Meadows (Sun)

Rudolf Eigenmann (Purdue U.)
Yoshiki Seo (NEC)
Matthijs van Waveren (Fujitsu)

Table of Contents

IV   Software

V   Applications

VI   ITBL

VII   Short Papers

VIII   International Workshop on OpenMP: Experiences and Implementations (WOMPEI 2003)

High Performance Computing Trends and Self Adapting Numerical Software

Jack Dongarra

University of Tennessee
dongarra@cs.utk.edu
and
Oak Ridge National Laboratory
dongarrajj@ornl.gov

1 Historical Perspective

In last 50 years, the field of scientific computing has undergone rapid change – we have experienced a remarkable turnover of technologies, architectures, vendors, and the usage of systems.   Despite all these changes, the long-term evolution of performance seems to be steady and continuous, following Moore's Law rather closely.   In 1965 Gordon Moore one of the founders of Intel, conjectured that the number of transistors per square inch on integrated circuits would roughly double every year. It turns out that the frequency of doubling is not 12 months, but roughly 18 months [8]. Moore predicted that this trend would continue for the foreseeable future. In Figure 1, we plot the peak performance over the last five decades of computers that have been called "supercomputers." A broad definition for a supercomputer is that it is one of the fastest computers currently available. They are systems that provide significantly greater sustained performance than that available from mainstream computer systems. The value of supercomputers derives from the value of the problems they solve, not from the innovative technology they showcase. By performance we mean the rate of execution for floating point operations. Here we chart KFlop/s (Kilo-flop/s, thousands of floating point operations per second), MFlop/s (Maga-flop/s, millions of floating point operations per second), GFlop/s (Gega-flop/s, billions of floating point operations per second), TFlop/s (Tera-flop/s, trillions of floating point operations per second), and PFlop/s (Peta-flop/s, 1000 trillions of floating point operations per second). This chart shows clearly how well this Moore's Law has held over almost the complete lifespan of modern computing – we see an increase in performance averaging two orders of magnitude every decade.

In the second half of the seventies, the introduction of vector computer systems marked the beginning of modern supercomputing. A vector computer or vector processor is a machine designed to efficiently handle arithmetic operations on elements of arrays, called vectors. These systems offered a performance advantage of at least one order of magnitude over conventional systems of that time.   Raw performance was the main, if not the only, selling point for supercomputers of this variety. However, in the first half of the eighties the integration of vector systems into conventional computing environments became more important.   Only those manufacturers that provided standard programming environments, operating systems and key applications were successful in getting the industrial customers that became

A. Veidenbaum et al. (Eds.): ISHPC 2003, pp. 1–9, 2003.

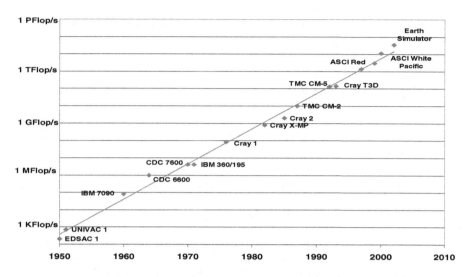

Fig. 1. Moore's Law and Peak Performance of Various Computers over Time.

essential for survival in the marketplace. Performance was increased primarily by improved chip technologies and by producing shared-memory multiprocessor systems, sometimes referred to as symmetric multiprocessors or SMP's. A SMP is a computer system which has two or more processors connected in the same cabinet, managed by one operating system, sharing the same memory, and having equal access to input/output devices. Application programs may run on any or all processors in the system; assignment of tasks is decided by the operating system. One advantage of SMP systems is scalability; additional processors can be added as needed up to some limiting factor determined by the rate at which data can be sent to and from memory.

Scalable parallel computing using distributed memory became the focus of interest at the end of the eighties. A distributed memory computer system is one in which several interconnected computers share the computing tasks assigned to the system. Overcoming the hardware scalability limitations of shared memory was the main goal of these new systems. The increase of performance of standard microprocessors after the Reduced Instruction Set Computer (RISC) revolution, together with the cost advantage of large-scale parallelism, formed the basis for the "Attack of the Killer Micros". The transition from Emitted Coupled Logic (ECL) to Complementary Metal-Oxide Semiconductor (CMOS) chip technology and the usage of "off the shelf" commodity microprocessors instead of custom processors for Massively Parallel Processors or MPPs was the consequence. The strict definition of MPP is a machine with many interconnected processors, where "many" is dependent on the state of the art. Currently, the majority of high-end machines have fewer than 256 processors, with the most on the order of 10,000 processors. A more practical definition of an MPP is a machine whose architecture is capable of having many processors–that is, it is scalable. In particular, machines with a distributed memory design (in comparison with shared memory designs) are usually synonymous with MPPs since they are not limited to a certain number of processors. In this sense, "many" is a number larger than the current largest number of processors in a shared-memory machine.

2   State of Systems Today

The acceptance of MPP systems not only for engineering applications but also for new commercial applications especially for database applications emphasized different criteria for market success such as stability of system, continuity of the manufacturer and price/performance. Success in commercial environments is now a new important requirement for a successful supercomputer business. Due to these factors and the consolidation in the number of vendors in the market hierarchical systems build with components designed for the broader commercial market are currently replacing homogeneous systems at the very high end of performance. Clusters build with components of the shelf also gain more and more attention. A cluster is a commonly found computing environment consists of many PCs or workstations connected together by a local area network. The PCs and workstations, which have become increasingly powerful over the years, can together, be viewed as a significant computing resource. This resource is commonly known as cluster of PCs or workstations, and can be generalized to a heterogeneous collection of machines with arbitrary architecture.

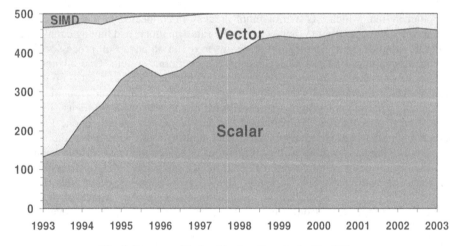

Fig. 2. Processor Design Used as Seen in the Top500.

At the beginning of the nineties, while the multiprocessor vector systems reached their widest distribution, a new generation of MPP systems came on the market, claiming to equal or even surpass the performance of vector multiprocessors. To provide a more reliable basis for statistics on high-performance computers, the Top500 [4] list was begun. This report lists the sites that have the 500 most powerful installed computer systems. The best LINPACK benchmark performance [9] achieved is used as a performance measure to rank the computers. The TOP500 list has been updated twice a year since June 1993. In the first Top500 list in June 1993, there were already 156 MPP and SIMD systems present (31% of the total 500 systems).

The year 1995 saw remarkable changes in the distribution of the systems in the Top500 according to customer types (academic sites, research labs, industrial/commercial users, vendor installations, and confidential sites). Until June 1995, the trend in the Top500 data was a steady decrease of industrial customers, matched by an increase in the number of government-funded research sites. This trend reflects the influence of governmental High Performance Computing (HPC) programs that made it possible for research sites to buy parallel systems, especially systems with distributed memory. Industry was understandably reluctant to follow this path, since systems with distributed memory have often been far from mature or stable. Hence, industrial customers stayed with their older vector systems, which gradually dropped off the Top500 list because of low performance.

Beginning in 1994, however, companies such as SGI, Digital, and Sun began selling symmetric multiprocessor (SMP) models in their workstation families. From the very beginning, these systems were popular with industrial customers because of the maturity of the architecture and their superior price/performance ratio. At the same time, IBM SP systems began to appear at a reasonable number of industrial sites. While the IBM SP was initially intended for numerically intensive applications, in the second half of 1995 the system began selling successfully to a larger commercial market, with dedicated database systems representing a particularly important component of sales.

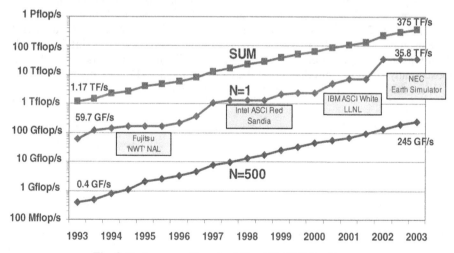

Fig. 3. Performance Growth at Fixed Top500 Rankings.

It is instructive to compare the growth rates of the performance of machines at fixed positions in the Top 500 list with those predicted by Moore's Law. To make this comparison, we separate the influence from the increasing processor performance and from the increasing number of processors per system on the total accumulated performance. (To get meaningful numbers we exclude the SIMD systems for this analysis, as they tend to have extremely high processor numbers and extremely low processor performance.) In Figure 3 we plot the relative growth of the total number of processors and of the average processor performance, defined as the ratio of total accumulated performance to the number of processors. We find that these two factors contribute almost equally to the annual total performance growth – a factor of 1.82. On average, then number of processors grows by a factor of 1.30 each year and the processor performance by a factor 1.40 per year, compared to the factor of 1.58 predicted by Moore's Law.

3 Future Trends

Based on the current Top500 data (which cover the last 13 years) and the assumption that the current rate of performance improvement will continue for some time to come, we can extrapolate the observed performance and compare these values with the goals of government programs such as the Department of Energy's Accelerated Strategic Computing Initiative (ASCI), High Performance Computing and Communications and the PetaOps initiative. In Figure 4 we extrapolate the observed performance using linear regression on a logarithmic scale. This means that we fit exponential growth to all levels of performance in the Top500. This simple curve fit of the data shows surprisingly consistent results. Based on the extrapolation from these fits, we can expect to see the first 100 TFlop/s system by 2005. By 2005, no system smaller then 1 TFlop/s should be able to make the Top500 ranking.

Looking even further in the future, we speculate that based on the current doubling of performance every year to fourteen months the first PetaFlop/s system should be available around 2009. Due to the rapid changes in the technologies used in HPC systems, there is currently no reasonable projection possible for the architecture of the PetaFlops systems at the end of the decade. Even as the HPC market has changed substantially since the introduction of the Cray 1 three decades ago, there is no end in sight for these rapid cycles of architectural redefinition.

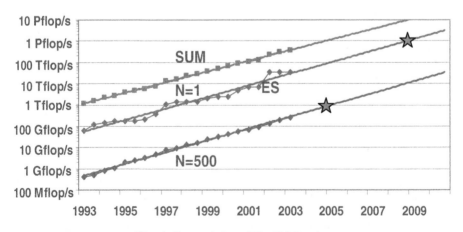

Fig. 4. Extrapolation of Top500 Results.

There are two general conclusions we can draw from these figures. First, parallel computing is here to stay. It is the primary mechanism by which computer performance can keep up with the predictions of Moore's law in the face of the increasing influence of performance bottlenecks in conventional processors. Second, the architecture of high-performance computing will continue to evolve at a rapid rate. Thus, it will be increasingly important to find ways to support scalable parallel programming without sacrificing portability. This challenge must be met by the development of software systems and algorithms that promote portability while easing the burden of program design and implementation.

The Holy Grail for software is *portable performance*. That is, software should be reusable across different platforms and provide significant performance, say, relative to peak speed, for the end user. Often, these two goals seem to be in opposition each to the other. Languages (e.g. Fortran, C) and libraries (e.g. Message Passing Interface (MPI) [7], Linear Algebra Libraries i.e., LAPACK [3]) allow the programmer to access or expose parallelism in a variety of standard ways. By employing standards-based, optimized libraries, the programmer can sometimes achieve both portability and high-performance. Tools (e.g. svPablo [11], Performance Application Programmers Interface (PAPI) [6]) allow the programmer to determine the correctness and performance of their code and, if falling short in some ways, suggest various remedies.

4   Self Adapting Numerical Software

As modeling, simulation, and data intensive computing become staples of scientific life across nearly every domain and discipline, the difficulties associated with scientific computing are becoming more acute for the broad rank and file of scientists and engineers. While access to necessary computing and information technology has improved dramatically over the past decade, the efficient application of scientific computing techniques still requires levels of specialized knowledge in numerical analysis, computer architectures, and programming languages that many working researchers do not have the time, the energy, or the inclination to acquire.

The classic response to this situation, introduced over three decades ago, was to encode the requisite mathematical, algorithmic and programming expertise into libraries that could be easily reused by a broad spectrum of domain scientists. In recent times, however, the combination of a proliferation in libraries and the availability of a wide variety of computing platforms, including varieties of parallel platforms, have made it especially hard to choose the correct solution methodology for scientific problems. The advent of new grid-based approaches to computing only exacerbates this situation. Since the difference in performance between an optimal choice of algorithm and hardware, and a less than optimal one, can span orders of magnitude, it is unfortunate that selecting the right solution strategy requires specialized knowledge of both numerical analysis and of computing platform characteristics.

What is needed now, therefore, is a way of guiding the user through the maze of different libraries so that the best software/hardware combination is picked automatically.

We propose to deal with this problem by creating *Self-adapting Numerical Software (SANS)* systems that not only meet the challenges of scientific computing today, but are designed to smoothly track the state of the art in scientific computing tomorrow

While we are focused on linear algebra computations, the ideas and innovations SANS systems embody will generalize to a wide range of other operations. Like the best traditional libraries, such system can operate as "black box" software, able to be used with complete confidence by domain scientists without requiring them to know the algorithmic and programmatic complexities it encapsulates. But in order to self-adapt to maximize their effectiveness for the user, SANS must encapsulate far more intelligence than standard libraries have aspired to.

5 Optimization Modes

The components of a SANS system can operate in several optimization modes, each being more or less appropriate depending on the component's level in the hierarchy and the nature of the data it's dealing with.

Completely Off-Line Optimization
This scenario is used in PHIPAC [1] and ATLAS [2], and it works well for the dense BLAS because the computational pattern is nearly independent of the input: matrix multiplication does the same sequence of operations independent of the values stored in the matrices. Because optimization can be done offline, one can in principle take an arbitrary amount of time searching over many possible implementations for the best one on a given micro-architecture.

Hybrid Off-Line-Time Optimization
This is the scenario in which Sparsity [5] can work (it can be run completely off-line as well) In both cases, some kernel building blocks are assembled off-line, such as matrix-vector or matrix-matrix multiply kernels for very small dimensions. Then at run time the actual problem instance is used to choose an algorithm. For Sparsity, the problem instance is described by the sparsity pattern of the matrix A . Any significant processing of this will also overwhelm the cost of a single matrix-vector multiplication, so only when many are to be performed is optimization worthwhile.

Completely Run-Time Optimization
This is the scenario to be followed when the available choices depend largely on the nature of the user data. The algorithmic decision making Intelligent Agent follows this protocol of inspecting the data and basing the execution on it. A standard example of inspector-executor is to examine the sparsity pattern of a sparse matrix on a parallel machine at run, and automatically run a graph partitioner to redistribute it to accelerate subsequent matrix-vector multiplications.

Feedback Directed Optimization
This scenario, not disjoint of the last, involves running the program, collecting profile and other information and recompiling with this information, or saving it for future

reference when similar problems are to be solved. We will make use of this mode through the explicit incorporation of a database of performance history information.

6   Optimized Libraries

Automation of the process of architecture-dependent tuning of numerical kernels can replace the current hand-tuning process with a semiautomated search procedure. Current limited prototypes for dense matrix-multiplication (ATLAS [2] and PHiPAC [1]) sparse matrix-vector-multiplication (Sparsity [5], and FFTs (FFTW [10]) show that we can frequently do as well as or even better than hand-tuned vendor code on the kernels attempted.

Current projects use a hand-written *search-directed code generator* (SDCG) to produce many different C implementations of, say, matrix-multiplication, which are all run on each architecture, and the fastest one selected. Simple performance models are used to limit the search space of implementations to generate and time. Since C is generated very machine specific optimizations like instruction selection can be left to the compiler. This approach can be extended to a much wider range of computational kernels by using compiler technology to automate the production of these SDCGs.

Sparse matrix algorithms tend to run much more slowly than their dense matrix counterparts. For example, on a 250 MHz Ultrasparc II, a typical sparse matrix vector multiply implementation applied to a document retrieval matrix runs at less than 10 MFlops, compared to 100 MFlops for a vendor-supplied dense matrix-vector multiplication routine, and 400 MFlops for matrix-matrix. Major reasons for this performance difference include indirect access to the matrix and poor data locality in access to the source vector x in the sparse case.

It is remarkable how the type of optimization depends on the matrix structure: For a document retrieval matrix, only by combining both cache blocking and multiplying multiple vectors simultaneously do we get a five fold speed; cache blocking alone yields a speedup of 2, and multiple vectors alone yield no speedup. In contrast, for a more structured matrix arising from face recognition, the unoptimized performance is about 40 Mflops, cache blocking yields no speedup, and using multiple vectors yields a five fold speedup to 200 Mflops In other words, the correct choice of optimization, and the performance, depends intimately on the matrix structure.

We will use these techniques as well as others and incorporate them into SANS: *register blocking*, where a sparse matrix is reorganized into fixed-size blocks that may contain some zeros, and *matrix reordering*, where the order of rows and columns are changed to reduce cache misses and memory coherence traffic on an SMP systems.

References

1. Optimizing Matrix Mulitply Using PHiPAC: A Portable, High-Performance ACSI C Coding Methodology, In Proceedings of the International Conference on Supercomputing, Vienna, Austria, July 1997.
2. Automated Empirical Optimizations of Software and the ATLAS Project, R.C. Whaley, A. Petitet, and J. Dongarra, Parallel Computing, 27 (1-2): 3-35, January 2001.

3. LAPACK Users' Guide – Third Edition, E. Anderson, Z. Bai, C. Bischof, S. Blackford, J. Demmel, J. Dongarra, J. Du Croz, A. Greenbaum, S. Hammaring, A. McKenney, and D. Sorensen, SIAM Publication, Philadelphia, 1999, ISBN 0-89871-447-8.
4. Top500 Report http://www.top500.org/
5. Optimizing sparse matrix computations for register reuse in SPARSITY, Eun-Jin Im and Katherine Yelick, In Vassil N. Alexandrov, Jack J. Dongarra, Benjoe A. Juliano, Ren´e S. Renner, and C.J. Kenneth Tan, editors, Computational Science – ICCS 2001. International Conference, San Francisco, CA, USA, May 2001, Proceedings, Part I., pages 127–136. Springer Verlag, 2001.
6. A Portable Programming Interface for Performance Evaluation on Modern Processors, S. Browne, J. Dongarra, N. Garner, G. Ho, and P. Mucci, International Journal of High Performance Computing Applications, Vol. 14, No. 3, pp 189--204, 2000.
7. MPI: The Complete Reference, Marc Snir, Steve Otto, Steven Huss-Lederman, David Walker, and Jack Dongarra, MIT Press, Boston, 1996.
8. Cramming More Components onto Integrated Circuits, Gordon E. Moore, Electronics (Volume 38, Number 8), April 19, pp. 114-117.
9. Performance of Various Computers Using Standard Linear Equations Software, (Linpack Benchmark Report), Jack J. Dongarra, University of Tennessee Computer Science Technical Report, CS-89-85, 2003. http://www.netlib.org/benchmark/performance.pdf
10. Fftw: An Adaptive Software Architecture for the FFT. Matteo Frigo and Stephen Johnson. In Proceedings of the International Conference on Acoustics, Speech, and Signal Processing, Seattle, Washington, May 1998.
11. SvPablo: A Multi-Language Architecture-Independent Performance Analysis System, Luiz DeRose and Daniel A. Reed, *Proceedings of the International Conference on Parallel Processing (ICPP'99)*, Fukushima, Japan, September 1999.

Kilo-instruction Processors

Adrián Cristal, Daniel Ortega, Josep Llosa, and Mateo Valero

Departamento de Arquitectura de Computadores,
Universidad Politécnica de Cataluña,
Barcelona, Spain,
{adrian,dortega,josepll,mateo}@ac.upc.es

Abstract. Due to the difference between processor speed and memory speed, the latter has steadily appeared further away in cycles to the processor. Superscalar out-of-order processors cope with these increasing latencies by having more in-flight instructions from where to extract ILP. With coming latencies of 500 cycles and more, this will eventually derive in what we have called Kilo-Instruction Processors, which will have to handle thousands of in-flight instructions. Managing such a big number of in-flight instructions must imply a microarchitectural change in the way the re-order buffer, the instructions queues and the physical registers are handled, since simply up-sizing these resources is technologically unfeasible. In this paper we present a survey of several techniques which try to solve these problems caused by thousands of in-flight instructions.

1   Introduction and Motivation

The ever increasing gap between processor and memory speed is steadily increasing memory latencies (in cycles) with each new processor generation. This increasing difference, sometimes referred to as the *memory wall* effect [33] affects performance by stalling the processor pipeline while waiting for memory accesses. These scenarios are very common in superscalar out-of-order microprocessors with current cache hierarchies.

In these processors, when a load instruction misses in L2 cache, the latency of the overall operation will eventually make this instruction the oldest in the processor. Since instructions commit in order, this instruction will disallow the retirement of newer instructions, which will fill entirely the re-order buffer (ROB) of the processor and halt the fetching of new instructions.

When this happens, performance suffers a lot. The amount of independent instructions to issue is limited due to the lack of new instructions. Eventually, all instructions independent on the *missing* load will get executed and the issuing of instructions will stop. In the left side of Fig. 1 we can see the different IPC for a 4 way processor different latencies to main memory (100, 500, and 1000 cycles) assuming a 128 entry ROB (architectural parameters will be described in section 5). For a current 100 cycle latency, the loss due to the lack of in-flight instructions is not enormous, but the loss for a 1000 cycle latency indeed it is. Notice that the slowdown observed in this figure from 100 cycle to 1000 cycle is almost 3 times.

A. Veidenbaum et al. (Eds.): ISHPC 2003, LNCS 2858, pp. 10–25, 2003.

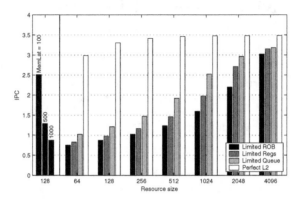

Fig. 1. Effect on IPC of unbounded resources for SPEC2000 fp applications

A well known microarchitectural solution to cope with this problem is to increase the amount of maximum number of in-flight instructions that the processor can handle. More instructions means that the halting will start later, thus affecting less overall performance, or not happen at all. Besides, increasing the availability of instructions to issue, increases the probability of finding independent instructions, thus also increasing IPC [11].

In order to increase the amount of in-flight instructions we must increase the capability of several resources that directly depend on the number of in-flight instructions, the most important ones being the ROB, the instruction queues and the physical registers. We will start our digression describing the problems of the re-order buffer (ROB). This structure keeps the state of each instruction since it gets prepared for execution (normally just after renaming) until it reaches commit stage and updates the architectural state of the machine. In the case of a 1000 cycle latency to memory and a 4 way processor, and assuming we would like not to halt the processor, this would mean at least a four thousand entry ROB. Several authors have pointed out the criticality of this resource [22]. Simply upsizing this structure will not be feasible technologically, which motivates us to look for other type of solutions. Unfortunately, this is not our only problem.

In-flight instructions not only live in the ROB. After the decode stage they also get inserted into Instruction Queues (IQ) where they wait until execution. Normally the instruction queues are divided depending on the type of the instruction, which leads to a more clustered and effective microarchitecture. Unfortunately, the handling of this IQs is even more critical than that of the ROB. Each cycle the tags of the destination registers of instructions that have finished must be compared against each of the source registers from each instruction in the IQ. This process, called wakeup logic, determines which instructions are ready to get executed. Another process, called selection logic, determines which ready instructions get selected and are effectively issued to their respective functional units and which remain in the IQ.

Having more in-flight instructions means having bigger and more complex IQs. This is specially true in our missing load scenario, since many of the instructions will depend on the missing load and will consume entries of their respective queues for a long time. Palacharla et al [23] showed how the wakeup and selection logic of instruction queues is on the critical path and may eventually determine processor cycle time. Several authors have presented techniques which take into account the complexity problem of IQs and handle it in innovative and intelligent ways [3, 29]. Lately [14] and [2] have proposed techniques that tackle the problem of large IQs derived from processors with a high number of in-flight instructions.

Another resource which is dependent on the amount of in-flight instructions is the amount of physical registers [8]. Just after fetching instructions, these go through a process of decoding and renaming. Renaming is the process of changing the logical names of the registers used by instructions for the names of physical registers [30]. Each destination logical register is translated into a new physical register and source registers get translated to whatever register was assigned to the destination register of their producer. This exposes more ILP since false dependencies[1] and certain hazards such as write after write (WAW) get handled in an elegant way.

Physical register lives start in the decode stage of the instruction that defines the value that that particular register will contain. The liberation of this resource must take place when all dependent instructions have already read the value. Since the determination of this particular moment in time is very difficult, microarchitectures conservatively free it when the following instruction that defines the same logical register becomes the oldest, since by then all dependent instructions from the prior definition of that logical register must have read their value. When a missing load stalls the commit of instructions, register lives get extended, thus increasing the necessity for physical registers.

Unfortunately, increasing physical registers is not simple nor inexpensive. Several studies [1, 24] show that the register file is in the critical path, and others [31] state that register bank accounts for a 15% to a 20% of overall power consumption.

At this point of our analysis it must be clear that simply up-sizing critical structures is not a feasible solution to tolerate memory latency. Nevertheless, we must not disdain the performance benefits that this increase brings. In Fig. 1 we can see the IPC obtained in our machine with a memory latency of 1000 cycles to main memory relative to the amount of resources. Each group of four bars presents rightmost the IPC obtained supposing perfect L2 cache, while the other three bars show the IPC when one of the three resources (ROB size, IQ size and number of physical registers) is limited to the amount shown on the X axis while the other two are supposed unbounded. The first conclusion we may draw from these results is how tightly coupled these three resources are.

---

[1] False dependencies occur when two instructions define the same logical register. In microarchitectures without renaming, both instructions get to use the same resource and the second instruction may not start until the first one and all its dependents have finished. Renaming solves this constraint by assigning different physical registers to both dependence chains.

Unbounding any two of them does not increase performance, since the remaining one becomes the bottleneck. Solutions to any of these three hurdles, although in many cases presented separately, will always need solutions to the other two to become effective.

From these numbers it is clear that the ratio between the amount of physical registers, the amount of IQ entries and the amount of ROB entries should not be the unity. Since not all instructions define registers and not all instructions in the ROB are actually using an IQ entry, the necessary amount of these two resources should be smaller than that of the ROB. This can be seen in the figure by noticing that the bar where the ROB is limited to the resource size is the worse of the three.

The most important conclusion that we can get from Fig. 1, which motivates our present discussion, is the fact that increasing the amount of in-flight instructions (which is achieved by augmenting the necessary resources) effectively allows to tolerate the latency from main memory, as expected. The results of Fig. 1 show how for a 1000 cycle latency, 4096 entries in the ROB practically achieves a performance close to the perfect L2 behavior.

The scenarios and the results explained in this paper are not exclusive of numerical applications, but also of integer applications. Unfortunately, integer applications also suffer from other bottlenecks such as branch mis-speculation and pointer chasing problems which are out of the scope of this paper. Discussion about these problems can be found in [13, 28].

Having reached this point in our discussion, it seems clear that although increasing in-flight instructions clearly delivers outstanding performance for numerical applications considering future memory latencies, it is not feasible to up size resources to achieve this goal. We need a different approach to this problem [4], and this approach is motivated from the results of Figs. 2 and 3.

In these figures we can see the breakdown of the resource usage on FP applications. In Fig. 2 we can see the relation between the number of in-flight instructions and the necessary entries in the FP instruction queue for a 4 way

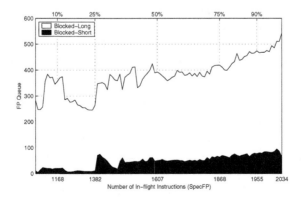

Fig. 2. Breakdown of FP Queue usage on FP applications

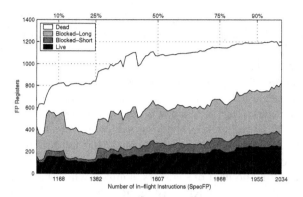

Fig. 3. Breakdown of FP Registers usage on FP applications

processor with 2048 entries in the ROB and a memory latency of 500 cycles. The x-axis is not linear. Instead we have also computed in average how much time of the total time did the number of in-flight instructions behave in a particular way. For example, we can see that on average a 25% of the time the number of in-flight instructions was less than 1382, on average a 50% of the time this number was less than 1607 and so on. The motivating part of figure is the fact that we have not only computed the amount of *to-be-issued* instructions with respect to the amount of in-flight instructions, but rather, we have divided the *to-be-issued* instructions into two different types, those depending on instructions which are going to be ready shortly, and those depending on instructions with very long latency. Of course, long latency instructions are those that have missed on L2 or any dependent instruction on them. As can be seen from Fig. 2, the amount of dependents on long latency instructions, tagged blocked long, in the figure is much bigger than that of the *soon to be ready*. The motivation behind our present discussion is obvious. Why consume critical resources for long latency instructions? It seems that a small number of entries in the FP queue will be able of handling the *soon to be ready* instructions.

Something similar can be read from Fig. 3. In it the x-axis follows the same behavior, indicating the amount of in-flight instructions and the percentile distribution over time. The y-axis accounts for the amount of necessary physical registers to cope with that amount of in-flight instructions. Again, we have classified physical register usage depending on their nearness to be used. Live registers are those that presently have values computed. Blocked-short are registers which do not yet have a value but depend on *soon to be ready* instructions and therefore their value is presumably going to be available soon. Blocked-long registers are those that will eventually have a value but that will remain for a long time without one. These are the registers associated with long latency instructions and their dependents. Dead registers are those registers which have their values computed, and that all their dependent instructions have already consumed it.

Here again we can see something motivating. At any particular point, Dead registers are not needed anymore[2] and Blocked-long registers will be needed in a future time. Therefore the amount of *necessary* registers at any point is much smaller, ranging less than 300 for nearly 2K in-flight instructions.

In this paper we shall present a survey of different techniques that address the problems of the ROB, the IQs and the physical registers. We will present summaries of our work and the work of others and the philosophy behind them. In section 2 we will address the problem of achieving very big ROBs. Following, in section 3 we will describe certain mechanisms that try to virtually enlarge the IQs with little cost. We shall also present in section 4 a survey of the different solutions proposed in the field of registers, some of which have been specifically been tailored to suit the case of Kilo-Instruction Processors. We will present brief results for a combination of these mechanisms in section 5 and we will conclude the paper in section 6.

2   Managing the Commit of Kilo-instruction Processors

The main purpose of the ROB is to enforce in-order commit, which is fundamental for microarchitectures with precise interrupts. The ROB allows the processor to recover the state after each specific instruction, and this means not only the state of the registers, but also the memory state, since stores only update memory when they get committed. Fortunately, recoveries are rare when compared to the amount of instructions. Recoveries may happen due to two causes, exceptions or mis-speculations. The former is highly unpredictable: a page miss, an illegal division, etc. The latter may only happen on speculated instructions, such as branches or memory operations incorrectly disambiguated[3]. Why not just try to honor the famous motto *Improve the common case* [11] and execute normal instructions as if they would never except nor be part of any mis-speculation?

This motivation is the one followed in [6], although initial work on checkpointing was made by [12]. As far as we know, [6] is the first approach to substitute the need for a large ROB with a small set of microarchitectural checkpoints.

Once instructions finish their execution, the processor can free the resources that the instruction is not going to need anymore. The processor does not need to wait for preceding instructions to finish. Of course, this heavily relies on the improbability of needing to return to this specific moment in the state of the machine. The processor is acting as if the state of the machine changed every group of executed instructions, instead of after each instruction as the Architecture mandates.

---

[2] In reality they are needed, since precise interrupts must still be taken care of. Nevertheless, we will speak freely using the verb *need* to mean that they are not to be treated with the same urge, i.e. they are not needed shall be understood as they are not required to be accessed every cycle.

[3] We recognise that many other types of speculation exist and have been published in the literature. Nevertheless, the two most common types are the ones we shall discuss and the philosophy can easily be extended to cover other cases.

Nevertheless, the functionality for returning to a precise state of the processor is still available. If any kind of error, i.e. mis-speculation, exception, ..., happened, the processor needs only to return to the previous checkpoint to the error and proceed after that, either inserting checkpoints after each instruction, in a step-by-step way, or if it remembered the excepting instruction, by taking a checkpoint at it. This re-execution of extra instructions is detrimental to performance. The total amount of extra re-execution can be minimised if the heuristics for taking checkpoints in the first place are driven to the most common mis-speculated instructions.

As instructions de-allocate their resources when they finish rather than waiting for all preceding instructions to finish, this mechanism has been called *Out-of-Order Commit* [5]. The name is somehow misleading, since the state of the machine is not modified at all. Only when the whole set of instructions belonging to a checkpoint finish, and the checkpoint is the last one, the state of the machine is modified. The checkpoints are thus committed in-order, respecting ISA semantics and interruption preciseness. The authors name this coupling of an out-of-order commit mechanism for instructions and an in-order commit mechanism for checkpoints as a *Hierarchical Commit* mechanism.

In [5, 6] different heuristics are given for taking checkpoints. In [6] the authors analyse the impact of having loads as the checkpoint heads. Later, in [5] branches are shown to be good places for taking checkpoints. This seems intuitive since the biggest number of mis-speculations happen at branches. Nevertheless taking a checkpoint at every branch is not necessary nor efficient. Many branches resolve very early. So, adjacent checkpoints can be merged to minimise the number of checkpoint entries and increase the average number of instructions per checkpoint. Another solution proposed is to delay the taking of the branch with some sort of simple ROB structure which holds the last n instructions. This pseudo-ROB structure would hold the state for the n newest instructions and decrease the need for so many checkpoints. Both papers have pointed out that the analysis and determination of where to take checkpoints is not a simple task, since it must combine the correct use of resources with the minimisation of recovery procedures. Further analysis should be expected in this field. In section 5 we will discuss the sensibility of certain parameters related to this mechanism.

3 Managing the Issue of Kilo-instruction Processors

Probably the simplest classification for instruction in Kilo-Instruction Processors is to consider that an instruction may be a *slow* one or a *fast* one (cf. Fig. 2). *Slow* instructions are those that depend on long latency instructions, such as loads that miss on L2 cache or long floating point operations. Issuing an instruction is not simple at all, and treating these *slow* instructions with unknown latencies increases the complexity for the whole set of instructions. Several studies have addressed this particular problem [3, 9, 20, 23, 29] and more specifically three of them [2, 5, 14] stand out with proposals on how to modify the instruction queues in order to tolerate the problems caused by bigger latencies.

[14] presents a mechanism for detecting long latency operations, namely loads that miss in L2. Dependent instructions from this load are then moved into a Waiting Instruction Buffer (WIB) where the instructions reside until the completion of the particular loads. In order to determine if an instruction depends on the load or not, the wakeup logic of the window is used to propagate the dependency. The selection logic is used to *select* the dependent instructions and send them into the WIB, which acts as a functional unit. This mechanism is very clever, but we believe that using the same logic for two purposes, puts more pressure on a logic which is on the critical path anyway [23]. Besides, the WIB contains all instructions in the ROB, which the authors consider not to be on the critical path. This way they supposedly simplify the tracking of which missing load instructions each of the WIB instructions depends on. [14] has been a pioneer in this field, among the first to propose solutions, but we believe that certain decisions in how the mechanism behaves will not be able to scale. The WIB structure is simpler than the IQ it replaces but still very complex to scale to the necessities of 10K in-flight instruction processors. Besides, the tracking of dependencies on the different loads is done orthogonally to all the outstanding loads which we believe may be another cause of not scaling correctly.

In [2] the authors propose a hierarchical instruction window (HSW) mechanism. This mechanism has two scheduling windows, one big and slow and another one smaller but faster. Each one of them has its own scheduler that schedules instructions indepently. This distribution forces the hardware to duplicate the number of functional units. Besides the managing of the register accesses is rather complex due to this clusterisation. The important thing about this mechanism is that all instructions are decoded and initially sent to the slow instruction window. Long latency instructions and their dependents are considered more critical and are then prioritised by moving them into the small and fast. The rationale behind this is that instructions on the critical path should finish as soon as possible. The rest of instructions can pay the penalty imposed by the slower and bigger instruction window. Notice how this mechanism is conceptually opposed to the one presented in [14] although both try to achieve the same.

Another reference that proposes a mechanism to handle the complexity and necessities of instruction queues for Kilo-Instruction Processors is [5]. This mechanism is conceptually nearer to that of [14] than to the one presented in [2]. It differs from it, however, in its complexity. This mechanism has been devised to scale further than the one presented in [14]. First of all, in order to simplify the detection of long latency loads and dependents it uses a pseudo-ROB structure which tracks the last n instructions. When leaving this structure (which can be easily coupled with the newest part of the ROB if the microarchitecture has one) instructions are analysed to see if they are a missing load or dependent on one. This dependency tracking is easier and does not need a wake-up or selection logic. If the instruction is a slow one, it is moved from the IQ[4] into a *Slow Lane*

---

[4] Microarchitecturally it would be simpler to invalidate entries in the IQ and to insert into the SLIQ from the pseudo-ROB, but for clarity in the explanation the instructions are assumed to *move* from the IQ to the SLIQ.

Instruction Queue (SLIQ) where it will wait until the missing load comes from main memory.

In order to manage several missing loads, instead of having more complexity in the SLIQ, the authors propose to treat instructions linearly. Each load has a starting point in this FIFO like structure and once it finishes it starts to move instructions from the SLIQ to the small and fast IQ. This is done at the maximum pace of 4 instructions per cycle with the starting penalty of 4 cycles, in our simulations. The instructions are re-introduced to the fast IQ when their source operands no longer depend on long latency instruction. However, it could happen that a second long latency load is resolved while the instructions dependent on the first one are still being processed. Two different situations can happen: first, if the new load is younger than the instructions being processed, it will be found by the wakening mechanism. After that, the mechanism could continue, placing the instructions dependent on any of the two loads in the instruction queues. Second, if the new load is older that the instructions being processed, it will not be found by the wakening process, so a new wakening process should be started for the second load. Extra penalty in the re-insertion seems not to affect performance, which leaves a margin for the consequences of the potential complexity of this SLIQ [5]. Finally, an additional advantage of the pseudo-ROB is reducing the misprediction penalty of branch instructions. Since our out-of-order commit mechanism removes the ROB, mispredicted branches force the processor to return up to the previous checkpoint, which is potentially some hundred instructions behind. The information contained in the pseudo-ROB allows to recover from branch mispredictions without needing to use a far checkpoint, whenever the branch instruction is still stored in the pseudo-ROB.

4 Managing the Dataflow Necessities of Kilo-instruction Processors

Decreasing the lifetime of registers has been a hot topic for over 10 years now. As explained in section 1, the motivation behind all this research is to diminish the impact that the register file and the renaming have on performance, which is very important as has been stated in several papers [1, 21, 23, 27, 32]. In many cases the authors propose to microarchitecturally modify the register structure to scale better for current or future processor generations [7, 16, 19, 24, 25]. In other cases, the authors propose to modify the renaming strategy to limit the amount of live registers at any moment and therefore decrease the long term necessities of the processor. All this can be accomplished due to the classification of different types of registers shown in Fig. 3.

A good example of this last type of research is presented in [21]. This paper is the first to our knowledge that proposes to recycle registers as soon as they have been fully and irreversibly superseded by the renaming logic, as opposed to waiting for instruction retirement. In order to be able of doing this, the microarchitecture keeps a per-register bookkeeping count of consumers of the register not yet executed. When a particular register has been unmapped, i.e. the logical has been mapped to another physical, and all its consumers have read the particular value, the mechanism is able to reuse this particular register for other purposes.

Solutions for delaying register allocation have also been presented in the literature prior to the research for Kilo-Instruction Processors, such as [10, 32]. These proposals require deadlock avoidance mechanisms. They also incur in a complexity penalty for Kilo-Instruction Processors[5] which makes them unsuitable for us. In the case that a particular instruction fails to secure a destination register and needs to be re-executed, two alternatives appear: either it must be kept in the IQ throughout its execution or the IQ must be augmented to admit re-insertion of instructions.

More recent papers have addressed partially the problem de-allocating resources in Kilo-Instruction Processors. One of them is [19] where the authors propose Cherry, a hybrid checkpoint/ROB-based mechanism that allows early recycling of multiple resources. They use the ROB to support speculative execution and precise interrupts, and state checkpointing for precise exceptions. They identify the instructions not subject to mis-speculation and apply early recycling of load/store queue entries and registers. This mechanism does not go beyond unresolved branches or speculative loads, which limits its scope.

Other papers have made proposals to recycle registers based on compiler support and/or special ISA instructions, most notably [15–17]. We find these mechanisms very interesting but in this paper we have limited our scope to microarchitectural solutions which do not need to change the ISA or the compiler.

A new proposal specifically targeted for Kilo-Instruction Processors is the one presented in [18], which is called Ephemeral Registers. This mechanism proposes an aggressive register recycling mechanism which combines delayed register allocation and early register recycling in the context of Kilo-Instruction Processors.

The mechanism has been tailored to support fast recovery from mis-speculations while supporting precise exceptions and interrupts. This is accomplished by combining these mechanisms with only one checkpointing, which allows the processor to non-conservatively deallocate resources. This is specially important in Kilo-Instruction Processors since these processors rely heavily on speculation mechanism to extract parallelism. This paper is also the first proposal that integrates both a mechanism for delayed register allocation and early register recycling and analyses the synergy between them. With respect to [10] they opt for assigning resources during the issue stage rather than in the writeback since this simplifies the design. The mechanism tries to secure a register within a few cycles after issue. During these cycles, the instruction keeps its IQ entry. If successful, the IQ entry is released; if unsuccessful, it is squashed and, since it retains its entry in the IQ, can be retried later.

---

[5] Of course the authors solution was not intended for Kilo-Instruction Processors and their solutions perfectly fit the problems addressed in their respective papers.

5   Experimental Results

In this section we shall present some experimental results of some of the techniques presented in the previous sections. The benchmark suite used for all the experiments is SPEC2000fp, averaging over all the applications in the set. All benchmarks have been simulated 300 million representative instructions, where representativeness has been determined following [26]. The simulator targets a superscalar out-of-order microarchitecture which fetches, issues and commits a maximum of 4 instructions per cycle. The branch predictor assumed is a 16K history gshare with a penalty in case of misprediction of 10 cycles. The memory hierarchy has separate instruction and data first level of cache (L1), each with 32Kb 4way and 2 cycles of hit access. The second level of cache (L2) is unified for data and instructions and its size is 512Kb with 10 cycles in case of a hit. Notice that in this description of the architecture we have intentionally omitted the most commons parameters, sizes of the IQs, size of the re-order buffer, latency to main memory and number of physical registers. Since our experiments deal exactly with these parameters, each one of the figures will detail the exact assumptions of the results presented.

In Fig. 4 we can see the IPC obtained with different numbers of checkpoints using the technique presented in [5] and explained in 2. The limit bar shows IPC results obtained when considering ROB of 4096 entries, which is microarchitecturally unfeasible. This limit helps us show how near we are of a pseudoperfect mechanism which does not take into account complexity. Having only 4 checkpoints produces just a 20% slowdown. If we increase to 8 checkpoints this slowdown decreases to only 9% and from 32 onwards the slowdown is just a 6%. This 6% slowdown with 256 times less entries is clearly spectacular.

Figure 5 presents results for the technique called *Slow Lane Instruction Queuing* presented in [5] and explained in section 3. The experiments show the IPC obtained on average by all the SPEC2000fp applications when executed under different architectural configurations. The three bars of each group show the IPC when the IQ size ranges from 32 to 128. It is remarkable that in each group

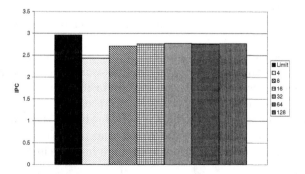

Fig. 4. Sensibility of our Hierarchical Commit mechanism to the amount of available checkpoints. With IQ size=2048 entries and 2048 Physical Registers.

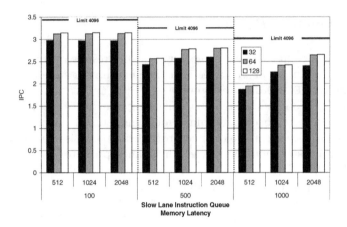

Fig. 5. IPC respective to the memory latency, the size of the SLIQ and the IQ size. With 16 Checkpoints and 4096 Physical Registers

of bars the 64 entry bar nearly achieves the same IPC as the 128 one. This is because with this mechanism having 64 entries is enough to extract enough parallelism with out forcing the processor to stall. It can also be seen that in order to tolerate higher latencies to main memory the SLIQ must be bigger.

Figure 6 presents a combination of the three different techniques from previous figures, i.e. SLIQ, Out-of-Order Commit and Ephemeral registers. These three different techniques were coupled together in a simulator which assumed a 64 entry pseudo-ROB and IQs of 128 entries. The figure is divided into three zones, each of them comprising the results for each of the memory latencies used throughout this paper (100, 500 and 1000). Each group is composed of three groups of two bars each. Each of the groups assumes an increasing amount of Virtual Registers. The two bars of each group represent the IPC with 256 or 512 physical registers. Besides each zone has two lines which represent the performance obtained with the baseline (with 128 ROB entries) and the performance obtained by a Limit microarchitecture where all the resources have been up-sized with no constraints. Of course this Limit microarchitecture is unfeasible, but it helps us understand how well our mechanisms are behaving, since it acts as an upper bound.

Figure 6 shows some interesting points. First of all it allows to see how the combination of these three orthogonal techniques work. Another conclusion that we can see is that there is still room for improvement with 1000 cycles or more. We can notice how the mechanisms nearly saturate at 500 cycles. For this latency having 1024 Virtual tags is nearly as having 2048, while in the 1000 latency zone we can see that the growth trend is far from saturating. This enforces us in the belief that Kilo-Instruction Processors with more in-flight instructions will still yield benefits.

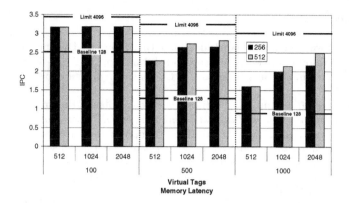

Fig. 6. IPC results of the combination of mechanisms (SLIQ, Out-of-Order Commit, Ephemeral registers) with respect to the amount of Virtual Registers, the memory latency and the amount of physical registers

6 Conclusions

The ever increasing difference between processor and memory speed is making the latter appear farther with each new processor generation. Future microarchitectures are expected to have 500 cycles to main memory or even more. In order to overcome this hurdle, more in-flight instructions will be needed to maintain ILP in the events of missing load instructions. Although recent history has shown how an increase in the number of in-flight instructions helps tolerate this increasing latency, it has also pointed out the limitations that simply up-sizing the critical resources has on cycle time and thus on performance.

The resources that directly depend on the amount of in-flight instructions are the re-order buffer, the instructions queues and the amount of physical registers available. In this paper we have shown that these critical resources are underutilised, and we have presented a survey of different techniques that accomplish the task of allowing thousands of in-flight instructions at reasonable cost. We have also shown that this pursuit is beneficial for performance and that it is attainable in a near future. We have also shown how a subset of all the techniques described act together forming a synergy.

Acknowledgments

The authors would like to thanks Sriram Vajapeyam and Oliver Santana for the help during the writting of this paper. This work has been supported by the Ministry of Science and Technology of Spain, under contract TIC-2001-0995-C02-01 by the CEPBA.

References

1. R. Balasubramonian, S. Dwarkadas, and D. Albonesi. Dynamically allocating processor resources between nearby and distant ilp. In *Proceedings of the 28th annual international symposium on on Computer architecture*, pages 26–37. ACM Press, 2001.
2. E. Brekelbaum, J. Rupley, C. Wilkerson, and B. Black. Hierarchical scheduling windows. In *Proceedings of the 35th annual ACM/IEEE international symposium on Microarchitecture*, pages 27–36. IEEE Computer Society Press, 2002.
3. M.D. Brown, J. Stark, and Y.N. Patt. Select-free instruction scheduling logic. In *Proceedings of the 34th annual ACM/IEEE international symposium on Microarchitecture*, pages 204–213. IEEE Computer Society, 2001.
4. A. Cristal, J.F. Martínez, J. Llosa, and M. Valero. A case for resource-conscious out-of-order processors. Technical Report UPC-DAC-2003-45, Universidad Politécnica de Cataluña, Department of Computer Architecture, July 2003.
5. A. Cristal, D. Ortega, J.F. Martínez, J. Llosa, and M. Valero. Out-of-order commit processors. Technical Report UPC-DAC-2003-44, Universidad Politécnica de Cataluña, Department of Computer Architecture, July 2003.
6. A. Cristal, M. Valero, A. Gonzalez, and J. LLosa. Large virtual robs by processor checkpointing. Technical Report UPC-DAC-2002-39, Universidad Politécnica de Cataluña, Department of Computer Architecture, July 2002.
7. J-L. Cruz, A. González, M. Valero, and N. P. Topham. Multiple-banked register file architectures. In *Proceedings of the 27th annual international symposium on Computer architecture*, pages 316–325. ACM Press, 2000.
8. Keith I. Farkas, Paul Chow, Norman P. Jouppi, and Zvonko Vranesic. Memory-system design considerations for dynamically-scheduled processors. In *Proceedings of the 24th annual international symposium on Computer architecture*, pages 133–143. ACM Press, 1997.
9. D. Folegnani and A. González. Energy-effective issue logic. In *Proceedings of the 28th Annual International Symposium on Computer Architecture*, pages 230–239, Göteborg, Sweden, June 30–July 4, 2001. IEEE Computer Society and ACM SIGARCH. *Computer Architecture News*, 29(2), May 2001.
10. A. González, J. González, and M. Valero. Virtual-physical registers. In *IEEE International Symposium on High-Performance Computer Architecture*, February 1998.
11. J.L. Hennessy and D.A. Patterson. *Computer Architecture. A Quantitative Approach. Second Edition.* Morgan Kaufmann Publishers, San Francisco, 1996.
12. W.M. Hwu and Y. N. Patt. Checkpoint repair for out-of-order execution machines. In *Proceedings of the 14th annual international symposium on Computer architecture*, pages 18–26. ACM Press, 1987.
13. N. P. Jouppi and P. Ranganathan. The relative importance of memory latency, bandwidth, and branch limits to performance. In *Workshop of Mixing Logic and DRAM: Chips that Compute and Remember.* ACM Press, 1997.
14. A.R. Lebeck, J. Koppanalil, T. Li, J. Patwardhan, and E. Rotenberg. A large, fast instruction window for tolerating cache misses. In *Proceedings of the 29th annual international symposium on Computer architecture*, pages 59–70. IEEE Computer Society, 2002.
15. J. Lo, S. Parekh, S. Eggers, H. Levy, and D. Tullsen. Software-directed register deallocation for simultaneous multithreaded processors. Technical Report TR-97-12-01, University of Washington, Department of Computer Science and Engineering, 1997.

16. L.A. Lozano and G.R. Gao. Exploiting short-lived variables in superscalar processors. In *Proceedings of the 28th annual international symposium on Microarchitecture*. IEEE Computer Society Press, November 1995.
17. M.M. Martin, A. Roth, and C.N. Fischer. Exploiting dead value information. In *Proceedings of the 30th annual ACM/IEEE international symposium on Microarchitecture*. IEEE Computer Society Press, December 1997.
18. J.F. Martínez, A. Cristal, M. Valero, and J. Llosa. Ephemeral registers. Technical Report CSL-TR-2003-1035, Cornell Computer Systems Lab, 2003.
19. J.F. Martínez, J. Renau, M.C. Huang, M. Prvulovic, and J. Torrellas. Cherry: checkpointed early resource recycling in out-of-order microprocessors. In *Proceedings of the 35th annual ACM/IEEE international symposium on Microarchitecture*, pages 3–14. IEEE Computer Society Press, 2002.
20. E. Morancho, J.M. Llabería, and A. Olivé. Recovery mechanism for latency misprediction. Technical Report UPC-DAC-2001-37, Universidad Politécnica de Cataluña, Department of Computer Architecture, November 2001.
21. M. Moudgill, K. Pingali, and S. Vassiliadis. Register renaming and dynamic speculation: an alternative approach. In *Proceedings of the 26th annual international symposium on Microarchitecture*, pages 202–213. IEEE Computer Society Press, 1993.
22. O. Mutlu, J. Stark, C. Wilkerson, and Y.N. Patt. Runahead execution: An alternative to very large instruction windows for out-of-order processors. In *Proceedings of the Ninth International Symposium on High-Performance Computer Architecture*, Anaheim, California, February 8–12, 2003. IEEE Computer Society TCCA.
23. S. Palacharla, N.P. Jouppi, and J.E. Smith. Complexity-effective superscalar processors. In *Proceedings of the 24th international symposium on Computer architecture*, pages 206–218. ACM Press, 1997.
24. I. Park, M. Powell, and T. Vijaykumar. Reducing register ports for higher speed and lower energy. In *Proceedings of the 35th annual ACM/IEEE international symposium on Microarchitecture*, pages 171–182. IEEE Computer Society Press, 2002.
25. A. Seznec, E. Toullec, and O. Rochecouste. Register write specialization register read specialization: a path to complexity-effective wide-issue superscalar processors. In *Proceedings of the 35th annual ACM/IEEE international symposium on Microarchitecture*, pages 383–394. IEEE Computer Society Press, 2002.
26. T. Sherwood, E. Perelman, and B. Calder. Basic block distribution analysis to find periodic behavior and simulation points in applications. In *Proceedings of the Intl. Conference on Parallel Architectures and Compilation Techniques*, pages 3–14, September 2001.
27. D. Sima. The design space of register renaming techniques. In *Micro, IEEE , Volume: 20 Issue: 5*, pages 70–83. IEEE Computer Society, September 1999.
28. K Skadron, P.A. Ahuja, M. Martonosi, and D.W. Clark. Branch prediction, instruction-window size, and cache size: Performance trade-offs and simulation techniques. In *IEEE Transactions on Computers*, pages 1260–1281. IEEE Computer Society, 1999.
29. J. Stark, M.D. Brown, and Y.N. Patt. On pipelining dynamic instruction scheduling logic. In *Proceedings of the 33rd Annual International Symposium on Microarchitecture*, pages 57–66, Monterey, California, December 10–13, 2000. IEEE Computer Society TC-MICRO and ACM SIGMICRO.
30. R.M. Tomasulo. An efficient algorithm for exploiting multiple arithmetic units. January 1967.

31. J. Tseng and K. Asanovic. Energy-efficient register access. In *XIII Symposium on Integrated Circuits and System Design*, September 2000.

32. S. Wallace and N. Bagherzadeh. A scalable register file architecture for dynamically scheduled processors. In *Proceedings: Parallel Architectures and Compilation Techniques*, October 1996.

33. W.A. Wulf and S.A. McKee. Hitting the memory wall: Implications of the obvious. In *Computer Architecture News*, pages 20–24, 1995.

CARE: Overview
of an Adaptive Multithreaded Architecture

Andrés Márquez and Guang R. Gao

University of Delaware, Newark DE 19716, USA,
{marquez,ggao}@capsl.udel.edu

Abstract. This paper presents the CARE (*Compiler Aided Reorder Engine* (CARE)) execution and architecture model. CARE is based on a decentralized approach for high-performance microprocessor architecture design – a departure from the mainly centralized control paradigm that dominated the traditional microprocessor architecture evolution. Under CARE, a processor is powered by a large number of fine-grain threads (called *strands)*, each enabled by individual events – such as those due to control dependency or data dependencies with unpredictable and/or long latencies. As a result, the CARE architecture consists of a large grid of small processing cells and their local memories. We outline the CARE architecture design as well as the related issues in strand *chaining* and synchronization support. Some experimental results are also presented.

In this paper, we study alternative approaches for microprocessor architectures as a departure from the mainly centralized control paradigm that dominated the traditional microprocessor architecture evolution. In other words, we can employ a fully decentralized control paradigm - that is the processor chip is controlled by a large number of fine-grain threads, each enabled by individual events due to program flow. As a result, the hardware architecture consists of a large grid of small processing cells and their local memories.

An important open question is: what execution and architecture model will be best suited to harness such a decentralized architecture paradigm? The CARE Execution and Architecture model introduced in this paper provides an entry point in developing an effective multithreaded execution model for such future massively parallel decentralized architectures. An important feature of CARE is the support for a multilevel fine-grain multithreading hierarchy. For example, the CARE model has two levels of threads: threaded procedures and strands. A threaded procedure is invoked asynchronously - forking a parallel thread of execution. A threaded procedure is statically divided into strands. A main feature of CARE is the effective synchronization between strands using only relevant dependencies, rather than global barriers. It also enables an effective overlapping of communication and computation, allowing a processor to execute any strand whose data and control is ready.

This paper provides an overview of the CARE execution and architecture model, explains how CARE supports explicit encoding of a partial order of strands – the scheduling quantum processed by the dispatch/issue section, and

A. Veidenbaum et al. (Eds.): ISHPC 2003, LNCS 2858, pp. 26–38, 2003.
© Springer-Verlag Berlin Heidelberg 2003

introduces the concept of strand chaining used to take advantage of a sequential fetching/decode mechanism by fetching instructions in successive strands in order. We describe the CARE scheduling window and how it is designed to support the strand enabling mechanism for register-based dependencies – data, control and memory dependencies – through the CARE architecture mechanism.

In Sec. 1 we make the case for a decentralized control paradigm with the Quicksort Kernel. Section 2 introduces the Execution Model of threads, strand chains and strands. In Sec. 3 a brief overview of the compiler phases is presented. Section 4 explains the major architectural components with an emphasis on the register based synchronization mechanism. Next, Sec. 5 gives a cross section of compiler derived experimental data. Related work and conclusions wrap the paper up.

1   Motivation

Consider the control flow graph (CFG) for the modified Quicksort Kernel in three-address code as described in its original form in the "dragon book" [1] in Fig. 1. The code has two inner parallel loops LR2 and LR3 that traverse a list. The list has guard nodes on each end to check for list overflows – not shown. A node in the list has as members: **(1)** an element value, **(2)** an unique identification (id) that is totally ordered, **(3)** a pointer to the next node and **(4)** a pointer to the previous node. LR2 starts from the first element in the list and searches for an element that is smaller than a predetermined pivot element v. v is the last element in the list. LR3 starts from the last element in the list and searches for an element that is larger than v. These two elements are swapped in B5 and the search continues from where LR2 and LR3 left off by starting a new iteration

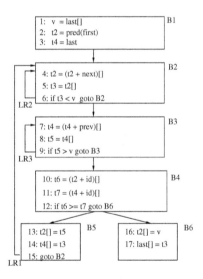

Fig. 1. CFG of Quicksort Kernel

of the outer loop LR1. Once the search spaces of LR2 and LR3 overlap each other, the outer loop LR1 exits. B6 performs a final swap between the pivot and the last fetched element in LR2. The kernel itself is invoked in a divide-and-conquer fashion – not shown – on the elements that were not traversed in the current call by their respective loops LR2 and LR3.

The main characteristics of this code are its small, independent inner loops LR2 and LR3 with unknown loop bounds. Each inner loop has two memory load instructions to fetch the address of a node and its corresponding value(4, 5, 7, 8). Memory is written in the swap instructions 13, 14, 16 and 17.

An architecture with a centralized control paradigm that progresses in its computation along one locus of control fetches these kernel instructions in the total dynamic order in which they appear in the program. Hence, an instruction overlap between the inner loops in the computation can only occur at the LR2 exit.

To provide more opportunities for instruction overlap, the inner loops are speculatively unrolled during runtime, yet the memory wall met in each iteration can render the speculation ineffective. Prefetching techniques to warm up the cache alleviate the penalties introduced by the memory wall but are less effective in the presence of reference indirections such as in list traversals due to the lack of a guaranteed spatial locality. Depending on the size and organization of the data cache there might be some temporal cache locality inside the kernel when search spaces of LR2 and LR3 overlap or when values fetched from memory are reused in case of a swap operation. There might be temporal locality in successive divide-and-conquer calls to the kernel.

Now consider the opportunities for instruction overlap in the kernel that arise from a decentralized architecture paradigm such as in an adaptive fine-grain multithreaded execution model: Several strands for each inner loop LR2 and LR3 can execute concurrently without the need for LR3 to wait for the conclusion of LR2. That is, assume each instruction in the inner loops is a strand and that instances of strands 4, 5 and 6 execute concurrently to strand instances 7, 8 and 9. Since twice the amounts of loads are issued as opposed to a centralized control paradigm, loads can be serviced at least twice as fast, provided there is interleaved memory with adequate bandwidth. Also, the lack of guaranteed spatial locality as in an array does not exclude "localized" spatial locality, depending on how heap memory is populated with nodes during construction of the list. This "localized" spatial locality is exploited by the multithreaded paradigm: Latency incurred by the traversal of one loop might be hidden by "localized" spatial locality in another loop and vice-versa as the computation progresses. Finally, executing both loops concurrently is similar to the "Unroll-and-Jam" compiler technique – without the unrolling and the static schedule. The benefits reside in the exploitation of temporal locality encountered as the search spaces of LR2 and LR3 overlap. This effect will be especially pronounced in the numerous small leaf nodes of the Quicksort divide-and-conquer sort tree.

Another important property: An adaptive system only fetches and dispatches strands into the scheduling window if their predecessors are issued from the scheduling window. Such a just-in-time policy in conjunction with the partitioning of strands along long latency boundaries prevents clogging of the scheduling window with instructions that will not be ready for issue in the proximate future. Hence, the opportunity to find instructions for overlap in the scheduling window increases even further.

2   CARE Execution Model

Hierarchy of Threads, Strand Chains and Strands: In the CARE execution model, a program is written as a set of *threaded functions*. Each instantiation of a *threaded function*, called a *thread*, has its own context called a *frame*, which is a block of memory holding function parameters and local variables. As in some other multithreaded models [2, 3], frames are dynamically allocated from a heap, and are typically linked to one another in a tree structure, rather than being confined to a linear stack. This allows threads to run concurrently, even if they are instances of the same procedure. Threads can be created and destroyed at runtime, and threads can create new threads.

Under CARE's execution model, a thread is divided into smaller units called *strands*. A strand is a sequence of instructions that, once activated, should execute to completion without any stalls due to long and/or unpredictable latencies: i.e. due to memory dependencies, control dependencies and explicit synchronization instructions. In CARE, this form of execution is called *ballistic execution*. A strand will not start executing until all data needed by that strand is produced (except for intermediate results produced within the strand itself) and all memory and control dependencies are satisfied.

Instructions are grouped into a strand according to the following *Split Phase Constraint*.

- Two instructions cannot be placed in the same strand if there is a control dependence between them;
- A load instruction and any instruction which *may* depend on the load cannot be placed in the same strand, except when a compiler can assure that enough useful instructions can be scheduled between them to hide the latency.

A strand is allocated a *private register pad* – a small number of registers that retain intermediate values generated and passed among instructions within the same strand.

To reduce the overhead of managing a large amount of small strands in hardware, strands are assembled at compile time to larger units called *strand chains* (SCs). One or more SCs compose a thread. Strands in SCs are fetched and dispatched according to the total order imposed by the chain. Hence, SCs are a way of grouping strands such that they can be fetched with *program counters* (PCs) and dispatched sequentially from memory. SCs can be fetched concurrently if the processor supports *multiple PCs*. As with strands, the same SC can be fetched and/or dispatched multiple times as in the case of loops or function calls.

Figure 2(a) shows an example snapshot of a runtime structure of the two threads qs-l and qs-r – the left and right divide-and-conquer invocations of the Quicksort Kernel – with their code and data allocated in memory. Each thread also has a frame allocated on the heap. Both, the instructions and the frame of each thread are referenced through *context pointers*. Thread qs-l has two dispatched strands with register pads qs-l1 and qs-l3 allocated on a register file. Thread qs-r has one strand dispatched, qs-r3, allocated on a separate register file.

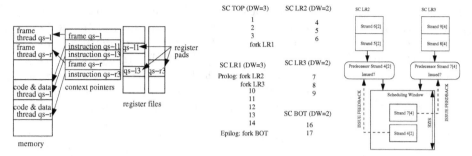

Fig. 2. (a) QS: Runtime Structure (b) SCs (c) Strand Fetch/Dispatch

Strand Fetch and Dispatch: As soon as a thread is invoked, the processor loads the addresses of the initial strands into PCs. PCs traverse SCs in sequential order, fetching and dispatching one strand after another until a *dispatch window* (DW) limit is reached or a resource limit is reached. A DW indicates how many strands in the SC should be monitored concurrently in the scheduling window.

For the Quicksort example in Fig. 2(b), the compiler generated 5 strand chains labelled TOP, BOT, LR1, LR2, LR3. SC TOP is the *Entry SC* that is enabled upon thread execution. Strands are labelled by their first instruction. Instances of a strand are denoted in brackets (e.g. strand 4, instance 3 is 4[3]). SC TOP has one strand 1 with instruction 1, 2 and 3. All other instructions form their own strands. E.g., SC LR2 has 3 strands: 4, 5 and 6.

The compiler augmented SCs with "FORK SC x" instructions that indicate to the processor that a new SC x is available for fetching. E.g., SC TOP forks SC LR1. LR1 is augmented with FORKs in its Prolog and Epilog. The Prolog forks the two inner loops LR2 and LR3 and the Epilog forks SC Bottom to conclude the Quicksort Kernel.

To support adaptive multithreading, each SC has an associated DW parameter e.g., SC A has a dispatch window of 3 – $DW = 3$. Note that SC LR1 has a $DW = 3$. Strands 10, 11, and 12 are probably fetched before loops SC LR2 and SC LR3 might finish execution. A less aggressive dispatch policy that tries to conserve scheduling window space could chain strands 10, 11 and 12 in a separate SC that is forked by SC LR2 and LR3.

Figure 2(c) shows a snapshot of a CARE processor with two active SCs – the inner loops LR2 (strands 4, 5, 6) and LR3 (strands 7, 8, 9)– fetching and dispatching instances concurrently into the scheduling window. In the example, each SC has a DW of 2. Strand 4 and strand 7 are dispatched to the scheduling window. Hence, each SC is allowed to fetch and dispatch another strand – 5 and 8. At this point, a *feedback* mechanism prevents any further strand dispatch until a strand is issued from the scheduling window.

Once the scheduling window issues a strand, the processor can estimate the time it will take for this strand to complete since its execution is *ballistic*. This information can prompt the fetch unit to begin *prefetching* successor strands to keep the scheduling window filled at capacity. Consequently, *predecoding* starts early as well.

Strand Enabling: A strand's enabling state is monitored by associating the strand with two firing rules maintained in hardware detailed in the next section: **(1)** The data firing rule maintained in the *Operand Synchronization Vector* (OSV). This rule tracks the data registers on which the strands depends. **(2)** The control and memory firing rule, also called the CSRF firing rule because it is maintained in the *Computation Structure Register File* (Sec. 4). This rule tracks the strand's outstanding control and memory dependencies that are not subsumed by a preestablished fetching order.

In the Quicksort example, the OSV for an instantiation of strand 5[·] is the new t2 operand produced by the load instruction 4[·]. The CSRF of strand 5[·] is the control predicate produced by the previous iteration. The CSRF of strand 5[·] is shared with strands 4[·] and 6[·].

3   CARE Compiler (CAREgiver)

CARE's compiler goal is to automatically generate the hierarchy of threads, SCs and strands introduced in the Execution Model Sec. 2. Flowchart Fig. 3 shows the steps involved in the compilation process. Inter-Thread data partitioning and synchronization, closely related to this work, has been elaborated in [4]. Intra-Thread compilation, detailed in the thesis [5], is divided into the following major phases: **(1)** The *attractor graph* does a high level partitioning of a program unit following the split phase constraint and some hight level policies regarding speculation. **(2)** The next step is a low level program unit partitioning that considers CARE's architectural features such as the handling of control transfer. The outcome is a *strand graph* (SG). **(3)** and **(4)** The next two phases establish the control and memory dependencies in the SG. **(5)** Now the compiler computes register allocation, control and memory predicates. This phase is specific to the architecture of the CSRF (Sec. 4). In case the register allocation fails, phases 2-5 are repeated with a register allocator constrained SG. **(6)** The CARE compiler decides when to fetch a strand relatively to other strands based on information such as available parallelism in the thread, number of PCs and scheduling window size.

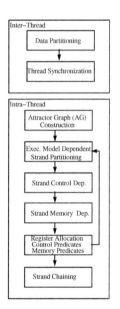

Fig. 3. Flowchart

4   CARE Architecture (CAREtaker)

Figure 4(a) presents the generic CARE architecture and its processor core. The processor core is divided along its *macro pipeline stages* into *a Fetch Unit, a Decoder Unit, a Dispatch/Issue Unit, some buffers* followed by *Functional Units* (FUs) and finally a *Write Back Unit*. Variables and constants that do not change across strands in a function are allocated to the Local Register File. Variables

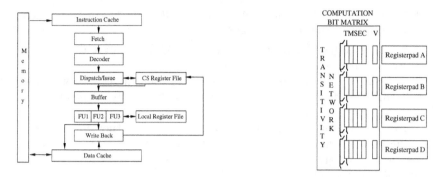

Fig. 4. (a) CARE Architecture: Core (b) CSRF

that are changed in a function and live across strand boundaries are allocated to the *Computation Structure Register File (CSRF)*.

CARE's processor core has an out-of-order *(OOO)-execution machine*, allowing different instructions of the program to execute concurrently at different rates. CARE also has an *OOO-Dispatch/Issue Unit* allowing strands to compete for decode time and reorder space. Finally, CARE has an *OOO-Fetch Unit* since several SCs in a thread are allowed to fetch concurrently. This last property alone changes the characteristic structure of a traditional OOO machine: There is no commit unit to sequentialize the concurrent instruction streams from the Functional Units since there is no single sequential fetch stream. The contract between the compiler and the architecture guarantees that there will not be any control or memory hazards between dependent instructions and therefore instruction results can immediately update the Register File as they become available. The contract stipulates that any false data dependencies – such as anti and output data dependencies – that might occur during runtime are to be avoided by the compiler: either through renaming, through synchronization at the fetch stage with a *FORK* instruction or synchronization at the dispatch/issue stage through the CSRF. A full description of the CARE architecture can be found in thesis [5].

The scheduling window in the Dispatch/Issue Unit is designed to support the strand enabling mechanism for register-based dependencies. It is assisted by a Computation Structure Register File (CSRF). Each element in the CSRF is uniquely named and contains status information required to implement CARE strand enabling rules. The CSRF is also equipped to hold dependence information related to predication and and to register map those memory accesses that cannot be resolved statically. A CSRF is partitioned at compile time, encoding different phases of the computation. The CSRF partition reflects precedence relations – control dependencies and memory dependencies – established at compile time.

In the example in Fig. 4(b), the CSRF is partitioned into register pads named A to D. Each register pad has a set of registers. Each register holds data and is associated with additional information in a *computation bit matrix* describing the data with respect to the computation state:

- The *valid* bit V. When disabled, identifies that the contents of the register depend on an outstanding operation.
- The *control* bit C. C acts as a control predicate. When enabled, — with the E bit disabled — identifies that the contents of the register depend on an outstanding control predicate evaluation.
- The *control evaluation* bit E. When enabled, signals that the control predicate evaluation is performed. The result of the control evaluation performed by a predicate evaluation instruction is stored in C.
- The *control speculate* bit S. When enabled, allows compiler assisted, dynamic control speculative execution of the strand. The S bit is determined by the compiler. The machine decides at runtime if speculation is necessary.
- The *memory precedence* bit M. When enabled, identifies that the contents of the register depend on an outstanding memory write operation. The M bit is determined by the compiler and reset by a memory write instruction.
- The *transitivity* T bit. When enabled, signals that the partial order control and memory precedence relations leading up to the contents of the register are satisfied. The register itself might still have the M bit enabled and/or an unevaluated control bit C.

Each register pad has a transitivity network as shown in Fig. 4(b). Each member register passes an enabled T bit to the next higher register if its own T bit is set, its M is not set and C is either unset, evaluated or S is set.

Each time the status in the computation bit vector for a register changes, the status is broadcast to the scheduling window. The strand firing rules with a CSRF are:

- OSV firing rule: All operands external to a strand that are allocated to registers are produced.
- CSRF firing rule: The CSRF firing rule for a strand is satisfied when the last result register in the strand (LRR) has its T bit set, the M bit is not enabled and the C bit is not enabled or an enabled C bit is positively evaluated or speculated. The computation status corresponds to the strand firing rule requirement that all control — possibly speculatively — and memory precedence relations for the strand are satisfied.

The transitivity network can wrap around to support loop execution with rotating register renaming. If deemed necessary by the compiler, neighboring register pads can interconnect their transitivity networks, effectively forming larger register pads.

Rotating register renaming requires the compiler to determine a tuple $L = (\omega, \alpha)$ of two parameters for each SC member of a loop (called *$\alpha\omega$-loop-bounding*): Parameter ω determines how many completed past iterations have to be preserved to execute the *current iteration*. The current iteration is the earliest iteration that is not yet completed. Parameter α determines how many iterations not completed — current and future iterations — are allowed to execute concurrently.

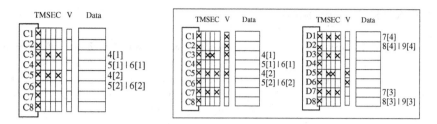

Fig. 5. QS: CSRF LR2, LR3 Allocation (a) Overall (b) Possible Snapshot

As an illustration, consider the inner loop LR2 – LR3 is analogous: The compiler determined an $L = (1, 2)$. The first parameter, ω, is 1 because the worst case loop carried dependency is 1. The second parameter, α, is 2 because the compiler decided to execute several overlapped instances of LR2 fitting one register pad: 1 current and 1 future instance – 2 futures would have fitted also.

Figure 5(a) shows an excerpt of a rotating register allocation for LR2, mapped onto CSRF pad C– with registers C1-C8 – with a transitivity network that loops T around the pad. Continuing with the LR2 example – LR3 is analogous, each loop instance is allocated to two registers: e.g., instance 1 has strand 4 allocated to C3 and strands 5 and 6 allocated to C4. Similarly, 4[2], 5[2] and 6[2] are allocated to C5 and C6. Eventually, 4[4], 5[4] and 6[4] allocate to registers C1 and C2. Note that strand 6[·] is also allocated to a CSRF. The predicate evaluation instruction in strand 6[·] does not write any data register, only C bits in the computation bit matrix. Yet, it still requires a CSRF allocation to track its partial order in the computation. Each register allocation also requires presetting the computation bit matrix with special *CARE Maintenance* instructions that also support repetitive tasks as in loops. That means for the example that for each newly allocated instance, the V and E bits are reset. Also, the S and C bits are set for each odd numbered register in the C pad. The C bit identifies a control predicate to enable the strand 4[·] since it is the first strand of a new iteration. The S bit identifies speculation. The transitivity bit T would not propagate past an unevaluated C bit unless the S bit is set. Since all C bits are masked with an S bit, all T bits in the C pad are set. Eventually, a control evaluation will exit the loop and enter a loop epilog – not shown – that copies the live-out results produced in the loop to compiler known registers.

Figure 5(b) shows a possible CSRF snapshot of LR2 and LR3 concurrent execution: LR2 is executing instances 1 and 2. The control dependence of instance 1 is evaluated to continue iterating. 4[1] produced a result. 4[2] which depends on 4[1] also produced a result. The elements 5[1] and 5[2] are still being loaded. Concurrently, LR3 is already executing instances 3 and 4. Both are being executed speculatively and no values have been produced so far for these instances.

5   CARE Experiments

In this section we give an overview of the results we gathered with our CAREgiver compiler framework. The parameters that are of most interest are the strand sizes, the exposed parallelism and the dependencies our particular CAREtaker architecture has to manage. Since the compiler has not been presented in more detail; only a cross section of strand parameters will be presented here. We chose the industrial strength MIPSPro compiler as the platform upon CAREgiver has been built. All the compiler phases are briefly presented in Sec. 3 were implemented almost in their entirety in the compiler. The benchmarks we ran are the SPEC2000 programs coded in C. Each function in a benchmark is a compilation unit whose individual experimental results contribute to the overall trends shown in the histograms and tables. Except for the control dependence histogram of Fig. 6(f), all charts depict *clustered histograms*: The histograms of each function in a benchmark are combined to form an overall histogram for the gzip benchmark. **1:** Each entry in a clustered histogram is plotted as a vertical line and is derived from an entry of a function histogram. **2:** The y-axis is plotted logarithmically, with the origin at 0.1 and increments to the power

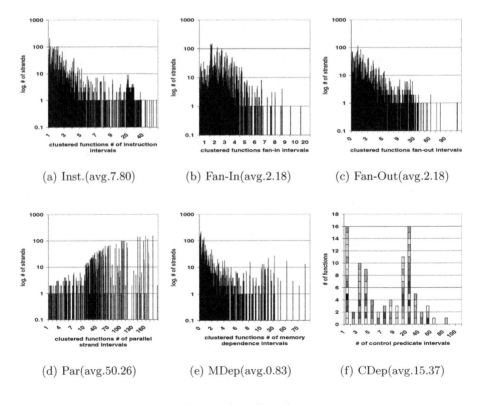

(a) Inst.(avg.7.80)   (b) Fan-In(avg.2.18)   (c) Fan-Out(avg.2.18)

(d) Par(avg.50.26)   (e) MDep(avg.0.83)   (f) CDep(avg.15.37)

Fig. 6. Gzip Strands

of 10. Figure 6(f) is an exception which will be described later. **3:** The x-axis depicts intervals, with the origin at either 0 or 1. It is incremented by 1 until the 10th interval, after which the increments are by decades. Numbers '16-24' would e.g. fall into the '20 interval' and so forth. The x-axis extends until the last populated interval and no further.

Histogram 6(a) plots the strand sizes: The maximum strand size is around 50 instructions with an average strand size of 7.80 instructions.

For the rest of the benchmarks, the average strand size ranges between 4.28 instructions for *mcf* to 10.74 instructions for *bzip2*. The median is either 1 or 2 instructions evenly distributed over all benchmarks. At the 3 quartile mark, the strand sizes range from 2-3 instructions. For the 4 floating point benchmarks coded in C, the average strand size ranges from 3.62 instructions for *art* to 7.75 instructions for *mesa*. The scheduling window will yield more than 1 instruction per entry due to one third to one quarter of the strands of a program.

Figure 6(b) plots the number of strand predecessors over all functions. The strand bulk has 1-5 predecessors. The "longest_match" function even has a strand with 17 predecessors (20 interval). The average number of predecessors for a strand in is 2.18. Figure 6(c) shows the number of successors ranging from 0 to some strands with successors in the 120 range. Most of the strands have successors in the interval from 0-30. The average number of successors is 2.18. These numbers are also confirmed for all other benchmarks: The average Fan-In, Fan-Out numbers for all benchmarks are ∼ 2. Fan-In and Fan-Out data identifies synchronization patterns. Architecturally speaking, the Fan-In number indicates how many tags have to be checked concurrently in the worst case in the scheduling window if speculation is allowed and firing delay is to be avoided. The number of concurrent tags impacts negatively on the scheduling window's complexity. Also of interest is the large number of single predecessor nodes, suggesting a large number of single sourced monadic instructions. The Fan-Out histogram is insightful as well: A synchronization mechanism solely based on synchronization counters would not fare well in a microarchitecture with such a large number of nodes exhibiting a large number of successors. This calls for a mechanism that shares synchronization events among several recipients such as broad or multicasting.

Strand parallelism before register allocation is plotted in the Fig. 6(d). It ranges from no parallelism in interval 1 to strands that have up to 180 parallel strands. In the histogram, there is a concentration of strands in the interval range from 10 to 60 strands. The average parallelism per function is 50.26 strands. The Tables 1 and 2 show max. and avg. strand parallelism for each benchmark. The average strand parallelism lies between 43.24 and 237.52 strands for the SPEC2000 integer benchmarks, and between 52.58 and 76.80 strands for the 4 floating point benchmarks. A topic to be fuller explored in the future is the strand parallelism that may be curtailed by the register allocator. The parallelism data is important in order to assess the required scheduling window size. To avoid delayed firing of strands due to a limited instruction window size, the window should be capable of holding up to 250 strands and be capable of in-

Table 1. SPEC00INT

	max.PAR	avg.PAR
gzip	316.44	50.26
vpr	596.09	52.94
gcc	5735.66	84.17
mcf	158.32	43.24
crafty	5316.7	237.52
parser	664.12	32.16
perlbmk	2154.41	58.86
gap	1468.3	77.95
vortex	1405.51	58.54
bzip2	484.20	42.56
twolf	1455.37	145.24

Table 2. SPEC00FL

	max.Par	avg.Par
mesa	1249.47	52.58
art	198.76	50.94
equake	733.14	67.74
ammp	980.94	76.80

specting the firing rules of those strands concurrently. The parallelism displayed in the scheduling window does not necessarily carry through to the execution units. Control- and/or memory-dependencies in conjunction with the machine's speculation policy might squash some of the parallelism exposed in the window.

Figure 6(e) depicts memory dependencies. The experiments did not perform alias analysis and therefore these numbers are to be interpreted as a worst case scenario. Most of the strands exhibit 0 or 1 memory dependencies. One pathological function has over 10 strands with more than 80 memory dependencies each. Experiments on the other benchmarks underscore the *gzip* example with average strand memory synchronization ranging from 0.62 for *bzip* to 1.45 for *twolf*.

Figure 6(f) is not clustered as previous histograms but additive. Each function in *gzip* contributes one unit in the interval. In the plot, most functions require checking 1-4 or 10-20 control predicates. The average number of control predicates is 15.37 for gzip. The plot confirms the trend observed in other benchmarks. The range of required average control predicates lies between 10.92 and 44.37 for all the benchmarks. Plots 6(e)(f) are important to assess the requirements for the CSRF in terms of M and C bits.

6   Related Work

This paper touches a vast body of research by other groups accumulated in recent years in the areas of multithreaded and decentralized architectures as well as compilers that is just too numerous to be mentioned individually. We therefore just highlight, without any claim of completeness, a few papers that more directly influenced the CARE architecture.

As a comprehensive background on multithreading, consult [6–9]. Survey articles on the evolution of fine-grain multithreaded architectures are in [10]. For an overview of fine-grain multithreading, in the context of EARTH, refer to [3, 11]. CARE's threaded function invocation is very similar to EARTH's. A recent predecessor to CARE is the Superstrand architecture [12].

7   Summary and Conclusions

This paper has given an overview of architecture features that allow fine-grain fetch, dispatch and synchronization mechanisms in the framework of a decentralized control paradigm. The Quicksort kernel showcases how such an architecture can be beneficial. CARE relies on compiler technology to generate threads, strand chains and strands. The preliminary experimental compiler data suggests that there is parallelism to be exploited that has to be managed judiciously in the scheduling window and synchronized e.g., in a CSRF.

Acknowledgements

We acknowledge the support of the current research by NSF: grants NSF-NGS-0103723, NSF-CCR-0073527, and DOE: grant DE-FC02-01ER25503. We also acknowledge support from the Defense Advanced Research Projects Agency (DARPA) and the National Security Agency (NSA). We also wish to thank many present and former members of our group at CAPSL (University of Delaware) as well as ACAPS (McGill) who have contributed in many ways to the essential ideas of decentralized architectures.

References

1. Aho, A.V., Sethi, R., Ullman, J.D.: Compilers — Principles, Techniques, and Tools. Corrected edn. Addison-Wesley Publishing Company, Reading, MA (1988)
2. Culler, D.E., Goldstein, S.C., Schauser, K.E., von Eicken, T.: TAM – a compiler controlled threaded abstract machine. JPDC **18** (1993) 347–370
3. Hum, H.H.J., Maquelin, O., Theobald, K.B., Tian, X., Gao, G.R., Hendren, L.J.: A study of the EARTH-MANNA multithreaded system. JPP **24** (1996) 319–347
4. Tang, X., Gao, G.R.: How "hard" is thread partitioning and how "bad" is a list scheduling based partitioning algorithm? Proceedings: SPAA '98 (1998) 130–139
5. Márquez, A.: Compiler Aided Reorder Engine (CARE) Architectures. PhD thesis, University of Delaware, Newark, Delaware (TBA 2003)
6. Agarwal, A., Lim, B.H., Kranz, D., Kubiatowicz, J.: APRIL: A processor architecture for multiprocessing. Proceedings: ISCA-17, (1990) 104–114
7. Alverson, R., Callahan, D., Cummings, D., Koblenz, B., Porterfield, A., Smith, B.: The Tera computer system. Proceedings: ICS (1990) 1–6
8. Gao, G.R., Bic, L., Gaudiot, J.L., eds.: Advanced Topics in Dataflow Computing and Multithreading. IEEE Computer Society Press (1995)
9. Tullsen, D.M., Eggers, S.J., Levy, H.M.: Simultaneous multithreading: Maximizing on-chip parallelism. Proceedings: ISCA-22 (1995) 392–403
10. Najjar, W. A., E.A.L., Gao, G.R.: Advances in dataflow computational model. Parallel Computing (1999), North-Holland, Vol. 25, No. 13-14, 1907–1929
11. Theobald, K.B.: EARTH: An Efficient Architecture for Running Threads. PhD thesis, McGill University, Montréal, Québec (1999)
12. Márquez, A., Theobald, K.B., Tang, X., Gao, G.R.: A Superstrand architecture and its compilation. Proceedings: Workshop on Multithreaded Execution, Architecture and Compilation (1999). Held in conjuction with HPCA-V

Numerical Simulator III –
A Terascale SMP-Cluster System
for Aerospace Science and Engineering:
Its Design and the Performance Issue

Yuichi Matsuo

National Aerospace Laboratory of Japan,
7-44-1, Jindaijihigashi, Chofu, Tokyo 182-8522, Japan
matsuo@nal.go.jp

Abstract. National Aerospace Laboratory of Japan has introduced a new terascale SMP-cluster-type parallel supercomputer system as the main engine of Numerical Simulator III, which is used for aerospace science and engineering purposes. It started its full operation in October 2002. The system has computing capability of 9.3Tflop/s peak performance and 3.6TB user memory, with about 1800 scalar processors for computation. It has a mass storage consisting of 57TB disk and 620TB tape library, and a visualization system tightly integrated with the computation system. In this paper, after reviewing the history of Numerical Simulator I and II, we will describe the requirements and design of the third generation system, focusing on its characteristic points. We will also show some performance measurement results and discuss the system effectiveness for typical applications at NAL.

1 Introduction

Remarkable development in processor technology is pushing up computing power day by day. However, parallel processing is inevitable for large-scale numerical simulation such as in Computational Fluid Dynamics (CFD). Meanwhile, shared-memory architecture is becoming a trend because of its programming ease. National Aerospace Laboratory of Japan (NAL) had been operating a distributed memory parallel supercomputer system consisting of 166 vector processors with 280Gflop/s peak performance for aerospace science and engineering since 1993. We have at last replaced it with an SMP-cluster-type distributed parallel system, consisting of about 1800 scalar processors, with a peak performance of 9.3Tflop/s, and 3.6TB of user memory. The new system is called Numerical Simulator III (NSIII). In this paper, after briefly reviewing the historical background of the previous Numerical Simulators, we will describe the new terascale SMP-cluster parallel computing system, and finally present performance evaluation result and discuss the system characteristics.

A. Veidenbaum et al. (Eds.): ISHPC 2003, pp. 39–53, 2003.

2   Numerical Simulator

The idea of Numerical Simulator (NS) has its origin in 1987 when 1Gflop/s vector-type supercomputer Fujitsu VP400 was installed in NAL, the first in Japan. In the NS project, we were aiming at making numerical simulation, particularly CFD, a useful tool for the R&TD of aerospace vehicles, offering less cost and less time when compared to traditional methods such as wind tunnel experiments or flight tests. At the first stage, we tried to perform numerical simulation on supercomputer to obtain higher-resolution results and more accurate aerodynamic data than that previously achieved. We succeeded in analyzing flow around a three-dimensional wing by numerically solving the Navier-Stokes equations (Navier-Stokes-based CFD).

In the second generation system (NSII), we intended to have an over-a-hundred-times faster computer system than NSI in order to apply Navier-Stokes-based CFD to more practical subjects such as parameter study for designing aircraft elements and aerodynamic analysis of a complete aircraft configuration. Eventually a distributed parallel supercomputer called Numerical Wind Tunnel (NWT) was installed at 1993. The name of NWT came from the concept that it would surpass wind tunnel tests in terms of both time and cost saving. NWT was a vector-parallel supercomputer, which had 166 vector processors with 280Gflop/s peak performance, and was the world's fastest supercomputer for the first three years of its service. With NWT, we have entered 100Gflop/s performance era, while at the same time completed the Navier-Stokes-based CFD tools. Figure 1 shows typical results in both engineering applications and scientific studies. In engineering applications, with the advent of multi-block CFD code, the objects to be solved were becoming larger in scale and more complicated, requiring 5-10M (1M denotes one million) grid points . Then, it took 10 hours or more to get a steady solution on NWT. In science studies, Direct Numerical Simulation (DNS) was becoming a more common tool for solving fundamental fluid dynamics problems concerning turbulence and combustion, where 10-50M grid points were used. In those days, parallelism was widely used by NWT users. The parallel code development and its execution was conducted with either a message-passing model via MPI or a data-parallel model via NWT-Fortran. However, NWT was a traditional supercomputer, and thus its interoperability and manageability was a concern. In addition, its operation rate was reaching about 90% in its last few years, signaling that it was time to replace it with a new machine.

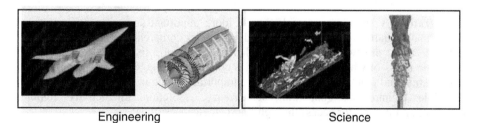

Engineering   Science

Fig. 1. Typical results from NWT.

3   Requirements to NSIII

In this section, we discuss about what we expect from the third generation Numerical Simulator system (NSIII). There are two main target applications of NSIII: (1) a practical use of the Navier-Stokes-based CFD for aerospace engineering R&TD; e.g. coupled analysis of multiple disciplines such as fluid dynamics and structural dynamics, and high-fidelity optimum design of aerospace vehicles by utilizing Navier-Stokes-based CFD, and (2) a very large-scale scientific simulations; e.g. detailed investigation of turbulence or combustion phenomena, and meteorological forecast and environmental issue concerning aircraft flight. For application (1), in particular, 'practical use' requires the system to be able to obtain the final solution within one day. In application (2), 'very large-scale' corresponds to the grid size of over 1G (one billion) points. In addition, we wish to standardize the use the numerical simulations as tools in aerospace R&TD.

Performance and Memory: In order to carry out more complex, larger-scale Navier-Stokes-based CFD than those performed on NWT, with the frequency of dozens of times per day, NSIII needs to perform one hundred times faster (about 10Tflop/s). Meanwhile, the system memory should be large enough to handle 1G-point scientific simulation, requiring NSIII to have memory capacity of one hundred times larger than NWT (about 5TB).

Data Management: In terms of developing a practical and useful system, we focus on how to manage and manipulate large amount of output data. One of the lessons learned from the operation of NWT is the importance of mass storage for large-scale simulations. As simulations become more complicated, data I/O becomes larger and more frequent, and lack of storage quantity might even lead to a bottleneck in the overall simulation work-flow. Therefore, we concluded that we need some hundreds TB of mass-storage for NSIII.

System Area Network: System area network, which connects subsystems to each other (e.g. computing system to mass-storage, or to visualization system), is another important issue when it comes to a system with Tflop/s performance and TB memory. The network speed between subsystems should be fast enough to manage large amount of data. Furthermore, the architecture including the file system should be a standard one for a secure RAS. Thus we required the target speed (bandwidth) to be 1GB/s.

Visualization: Visualizing terascale data is more important and more difficult than the normal visualization because data reduction and comprehension techniques are required. There are two aspects in visualization; what to see, and how to make it visible. In terms of what to see, technologies concerning large data manipulation, fast visualization and feature extraction, etc. are important, while in terms of how to make it visible, technologies concerning screen size, resolution, interactivity, remote visualization function for access through internet, etc. are critical.

System Availability Aspects: Compatibility of resources (i.e. programs and data) with the previous NWT system is one of the most important issues for NSIII. If not taken into account, a lot of time and labor would be required for system transition. Another important issue is standard and common characteristics of software to make

NSIII usable by a much larger community. Single system image is also critical from the usability viewpoint. In addition, possible future extension to grid computing or web computing will be included. The compiled requirements of NSIII are summarized in Table 1.

Table 1. Summary of requirements to NSIII.

Issue	Requirements
Performance	1) 10Tflop/s 2) 5TB user memory 3) 500TB mass storage 4) 1GB/sec data transfer rate
Functionality	5) Resource compatibility with NWT 6) Standard and common characteristics of software 7) Single system image 8) Extension to grid computing and web computing

4   Basic Design of NSIII

How do we realize a scalable computer system with 10Tflop/s of peak performance? There are only a few systems with performance of 10Tflop/s or more in the world [1,2], so we have no general discipline to follow in order to construct this type of machine. At present, no processor chip, either vector or scalar, is faster than 10Gflop/s. Therefore, to obtain 10Tflop/s in total performance, 1000 or more chips are needed. We decided that the current PC-cluster systems are not appropriate as our choice, due to the uncertainty in reliability and manageability for a system with 1000 or more CPUs. According to our experience with NWT, we have the following empirical formulae for the system configuration: 1) shared memory architecture is convenient within a node in terms of programming ease, and 2) interconnect topology between nodes should be of crossbar-switch type due to better scalability. Furthermore, from the cost viewpoint, the total number of nodes should be less than 100. Thus, if the system is to satisfy the requirement mentioned above, each node should have over 100Gflop/s performance with over 50GB of shared memory.

The next issue is the mass storage of over 500TB. It is difficult for us to obtain over 500TB of storage using only disk, due to the cost overhead. Thus, the storage system should be a combination of disk and tape, with a data management system between them for an effective usage. Another related issue is how to achieve 1GB/s data transfer bandwidth between subsystems. We did not adopt so-called parallel I/O through multiple nodes because the file system will become complicated. Instead, we adopted a strategy to use conventional I/O through a specific I/O node.

The I/O node is a special-purpose node that handles I/O process only. The node is connected to mass storage. It is impossible to achieve 1GB/s speed using only single connection with the current technology, so we introduced a striping technique. In addition, in order to achieve single system image, a high-speed and transparent network file system is introduced. Shown in Fig. 2 is the conceptual design of NSIII, which is discussed in more detail in Ref [3].

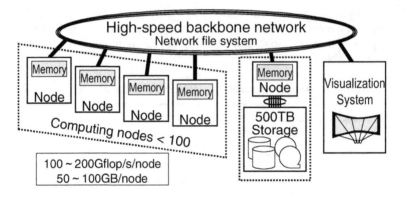

Fig. 2. Conceptual design of NSIII.

5   System Overview of NSIII

5.1   Hardware

After the detailed considerations described above, we introduced the new terascale SMP cluster-type supercomputer system as NSIII in 2002. Its configuration is described in Fig. 3.

The whole system consists of 4 subsystems; computing subsystem, mass storage subsystem, visualization subsystem, and network subsystem.

The computing subsystem is called Central Numerical Simulation System (CeNSS). We have 18 cabinets (each is Fujitsu PRIMEPOWER HPC2500), where a cabinet is the physical unit of hardware. A cabinet includes 128 CPUs with 256GB shared memory, and can act as a 128-way symmetric-multi-processor (SMP) system

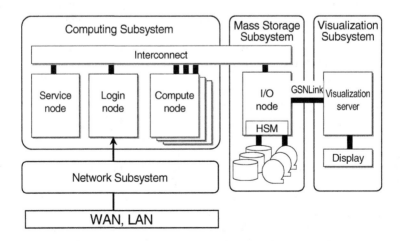

Fig. 3. System configuration of NSIII.

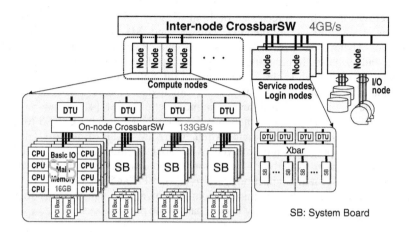

Fig. 4. CeNSS configuration in detail.

in its maximum configuration. The CPU is SPARC64 V scalar chip with 1.3GHz clock, implementing conventional scalar speed-up technologies such as prefetch, out-of-order instructions and 4 floating-point operations per cycle. Thus, the theoretical peak performance of one cabinet becomes 665.6Gflop/s. 14 cabinets are dedicated for computation, giving a total peak computing performance of 9.3Tflop/s and 3.6TB of memory. A cabinet can be partitioned into either 2 or 4 nodes according to need, where a node is the logical unit in terms of operating system. Here, each cabinet is partitioned into 4 nodes where each node is 32-way SMP, giving a total of 56 computation nodes. For the remaining cabinets, 3 are partitioned into 2 nodes each, to give 6 nodes with 64-way SMP. Four of these are used for service nodes and 2 are for login nodes. Note that the service nodes are used for multiple purposes, such as compilation, debugging, TSS, and ISV-application execution, while the login nodes are used in a round-robin fashion for entrance management. The remaining cabinet is a 128-way SMP, used as a specific I/O node. All nodes are connected to a crossbar interconnect network through one or two data-transfer-units (DTUs) per node with 4GB/s bi-directional bandwidth. The detailed structure of CeNSS is shown in Fig. 4.

The mass storage subsystem is called Central Mass Storage System (CeMSS), which is composed of 57TB FC RAID disk (Fujitsu PW-D500B1) and 620TB LTO tape library (IBM3584). The disk system is connected to I/O node through 80 Fiber Channel (FC) links, while the tape library through 40 FC links. RAID5 was constructed for the disk storage, securing a total of 50TB of user area. The tape library consists of 40 drives.

We introduced striping technique to attain 1GB/s data transfer bandwidth. From a benchmark test, we decided to apply 16 FC stripes. To manage large amount of data transfer between disk and tape, we introduced Hierarchical Storage Management (HSM) software, so user does not have to worry about whether file location is on disk or tape.

The visualization system is called Central Visualization System (CeViS), consisting of a visualization server, a large-sized flat-screen tiled display system, and a number of graphics terminals. The visualization server (SGI Onyx3400) has 32CPUs and 64GB of shared memory available. The large-sized tiled display system is called Aerovision, with a screen size of 4.6m wide by 1.5m high, and a resolution of 3300 by 1028 dots. Edge-blending, stereo, and 3D pointing functions are implemented. Using the visualization server and Aerovision, we can easily manipulate and visualize large amount of output data from the CeNSS system. The CeViS system is connected to the CeNSS system through GSNLink, which is a high speed connection employing 4 stripes of Gigabyte System Network (GSN) connection. The CeViS system is described in more detail in Ref. [4].

The network system is called NS network; NSnet, which connects CeNSS to LAN and WAN through Gigabit Ethernet and switch. There are 1.5TB of network-attached disk, in which user programs are archived. A number of important servers such as DNS and NIS are also available. Connection to the internet is provided through firewall, using only secured connections such as SSH, VPN and https.

In Fig. 5, the layout of the computing room is shown. It is worth noting that the operational cost in terms of electricity is only half of the previous system.

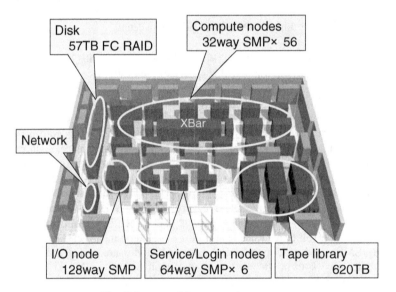

Fig. 5. Layout of the computing room.

5.2   Software

Various factors concerning software should be taken into account in order to achieve efficient and reliable operation in such a large computer system as NSIII. As far as CeNSS is concerned, the software system consists of three layers, as shown in Fig. 6. As the operating system, we adopted Solaris8, which is a mature and standard 64-bit UNIX operating system, considering its portability, interoperability and scalability. With Solaris8 operating environment, most of the standard facilities associated with a UNIX workstation can be made available to users.

Over the operating system, we have a middleware layer, which provides various environments to users. High-performance parallel environment provides function-alities concerning memory usage and file system for high-speed and large-scale computing. Large page is an option to enhance the performance of memory access for huge programs. Shared rapid file system (SRFS) is a high-speed network file system, which can utilize interconnect and ensure file compatibility. Using SRFS, users will see only one unique file system from any node.

Job launching/control environment provides job-management-related system. We are accommodating various job types, from conventional batch job to interactive job, group job, and so on. Consequently, job scheduling is a very important issue to operate a huge computing resource efficiently. We have developed our own intelligent job scheduler called NSJS, which can smoothly manage all types of job using a generic job class, instead of defining many different job classes.

Programming developing environment is also crucial for parallel system. In principal, we use the 'thread parallel' model within a node, with automatic parallelism or OpenMP, whereas between nodes we use the 'process parallel' model, with MPI or XPFortran, as shown in Fig.7. Codes that use 'process parallel' only can also be run. Since we have already been accustomed to our own data parallel language, NWT-Fortran (similar to HPF), we will also use its successor, XPFortran. This programming style is very convenient for us, because if we think of one NWT processor mapped onto a node of CeNSS, the transition in programming from NWT to CeNSS becomes quite natural, particularly in parallel coding.

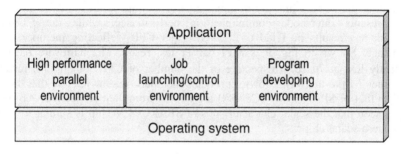

Fig. 6. Software configuration for CeNSS.

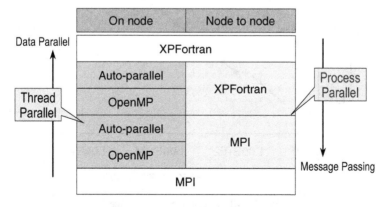

Fig. 7. Programming paradigm in NSIII.

As for application softwares, we have introduced NASTRAN into the system, a standard structural analysis package in aerospace engineering, and some other tools for visualization or web-based access. Online visualization is considered very important, so we developed a connection-software operating on GSNLink. Web-based access to NSIII will become inevitable in the future, and this leads to the development of web-based access software called WANS (Web-Access to NS). Using WANS, we can execute a number of operations concerning file and job through internet.

As for CeViS, we have installed well-known commercial visualization software packages, such as AVS, EnSightGold, FIELDVIEW, and graphics libraries such as OpenGL, Performer and Volumizer. In order to visualize very large amount of data, we have developed a specific volume visualization tool using volume rendering. In addition, because the software runs on GSNLink, we developed a communication library based on the ST protocol, called STF, to enable faster data communication between CeNSS and CeVIS, as compared to TCP/IP protocol.

6   Performance Evaluation

On the basic performance of CeNSS, the MPI ping-pong test result shows 3.58GB/s bandwidth at the peak, while the minimal latency is 4.3μsec, as indicated in Fig. 8. This represents inter-node communication performance, while the STREAM benchmark test results in Triad is 2.254GB/s per CPU, reflecting memory access performance. Shown in Fig. 9 is MPI barrier test result. The hardware barrier is consistently low at 7μsec, irrespective of the number of CPU. In Fig. 10, result of basic kernel 7 from the well-known Euroben benchmark is shown. Note that the data other than that of SPARC64 V (CeNSS) are quoted from their respective web pages. It can be seen that the scalar characteristics of SPARC64 V chip is similar to that of other modern scalar chips.

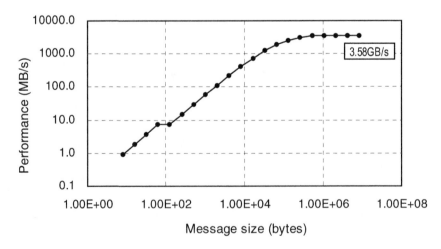

Fig. 8. MPI ping-pong performance.

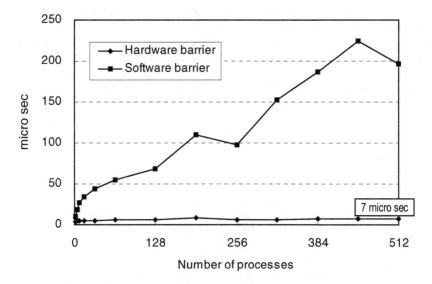

Fig. 9. MPI barrier performance.

Kernel 7: DotProd -- s = s + x1(i)*x2(i), i = 1, n

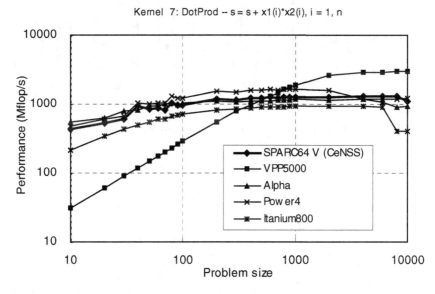

Fig. 10. Euroben benchmark Kernel7 performance.

In a standard LINPACK benchmark test, the sustained peak performance $R_{max}=$ 5.406Tflop/s is obtained with $N_{max}=658,800$. For the measurement, the system is reconfigured such that each node becomes 64-way SMP, and the job configuration of 63 threads by 36 processes is used, giving $R_{peak}=11.98$Tflop/s.

As for NAL's applications, first the performance of single CPU is measured. We examined 5 typical CFD codes for single CPU; four are Navier-Stokes-based CFD codes and one is a simple panel-method code. In Fig. 11, the sustained performance results between VPP300 and CeNSS are compared, where 'VPP300' represents the result obtained from VPP300 with one processor, 'CeNSS' is for CeNSS without tuning, and 'CeNSS-T' is with tuning. VPP300 has a vector processor whose peak performance is 2.2Gflop/s. Generally, after tuning, around 400 – 700Mflop/s performance can be obtained on CeNSS, which is comparable to the VPP300. It should be noted that only typical types of tuning are applied; for example, exchange of array index and loop rearrangement in order to decrease L2-cache miss. OpenMP is employed if auto-parallelization does not work. Computation is overlapped with communication to cancel the communication overhead.

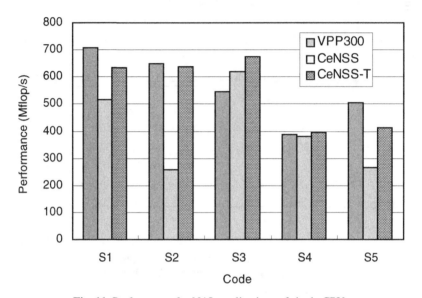

Fig. 11. Performance for NAL applications of single CPU.

Table 2. Specifications of NAL parallel applications.

Code	Application		Simulation Model	Numerical Method	Parallel strategy	Language
P1	Aircraft	(Engineering)	LES	FDM	OpenMP+MPI	F77
P2	Combustion	(Science)	DNS	FDM	OpenMP+MPI	F77
P3	Turbulence	(Science)	DNS	FDM+FFT	OpenMP+XPF	F77
P4	Helicopter rotor	(Engineering)	URANS	FDM	AutoParallel+XPF	F77

LES: Large-Eddy Simulation
DNS: Direct Numerical Simulation
URANS: Unsteady Reynolds-Averaged Navier-Stokes

FDM: Finite Difference Method
FFT: Fast Fourier Transform

Next, the performance of NAL's parallel applications is measured. As indicated in Table 2, four types of applications with different memory-access/data-transfer features are chosen. All codes are tuned. In Fig. 12, the code characteristic is plotted schematically. Note that the figure shows only the relative position of the codes, not their absolute values. Code P2 is the most CPU intensive application in which arithmetic operations are dominant, while Code P4 is located at the opposite side where both memory access and data transfer are highly loaded, a severe situation for the system.

In Figs. 13 and 14, scaled-up test results for Codes P1 and P2 are shown, respectively. In both cases, the performance is measured at the condition of fixed 30 threads using OpenMP, with the number of processes using MPI is ranging from 1 to 72 . The scalability is very good up to 72 processes. Both codes exceed 1 Tflop/s of sustained performance. In the case of Code P2, which is used for hydrogen-combustion simulation, over 1.2Tflop/s performance is obtained. In both applications, the simulation size and the time taken are very large, so fine scalability is important.

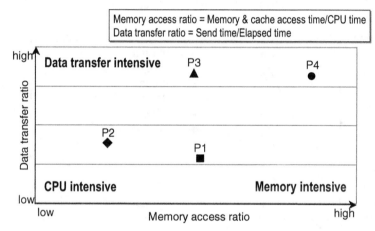

Fig. 12. Comparison of the NAL parallel application characteristics.

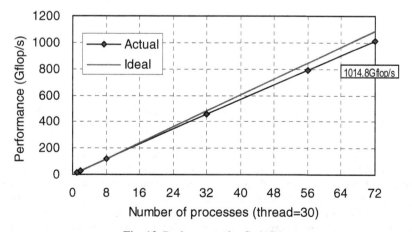

Fig. 13. Performance for Code P1.

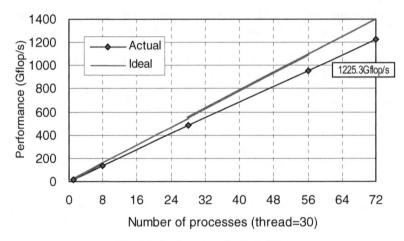

Fig. 14. Performance for Code P2.

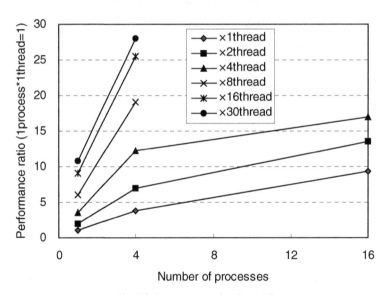

Fig. 15. Performance for Code P3.

Speed-up test results for Codes P3 and P4 are shown in Figs. 15 and 16, respectively. It can be seen in Fig. 15, that even with the same number of CPUs, the performance is better with 4 threads by 4 processes than that with 1 thread by 16 processes, which means that in this case the data communication through shared memory within a node is faster than that through the crossbar interconnect. We can also find by comparing Figs. 15 and 16, that the process scalability is better for Code P4 than for Code P3. After analyzing the data communication, we find that this is because data transfer packet for one communication cycle is larger in Code P3 than in Code P4, resulting in heavier load on the crossbar interconnect. Code P3 contains FFT algorithm, which causes relatively large data transfer between nodes. Improving the performance of Code P3 will be left for future work, due to its long computation time. The reason of worse thread scalability in Code P4 is large L2-cache miss. We concluded that a job needs to be designed carefully, taking into consideration the appropriate combination between the numbers of threads and processes. To establish the simplest way to optimize job design is another future work.

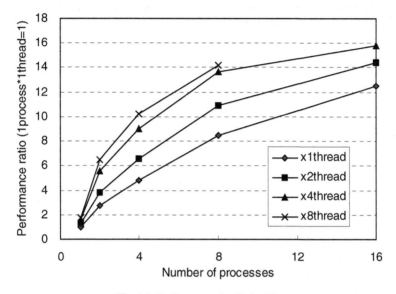

Fig. 16. Performance for Code P4.

7   Summary

NAL's Numerical Simulator III has been described and its performance measurement results have been presented and discussed. At present, although cluster-type systems, particularly PC clusters, are gaining popularity even for HPC systems, due to its cost-performance, we think that the system RAS capabilities are still not sufficient for a terascale system. We have designed the NSIII system based on our experience with large parallel system, and believe that our choice is the best for the current technology trend. The NSIII has started its operation in October 2002, and it has been playing its role as the main computing engine for aerospace science and engineering at NAL. We have confirmed Tflop/s of sustained performance in real-life applications. Further performance improvement on the software side remains as future work.

Acknowledgement

The author would like to thank NSIII development and operation teams for their invaluable support. The author would also like to thank Fujitsu HPC team for their cooperation in this work.

References

1. ASCI Program, `http://www.llnl.gov/asci/`
2. Earth Simulator, `http://www.es.jamstec.go.jp/esc/eng/index.html`
3. Matsuo, Y., et. al, "Numerical Simulator III – Building a Terascale Distributed Parallel Computing Environment for Aerospace Science and Engineering," *Parallel Computational Fluid Dynamics, New Frontiers and Multi-Disciplinary Applications*, Elsevier (2003), 187-194.
4. Matsuo, Y., "An Immersive and Interactive Visualization System for Large-Scale CFD," In *Proceedings of 4th ASME/JSME Joint Fluids Engineering Conference*, FEDSM2003-45201, June 2003.

Code and Data Transformations for Improving Shared Cache Performance on SMT Processors

Dimitrios S. Nikolopoulos

Department of Computer Science, The College of William & Mary,
McGlothlin–Street Hall, Williamsburg, VA 23187–8795, U.S.A.,
dsn@cs.wm.edu

Abstract. Simultaneous multithreaded processors use shared on-chip caches, which yield better cost-performance ratios. Sharing a cache between simultaneously executing threads causes excessive conflict misses. This paper proposes software solutions for dynamically partitioning the shared cache of an SMT processor, via the use of three methods originating in the optimizing compilers literature: dynamic tiling, copying and block data layouts. The paper presents an algorithm that combines these transformations and two runtime mechanisms to detect cache sharing between threads and react to it at runtime. The first mechanism uses minimal kernel extensions and the second mechanism uses information collected from the processor hardware counters. Our experimental results show that for regular, perfect loop nests, these transformations are very effective in coping with shared caches. When the caches are shared between threads from the same address space, performance is improved by 16–29% on average. Similar improvements are observed when the caches are shared between threads from different address spaces. To our knowledge, this is the first work to present an all-software approach for managing shared caches on SMT processors. It is also one of the first performance and program optimization studies conducted on a commercial SMT-based multiprocessor using Intel's hyperthreading technology.

Keywords: multithreaded processors, compilers, memory hierarchies, runtime systems, operating systems.

1 Introduction

Simultaneous Multithreaded Processors (SMT) were introduced to maximize the ILP of conventional superscalars via the use of simultaneous instruction issue from multiple threads of control [16]. SMT architectures introduce modest extensions to a superscalar processor core, while enabling significant performance improvements for both parallel shared-memory programs and server workloads. To improve cost-effectiveness, threads on SMT processors share all processor resources, except from the registers that store the private architectural state of each thread. In particular, SMT processors use shared caches at all levels. From a performance perspective, a shared cache used by a multithreaded program accelerates inter-thread communication and synchronization, because threads

A. Veidenbaum et al. (Eds.): ISHPC 2003, LNCS 2858, pp. 54–69, 2003.

that share an address space can prefetch data in the cache for each other [8]. However, for scientific programs which are organized to fully utilize the cache, a shared cache organization introduces significant problems. The working sets of different threads that share a cache may interfere with each other causing an excessive amount of conflict misses. Unless the cache space of the processor is somehow partitioned and "privatized", the problem of coping with thread conflicts on shared caches can become a serious performance limitation.

In this paper, we present software solutions for partitioning shared caches on SMT processors, based on standard program transformations and additional support from the OS, or information collected from the processor hardware counters. Our solutions work in scientific codes with perfect loop nests, which are tiled for improved temporal locality. Tiling is perhaps the most important locality optimization in codes that operate on dense multidimensional arrays. It is used extensively in scientific libraries such as LAPACK. It is also the target of numerous compiler optimizations for memory hierarchies [3, 4, 6, 9, 12, 17].

Our solutions combine dynamic tiling with either copying, or modification of the array layouts in memory. Dynamic tiling amounts to changing the tile size of the program at runtime, if a program senses that its threads are competing for shared cache space with other threads. Competition can be self-interfering (i.e. multiple threads of the same address space compete for cache space on the same SMT), or cross-interfering (i.e. threads from different address spaces compete for shared cache space on the same SMT). We opt for dynamic tiling instead of static adjustment of tile sizes for a shared cache, to give the program an opportunity to adapt its tile size without a-priori knowledge of the assignment of processors and intra-processor threads to the program. The program can adapt its execution to configurations that use one thread per processor or multiple threads per processor, as well as to multiprogrammed execution with threads from multiple address spaces sharing a processor.

Copying is a method for reducing the intra-tile and inter-tile interference in the cache [15]. Copying has been used along with other compiler optimizations for reducing conflict misses. We use copying as a means to confine the tiles used by a thread to a fixed set of cache locations. Coupled with dynamic adjustment of the tile size, copying achieves an implicit spatial partitioning of the cache between the working sets of competing threads to improve performance.

For codes that do not lend themselves to copying due to either correctness or performance implications, we use block layouts of arrays in memory and interleaved assignment of blocks to processors, to achieve a similarly effective spatial partitioning of the cache between tiles belonging to different threads. This solution is expected to be more widely applicable than copying. However, it must cope with restrictions posed by the programming language on the layout of arrays in memory.

The proposed methods have some technical implications. They require a mechanism to detect cache interference between threads at runtime, as well as mechanisms to control the tile size and allocate memory blocks for copying tiles to. We provide solutions to these problems, using either additional support from the operating system or information collected from the processor hardware counters.

To the best of our knowledge, this paper is the first to propose the use of standard code and data layout transformations to improve the management of shared caches on SMT processors without modifications to the hardware. It is also among the first to measure the impact of the proposed transformations on an actual commercial multiprocessor with SMT processors. We use two simple kernels, a tiled parallel matrix multiplication and a tiled point SOR stencil to illustrate the effectiveness of the proposed optimizations. We present experiments from a four-processor Dell PowerEdge 6650 with Xeon MP processors. The processors use Intel's hyperthreading technology and allow simultaneous execution of instructions from two threads. We illustrate the benefits of dynamic tiling, copying and block data layouts using both stand-alone executions of the programs with multiple threads per processor and multiprogrammed workloads with threads from different programs sharing processors.

The rest of this paper is organized as follows. Section 2 details the proposed transformations. Section 3 presents techniques to detect cache contention due to sharing between threads at runtime. Section 4 presents our experimental setting and results. Section 5 provides an abbreviated review of related work. Section 6 concludes the paper.

2   Dynamic Tiling, Copying and Block Data Layout

Figure 1 illustrates the use of the cache if two threads that run disjoint parts of a tiled loop nest share the cache, assuming a perfectly tiled code with a tile size that fits exactly in the cache. Each thread individually tries to use the entire cache space C, by selecting the appropriate tile size. Since the cache can only fit one tile of size $\leq C$, there are excessive conflict misses, while each thread is trying to reuse data in the tile on which it is currently working. These conflict misses lead to poor cache utilization and poor performance.

The immediate solution that comes to mind is to reduce the tile size to a fraction of the cache, e.g. $C/2$ in the case of two threads sharing the cache. The problem with this solution, besides a potential reduction of spatial locality due to smaller tile sizes [4], is that it does not reduce cache conflicts, as shown in Figure 2. Despite the smaller tile size, each thread tries to load two tiles in the cache. Even if each thread individually avoids self-interference between its own tiles, there are still conflicts between tiles belonging to different threads.

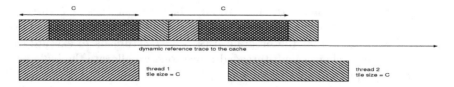

Fig. 1. Cache use when full-size tiles from two threads are loaded in the cache.

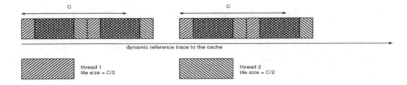

Fig. 2. Cache use when half-size tiles from two threads are loaded in the cache.

```
for (jj = 0; jj < n; jj += tj)
    for (kk = 0; kk < n; kk += tk)
        for (k = kk; k < MIN(n,kk+tk); k++) {
            for (j = jj; j < MIN(n,jj+tj); j++)
                b_block[(k-kk)*tj+(j-jj)] = b[k*n+j];
            for (i = 0; i < n; i++)
                for (k = kk; k < MIN(n,kk+tk); k++)
                    for (j = jj; j < MIN(n,jj+tj); j++)
                        c[i*n+j] += a[i*n+k]*b_block[(k-kk)*tj+(j-jj)];}
```

Fig. 3. A tiled matrix multiplication with copy.

One way to solve this problem is to use smaller tiles and at the same time ensure that tiles from the same thread occupy always the same set of cache blocks. A simple method to accomplish this is to use tile copying [15]. Each thread allocates a memory block equal to the tile size. During the execution of the loop nest, the thread copies the working tile into the memory block, executes the loop nest iterations that work on the tile and then copies the tile back to the array, if the tile is modified during the iterations. Figure 3 shows the tiled matrix multiplication with copy.

Copying fixes the virtual address range and the set of cache locations occupied by a tile, while the tile is being reused in the loop nest. If the thread selects a tile size which is a fraction of C, copying guarantees that during the loop, all tiles use mostly the specified fraction of the cache. Care must still be taken so that tiles from different threads do not interfere in the cache. To address this problem we consider two cases, tiles belonging to different threads in the same address space and tiles belonging to threads in different address spaces.

For tiles belonging to the same address space, the problem can be solved with proper memory allocation of the buffers to which tiles are copied. In practical cases, virtual addresses are mapped to cache lines modulo the number of sets (for set-associative caches) or the number of cache lines (for direct-mapped caches). Assume that the cache needs to be partitioned between the tiles of T threads. We select a tile size which is the best size for a cache of size C/T and apply copying. If we assume that the program uses $N = PT$ threads to execute the loop nest, P the number of multithreaded processors, we allocate the buffers to which tiles are copied in virtual addresses $r_0, r_0 + \frac{C}{T}, \ldots, r_0 + (N-1)\frac{C}{T}$, where r_0 is a starting address aligned to the cache size boundary. After memory allocation is performed, the runtime system schedules threads so that two threads the tiles of which conflict in the cache are never scheduled on the same processor. More formally, if two tiles with starting addresses $r_0 + i\frac{C}{T}, r_0 + j\frac{C}{T}, 0 \leq i < j \leq N - 1$, are used by two different threads, the threads should be scheduled on the same SMT processor only if $mod(j, i) \neq 0$. For the case of 2 threads per processor, this means that threads that work on even-numbered tiles should be scheduled on even-numbered hardware contexts and threads that work on odd-numbered tiles should be scheduled on odd-numbered hardware contexts. In simpler words, tiles should be interleaved between hardware contexts. In the general case of multithreaded processors with T threads per processor, two threads using tiles with starting addresses $r_0 + i\frac{C}{T}, r_0 + j\frac{C}{T}$ should not be scheduled on the same processor if $mod(i, T) = mod(j, T)$.

For tiles belonging to different address spaces, guaranteeing conflict avoidance is difficult because a thread running tiled code does not have any knowledge of how a competing thread uses the shared cache. Copying can only guarantee that each thread that runs tiled code reuses a fixed set of locations in the cache. If all the threads that share the cache run tiled computations, the conflict problem will be alleviated with dynamic tiling and copying if the cache is set-associative, which is the case in all modern microprocessors. Otherwise, if we assume that each program can use only local information to control its caching strategy, the problem is intractable.

If threads need to change their tile sizes at runtime, they need to allocate a new memory block for each tile size used. The additional memory consumption is therefore equal to the sum of the different tile sizes that a thread uses at runtime, times the number of threads in the program. For the transformation considered here, each thread uses at most two tile sizes and the space overhead is $PT(C + \frac{C}{T})$. The number of memory allocations is equal to the number of different tile sizes used at runtime, times the number of threads in the program i.e. $2PT$. Since memory allocations of buffers can be executed in parallel, the actual overhead is 2 memory allocations.

Copying can not be used for free. It incurs instruction overhead and additional cache misses while data is copied from arrays to buffers. We use the analysis of [15] to balance the earnings with the overhead of copying, assuming that all accesses to the reused tile will conflict with probability $1/T$. We estimate the number of cache accesses and cache misses during innermost loop nest iterations that reuse the tile, as well as the number of accesses and instructions executed while copying a tile from an array to a buffer. Then, we compare the cost of conflict misses to the cost of copying, by associating a fixed latency for each cache miss and each copy instruction. The average latency of cache misses and instructions is obtained by microbenchmarking the target processors. We use PAPI [1] for this purpose.

Unfortunately, copying may still not be applicable because of its runtime cost. This is frequently the case when both copy-in and copy-out operations must be used to maintain data consistency. An alternative solution to copying is to use array layouts that achieve the same effect as copying on the cache. Block data layouts [10] can be used for this purpose. With a block data layout, an array

1	2	3	4
5	6	7	8
9	10	11	12
13	14	15	16

1	2	5	6
3	4	7	8
9	10	13	14
11	12	15	16

Fig. 4. The array on the right shows the block data layout of the 4×4 array on the left, using 2×2 blocks.

is first decomposed in blocks, which correspond to the tiles used when tiling is applied in the loop nest that accesses the array. The elements of each block are stored in contiguous locations in memory as shown in Figure 4. Previous work has shown that a block data layout can be equally or more efficient than a conventional data layout due to good use of the memory hierarchy and the TLB [10]. In our context, the block data layout has a second hidden advantage. It lets us control explicitly the location of the virtual address space in which each block is stored.

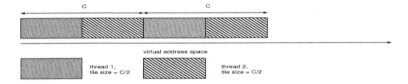

Fig. 5. Block data layout with interleaved assignment of blocks to threads.

If copying can not be used, we use the block data layout and position the blocks in the virtual address space so that blocks used by the same thread in the parallel execution are placed with a distance of C between each other, C being the cache size. As shown in Figure 5, blocks belonging to different threads are interleaved in memory and regions of the address space with size equal to the cache size C are partitioned between blocks. Conceptually, we try to force conflicts between blocks of the same thread and avoid conflicts between blocks of different threads. For caches shared between threads from different programs, interference may still occur, but this solution ensures that each thread will only compete for an equal share of the cache, instead of all cache locations.

To implement dynamic tiling and copying we create two versions of the tiled part of the loop nest. The first version executes the properly tiled loop, where the tile size is selected to utilize the entire cache space. The second version executes the loop nest with the reduced tile size and tile copy. More formally, if the loop nest has n levels, $L_1 \ldots L_n$, out of which $m < n$ levels are the outermost loop control levels and $n-m+1$ are the innermost tiled levels, the first version executes the tiled levels $L_m \ldots L_n$, whereas the second version executes $n - m + k, k > 1$ levels, including the tiled levels $L_m \ldots L_n$ using a small tile size that utilizes $\frac{1}{T}$ of the available cache space and $k-1$ outermost levels used to partition the original tiles into smaller tiles. The selection between the two versions is done with a conditional that checks if cache contention between multiple threads occurs at runtime. The implementation of this conditional is discussed in more detail in Section 3. Switching between the two versions is done at level L_m in the original loop nest, i.e. at tile boundaries. This simplifies the code significantly but may delay the adaptation to cache contention by as much as the time required for the processing of one tile. The buffer space used for copying is statically preallocated in the code.

If the block data layout is used, the array is transformed from the standard to the block layout and two versions of the loop nest are created again. Both versions use the block data layout. In this case, the base tile size is selected to utilize $\frac{1}{T}$ of the available cache space. In the first version, each thread accesses combined large tiles, each composed of T consecutive smaller tiles at level L_m. In the second version, each thread accesses the smaller tiles with a stride equal to T, so that tiles are interleaved between threads. The first version of the code executes the $n - m$ innermost tiled loops using the compound tile, whereas the second version executes the $n - m$ innermost tiled loops of the original code using the small tile size and one more loop level to control the interleaving of tiles between processors. Again, a conditional is used to select between the two versions.

We experimented with several compile-time algorithms for selecting the tile size and we ended up using the one proposed by Chame and Moon [3]. Chame and Moon's algorithm computes an initial set of tile sizes that avoids self-interference. It calculates a set of rectangular tile sizes starting with a tile of one row with size equal to the cache size and incrementally increasing the number of rows by one, while setting the maximum row size equal to the minimum distance between the starting addresses of any two tile rows in the cache. For each tile size of this set, the algorithm performs a binary search on all possible tile row sizes, to find the tile row size that minimizes capacity misses and cross-interference misses. The binary search is based on the assumption that capacity misses decrease monotonically with an increasing row size and cross-interference misses increase monotonically with an increasing row size. The estimated capacity misses are calculated analytically for each array reference from the parameters of the loop nest. The cross-interference misses are estimated probabilistically from the cache footprints of array references in each tile.

After extensive experimentation we found that Chame and Moon's algorithm tends to perform consistently better than other heuristics [3, 4, 12] when the tile size is reduced to cope with cache sharing. We note however that it is beyond the scope of this paper to investigate thoroughly the relative performance of tiling algorithms on SMT processors.

A point worth noting is that we dynamically change the tile size to the best tile size for a cache with size equal to $\frac{1}{T}$ of the actual cache size. This implies that each thread is expected to use as much cache space as possible, which is valid for tiled parallel codes. Nevertheless, in arbitrary workloads (for example, with multiprogramming, or if the application uses one thread for communication and one thread for computation), situations may arise in which one thread requires little or no cache space, whereas another thread running on the same processor requires all the available cache space. We leave this issue open for investigation in future work.

A drawback of using block data layouts is that if for any reason the standard language layouts (row-major or column-major) of the arrays must be preserved, additional copying must be used to create the block layout before the loop nest and restore the original layout after the loop nest. In the codes we used for the experiments presented in this paper, both of which are composed of a single loop nest, creating the block layout before the loop nest and restoring the original layout after the loop nest did not introduce significant overhead. We measured the overhead for data set sizes between 90 and 200 Megabytes and the arithmetic mean was 1.1% of the total execution time.

We have implemented the aforementioned transformations by hand in two regular tiled codes, a blocked matrix multiplication and a 2-D point SOR kernel. Work for formalizing and automating these transformations in a compiler is in progress. Since we are interested in the performance of parallel code, it is necessary to partition the tiles between processors. We use a simple blocked partitioning scheme in which each processor is assigned a stack of contiguous tiles. We used OpenMP and the default static scheduling algorithm for parallelization. This simple scheduling scheme leads often to load imbalance, but this effect is mitigated when the number of tiles per thread is increased.

3   Detecting Cache Contention

If the compiler could know in advance the number of threads used by a program and the processors on which these threads are going to execute, it could selectively generate code with or without the transformations that optimize execution for shared caches. Unfortunately, the number of threads is usually a runtime parameter and the processors on which threads execute is unknown until runtime. In most shared-memory multiprocessors, where programmers use either POSIX threads or OpenMP to parallelize, the number of threads is set by either an environment variable or a call to a library function. For a P-processor machine with T-way multithreaded processors, and a program that uses $2 \leq t < PT$ threads, the operating system may decide to co-schedule threads on the same processor(s) or not, depending on the load and the scheduling policy.

Furthermore, although a single program running on an idle machine is most likely going to execute with a fixed number of threads pinned to processors and the OS can avoid cache sharing simply by not co-scheduling multiple threads of the same program on the same processor, if the machine runs multiple programs and/or system utilities at the same time, the scheduler may decide to change the number and the location of threads dynamically. Last but not least, a single parallel program may wish to utilize all processors and threads within processors. In this case, cache sharing is inevitable. In general, it is desirable for a program to identify locally at runtime whether there is cache sharing or not. We examine two mechanisms to accomplish this goal, one that uses an extension to the operating system and one that uses the processor hardware counters.

The first solution is to extend the operating system with an interface that exports the placement of threads on processors to a page in shared memory, which is readable by user programs. Earlier work has coined the term shared arena to characterize this interface [5]. A shared arena can be used by a user-level threads library in many ways for improving thread scheduling and synchronization algorithms under multiprogrammed execution. In our case, we use the shared arena to export a bit vector that shows which hardware contexts are busy running either user-level code, or operating system handlers. The OS sets a bit whenever it assigns a thread or an interrupt handler to run on a hardware context. We used Linux, in which the hardware contexts are treated as virtual CPUs, with multiple virtual CPUs per physical processor. Each virtual CPU is seen as a separate processor by the software. The tiled code that we generate checks this vector each time it picks a new tile to work on. The problem with an OS extension, even a very simple one, is that it is not portable and requires kernel modifications which may or may not be possible. We investigated a second option, which is to try to implicitly detect cache sharing by measuring the cache misses with hardware performance counters while the program is executing. Hardware performance counters have become a standard feature of modern microprocessors, therefore a solution based on them would be significantly more portable. Using hardware performance counters, we compare the actual number of cache misses per tile, against the expected number of cache misses, which is calculated analytically, using stack distances [2].If the measured number of cache misses exceeds the expected number by a threshold, we assume interference due to cache sharing. The threshold is allowed to compensate for potential self and cross-interference between data of the same thread. We have experimented with several threshold values and found that a threshold equal to the expected number of cache misses is always an accurate indication of contention for cache space.

4   Experimental Setting and Results

We experimented on a 4-processor Dell PowerEdge 6650 Server. Each processor is a 1.4 GHz Xeon MP with hyperthreading technology. The processor can execute simultaneously instructions from two threads, which appear to the software as two different processors. We call these processors virtual processors onwards, to differentiate from the actual physical processors, which are multithreaded. Each physical processor has an 8 KB L1 data cache integrated with a 12 KB execution trace cache, an 8-way associative 256 KB L2 cache and an integrated 512 KB L3 cache. The system runs Linux 2.4.20, patched with a kernel module that binds threads[1] to virtual processors. The processor binding module is not a scheduler. We use it only to specify the placement of threads on virtual processors and

---

[1] In Linux, threads and processes are synonyms, each thread has a unique PID, threads can share an address space and thread switching is performed in the kernel.

enforce cache sharing, in order to run controlled experiments. Linux provides no other way to achieve the same effect at user level. Nevertheless, we note that the dynamic tiling algorithm is scheduler-independent. It is designed to detect cache contention and sharing, regardless of the processor allocation policy of the OS.

We used two codes, a tiled matrix multiplication and the 5-point stencil loop of a 2-D point SOR code, both parallelized with OpenMP. Copying was used in the matrix multiplication, whereas block data layout with proper block alignment was used in the SOR code. Copying is affordable in matrix multiplication but not in SOR. The dynamic tiling transformation was applied for the L1, the L2 and the L3 cache recursively. We used large data sets, so that the performance of the codes becomes more bound to the L2 and L3 cache latencies. Note that the caches maintain the inclusion property. The programs were compiled with the Intel C compiler, version 7.1, using the O3 optimization level. Besides standard code optimizations, the compiler performs profile-guided interprocedural optimization and function inlining, partial redundancy elimination, automatic vectorization of inner loops and automatic padding of arrays for improved cache performance. Padding was enabled in matrix multiplication but disabled in SOR to control explicitly the data layout. Cache contention was detected by measuring L1 cache misses using PAPI.

Figure 6 shows the execution time of the tiled matrix multiplication using fixed tile sizes and dynamic tiling with copy. The matrix multiplication is executed as a stand-alone program on an otherwise idle machine. We run the code with 1 thread on one processor, 2 threads on 2 physical processors (labelled 2×1), 4 threads on two physical processors (labelled 2×2), 4 threads on 4 physical processors (labelled 4×1) and 8 threads on 4 physical processors (labelled 4×2).

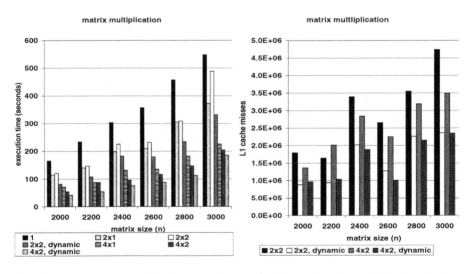

Fig. 6. Execution times (left chart) and number of L1 cache misses (right chart) of an $n \times n$, $2000 \leq n \leq 3000$, matrix multiplication with fixed tile sizes and dynamic tiling with copying.

The label suffix *dynamic* indicates the use of tiling with dynamic reduction of the tile size when cache contention is detected, instead of fixed tile sizes that use the entire cache space. The execution time savings for two threads running on two processors average 26% (standard deviation is 5.29) and for eight threads running on four processors 15.66% (standard deviation is 8.53). The average speedup (measured as the arithmetic mean of the parallel execution time for all matrix sizes, divided by the execution time with 1 thread), is improved from 1.41 to 1.91 on 2 SMTs running 4 threads and from 2.95 to 3.89 on 4 SMTs running 8 threads. The sublinear speedups are attributed to load imbalance due to uneven distribution of tiles between threads and memory hierarchy effects, including limitations of bus bandwidth and bus contention.

It is interesting to notice that simultaneous multithreading on a single processor does not always improve performance. With fixed tile sizes, using 4 threads on 2 processors leads to a performance drop of 9.3% (arithmetic mean across all matrix sizes), compared to the performance achieved with 2 threads running on 2 processors. On the contrary, with dynamic tiling, performance is improved by 17.95% on average in the same setting. With 8 threads on 4 processors, performance is always improved, compared to the performance achieved with 4 threads on 4 processors. The arithmetic mean of the improvement is 15.44% with fixed tile sizes and 35.44% with dynamic tiling.

The right chart in Figure 6 shows the number of L1 cache misses throughout the execution of the programs with fixed tile sizes and dynamic tiling, using 4 threads on 2 SMTs and 8 threads on 4 SMTs. Each bar is the arithmetic mean of the number of cache misses of all threads, during the execution of the matrix multiplication loop nest, excluding initialization.On two processors with two threads each, L1 cache misses are reduced by 54.63% on average (standard deviation is 6.34). On four processors with two threads each, L1 cache misses are reduced by 61.15% on average (standard deviation is 10.57). L2 cache misses (not shown in the charts) are reduced by 51.78% and 55.11% respectively. Although cache misses are reduced more with 8 threads than with 4 threads, the parallel execution time is not improved similarly.

Figure 7 shows the execution times and L1 cache misses of the tiled SOR code. The observed results are very similar to matrix multiplication and show that the block data layout transformation is almost equally effective. Execution time improvements average 19.45% with 4 threads (standard deviation is 11.32) and 29.19% with 8 threads (standard deviation is 13.24). Speedup is improved from 1.76 to 2.17 with 4 threads on 2 SMTs and from 3.28 to 4.65 with 8 threads on 4 SMTs. Using 2 threads instead of 1 thread on the same SMT leads to marginal performance losses or improvements with fixed tile sizes (in the range of ± 5%) and more significant improvements (ranging from 12.23% to 32.42%) with dynamic tiling. L1 cache misses are reduced by 50.55% on average (standard deviation is 10.86). L2 cache misses (not shown in the charts) are reduced by 59.44% on average (standard deviation is 16.12).

Figure 8 shows the average execution time of two tiled matrix multiplications running simultaneously with four threads each, using a stride-2 allocation of

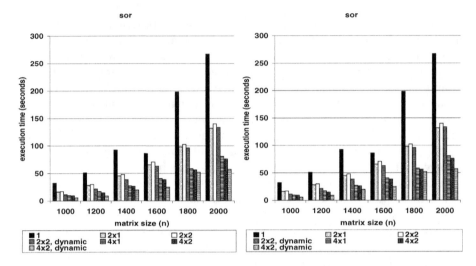

Fig. 7. Execution times (left) and number of L1 cache misses (right) of an $n \times n$, $1000 \leq n \leq 2000$, point SOR code tiled in two dimensions, with fixed tile sizes and dynamic tiling with block data layout.

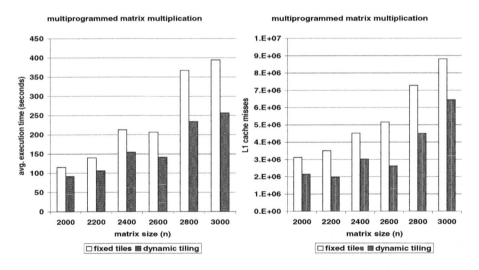

Fig. 8. Average execution time and number of L1 cache misses of two tiled matrix multiplications executed with four threads each, using stride-2, odd-even allocation of virtual processors to threads.

virtual processors to threads. This means that one matrix multiplication runs on even-numbered virtual processors and another matrix multiplication runs on odd-numbered virtual processors simultaneously. This experiment evaluates if dynamic tiling and copying can alleviate cache contention between independent tiled codes sharing the cache. The arithmetic mean of the reduction of execution time is 28.9% (standard deviation is 5.83). L1 cache misses are reduced by 61.24% on average. L2 cache misses (not shown) are reduced by 71.79% on average.

In the specific experiment, cache contention is detected by the hardware performance counters. We repeated the experiment and used sharing notifications from the OS instead of the processor hardware counters. We observed measurable but not significant improvements. We ran one more experiment to test whether our techniques work equally well with a more dynamic workload competing for cache space with the transformed programs. We executed the tiled matrix multiplication with four threads, using a stride-2 processor allocation. Concurrently, we launched a script with a blend of integer, floating point, system call, disk I/O and networking microbenchmarks, adapted from the Caldera AIM benchmark (version s9110). The script was modified so that individual microbenchmarks were distributed between the odd-numbered virtual processors, while the even-numbered virtual processors were executing one parallel matrix multiplication. The script creates significant load imbalance between the processors, as some of the benchmarks (e.g. I/O and networking) make intensive use of caches due to buffering, while others (e.g. system calls) make little or no use of caches.

Figure 9 shows the execution time and L1 cache misses of matrix multiplication with this workload, for various matrix sizes. Performance is improved by 24.69% on average (standard deviation is 4.47). L1 cache misses are reduced by 64.35% on average (standard deviation is 9.89). During the experiments, we instrumented the code to check how long the code was executing with the reduced tile size and how long it was executing with the larger tile size. A smaller tile size was used during 94.53% of the execution time. The code switched twice

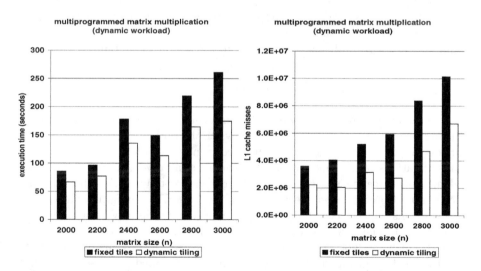

Fig. 9. Execution time and L1 cache misses of one tiled matrix multiplication with four threads running on even-numbered virtual processors, together with the multiprogramming workload of Caldera's AIM benchmark, spread among the four odd-numbered virtual processors.

from bigger to smaller tiles and vice versa. Some further testing has shown that theoretically, the adaptive code should increase more frequently the tile size and reduce execution time even further, because the idle intervals, as indicated by the load of each processor, account for more than 15% of the execution time of the tiled code. On the other hand, when cache misses were used as the sole indication of contention, the adaptation of the code was correct and the proper tile size was used in all cases. We plan to examine this issue in more detail in future work.

5   Related Work

Simultaneous multithreaded processors have been a focal topic in computer systems research for almost 8 years, initially along the hardware design space [16] and later, along the operating system design space [8, 11]. From the hardware design perspective, all studies concluded that shared caches at all levels of the cache hierarchy yield the most cost-effective design. From the software perspective, the related work has focused on parallel programming issues such as synchronization and memory allocation, but not on program optimizations for shared caches.

Tiling and multilevel blocking of array codes for sequential and parallel machines has been investigated in an enormous number of publications, only a sample of which is referenced here [3, 4, 6, 7, 9, 12, 15, 17, 18]. Recently, researchers have turned their attention to managing space-shared caches on multiprogrammed systems, via a number of hardware and software techniques, including static cache partitioning [13] and memory-aware job scheduling [14]. These studies place the burden of effective management of shared caches on the operating system. Otherwise, they propose hardware cache partitioning. We have presented a software method based on standard compiler transformations, which requires little or no additional support from the operating system.

6   Conclusions and Future Work

This paper presented a set of program transformations, namely dynamic tiling, copying and block data layouts, which alleviate the problem of conflicts between threads on the shared caches of simultaneous multithreaded processors. These transformations result to significant performance improvements, both for standalone multithreaded programs and for programs competing for shared cache space in multiprogrammed execution environments. The strength of the contribution of this paper is the ability to use standard and well-known code and data layout transformations to cope with a problem (thread interference in shared caches) which has not been addressed and has been merely investigated outside the hardware context. Given that both multithreaded processors and chip multiprocessors with shared caches are gaining more ground in the market, investigating program optimizations in the software context will contribute to the development of better compilers and other system software tools for these systems.

The weaknesses of the contribution of this paper parallel the weaknesses of practically all transparent locality optimization frameworks. They are effective in limited cases, mostly numerical scientific codes with "easy to analyze" loop nests and data access patterns. One direction of future work is the investigation of the development of more generic cache-conscious program and data layout schemes for processors with shared on-chip caches. We are also seeking a convenient analytical framework for quantifying conflict misses in shared caches, to derive automatic program transformations. One more unexplored issue which arises from this work is the use of dynamic instead of static cache partitioning methods, based on the cache footprints of threads.

Acknowledgments

Xavier Martorell implemented the kernel module for binding threads to virtual processors, which was used in this paper after some minor modifications. Christos Antonopoulos implemented C macros for interfacing with PAPI and solved several technical problems of PAPI on the Xeon MP processors. This research was supported by a startup research grant from the College of William&Mary and NSF Research Infrastructure Grant EIA-9972853.

References

1. S. Browne, J. Dongarra, N. Garner, K. London, and P. Mucci. A Scalable Cross-Platform Infrastructure for Application Performance Tuning Using Hardware Counters. In *Proc. of Supercomputing'2000: High Performance Networking and Computing Conference*, Dallas, TX, Nov. 2000.
2. C. Cascaval and D. Padua. Estimating Cache Misses and Locality using Stack Distances. In *Proc. of the 17th ACM International Conference on Supercomputing (ICS'2003)*, pages 150–159, San Francisco, CA, June 2003.
3. J. Chame and S. Moon. A Tile Selection Algorithm for Data Locality and Cache Interference. In *Proc. of the 13th ACM International Conference on Supercomputing (ICS'99)*, pages 492–499, Rhodes, Greece, June 1999.
4. S. Coleman and K. McKinley. Tile Size Selection Using Cache Organization and Data Layout. In *Proc. of the 1995 ACM SIGPLAN Conference on Programming Languages Design and Implementation (PLDI'95)*, pages 279–290, San Diego, CA, June 1995.
5. D. Craig. An Integrated Kernel and User-Level Paradigm for Efficient Multiprogramming. Technical Report CSRD No. 1533, University of Illinois at Urbana-Champaign, June 1999.
6. I. Kodukula, N. Ahmed, and K. Pingali. Data-Centric Multilevel Blocking. In *Proc. of the 1997 ACM SIGPLAN Conference on Programming Languages Design and Implementation (PLDI'97)*, pages 346–357, Las Vegas, Nevada, June 1997.
7. N. Mateev, N. Ahmed, and K. Pingali. Tiling Imperfect Loop Nests. In *Proc. of the IEEE/ACM Supercomputing'2000: High Performance Networking and Computing Conference (SC'2000)*, Dallas, TX, Nov. 2000.

8. L. McDowell, S. Eggers, and S. Gribble. Improving Server Software Support for Simultaneous Multithreaded Processors. In *Proc. of the 2003 ACM SIGPLAN Symposium on Principles and Practice of Parallel Programming (PPoPP'2003)*, San Diego, CA, June 2003.
9. K. McKinley, S. Carr, and C. Tseng. Improving Data Locality with Loop Transformations. *ACM Transactions on Programming Languages and Systems*, 18(4):424–453, July 1996.
10. N. Park, B. Hong, and V. Prasanna. Analysis of Memory Hierarchy Performance of Block Data Layout. In *Proc. of the 2002 International Conference on Parallel Processing (ICPP'2002)*, pages 35–42, Vancouver, Canada, Aug. 2002.
11. J. Redstone, S. Eggers, and H. Levy. Analysis of Operating System Behavior on a Simultaneous Multithreaded Architecture. In *Proc. of the 9th International Conference on Architectural Support for Programming Languages and Operating Systems (ASPLOS'IX)*, Cambridge, MA, Nov. 2000.
12. G. Rivera and C. Tseng. A Comparison of Tiling Algorithms. In *Proc. of the 8th International Conference on Compiler Construction (CC'99)*, pages 168–182, Amsterdam, The Netherlands, Mar. 1999.
13. G. Suh, S. Devadas, and L. Rudolph. Analytical Cache Models with Applications to Cache Partitioning. In *Proc. of the 15th ACM International Conference on Supercomputing (ICS'01)*, pages 1–12, Sorrento, Italy, June 2001.
14. G. Suh, L. Rudolph, and S. Devadas. Effects of Memory Performance on Parallel Job Scheduling. In *Proc. of the 8th Workshop on Job Scheduling Strategies for Parallel Processing (JSSPP'02)*, pages 116–132, Edinburgh, Scotland, June 2002.
15. O. Temam, E. Granston, and W. Jalby. To Copy or Not to Copy: A Compile-Time Technique for Assessing when Data Copying Should be Used to Eliminate Cache Conflicts. In *Proc. of the ACM/IEEE Supercomputing'93: High Performance Networking and Computing Conference (SC'93)*, pages 410–419, Portland, OR, Nov. 1993.
16. D. Tullsen, S. Eggers, and H. Levy. Simultaneous Multithreading: Maximizing On-Chip Parallelism. In *Proceedings of the 22nd International Symposium on Computer Architecture (ISCA'95)*, pages 392–403, St. Margherita Ligure, Italy, June 1995.
17. M. Wolf and M. Lam. A Data Locality Optimizing Algorithm. In *Proc. of the 1991 ACM SIGPLAN Conference on Programming Language Design and Implementation (PLDI'91)*, pages 30–44, Toronto, Canada, June 1991.
18. J. Xue. *Loop Tiling for Parallelism*. Kluwer Academic Publishers, Aug. 2000.

Improving Memory Latency Aware Fetch Policies for SMT Processors

Francisco J. Cazorla[1], Enrique Fernandez[2], Alex Ramírez[1], and Mateo Valero[1]

[1] Dpto. de Arquitectura de Computadores, Universidad Politécnica de Cataluña,
{fcazorla,aramirez,mateo}@ac.upc.es
[2] Dpto. de Informática y Sistemas, Universidad de Las Palmas de Gran Canaria,
efernandez@dis.ulpgc.es

Abstract. In SMT processors several threads run simultaneously to increase available ILP, sharing but competing for resources. The instruction fetch policy plays a key role, determining how shared resources are allocated.
When a thread experiences an L2 miss, critical resources can be monopolized for a long time choking the execution of the remaining threads. A primary task of the instruction fetch policy is to prevent this situation. In this paper we propose novel improved versions of the three best published policies addressing this problem. Our policies significantly enhance the original ones in throughput, and fairness, also reducing the energy consumption.

Keywords: SMT, multithreading, fetch policy, long latency loads, load miss predictors

1   Introduction

Multithreaded and Simultaneous Multithreaded Processors (SMT) [3], [8], [9], [10] concurrently run several threads in order to increase available parallelism. The sources of this parallelism come from the instruction level parallelism (ILP) of each thread alone, from the additional parallelism that provides the freedom of fetching instructions from different independents threads, and from mixing them in appropriate way to the processor core. Problems arise because shared resources have to be dynamically allocated between these threads. The responsibility of the fetch policy will be to decide which instructions (and from which thread) come into the processor, hence it determines how this allocation is done, playing a key role in obtaining performance.

When a thread experiences an L2 cache miss, following instructions spend resources for a long time while making little progress. Each instruction occupies a ROB entry and a physical register (not all) from the rename stage to the commit stage. It also uses an entry in the issue queues while any of its operands is not ready, and a functional unit (FU). Neither the ROB nor the FUs represent a problem, because the ROB is not shared and the FUs are pipelined. The issue queues and the physical registers are the actual problems, because they are used

A. Veidenbaum et al. (Eds.): ISHPC 2003, LNCS 2858, pp. 70–85, 2003.

for a variable, long period. Thus, the instruction fetch (I-fetch) policy must prevent an incorrect use of these shared resources to avoid significant performance degradation.

Several policies have been proposed to alleviate the previous problem. As far as we know, the first proposal to address it was mentioned in [11]. The authors suggest that a load miss predictor could be used to predict L2 misses switching between threads when one of them is predicted to have an L2 miss. In [4] the authors propose two mechanisms to reduce load latency: data pre-fetch and a policy based on a load miss predictor (we will explain these policies in the related work section). STALL [7], fetch-stalls a thread when it is declared to have an L2 missing load until the load is resolved. FLUSH [7] works similarly and additionally flushes the thread which the missing load belongs to. DG [2] and PDG [2] are two recently proposed policies that try to reduce the effects of L1 missing loads. Our performance results show that FLUSH outperforms both policies, hence we will not evaluate DG and PDG in this paper.

In the first part of this paper we analyze the space of parameters of STALL, FLUSH and a policy based on the usage of load miss predictors (L2MP), and compare their effectiveness. As we will see none of them outperforms all other in all cases, but each behaves better depending on the metric and on the workload. Based on this initial study, in the second part we propose improved versions of each of them. Throughput and fairness [5] results show that in general improved versions achieve important performance increments over the original versions for a wide range of workloads, ranging from two to eight threads.

The remainder of this paper is structured as follows: we present related work in Section 2. Section 3 presents the experimental environment and the metrics used to compare the different policies. In Section 4 we explain the current policies. Section 5 compares the effectiveness of those policies. In Section 6 we propose several improvements for the presented polices. Section 7 compares the improved policies. Finally Section 8 is devoted to the conclusions.

2   Related Work

Current I-fetch policies address the problem of L2 missing loads latency in several ways. Round Robin [8] is absolutely blind to this problem. Instructions are alternatively fetched from available threads, even when any of them has in-flight L2 misses. ICOUNT [8] only takes into account the occupancy of the issue queues, and disregards that a thread can be blocked on an L2 miss, while making no progress for many cycles. ICOUNT gives higher priority to those threads with fewer instructions in the queues (and in the pre-issue stages). When a load misses in L2, dependent instructions occupy the issue queues for a long time. If the number of dependent instructions is high, this thread will have low priority. However, these entries cannot be used by the other threads degrading their performance. On the contrary, if the number of dependent instructions after a load missing in L2 is low, the number of instructions in the queues is also low, so this thread will have high priority and will execute many instructions that cannot be committed for a long time. As a result, the processor can run out of registers. Therefore, ICOUNT only has a limited control over the issue queues, because it cannot prevent threads from using the issue queues for a long time. Furthermore, ICOUNT ignores the occupancy of the physical registers.

More recent policies, implemented on top of ICOUNT, focus in this problem and add more control over issue queues, as well as control over the physical registers.

In [11] a load hit/miss predictor is used in a super-scalar processor to guide the dispatch of instructions made by the scheduler. This allows the scheduler to dispatch dependent instructions exactly when the data is available. The authors propose several hit/miss predictors that are adaptations of well known branch miss predictors. The authors suggest adding a load miss predictor in an SMT processor in order to detect L2 misses. This predictor would guide the fetch, switching between threads when any of them is predicted to miss in L2.

In [4] the authors propose two mechanisms oriented to reduce the problem associated to load latency. They use data prefetching and conclude that it is not effective because, although the latency of missing loads is reduced, this latency is still significant. Furthermore, as the number of threads increases, the gain decreases due the pressure put on the memory bus. The second mechanism uses a load miss predictor, and when a load is predicted to miss the corresponding thread is restricted to use a maximum amount of available resources. When the missing load is resolved the thread is allowed to use the whole resources.

In [7] the authors propose several mechanisms to detect an L2 miss (detection mechanism) and different ways of acting on a thread once it is predicted to have an L2 miss (action mechanism). The detection mechanism that presents the best results is to predict miss every time that a load spends more cycles in the cache hierarchy than needed to access the L2 cache, including possible resource conflicts (15 cycles in the simulated architecture). Two action mechanism present good results: the first one, STALL, consists of fetch-stalling the offending thread. The second one, FLUSH, flushes the instructions after the L2 missing load, and also it stalls the offending thread until the load is resolved. As a result, the offending thread temporarily does not compete for resources, and what is more important, the resources used by the offending thread are freed, giving the other threads full access to them. FLUSH results show performance improvements over ICOUNT for some workloads, especially for workloads with a few number of threads. However, FLUSH requires complex hardware, and increase the pressure on the front-end of the machine because it requires squashing all instruction after a missing load. Furthermore, due to the squashes, many instructions need to be re-fetched and re-executed. STALL is less aggressive than FLUSH, does not require hardware as complex as FLUSH, and does not re-execute instructions. However, in general its performance results are worse.

In this paper we present improved versions of FLUSH, STALL, and L2MP that clearly improves the original ones in both throughput and fairness.

3   Metrics and Experimental Setup

We have used three different metrics to make a fair comparison of the policies: the IPC throughput, a metric that balances throughput and fairness (Hmean), and a metric that takes into account the extra energy used due the re-execution of instructions (extra fetch or EF).

We call the fraction IPC_{wld}/IPC_{alone} the relative IPC, where the IPC_{wld} is the IPC of a thread in a given workload, and the IPC_{alone} is the IPC of a thread when it runs isolated. The Hmean metric is the harmonic mean of the relative IPC of the threads in a workload [5]. Hmean is calculated as shown in Formula 1.

$$Hmean = \frac{\#threads}{\sum_{threads} \frac{IPC_{alone}}{IPC_{wld}}} \, . \tag{1}$$

The extra fetch (EF) metric measures the extra instructions fetched due to the flush of instructions (see Formula 2). Here we are not taking into account the flushed instructions due to branch mispredictions, but only those related with the loads missing in L2. EF compares the total fetched instructions (flushed and not flushed) with the instructions that are fetched and not flushed. The higher the value of the EF the higher the number of squashed instructions respect to the total fetched instructions. If no instructions is squashed, EF is equal to zero because the number of fetched instructions is equal to the number of fetched and not squashed instructions.

$$EF = \frac{TotalFetched * 100}{Fetched \ and \ not \ squashed} - 100 \ (\%) \, . \tag{2}$$

We have used a trace driven SMT simulator, based on SMTSIM [9]. It consists of our own trace driven front-end and a modified version of SMTSIM's back-end. Baseline configuration is shown in Table 1 (a).

Traces were collected of the most representative 300 million instruction segment following the idea presented in [6]. The workload consists on all programs from the SPEC2000 integer benchmark suite. Each program was executed using the reference input set and compiled with the $-O2 - non\_shared$ options using DEC Alpha AXP-21264 C/C++ compiler. Programs are divided in two groups based on their cache behavior (see Table 1 (b)): those with an L2 cache miss rate higher than 1%[1] are considered memory bounded (MEM), the rest are considered ILP. From these programs we create 12 workloads, as shown in Table 2, ranging from 2 to 8 threads. In the ILP workloads all benchmarks have good cache behavior. All benchmarks in the MEM workloads have an L2 miss rate higher than 1%. Finally, the MIX workloads include ILP threads as well as MEM threads. For MEM workloads some benchmarks were used twice, because there are not enough SPECINT benchmarks with bad cache behavior. The replicated benchmarks are boldfaced in Table 2. We have shifted second instances of replicated benchmarks by one million instructions in order to avoid both threads accessing the cache hierarchy at the same time.

---

[1] The L2 miss rate is calculated with respect to the number of dynamic loads

Table 1. From left to right. (a) Baseline configuration; (b) L2 behavior of isolated benchmarks

Processor Configuration	
Fetch /Issue /Commit Width	8
Fetch Policy	ICOUNT 2.8
Queues Entries	32 int, 32 fp, 32 ld/st
Execution Units	6 int, 3 fp, 4 ld/st
Physical Registers	384 int, 384 fp
ROB Size / thread	256 entries
Branch Predictor Configuration	
Branch Predictor	2048 entries gshare
Branch Target Buffer	256 entry, 4 -way associative
RAS	256 entries
Memory Configuration	
L1 Icache, Dcache	64K bytes, 2 -way, 8-banks, 64-byte lines , 1 cycle access
L2 cache	512K bytes, 2 -way, 8-banks, 10 cycles lat., 64-byte lines
Main Memory latency	100 cycles
TLB miss penalty	160 cycles

	L2 miss rate	Thread typ e
mcf	29.6	MEM
twolf	2.9	
vpr	1.9	
parser	1.0	
gap	0.7	ILP
vortex	0.3	
gcc	0.3	
perlbmk	0.1	
bzip2	0.1	
crafty	0.1	
gzip	0.1	
eon	0.0	

Table 2. Workloads

Num. of threads	Thread type	Benchmarks
2	ILP	gzip, bzip2
	MIX	gzip, twolf
	MEM	mcf, twolf
4	ILP	gzip, bzip2, eon, gcc
	MIX	gzip, twolf , bzip2, mcf
	MEM	mcf, twolf, vpr, **twolf**
6	ILP	gzip, bzip2, eon, gcc crafty, perlbmk
	MIX	gzip, twolf, bzip2, mcf, vpr, eon
	MEM	mcf, twolf, vpr, parser, **mcf, twolf**
8	ILP	gzip, bzip2, eon, gcc crafty, perlbmk, gap, vortex
	MIX	gzip, twolf, bzip2, mcf, vpr, eon, parser, gap
	MEM	mcf, twolf, vpr, parser, **mcf, twolf, vpr, parser**

4   Current Policies

In this section we discuss several important issues about the implementation of the policies that we are going to evaluate: L2MP, STALL and FLUSH.

In this paper we evaluate a policy that uses predictors to predict L2 misses. We call this policy L2MP. The L2MP mechanism is shown in Figure 1. The predictor acts in the decode stage. It is indexed with the PC of the loads: if a load is not predicted (1) to miss in L2 it executes normally. If a load is predicted (1) to miss in L2 cache, the thread it belongs to is stalled (2). This load is tagged indicating that it has stalled the thread. When this load is resolved (either in the Dcache (3), or in the L2 cache(4)) the corresponding thread is continued.

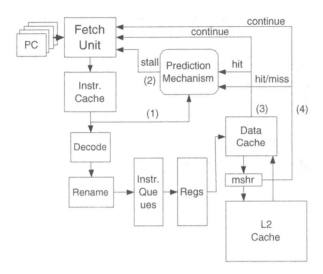

Fig. 1. L2MP mechanism

We have explored a wide range of different load miss predictors [1]. The one that obtains the best results is the predictor proposed in [4], what we call predictor of patterns.

About FLUSH and STALL, in [7] a load is declared to miss in L2 when it spends more than 15 cycles in the memory hierarchy. We have experiment different values for this parameter, and 15 presents the best overall results for our baseline architecture. Three additional considerations about FLUSH and STALL are: a data TLB miss also triggers a flush (or stall); a 2-cycle advance indication is received when a load returns from memory; and this mechanism always keeps one thread running. That is, if there is only one thread running, it is not stopped even when it experiences an L2 miss.

5   Comparing the Current Policies

In this section we will determine the effectiveness of the different policies addressing the problem of load latency. We will compare the STALL, FLUSH and L2MP policies using the throughput and the Hmean.

In Figure 2 we show the throughput and the Hmean improvements of STALL, FLUSH, and L2MP over ICOUNT. L2MP achieves important throughput increments over ICOUNT, mainly for 2-thread workloads. However, fairness results using the Hmean metric indicates that for the MEM workloads the L2MP is more unfair than ICOUNT. Only for 8-thread workloads L2MP outperforms ICOUNT in both throughput and Hmean. Our results indicate that it is because L2MP hurts MEM threads and boosts ILP threads, especially for few-thread workloads.

If we compare the effectiveness of L2MP with other policies addressing the same problem, like STALL, we observe that L2MP only achieves better through-

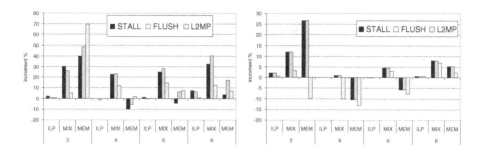

Fig. 2. Comparing current policies. (a) throughput increment over ICOUNT; (b) Hmean increment over ICOUNT

put than STALL for MEM workloads. However, L2MP heavily affects fairness. We will explain why L2MP does not obtain results as good as STALL soon.

The results of FLUSH and STALL are very similar. In general FLUSH slightly outperforms STALL, especially for MEM workloads and when the number of threads increases. This is because when the pressure on resources is high it is preferable to flush delinquent threads, and hence free resources, than stall these threads holding resources for a long time.

As stated before, no policy outperforms the other neither for all workloads, nor for all metrics. Each one behaves better depending on the particular metric and workload.

6   Improved Policies

6.1   Improving L2MP

We have seen that L2MP alleviates the problem of load latency, but it does not achieve results as good as other policies addressing the same problem. The main drawback of L2MP are the loads missing in L2 cache that are not detected by the predictor. These loads can seriously damage performance because following instructions occupy resources for a long time. Figure 3 depicts the percentage of missing loads that are not detected by the predictor of patterns. This percentage is quite significant (from 50% to 80%), and thus the problem still persists.

We propose to add a safeguard mechanism to "filter" this undetected loads. That is, this mechanism acts on loads missing on L2 that are not detected by the predictor. The objective is to reduce the harmful effects caused by these loads. In this paper we have used STALL [7] as safeguard mechanism.

Our results show that, when using L2MP, the fetch is absolutely idle for many cycles (i.e. 15% for the 2-MEM workload) because all threads are stalled by the L2MP mechanism. Another important modification that we have made to the original L2MP mechanism is to maintain always one thread running in order to avoid idle cycles of the processor.

The Figures 4 (a) and (b) show the throughput and the Hmean increment of L2MP+ over L2MP. Throughput results show that for MEM workloads L2MP

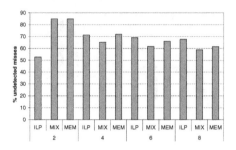

Fig. 3. Undetected L2 misses

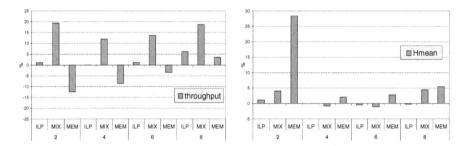

Fig. 4. L2MP+ vs. L2MP. (a) throughput results; (b) Hmean results

outperforms L2MP+ (5.1% on average), and for MIX L2MP+ outperforms L2MP (16% on average).

We have investigated why L2MP improves L2MP+ for MEM workloads. We detected that L2MP+ significantly improves the IPC of *mcf* (a thread with a high L2 miss rate), but this causes an important reduction in the IPC of the remaining threads. And given that the IPC of *mcf* is very low, the decrement in the IPC of the remaining threads affects more the overall throughput than the increment in the IPC of *mcf*. Table 3 shows the relative IPC of *mcf* and the remaining threads for each MEM workload. In all cases L2MP+ improves the IPC of *mcf* and degrades the IPC of the remaining threads. This indicates that the

Table 3. Relative IPCs

RELATIVE IPCs		L2MP	L2MP+	Increment
2 – MEM	mcf	0.32	0.64	100
	remaining	0.72	0.60	-17.01
4 – MEM	mcf	0.30	0.48	61.36
	remaining	0.39	0.35	-10.07
6 – MEM	mcf	0.29	0.39	33.72
	remaining	0.30	0.27	-7.63
8 – MEM	mcf	0.28	0.34	21.69
	remaining	0.19	0.19	0.48

original policy favors ILP threads but it is at the cost of hurting MEM threads. Hmean results, see Figure 4 (b), confirm that the L2MP+ policy presents a better throughput-fairness balance than the L2MP policy. L2MP+ only suffers slowdowns lower than 1%.

6.2   Improving FLUSH

The FLUSH policy always attempts to leave one thread running. In doing so, it does not flush and fetch-stall a thread if all remaining threads are already fetch-stalled. The Figure 5 shows a timing example for 2 threads. In the cycle c0 the thread T0 experiences an L2 miss and it is flushed and fetch-stalled. After that, in cycle c1, thread T1 also experiences an L2 miss, but it is not stalled because it is the only thread running. The main problem of this policy is that by the time the missing load of T0 is resolved (cycle c2), and this thread can proceed, the machine is presumably filled with instructions of thread T1. These instructions occupy resources until the missing load of T1 is resolved in cycle c3. Hence, performance is degraded.

The improvement we propose is called Continue the Oldest Thread (COT), and it is the following: when there are N threads, N-1 of them are already stalled, and the only thread running experiences an L2 miss, it is effectively stalled and flushed, but the thread that was first stalled is continued. In the previous example the new timing is depicted in Figure 6. When thread T1 experiences an L2 miss it is flushed and stalled, and T0 is continued. Hence, instructions of T0 consume resources until cycle c2 when the missing load is resolved. However, this does not affect to the thread T1 because it is stalled until cycle c3. In this example COT improvement has been applied to FLUSH, but it can be applied to any of the fetch-gating policies. We have applied it also to STALL. We call the new versions of FLUSH and STALL, FLUSH+ and STALL+.

Figure 7 (a) shows the throughput and the Hmean increments of FLUSH+ over FLUSH. We observe that FLUSH+ improves FLUSH for all workloads. We

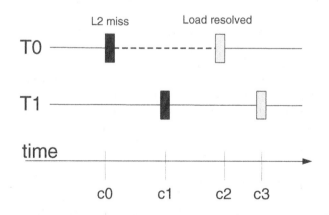

Fig. 5. Timing of the FLUSH policy

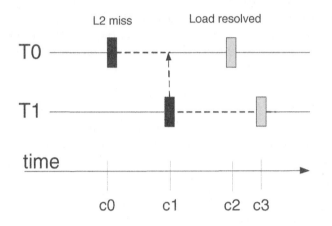

Fig. 6. Timing of the improved FLUSH policy

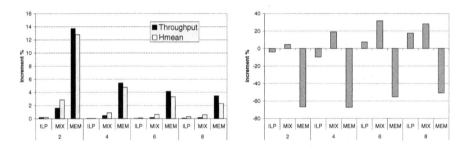

Fig. 7. FLUSH+ vs. FLUSH. (a) throughput and Hmean results; (b) EF results

also observe that for MEM workloads FLUSH+ clearly outperforms FLUSH, for both metrics, and the improvement decreases as the number of thread increases. This is because as the number of threads increases the number of time that only one thread is running and the remaining are stopped is lower. For MIX and ILP workloads the improvement is lower than for the MEM workloads because the previous situation is also less frequent. Concerning to flushed instructions, in Figure 7(b) we see the EF increment of FLUSH+ over FLUSH (remember the lower the value the better the result). We observe that, on average, FLUSH+ decrements by 60% FLUSH for MEM workloads and only increments by 20% FLUSH for MIX workloads. These results effectively indicate that FLUSH+ presents a better throughput-fairness balance than FLUSH, and also reduces extra fetch.

6.3   Improving STALL

The improved STALL policy, or STALL+, consists of applying the COT improvement to STALL.

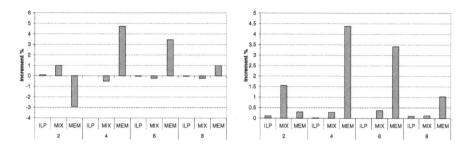

Fig. 8. STALL+ vs. STALL. (a) throughput results; (b) Hmean results

Figures 8 (a) and (b) show the throughput and the Hmean increment of STALL+ over STALL. We observe that the improvements of STALL+ over STALL are less pronunced than the improvements of FLUSH+ over FLUSH. Throughput results show that in general STALL+ improves STALL, and only for the 2-MEM workload there is a remarkable slowdown of 3%. The Hmean results show that STALL+ outperforms STALL for all workloads, and especially for MEM workloads. The EF results are not shown because the STALL and STALL+ policies do not squash instructions.

We analyzed why STALL outperforms STALL+ for the 2-MEM workload. We observe that the cause is the benchmark mcf. The main characteristic of this benchmark is its high L2 miss rate. On average, one of every eight instructions is a load failing in L2. In this case the COT improvement behaves as show in Figure 9: in cycle c0 the thread T0 (mcf) experiences an L2 miss and it is fetch-stalled. After that, in cycle c1, T1 experiences an L2 miss it is stalled and T0 (mcf) is continued. Few cycles after that, mcf experiences another L2 miss, thus the control is returned to thread T1. The point is that with FLUSH every time a thread is stalled it is also flushed. With STALL this is not the case, hence from cycle c1 to c2 mcf allocates resources that are not freed for a long time degrading the performance of T1. That is, COT improvement for STALL policy improves the IPC of benchmarks with high L2 miss rate, mcf in this case, but it hurts the IPC of the remaining. This situation is especially acute for 2-MEM workloads. To solve this problem, and other ones, we develop a new policy called FLUSH++.

6.4   FLUSH++ Policy

This policy tries to obtain the advantages of both policies, STALL+ and FLUSH+, and it focuses in the following points:

- For MIX workloads STALL+ presents good results. It is an alternative to FLUSH+ avoiding instruction re-execution.
- Another objective is to improve the IPC of STALL+ for MEM workloads with a moderate increment in the re-executed instructions.

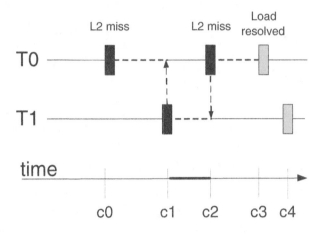

Fig. 9. Timing when the *mcf* benchmark is executed

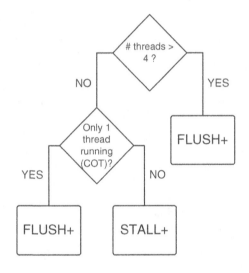

Fig. 10. FLUSH++ policy

- The processor knows every cycle the number of threads that are running. This information can be easily obtained for any policy at runtime.

FLUSH++ works differently depending on the number of running threads, see Figure 10.

- If the number of running threads is less than four, it combines STALL+ and FLUSH+. In a normal situation it behaves like STALL+, but when the COT improvement is triggered it acts as FLUSH+. That is, the flush is only activated when there is only one thread running and it experiences an L2 miss. In the remaining situations threads are only stalled.

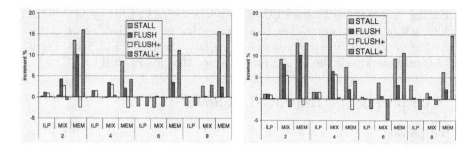

Fig. 11. FLUSH++ vs. FLUSH, STALL, FLUSH+ and STALL+. (a) throughput results; (b) Hmean results

- When there are more than four threads running, we must consider two factors. On the one hand, the pressure on resources is high. In this situation is preferable to flush delinquent threads instead of stalling it because freed resources are highly profited by the other threads. On the other hand, FLUSH+ improves FLUSH in both throughput and fairness for four-or-more thread workloads. For this reason, if the number of threads is greater than four, we will use the FLUSH+ policy.

In Figure 11 we compare FLUSH++ with the original STALL and FLUSH policies, as well as with the improved versions STALL+ and FLUSH+. The Figure (a) shows the throughput results, and the Figure (b) the Hmean results. We observe that FLUSH++ outperforms FLUSH in all cases in throughput as well as in Hmean. Furthermore, in Figure 12 it can been seen that for 2-, and 4-thread workloads FLUSH++ clearly reduces the EF. Concerning STALL, throughput results show that FLUSH++ only suffers slight slowdowns lower than 3% for the 6-MIX workload. Hmean results show that FLUSH++ always outperforms STALL.

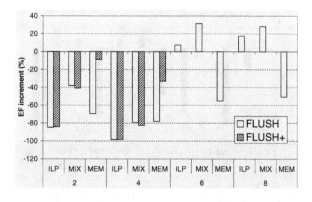

Fig. 12. FLUSH++ increments in EF over FLUSH, and FLUSH+

For ILP and MIX workloads FLUSH++ outperforms FLUSH+ and for MEM workloads it is slightly worse. The most interesting point is that, as we can see in the Figure 12 FLUSH++ considerably reduces the EF of FLUSH+. For 6-, and 8-thread workloads the results are the same that for FLUSH+.

7 Comparing the Improved Policies

In the previous section we saw that FLUSH++ outperforms FLUSH+ and STALL+. In this section we will compare FLUSH++ with L2MP+.

Figure 13 depicts the throughput and Hmean increments of L2MP+ over FLUSH++. The throughput results show that L2MP+ improves FLUSH++ for MIX workloads, and that FLUSH++ is better than L2MP+ for MEM workloads. The Hmean results indicate that only for 6-, and 8-thread workloads L2MP+ is slightly more fair than FLUSH++ for ILP and MIX workloads, for the remaining workloads FLUSH++ is more fair.

In general, FLUSH++ outperforms L2MP+, however for some configurations this is not the case. This confirms that each policy presents better results than the remaining depending on the particular workload and metric.

8 Conclusions

SMT performance directly depends on how the allocation of shared resources is done. The instruction fetch mechanism dynamically determines how the allocation is carried out. To achieve high performance it must avoid the monopolization of a shared resource by any thread. An example of this situation occurs when a load misses in the L2 cache level. Current instruction fetch policies focus on this problem and achieve significant performance improvements over ICOUNT.

A minor contribution of this paper is that we compare three different policies addressing this problem. We show that none of the presented policies clearly out-

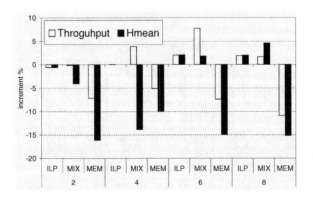

Fig. 13. Improved policies. Throughput and Hmean increments of L2MP+ over FLUSH++

performs all others for all metrics. The results vary depending on the particular workload and the particular metric (throughput, fairness, energy consumption, etc). The main contribution is that we have presented four improved versions of the three best policies addressing the described problem. Our results show that this enhanced versions achieve a significant improvement over the original ones:

- The throughput results indicate that L2MP+ outperforms L2MP for MIX workload (16% on average) and is worse than L2MP only for 2-, 4- and 6-MEM workloads (8% on average). The Hmean results show that L2MP+ outperforms L2MP especially for 2-thread workloads.
- The FLUSH+ policy outperforms FLUSH in both throughput and fairness especially for MEM workloads. Furthermore, it reduces extra fetch by 60% for MEM workloads and only increments extra fetch by 20% for MIX workloads.
- Throughput results show that in general STALL+ improves STALL, and only for the 2-MEM workload there is a remarkable slowdown of 3%. The Hmean results show that STALL+ outperforms STALL for all workloads, and especially for MEM workloads.
- FLUSH++, a new dynamic control mechanism, is presented. It adapts its behavior to the dynamic number of "alive threads" available to the fetch logic. Due to this additional level of adaptability, it is remarkable that FLUSH++ policy fully outperforms FLUSH policy in both, throughput and fairness. FLUSH++ also reduces EF for the 2- and 4-thread workloads, and moderately increases EF for the 6-MIX and 8-MIX workloads. Regarding STALL+, FLUSH++ outperforms STALL+ policy, with just a slight exception in throughput in 6-MIX workload.

Acknowledgments

This work was supported by the Ministry of Science and Technology of Spain under contract TIC-2001-0995-C02-01, and grant FP-2001-2653 (F. J. Cazorla), and by CEPBA. The authors would like to thank Oliverio J. Santana, Ayose Falcon and Fernando Latorre for their comments and work in the simulation tool. The authors also would like to the reviewers for their valuable comments.

References

1. F.J. Cazorla, E. Fernandez, A. Ramirez, and M. Valero. Improving long-latency-loads-aware fetch policies for SMT processors. Technical Report UPC-DAC-2003-21, Universitat Politecnica de Catalunya, May 2003.
2. A. El-Moursy and D.H. Albonesi. Front-end policies for improved issue efficiency in SMT processors. *Proceedings of the 9th Intl. Conference on High Performance Computer Architecture*, February 2003.
3. M. Gulati and N. Bagherzadeh. Performance study of a multithreaded superscalar microprocessor. *Proceedings of the 2nd Intl. Conference on High Performance Computer Architecture*, pages 291–301, February 1996.

4. C. Limousin, J. Sébot, A. Vartanian, and N. Drach-Temam. Improving 3d geometry transformations on a simultaneous multithreaded SIMD processor. *Proceedings of the 13th Intl. Conference onSupercomputing*, pages 236–245, May 2001.
5. K. Luo, J. Gummaraju, and M. Franklin. Balancing throughput and fairness in SMT processors. *Proceedings of the International Symposium on Performance Analysis of Systems and Software*, pages 164–171, November 2001.
6. T. Sherwood, E. Perelman, and B. Calder. Basic block distribution analysis to find periodic behavior and simulation points in applications. *Proceedings of the Intl. Conference on Parallel Architectures and Compilation Techniques*, September 2001.
7. D. Tullsen and J. Brown. Handling long-latency loads in a simultaneous multi-threaded processor. *Proceedings of the 34th Annual ACM/IEEE Intl. Symposium on Microarchitecture*, December 2001.
8. D. Tullsen, S. Eggers, J. Emer, H. Levy, J. Lo, and R. Stamm. Exploiting choice: Instruction fetch and issue on an implementable simultaneous multithreading processor. *Proceedings of the 23th Annual Intl. Symposium on Computer Architecture*, pages 191–202, April 1996.
9. D.M. Tullsen, S. Eggers, and H. M. Levy. Simultaneous multithreading: Maximizing on-chip parallelism. *Proceedings of the 22th Annual Intl. Symposium on Computer Architecture*, 1995.
10. W. Yamamoto and M. Nemirovsky. Increasing superscalar performance through multistreaming. *Proceedings of the 1st Intl. Conference on High Performance Computer Architecture*, pages 49–58, June 1995.
11. A. Yoaz, M. Erez, R. Ronen, and S. Jourdan. Speculation techniques for improving load related instruction scheduling. *Proceedings of the 26th Annual Intl. Symposium on Computer Architecture*, May 1999.

Tolerating Branch Predictor Latency on SMT

Ayose Falcón, Oliverio J. Santana, Alex Ramírez, and Mateo Valero

Departament d'Arquitectura de Computadors, Universitat Politècnica de Catalunya,
Barcelona, Spain,
{afalcon,osantana,aramirez,mateo}@ac.upc.es

Abstract. Simultaneous Multithreading (SMT) tolerates latency by executing instructions from multiple threads. If a thread is stalled, resources can be used by other threads. However, fetch stall conditions
caused by multi-cycle branch predictors prevent SMT to achieve all its
potential performance, since the flow of fetched instructions is halted.
This paper proposes and evaluates solutions to deal with the branch
predictor delay on SMT. Our contribution is two-fold: we describe a decoupled implementation of the SMT fetch unit, and we propose an inter-
thread pipelined branch predictor implementation. These techniques prove to be effective for tolerating the branch predictor access latency.

keywords: SMT, branch predictor delay, decoupled fetch, predictor
pipelining.

1 Introduction

Superscalar processors take advantage of ILP by executing multiple instructions
of a single program during each cycle. They require accurate branch predictions
to feed the execution engine with enough instructions, mainly from correct paths.
Simultaneous multithreaded processors (SMT) [15, 16] take one step further by
exploiting TLP, i.e. executing multiple instructions from multiple threads during
each cycle. Multithreading adds pressure to the branch predictor, increasing the
total number of predictor accesses per cycle.

Branch predictors consist of one or more large tables that store prediction
information. Each cycle, they must provide a prediction in order to keep on
fetching instructions to feed the execution core. However, as feature sizes shrink
and wire delays increase, it becomes infeasible to access large memory structures in a single cycle [1, 2]. This involves that small branch prediction tables
are needed to generate a branch prediction each cycle. However, the low accuracy of small branch predictors degrades the processor performance. This forces
superscalar processors to use big and accurate branch predictors in combination
with mechanisms for tolerating their access latency [5, 9].

The impact of branch predictor latency also affects SMT considerably. The
problem gets even worse because there are several threads trying to access the
shared branch predictor. If the branch predictor is unable to give a response in
one cycle, multiple threads could be stalled at the fetch stage for several cycles.

A. Veidenbaum et al. (Eds.): ISHPC 2003, LNCS 2858, pp. 86–98, 2003.

In this paper, we evaluate the impact of the branch predictor latency in the context of SMT processors. We show that the increased access latency of branch predictors degrades the overall processor performance. However, reducing the predictor size is not a solution, since the lower latency does not compensate the lower accuracy of a smaller predictor. We also evaluate the effect of varying the number of branch predictor access ports. Since an SMT processor can fetch from several threads each cycle, the number of access ports has an important effect on performance. In this context, mechanisms for tolerating the branch predictor access latency can provide a worthwhile performance improvement.

We show that decoupling the branch predictor from the instruction cache access [9] is helpful for tolerating the branch predictor delay on SMT. Branch predictions are stored in intermediate queues, and later used to drive the instruction cache while new predictions are being generated. This technique allows to increase the latency tolerance of the branch predictors. The decoupled scheme, combined with the ability of predicting from four different threads, allows a 4-port branch predictor to achieve a performance similar to an ideal 1-cycle latency branch predictor.

Finally, we propose an inter-thread pipelined branch predictor design, which interleaves prediction requests from different threads each cycle. Although a particular thread should wait for its previous prediction to start a new one, a different thread can start a new prediction. This allows to finish a branch prediction each cycle at a relative low complexity. Using the inter-thread pipelining technique, a 1-port branch predictor can achieve a performance close to a 4-port branch predictor, but requiring 9 times less chip area.

The rest of this paper is organized as follows. Section 2 exposes previous related work. Our experimental methodology is described in Section 3. Section 4 analyzes the impact of branch predictor delay on SMT. Section 5 shows the effect of varying the number of branch predictor access ports. In Section 6 we evaluate a decoupled design of an SMT fetch engine. Section 7 describes our inter-thread pipelined branch predictor design. Finally, Section 8 exposes our concluding remarks.

2   Related Work

The increase in processor clock frequency and the slower wires in modern technologies prevent branch predictors from being accessed in a single cycle [1, 5]. In the last years, research effort has been devoted to find solutions for this problem.

A first approach to overcome the branch predictor access latency is to increase the length of the basic prediction unit. Long prediction units allow to feed the execution engine with instructions during several cycles, hiding the latency of the following prediction. The FTB [9] extends the classic concept of BTB, enlarging the prediction unit by ignoring biased not taken branches. The next stream predictor [8] takes this advantage one step further, ignoring all not taken branches. The use of path correlation allows the stream predictor to provide accurate prediction of large consecutive instruction streams. The next trace predictor [4] also tries to enlarge the prediction unit by using instruction traces, which are stored in a trace cache. In practice, instruction streams are longer than traces [8], being more tolerant to access latency at a lower cost and complexity.

Decoupling branch prediction from the instruction cache access [9] helps to take advantage of large basic prediction units. The branch predictor generates requests that are stored in a fetch target queue (FTQ) and used to drive the instruction cache. When fetching instructions is not possible due to instruction cache misses or resource limitations, the branch predictor can still make new predictions. Therefore, the presence of an FTQ makes less likely for the fetch engine to stay idle due to the branch predictor access latency.

Prediction overriding [5] is a different approach for tolerating the predictor access latency. It uses a small branch predictor that provides a prediction in a single cycle, and a slower but more accurate predictor which overrides the first prediction some cycles later if they differ. Another promising idea is pipelining the branch predictor [6, 10]. Using a pipelined predictor, a new prediction can be started each cycle. However, this is not trivial, since the outcome of a branch prediction is needed to start the next one. Therefore, a branch prediction can only use the information available in the cycle it starts, which forces the branch predictor to use in-flight speculative information [10]. In general, this kind of pipelined predictors, as well as the prediction overriding mechanisms, require an increase in the complexity of the fetch engine.

3   Simulation Setup

We use a trace-driven version of the SMTSIM [13] simulator. We allow wrong path execution by having a separate basic block dictionary which contains information of all static instructions. We have modified the fetch stage of the simulator, by dividing it into two different stages: a prediction and a fetch stage.

The baseline processor configuration is shown in Table 1. We provide simulation results obtained by the FTB fetch architecture [9], using a gskew conditional branch predictor [7], as well as the stream fetch architecture [8]. Both fetch architectures use the ICOUNT.2.8 [14] fetch policy (up to 8 instructions from up to 2 threads). ICOUNT gives priority to threads with the fewest number of instructions in the decode, rename and dispatch stages of the processor.

Table 1. Simulation parameters

Fetch Policy	ICOUNT.2.8
Branch Predictor	FTB + gskew predictor, or stream predictor
Return Address Stack	64 entries (per thread)
Decode and Rename Width	8 instructions
Issue Queues	32-entry int, fp and ld/st
Reorder Buffer	256 entries (per thread)
Physical Registers	384 int and 384 fp
Functional Units	6 int, 3 fp, 4 ld/st
L1 I-Cache, L1 D-Cache	64KB, 2-way, 64B/line, 8 banks
Unified L2 Cache	1MB, 2-way, 64B/line, 8 banks, 10 cycle lat.
Main Memory Latency	100 cycles

The workloads used in our simulations are composed of benchmarks selected from the SPECint2000 suite. We have chosen eight benchmarks with a high instruction-level parallelism because our study is focused on the fetch engine architecture[1]. Benchmarks were compiled on a DEC Alpha AXP-21264 using Compaq's C/C++ compiler with '-O2' optimization level. Due to the large simulation time of SPECint2000 benchmarks, we collected traces of the most representative 300 million instruction slice, following the idea presented in [11]. These traces were collected executing the *reference* input set.

4   Motivation

We have measured the access time for the branch prediction structures evaluated in this paper using CACTI 3.0 [12]. We have modified it to model tagless branch predictors and to work with setups expressed in bits instead of bytes. Data we have obtained corresponds to a 0.10μm process, which is expected to be used in a near future. We assume an aggressive 8 fan-out-of-four (FO4) delays clock period, that is, a 3.47 GHz clock frequency as reported in [1]. It has been claimed in [3] that 8 FO4 delays is the optimal clock period for integer benchmarks in a high performance processor implemented in 0.10μm technology.

Figure 1 shows the prediction table access time obtained using CACTI. We have measured the access time for 2-bit counter tables ranging from 8 to 64K entries. We have also measured the access time for an FTB and a stream predictor ranging from 32 to 4K entry tables. These tables are assumed to be 4-way associative. Data in Fig. 1 is presented for branch predictors using 1, 2, and 4 access ports[2].

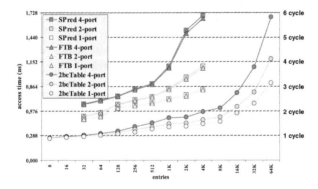

Fig. 1. Access time to branch prediction tables at 0.10μm and 3.47 GHz

---

[1] The eight benchmarks selected are *gzip*, *bzip2*, *eon*, *gcc*, *crafty*, *vortex*, *gap*, and *parser*. To simulate the effect of different numbers of threads in the evaluated mechanisms, we use workloads with 2, 4, 6, and 8 threads.

[2] Along this paper, when we refer to a predictor using n ports, we mean a predictor using n read ports and n write ports.

Table 2. Configuration of the evaluated branch predictors

		0.5KB	32KB
FTB	FTB	64 entry, 4-way	4096 entry, 4-way
	gskew	three 128 entry tables 7 bit history	three 8192 entry tables 13 bit history
stream predictor		16 entry, 4-way 1st level 64 entry, 4-way 2nd level DOLC 0-0-0-4	1024 entry, 4-way 1st level 4096 entry, 4-way 2nd level DOLC 12-2-4-10
access latency (both predictors)		1-port: 2 cycles 2-port: 2 cycles 4-port: 3 cycles	1-port: 3 cycles 2-port: 4 cycles 4-port: 6 cycles

This data shows that devoting multiple cycles to access the branch predictor is unavoidable. Even single-cycle conditional branch predictors, composed of a small 2-bit counters table, should use a multi-cycle FTB to predict the target address of branch instructions. On the other hand, although an FTB with a size below the evaluated range could be accessed in a single cycle, its poor prediction accuracy will be more harmful for the processor performance than the increased latency of a larger FTB.

In order to analyze the tradeoff between fast but inaccurate and slow but accurate predictors, we have chosen two setups for the two evaluated branch predictors: a 0.5KB setup and a 32KB setup. We have explored a wide range of history lengths for the gskew predictor, as well as DOLC index configurations [8] for the stream predictor, and selected the best one found for each setup. The four evaluated predictor setups are shown in Table 2.

Impact of Branch Predictor Delay on SMT Performance

As discussed previously, the branch predictor delay is a key topic in an SMT fetch engine design. On the one hand, a fast but small branch predictor causes a high number of mispredictions which degrades processor performance. On the other hand, a big and accurate predictor requires multiple cycles to be accessed, which also degrades performance. In this section we explore this tradeoff.

Figure 2.a shows the prediction accuracy of the four evaluated branch predictor setups. Clearly, small 0.5KB predictors provide the worst prediction accuracy. The exception is the 2-thread workload. It contains two benchmarks (*gzip* and *bzip2*) with few static branches, which allows a small branch predictor to predict branches accurately. The larger number of total static branches in the rest of workloads increases aliasing, degrading the accuracy of the 0.5KB predictors. Moreover, the larger number of threads also increases aliasing, degrading even more the accuracy of these predictors. Nevertheless, the 32KB predictors provide an accuracy over 95% for all the evaluated workloads due to their larger size.

According to this data, a 32KB predictor should be used. This predictor requires two access ports in order to generate two predictions each cycle, since the ICOUNT.2.8 fetch policy can fetch from two different threads each cycle. However, such a predictor needs 4 cycles to be accessed, which could degrade processor performance. An alternative is to use a smaller 0.5KB predictor, which

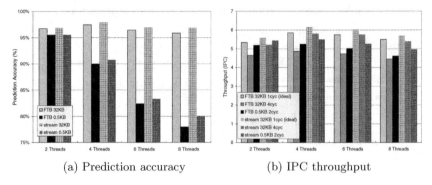

(a) Prediction accuracy   (b) IPC throughput

Fig. 2. Comparison of ideal (1-cycle latency) and realistic branch predictors. Figure (a) shows the prediction accuracy obtained by the four evaluated branch predictor setups. Figure (b) shows the IPC throughput achieved by ideal 32KB predictor setups and realistic predictor setups. All predictors have two access ports

is faster but also less accurate. Figure 2.b shows a comparison of the performance achieved by ideal 32KB predictor setups, i.e. with 1-cycle latency, and realistic predictor setups: 32KB predictors with 4-cycle latency and 0.5KB predictors with 2-cycle latency.

In the 2-thread workload, the 2-cycle latency 0.5KB predictors provide a performance better than the 4-cycle latency 32KB predictors. This is caused by the good prediction accuracy achieved by the 0.5KB predictors for this workload. In the rest of workloads, its higher speed enables the 0.5KB FTB to achieve a better performance than the 4-cycle 32KB FTB, but the 4-cycle latency 32KB stream predictor achieves a better performance than the smaller one. This happens because streams are longer than FTB fetch blocks [8], making the stream predictor more latency tolerant. In the remainder of this paper we describe techniques to tolerate the branch predictor access latency, trying to reach the ideal performance. For the purpose of brevity, we only show data for the 32KB branch predictor setups, since the 0.5KB predictors behave in a similar way.

5   A First Approach: Varying the Number of Ports

A first solution to alleviate the effect of the branch predictor latency is modifying the number of access ports. The branch predictors evaluated in the previous section use two access ports because the fetch policy used by the processor is ICOUNT.2.8 [14], which can fetch from two different threads each cycle. We assume that these predictions are stored in intermediate buffers and can be used in following cycles, while new predictions are being generated.

Reducing the number of access ports involves a reduction in the predictor access latency. Therefore, it is interesting to explore the tradeoff between using a single access port and having a higher latency. On the other hand, increasing the number of ports involves an increment in the access latency. Despite being

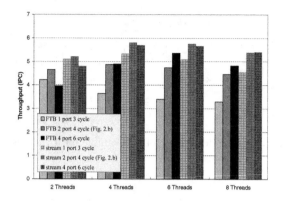

Fig. 3. IPC throughput achieved by the 32KB branch predictor setups using 1, 2, and 4 access ports

slower, a 4-port predictor can provide 4 predictions each cycle, helping to hide the access latency.

Figure 3 shows the performance achieved by the 32KB predictor setups using 1, 2, and 4 access ports. Data using 2 ports is the same data shown in Fig. 2.b. The first observation in this figure is that reducing the number of ports harms the processor performance. The reduction in the access time does not compensate losing the ability to make two predictions in each access. The second observation is that increasing the number of access ports allows to achieve a performance similar to the faster 2-port predictors, even achieving a higher performance in some cases. This is caused by the ability of making four predictions in each access. The exception is the 2-thread workload, where a maximum of two predictions can be generated each cycle.

Nevertheless, there is still room for improvement. The slowdown of the different configurations, varying the number of ports, against the 1-cycle latency ideal 32KB predictors using 2 access ports, shown in Fig. 2.b, ranges from 19% to 56% using the FTB, and from 9% to 17% using the stream predictor.

6   A Decoupled SMT Fetch Engine

A solution to alleviate the effect of the branch predictor latency is decoupling the fetch address generation from the fetch address consumption, as proposed in [9]. Our proposal of decoupled SMT fetch engine is shown in Fig. 4. For each thread, a fetch target queue (FTQ) stores predictions done by the branch predictor, which are consumed later by the fetch unit. It is important to note that, while the branch predictor and the fetch unit are shared among all the threads, there is a separate FTQ for each thread. The number of threads which can use the branch predictor depends on the number of access ports. These threads are selected using a *Round Robin* strategy.

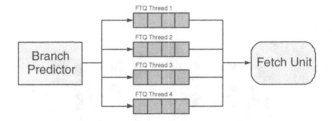

Fig. 4. SMT decoupled fetch (example for 4 threads)

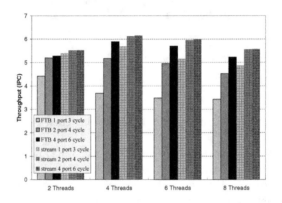

Fig. 5. IPC throughput using a decoupled fetch with a 4-entry FTQ. Data is shown for 32KB predictors using 1, 2, and 4 access ports

Figure 5 shows performance results using a decoupled fetch with a 4-entry FTQ per thread. For all workloads, decoupling the fetch involves a performance improvement with respect to the performance results shown in the previous section (Fig. 3). The FTQs allow the branch predictor to work at a different rate than the instruction cache, introducing requests in the queues even while the fetch engine is stalled due to cache misses or resource limitations. The information contained in the FTQs can be used to drive the instruction cache during the following branch predictor accesses, hiding its access latency.

Table 3 shows the speedup achieved by decoupling the branch predictor. The best average speedups are achieved by the 4-port predictors. This makes sense because they have a higher latency, and thus they can benefit the most from a decoupled scheme. It is also interesting to note that the FTB achieves higher speedups than the stream predictor. The stream fetch engine uses longer prediction units than the FTB fetch architecture [8]. Therefore, the use of an FTQ is more beneficial for the FTB fetch architecture than for the stream fetch engine, since a larger prediction unit makes it easier to hide the branch predictor access latency.

The last column of Table 3 shows the average slowdown of the decoupled scheme against the ideal 1-cycle 32KB predictors using 2 access ports and a

Table 3. IPC speedup (%) using a decoupled SMT fetch (4-entry FTQ) over a coupled SMT fetch (data in Fig. 3). Data is shown for 32KB predictors. The last column shows the average slowdown against ideal 1-cycle 32KB predictors using 2 access ports in a coupled SMT fetch (shown in Fig. 2.b)

		2 Threads	4 Threads	6 Threads	8 Threads	AVG	AVG slowdown
FTB	1 port	5%	1%	3%	4%	3%	50%
	2 ports	12%	6%	5%	2%	6%	14%
	4 ports	34%	20%	6%	8%	17%	2%
stream predictor	1 port	5%	6%	1%	7%	5%	11%
	2 ports	6%	5%	4%	3%	5%	1%
	4 ports	15%	8%	6%	3%	8%	1%

coupled fetch engine (shown in Fig. 2.b). There is little room for improvement in the 4-port predictors, since their ability of making four predictions combined with the FTQ allows the fetch engine to hide the access latency. The stream predictor using 2 access ports also achieves a performance close to the ideal 1-cycle predictor due to the longer size of instruction streams. However, the rest of setups can still be improved. This is specially true for the FTB using a single access port, which achieves half the performance of the ideal FTB.

7 An Inter-thread Pipelined Branch Predictor

Pipelining is a technique that allows the branch predictor to provide a prediction each cycle [6, 10]. The branch predictor access is divided into several stages, the last of which provides the final branch outcome. With this technique, large branch predictors can be used without impacting performance due to their large access time. However, pipelining the branch predictor implies that the next prediction is initiated before the current one has finished. Therefore, the next prediction cannot use the information generated by the current one (including next fetch address), which can harm prediction accuracy.

Figure 6 shows different implementations of a branch predictor. Figure 6.a shows a non-pipelined branch predictor. A branch prediction begins only when the previous one has finished. Hence, some bubbles are introduced in the pipeline. Figure 6.b shows a pipelined branch predictor as proposed in [10]. In order to

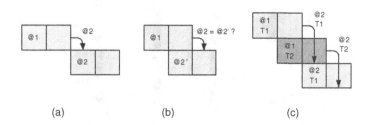

(a)   (b)   (c)

Fig. 6. Examples of accesses to a 2-cycle branch predictor: (a) Non-pipelined predictor; (b) Pipelined predictor [10]; (c) SMT pipelined predictor

initiate a new prediction without waiting for the previous one (@1), the new fetch address (@2') is calculated by using speculative dynamic information of pending predictions, as well as by predicting future decoding information. When a prediction is finished, the target address obtained (@2) is compared with the predicted target address (@2') used for generating the following prediction. If both target addresses are different, wrong speculative instructions should be discarded, resuming the instruction fetch with the correct prediction (@2).

Although using in-flight information and decoding prediction like in [10] allows accurate latency tolerant branch prediction, it also involves a high increase in the fetch engine complexity (predict the presence of branches, maintain in-flight prediction data, support recovery mechanisms, etc). To avoid this complexity, we propose a pipelined branch predictor implementation for SMT. Our proposal interleaves branch prediction requests for each thread, providing a prediction each cycle despite of the access latency. Each thread can initiate a prediction as long as there is no previous pending prediction for this thread. Thus, if there are enough threads being executed in the processor, the branch predictor latency can be effectively hidden.

Figure 6.c shows the behavior of our inter-thread pipelined branch predictor. This example employs a 2-cycle 1-port branch predictor and a 2-thread workload. Each cycle, a new prediction for a different thread can be initiated. Therefore, although each individual prediction takes 2 cycles, the branch predictor provides a prediction each cycle. As more threads are executed, the pipeline is always filled with new prediction requests, so branch predictor latency is totally masked.

Figure 7 shows performance results using an inter-thread pipelined branch predictor in a decoupled SMT fetch engine. The main observation is that the performance of all the evaluated setups is similar. Table 4 shows the IPC speedups achieved by using a decoupled inter-thread pipelined branch predictor over the decoupled fetch with a non-pipelined branch predictor. As stated in the previous section, there was little room for improving the 4-port predictors. Since

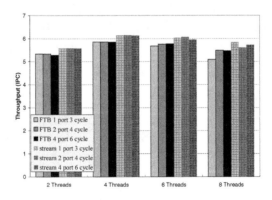

Fig. 7. IPC throughput using an inter-thread pipelined branch predictor and a decoupled fetch engine. Data is shown for 32KB predictors using 1, 2, and 4 access ports

Table 4. IPC speedup (%) using a pipelined branch predictor in a decoupled fetch engine over a non-pipelined decoupled predictor (data in Fig. 5). Data is shown for 32KB predictors. The last column shows the average slowdown against ideal 1-cycle 32KB predictors using 2 access ports in a coupled SMT fetch (shown in Fig. 2.b)

		2 Threads	4 Threads	6 Threads	8 Threads	A V G A V G	slowdown
	1 port	21%	59%	63%	48%	48%	2%
FTB	2 ports	2%	13%	16%	21%	13%	0%
	4 ports	0%	0%	1%	5%	1%	0%
stream predictor	1 port	3%	8%	17%	20%	12%	0%
	2 ports	1%	0%	2%	1%	1%	0%
	4 ports	1%	0%	0%	3%	1%	0%

decoupled branch predictors with four access ports can efficiently hide the access latency, pipelining them has little impact on processor performance. On the contrary, the single port predictors achieve an important performance improvement by using our inter-thread pipelining technique. The 1-port stream predictor achieves an average 12% speedup, while the 1-port FTB achieves a 48% speedup.

These results show that the inter-thread pipelined branch predictor is an efficient solution to tolerate the branch predictor latency on SMT. The last column of Table 4 shows the average slowdown of the decoupled inter-thread pipelined predictors against the ideal 1-cycle 32KB predictors using 2 access ports and a coupled fetch engine (shown in Fig. 2.b). All the evaluated setups almost achieve the performance of an ideal 1-cycle predictor. Therefore, the combination of an inter-thread pipelined predictor design and a decoupled SMT fetch engine constitutes a fetch unit that is able to exploit the high accuracy of large branch predictors without being penalized for their large access latency.

Finally, an important conclusion that can be drawn from these results is that an inter-thread pipelined branch predictor with a single access port provides a performance close to a branch predictor using four access ports, even if it is pipelined. A reduction in the number of access ports involves a large reduction in the chip area devoted to the branch predictor: Our 32KB 1-port inter-thread pipelined design requires 3 times less area than a similar predictor using 2 access ports, or even 9 times less area than a similar predictor using 4 access ports.

8   Conclusions

Current technology trends prevent branch predictors from being accessed in a single cycle. In this paper, we have shown that the branch predictor access latency is a performance limiting factor for SMT processors. Although the predictor latency degrades the potential performance, reducing its size is not a solution, since the lower prediction accuracy degrades the performance even more.

Using a branch predictor with a single port reduces the access latency, but only one prediction can be generated each time, limiting SMT performance. On the contrary, using four access ports increases the access latency, degrading performance, although generating four predictions each time partially compensates this degradation. Our results show that techniques for reducing the impact of prediction delay on SMT can provide a worthwhile performance improvement.

We propose to decouple the SMT fetch engine to tolerate the branch predictor latency. A decoupled SMT fetch provides larger improvements for those predictors with a higher latency, so the 4-port predictors benefit the most from this technique, achieving speedups ranging from 3% to 34%.

We also propose an inter-thread pipelined branch predictor for SMT. Our design maintains a high prediction generation throughput with a low complexity. Each cycle, a new branch prediction is initiated, but only from a thread that is not waiting for a previous prediction. Our inter-thread pipelined mechanism allows the 1-port predictors to achieve important speedups over decoupled non-pipelined predictors, ranging from 3% to 63%. Moreover, inter-thread pipelined 1-port predictors are able to achieve a performance close to 4-port predictors, but reducing the required chip area by a factor of nine.

In summary, SMT tolerates the latency of memory and functional units by issuing and executing instructions from multiple threads in the same cycle. The techniques presented in this paper allow to extend the latency tolerance of SMT to the processor front-end. Thus, multi-cycle branch predictors and instruction fetch mechanisms can be used without affecting SMT performance.

Acknowledgments

This work has been supported by the Ministry of Education of Spain under contract TIC–2001–0995–C02–01, CEPBA and an Intel fellowship. A. Falcón is also supported by the Ministry of Education of Spain grant AP2000–3923. O.J. Santana is also supported by the Generalitat de Catalunya grant 2001FI–00724–APTIND. We especially thank Francisco Cazorla for his help developing the simulation tool.

References

1. V. Agarwal, M. Hrishikesh, S. Keckler, and D. Burger. Clock rate versus IPC: The end of the road for conventional microarchitectures. *Procs. of the 27th Intl. Symp. on Computer Architecture*, June 2000.
2. R. Ho, K. Mai, and M. Horowitz. The future of wires. *Proceedings of the IEEE*, Apr. 2001.
3. M. Hrishikesh, D. Burger, S. Keckler, P. Shivakumar, N. Jouppi, and K. Farkas. The optimal logic depth per pipeline stage is 6 to 8 FO4 inverter delays. *Procs. of the 29th Intl. Symp. on Computer Architecture*, May 2002.
4. Q. Jacobson, E. Rottenberg, and J. Smith. Path-based next trace prediction. *Procs. of the 30th Intl. Symp. on Microarchitecture*, Dec. 1997.
5. D. Jiménez, S. Keckler, and C. Lin. The impact of delay on the design of branch predictors. *Procs. of the 33rd Intl. Symp. on Microarchitecture*, Dec. 2000.
6. D. Jiménez. Reconsidering complex branch predictors. *Procs. of the 9th Intl. Conf. on High Performance Computer Architecture*, Feb. 2003.

7. P. Michaud, A. Seznec, and R. Uhlig. Trading conflict and capacity aliasing in conditional branch predictors. *Procs. of the 24th Intl. Symp. on Computer Architecture*, June 1997.
8. A. Ramirez, O. J. Santana, J. L. Larriba-Pey, and M. Valero. Fetching instructions streams. *Procs. of the 35th Intl. Symp. on Microarchitecture*, Nov. 2002.
9. G. Reinman, B. Calder, and T. Austin. Optimizations enabled by a decoupled front-end architecture. *IEEE Trans. on Computers*, 50(4):338–355, Apr. 2001.
10. A. Seznec and A. Fraboulet. Effective ahead pipelining of instruction block address generation. *Procs. of the 30th Intl. Symp. on Computer Architecture*, June 2003.
11. T. Sherwood, E. Perelman, and B. Calder. Basic block distribution analysis to find periodic behavior and simulation points in applications. *Procs. of the Intl. Conf. on Parallel Architectures and Compilation Techniques*, Sept. 2001.
12. P. Shivakumar and N. Jouppi. CACTI 3.0, an integrated cache timing, power and area model. TR 2001/2, Compaq WRL, Aug. 2001.
13. D. Tullsen. Simulation and modeling of a simultaneous multithreading processor. *22nd Computer Measurement Group Conference*, Dec. 1996.
14. D. Tullsen, S. Eggers, J. Emer, H. Levy, J. Lo, and R. Stamm. Exploiting choice: Instruction fetch and issue on an implementable simultaneous multithreading processor. *Procs. of the 23rd Intl. Symp. on Computer Architecture*, May 1996.
15. D. Tullsen, S. Eggers, and H. Levy. Simultaneous multithreading: Maximizing on-chip parallelism. *Procs. of the 22nd Intl. Symp. on Computer Architecture*, June 1995.
16. W. Yamamoto and M. Nemirovsky. Increasing superscalar performance through multistreaming. *Procs. of the Intl. Conf. on Parallel Architectures and Compilation Techniques*, June 1995.

A Simple Low-Energy Instruction Wakeup Mechanism

Marco A. Ramírez [1, 4], Adrian Cristal[1], Alexander V. Veidenbaum [2],
Luis Villa [3], and Mateo Valero [1]

[1] Computer Architecture Department U.P.C. Spain,
{mramirez, adrian, mateo}@ac.upc.es
Phone: +34-93-401-69-79, Fax: +34-93-401-70-55
[2] University of California Irvine CA
alexv@ics.uci.edu
[3] Mexican Petroleum Institute, Mexico
lvilla@imp.mx
[4] National Polytechnic Institute, México

Abstract. Instruction issue consumes a large amount of energy in out of order processors, largely in the wakeup logic. Proposed solutions to the problem require prediction or additional hardware complexity to reduce energy consumption and, in some cases, may have a negative impact on processor performance. This paper proposes a mechanism for instruction wakeup, which uses a multi-block instruction queue design. The blocks are turned off until the mechanism determines which blocks to access on wakeup using a simple successor tracking mechanism. The proposed approach is shown to require as little as 1.5 comparisons per committed instruction for SPEC2000 benchmarks.

Keywords: Superscalar processors, Out of order execution, Instruction window, Instruction wake up, Low power, CAM.

1 Introduction

Modern high-performance processors issue instructions in order but allow them to be executed out of order. The out-of-order execution is driven by the availability of operands; i.e. an instruction waits until its operands are ready and then is scheduled for execution. An instruction queue (IQ) is a CPU unit used to store waiting instructions after they are issued and until all their operands are ready. Associated with the instruction queue are wakeup logic and selection logic [1].

The wakeup logic is responsible for waking up instructions waiting in the instruction queue as their source operands become available. When an instruction completes execution, a register number (tag) of its result register is broadcast to all instructions waiting in the instruction queue. Each waiting instruction compares the result register tag with its own source operand register tags. Three-address instructions are assumed in this paper and an instruction may have 0, 1 or 2 source operands. On a match, a source operand is marked as available. Thus, the instruction is marked ready to execute when all its operands are ready.

Selection logic is responsible for selecting instructions for execution from the pool of ready instructions. A typical policy used by the selection logic is oldest ready first.

A. Veidenbaum et al. (Eds.): ISHPC 2003, pp. 99–112, 2003.

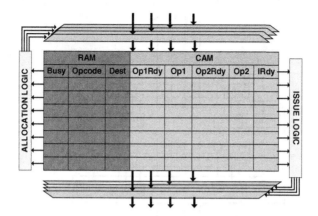

Fig. 1. An Instruction Queue Architecture using RAM-CAM array (4-wide issue).

A commonly used implementation of the instruction queue is shown in Fig. 1 and uses RAM-CAM arrays to store the necessary information [2]. The RAM section (solid area in the figure) stores the instruction opcode, the destination (result) register tag, and the busy bit indicating the entry is used. The CAM section stores the two source operand register tags and the corresponding ready bits.

One or more of its operands may be ready as the instruction is initially placed in the instruction queue. In this case they are immediately marked ready by setting the corresponding flag(s). If one or more of its operands are not ready, the instruction waits in the queue until it is generated by the producing instruction. As instructions complete execution and broadcast their destination register tag, each CAM entry performs a comparison of that tag with operand1 tag and/or operand2 tag. On a match, operand1 ready flag (Op1Rdy) and/or operand2 ready flag (OpRdy2) are set. When both Op1Rdy and Op2Rdy are set, the Instruction Ready (IRdy) flag is set.

Figure 2 shows a more detailed view of the CAM section of the queue assuming an issue width of four instructions. Eight comparators are required per each IQ entry. All comparisons are ideally completed in one clock cycle to allow up to four instructions to wake up per cycle. It is clear from the figure that the wakeup logic is both time and energy consuming. This paper focuses on reducing the latter.

The energy consumption of the wakeup logic is a function of the queue size, issue width, and the design used. Overall, the energy used in the IQ is dominated by the total number of comparisons performed in the CAM portion of the IQ. Consider the CAM cell organization shown in Fig. 2b. The precharge signal enables the comparison in a given entry. Most of the energy dissipated during a comparison is due to the precharge and the discharge of the match line ML. Almost no energy is dissipated when the precharge of an entry is disabled.

The actual energy dissipated in a CAM cell depends on the cell design and the process parameters. It is, however, proportional to the number of comparisons performed. Therefore, the rest of this paper uses the number of comparisons to evaluate and compare different wakeup mechanisms. In general, a larger instruction queue is highly desirable in a dynamically scheduled processor as it can expose additional instruction level parallelism. This increases the energy consumption in the IQ even further. Several solutions have been proposed to address the IQ energy consumption.

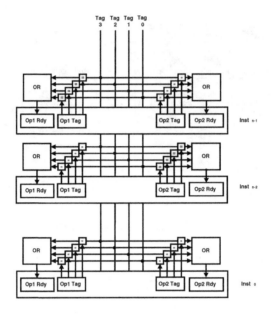

Fig. 2a. Organization of the CAM Section of the Instruction Queue.

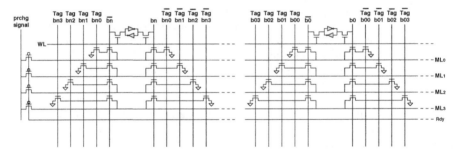

Fig. 2b. A transistor-level design of an n+1-bit Entry CAM.

This paper proposes a new solution at the microarchitecture level, which is simple, requires minimal CPU modification, and yet results in a very large reduction in the number of comparisons required and therefore the energy consumption of the instruction queue.

This solution can be used in conjunction with some of the previously proposed techniques to achieve even better results.

One of the commonly used IQ designs partitions the instruction queue into integer, floating point, and load/store queues. This reduces the destination tag fanout and the number of comparisons performed. The design proposed in this paper further divides each queue into a number of separate blocks. All queue blocks are normally disabled and perform no comparisons unless explicitly enabled (the precharge is turned on). A method for tracking which block(s) should be enabled upon an instruction completion is proposed. In addition, only valid entries are compared within each block as proposed in [3]. The proposed design is shown to lead to a large reduction in the total number of comparisons. The new design does not affect the IPC.

The rest of the paper is organized as following. Section 2 describes related research, Section 3 presents the proposed wakeup mechanism, and Section 4 shows the results obtained.

2 Related Work

The energy consumption of a modern dynamically scheduled superscalar processor is between 50 and 100 Watts. At the micro-architecture level, the issue logic is one of the main consumers of energy responsible for approximately 25% of the total energy consumption [3]. Many approaches to designing the wakeup logic have been proposed, both to reduce delay and to reduce energy consumption. [4] proposed a pointer-based solution, where each instruction has a pointer to its dependent instruction for direct wakeup. This was done for a singe-issue, non-speculative processor, however.

[5] extended the above solution to modern processors with wide issue and speculation. The effect of one, two, or three successor pointer entries per instruction was evaluated. Three approaches to deal with the case of an instruction with more successors than pointer entries were proposed. The first one stalled the instruction issue. The second one stopped recording successors and instead woke the instruction up when it reached the top of the instruction window. Both of these approaches lead to a loss of IPC. The third approach added a scoreboard to avoid stalls, an expensive approach to say the least. Finally, successor pointers were saved on speculated branches, which is quite expensive. Overall, the solution does not require the use of CAM and thus significantly reduces both the delay and the energy consumption of the wakeup logic.

[6] proposed a circuit design to adaptively resize an instruction queue partitioned into fixed size blocks (32 entries and 4 blocks were studied). The resizing was based on IPC monitoring. The use of self-timed circuits allowed delay reduction for smaller queue size. [7] further improved this design using voltage scaling. The supply voltage was scaled down when only a single queue block was enabled.

[8] used a segmented bit line in the IQ RAM/CAM design. Only selected segments are used in access and comparison. In addition, a form of bit compression was used and a special comparator design to reduce energy consumption on partial matches.

[3] proposed a design, which divided the IQ into blocks (16 blocks of 8 entries). Blocks which did not contribute to the IPC were dynamically disabled using a monitoring mechanism based on the IPC contribution of the last active bank in the queue. In addition, their design dynamically disabled the wake up function for empty entries and ready operands[1].

[9] split the IQ into 0-, 1-, and 2-tag queues based on operand availability at the time an instruction enters the queue. This was combined with a predictor for the 2-operand queue that predicted which of the two operands would arrive last. The wakeup logic only examined the operand predicted to arrive last. This approach reduces the IPC while saving energy. First, an appropriate queue with 0-, 1-, or 2-tags

must be available at issue, otherwise a stall occurs. Second, the last-operand prediction may be incorrect, requiring a flush.

[10] also used a single dependent pointer in their design. However, a tag comparison is still always performed requiring a full CAM. In the case of multiple dependent instructions a complex mechanism using *Broadcast* and *Snoop* bits reduces the total number of comparisons. The *Broadcast* bit indicates a producer instruction with multiple dependents (set on the second dependent). Each such dependent is marked with a *Snoop* bit. Only instructions with a *Snoop* bit on perform a comparison when an instruction with a *Broadcast* bit on completes. Pointer(s) to squashed dependent instructions may be left dangling on branch misprediction and cause unnecessary comparisons, but tag compare guarantees correctness.

[11] used a full bit matrix to indicate all successors of each instruction in the instruction queue. Optimizations to reduce the size and latency of the dependence matrix were considered. This solution does not require the use of CAM but does not scale well with the number of physical registers and the IQ size which keep increasing. [12] also described a design of the Alpha processor using a full register scoreboard.

[13] proposed a scalable IQ design, which divides the queue into a hierarchy of small, and thus fast, segments to reduce wakeup latency. The impact on energy consumption was not evaluated and is hard to estimate.

In addition, dynamic data compression techniques ([14], [15]) have been proposed as a way to reduce energy consumption in processor units. They are orthogonal to the design proposed here.

3   A New Instruction Wakeup Mechanism

The main objective of this work is to reduce the energy consumption in the wakeup logic by eliminating more unnecessary comparisons. The new mechanism is described in this section and is presented for a single instruction completing execution. The wakeup logic for each source operand in the IQ entry is replicated for each of the multiple instructions completing in the same cycle.

The approach proposed here takes advantage of the fact that in a distributed instruction queue with n blocks each source operand only requires one of the n blocks to be involved in instruction wake-up. In addition, a completing instruction typically has few dependent instructions waiting in the IQ for its operand [16]. For example, if an instruction has two successors then at most two out of the n IQ blocks need to be used for comparison. The remaining $n-2$ blocks can be "disabled" and do not need to be involved in the wakeup process. This can be accomplished by gating their precharge signal. Clearly, the benefits of this method will grow with an increase in n, but at some point the complexity of the design will become to high. Four or eight blocks per instruction queue are a practical choice of n used in this paper.

Consider an instruction queue partitioned into n blocks such that each block Bi can be enabled for tag comparison separately from all the others. All blocks are normally disabled and require an explicit enable to perform the tag search. A block mapping table, a key element of the proposed design, selects which block(s) hold the successor instructions and only these blocks are enabled to compare with the destination register tag of a completing instruction.

The block mapping table (BT) is a bookkeeping mechanism that records which of the *n* blocks contain successor instructions for a given destination register. It is organized as an *n* by *M* bit RAM, where *n* is the number of banks and *M* is the number of physical registers. It is shown in the upper left corner of Fig. 3. The table records for each destination register which IQ blocks contain its successor instructions. A BT row thus corresponds to a producing instruction and is checked when the instruction completes. It is read using the destination register tag of the completing instruction to find all the banks that contain its successor instructions. Such a row will be called a block enable vector (BE) for the register. An individual bit in BE will be called a block entry. For example, Fig. 3 shows BT[4], the BE for the physical register 4, with only entries for blocks 0 and 2 are set to '1'. The 1's indicate that successor instructions using register 4 are in blocks 0 and 2.

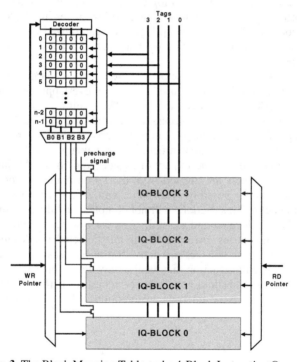

Fig. 3. The Block Mapping Table and a 4-Block Instruction Queue.

The operation of the proposed design is as follows. When an instruction is decoded it is entered into the instruction queue with the source operand designators specifying the physical register sources. An IQ block to enter the instruction is selected at this time. Several block assignment algorithms are possible; the results presented in this paper are based on the round-robin assignment algorithm. The destination and source physical register tags are then entered in the IQ as per Figure 1. At this point two operations on the BT take place concurrently.

BE Allocation. The enable vector for the instruction's destination register Ri, BT[i], is cleared. It is guaranteed that by this time all dependent instructions have obtained a previous copy of this register.

BE Entry Modification. A BE entry for each source register of the instruction that is not ready is set. This will indicate for each register what blocks contain its dependent instructions. The source operand register tag is used to select a BT row to modify. The modification consists of setting the block entry in the BE to a 1. For example, an instruction allocated to IQ block 'j' that uses a source ope   rand register 'i' sets entry 'j' in BT[i] to '1.' Note that multiple instructions in a block can all set the entry.

The BE is used when an instruction completes execution and produces a result. Its destination register is used to index the Block Table and to read out a corresponding block enable vector. Only the wakeup logic in blocks enabled by a '1' in the BE is going to perform comparisons. Furthermore, within an enabled block any of the previously proposed designs for wakeup logic in a centralized queue can be used to further reduce the number of comparisons. Only active entries are compared in each block in this paper.

The addition of the BT access in the path of the CAM precharge signal may lead to an increased delay. Pipelining can be used if a design overlapping the BT access with other parts of the CAM operation is impossible. The result tag can be broadcast to the BT one cycle earlier and the BT access completed and latched before the CAM access. This is possible in current and future processors with deep pipelines that take several cycles to select/schedule an instruction after wakeup.

A final issue to consider is the impact of branch prediction. All instructions that were executed speculatively are deleted from the IQ on branch misprediction. Their BEs remain in the table, possibly with some entries set. However, this is not a problem since the deleted instructions will not complete and their destination register will eventually be re-allocated and cleared. BEs for instructions prior to the branch may, however, contain entries corresponding to mispredicted and deleted instructions.

These entries may cause unnecessary activation of IQ banks for comparisons. In the worst case, a deleted instruction may be the only dependent instruction in a block and cause its activation. This does not affect correctness or performance of a program, but may lead to unnecessary energy consumption. The impact of this, in our experience, is negligible.

4   Results and Analysis

The proposed approach was simulated using an Alpha 21264-like microarchitecture. The SimpleScalar simulator was re-written to accomplish this. The processor configuration is shown in Table 1. The reorder buffer with 256 entries was used to avoid bottlenecks.

As in the R10K design, separate integer and floating-point instruction queues were used. In addition, Ld/St instructions were dealt with in the Ld/St queue as far as wakeup was concerned. Only one source operand was assumed necessary for memory access instructions, namely the address. All entries in the LSQ are compared to the integer result register tag. Data for stores is dealt with separately during

Table 1. Processor Configuration.

Element	Configuration
Reorder Buffer	256 entries
Load/Store Queue	64 entries
Integer Queue	32-64 entries
Floating Point queue	32-64 entries
Fetch/decode/commit width	4/4/4
Functional units	4 integer/address ALU, 2 integer mult/div, 4 fp adders, 2 mult/div and 2 memory ports.
Branch predictor	16 K-entry GShare
Branch penalty	8 cycles
L1 Data cache	32 KB, 4 way, 32 byte/line, 1 cycles
L1 Instruction cache	32 KB, 4 way, 32 byte/line, 1 cycles
L2 Unified cache	512 KB, 4 way, 64 byte/line, 10 cycles
TLB	64 entries, 4 way, 8KB page, 30 cycles
Memory	100 cycles
Integer Register file	128 Physical Registers
FP Register file	128 Physical Registers

commit when stores access memory. The integer and f.p. Instruction Queues were split into either 4 or 8 blocks.

SPEC2000 benchmarks were compiled for the DEC Alpha 21264 and used in simulation. A dynamic sequence of instructions representing its behavior was chosen for each benchmark and 200M committed instructions were simulated, with statistics gathered after a 100M-instruction warm-up period.

A queue design that disables comparison on ready operands and empty IQ entries (collectively called active entries), similar to [3] but without dynamic queue resizing is referred to as Model A. The impact of the proposed queue design, Model C, is evaluated using the number of wakeup comparisons per committed instruction and is compared with model A. Model C also uses only active entries in comparisons. The proposed design is utilized only in the integer and f.p. queues, the Ld/St queue are identical in both models. The number of comparisons performed in the Ld/St queue is thus the same in both models and is not reported.

Comparisons are divided into those that result in a match on each wakeup attempt (necessary) and all the rest (unnecessary) in the analysis. The former are labeled Match'in the graphs and the latter labeled No-Match."

Figs. 4 and 5 show comparisons per committed instruction for integer and f.p. benchmarks, respectively, and the Model A queue design. They show that most of comparisons are unnecessary and that, on average, the integer and f.p. queues have similar behavior. The number of comparisons per committed instruction is 13 and 15, on average, for integer queue size of 32 and 64, respectively. The averages are 12 and 17 for the same f.p. queue sizes.

Figs. 6 and 7 compare the two different queue designs and the effect of the queue size and of the number of blocks. The results are averaged over all integer and all f.p. benchmarks, respectively, and show the total number as well as the fraction of necessary comparisons per committed instruction. The results for Model C are always significantly better than Model A results for both 4- and 8–block queue organization. They are even better for a 1-block Model C design, which is a single queue, because some instructions do not use a destination register and the use of the BT allows wakeup to be completely avoided in this case. In general, results improve with the increase in the number of blocks.

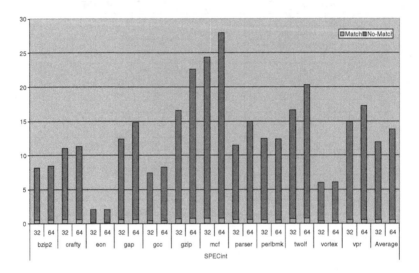

Fig. 4. Comparisons per Instruction for Model A and Integer Benchmarks.

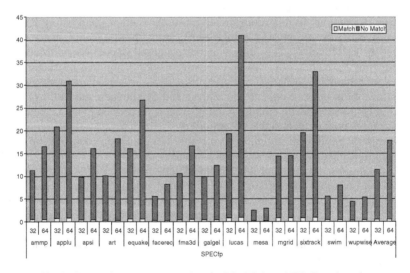

Fig. 5. Comparisons per Instruction for Model A and F.P. Benchmarks.

The 4-block IQ design achieves a 78% improvement, on average, over model A for the integer queue and 73% for the f.p. queue. The relative improvement is almost identical for both queue sizes. In absolute terms, the number of comparisons performed is higher for the larger queue size.

The 8-block design results in fewest comparisons. It reduces the number of comparisons in Model A by 87% for both the 32- and the 64-entry IQ for integer benchmarks. For f.p. benchmarks the reduction is 85%. More than one third of the total comparisons are necessary indicating the potential for further improvement.

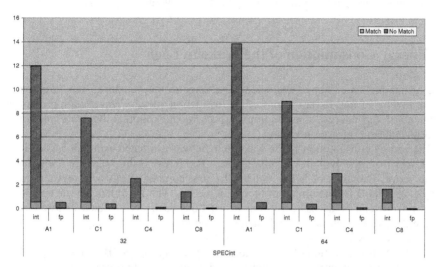

Fig. 6. Average Number of Comparisons per Instruction for Integer Codes

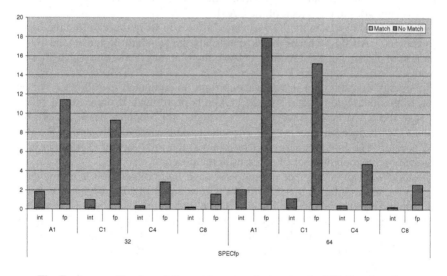

Fig. 7. Average Number of Comparisons per Instruction for F.P. Benchmarks.

The total number of comparisons per committed instruction is 1.45 and 1.72, on average, for integer queue sizes of 32 and 64, respectively, and eight blocks. The averages are 1.59 and 2.59 for f.p. benchmarks for the same queue sizes.

Overall, a larger instruction queue has more valid IQ entries, which explains the increase in the number of comparisons per instruction. For example, the average number of IQ entries per cycle is ~41% higher in the 64-entry/8-block queue for f.p. benchmarks as compared to the 32-entry queue. The difference is smaller for integer benchmarks, approximately 20%.

The associated IPC increase is small, only about 13% for f.p. codes and near 0 for integer codes (using a harmonic mean instead of an average).

5   Energy Consumption of the Wakeup Logic

The results presented above evaluated the change in the number of tag comparisons performed by various organizations of the instruction queue. This section presents an estimate of the energy consumption and shows the optimal IQ configuration to minimize the energy consumption.

The energy consumption of the wakeup logic was estimated using models of RAM and CAM from the Wattch [17] simulator. The RAM is the Mapping Table that is organized as R entries of B bits, where R is the number of physical registers in the CPU (128) and B is the number of blocks in the IQ. The RAM has 8 write and 4 read ports since it can be accessed by four instructions each cycle. Each instruction can update 2 entries when issued and each completing instruction reads an entry to find successor IQ blocks.

The CAM is the portion of the IQ storing tags and has (IQS / B) entries, where IQS is the number of IQ entries. An entry contains two separate source register tags of 7 bits each and has 4 write and 4 read ports. The total CAM size is 32 or 64 entries, but an access only goes to one block of the CAM.

The number of blocks in the IQ was varied from 1 to 64 in order to find the optimal organization. An IQ with one block represents the standard design. The IQ with block size of 1 does not even require a CAM; the Mapping Table entry identifies each dependent instruction in the IQ.

The energy consumption in the wakeup logic is computed as follows. For each access to the instruction queue a CAM lookup is performed. If more than 1 block is used than the RAM access energy is also included for Mapping Table access, but a smaller CAM is accessed in this case. The total lookup energy is the sum of RAM and CAM access. The lookup energy per program is the total lookup energy per access multiplied by the total number of accesses in the program.

The total wakeup energy per program for each IQ configuration is normalized to the energy consumption of the IQ with 64 entries and 1 block. Figure 8 shows the total normalized energy for the Integer Queue and Figure 9 for the Floating point Queue. The total normalized energy is averaged over all integer and all floating point benchmarks, respectively, for IQs with 32 and 64 entries. In each case the number of blocks is varied from 1 to 32 or 64. The RAM and CAM energy consumption are shown separately.

As the number of blocks is increased, the CAM energy consumption is reduced because only one block is accessed at a time and the number of entries accessed is reduced. At the same time, the Mapping Table RAM entry has more bits and requires more energy to access. The optimal configuration varies between 4 and 8 blocks depending on the total IQ size. The difference between the 4- and 8-block configurations is small.

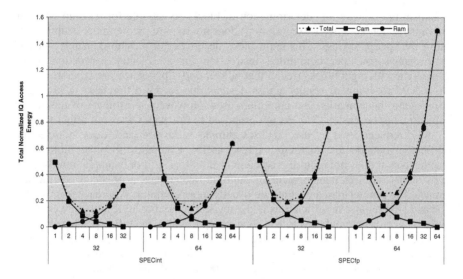

Fig. 8. Normalized IQ Energy Consumption for Integer Queue.

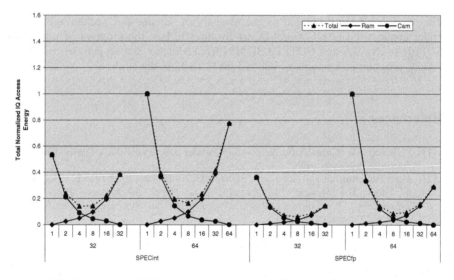

Fig. 9. Normalized IQ Energy Consumption for Floating Point Queue.

6 Conclusions

The instruction queue architecture presented in this paper uses a multi-block organization and delivers a large reduction in the number of comparisons per instruction. The major contribution of this work is the use of the block mapping mechanism that allows fast determination of blocks to activate for wakeup when an earlier instruction completes execution. It does not require prediction or adaptivity to achieve the energy savings. This multi-block design performs approximately one and a half comparisons per committed instruction for a 32-entry instruction queue organization with eight blocks (recall that integer *and* f.p. queues are separate).

It is hard to compare results across designs due to the differences in processor configuration, benchmarks, and compilers used. However, an estimate of the design complexity can be made. The design proposed in this paper does not affect the IPC and only reduces energy use. It uses simple hardware and does not require modification outside the instruction queue itself. Compared to other multi-block queue designs it does not require prediction or monitoring. For example, multi-queue design of [9] requires management of multiple queues and a predictor for the last operand. The pointer-based designs are usually more complex. Approaches presented in [5] and [10] require modifications in the register renaming logic and quite complex pointer/bit vector manipulation to add successors. Finally, the full dependence matrix is not scalable and has its own overheads [11] [12]. The design proposed here is a simpler alternative delivering a very large reduction in energy consumption. Its results can be further improved by using compression and other techniques.

Acknowledgements

This work has been supported by the Ministry of Science and Technology of Spain, under contract TIC-2001-0995-C02-01 by the CEPBA, the program of scholarships SUPERA-ANUIES and COTEPABE-IPN from Mexico, and, in part, by the DARPA PAC/C program under contract F33615-00-C-16 32 in the USA.

References

[1] S. Palacharla, "*Complexity effective Superscalar processors*" PhD Thesis, University of Wisconsin, Madison 1998.

[2] Alper Buyoktusunoglu, Stanley E. Shuster. David Brooks, Pradid Bose, Peter W. Cook, and David H. Albonesi, "*An Adaptive Issue Queue for Reduced Power at High Performance*", Workshop on Power Aware Computer Systems, in conjunction with ASPLOS-IX, November 2000.

[3] Daniel Folegnani and Antonio Gonzłez, " *Energy Effective Issue Logic*", Proceedings of 28th Annual of International Symposium on Computer Architecture, 2001. Page(s): 230-239, Gteborg Sweden.

[4] Shlomo Weiss, James E. Smith, "*Instruction Issue Logic for Pipelined Supercomputers*", Proceedings of 11th Annual International Symposium on Computer Architecture, 1984 Page(s): 110-118.

[5] Toshinori Sato, Yusuke Nakamura, Itsujiro Arita, "*Revisiting Direct Tag Search Algorithm on Superscalar Processors*," Workshop Complexity-Effective Design, ISCA 2001.

[6] Alper Buyoktusunoglu, Stanley E. Shuster. David Brooks, Pradid Bose, Peter W. Cook, and David H. Albonesi, "*An Adaptive Issue Queue for Reduced Power at High Performance*", Workshop on Power Aware Computer Systems, in conjunction with ASPLOS-IX, November 2000.

[7] Vasily G. Moshnyaga, "*Reducing Energy Dissipation of Complexity Adaptive Issue Queue by Dual Voltage Supply*", Workshop on Complexity Effective Design, June 2001.

[8] Gurhan Kukut, Kanad Ghose, Dmitry V. Ponomarev, Peter M. Kogge, "*Energy-Efficient Instruction Dispatch Buffer Design for Superscalar Processors*", Proceedings of International Symposium on Low Power Electronics and Design August 2001 Huntington Beach California, USA.

[9] Dan Ernst, Todd Austin, "*Efficient Dynamic Scheduling Through Tag Elimination,*" Proceedings of 29th Annual of International Symposium on Computer Architecture, 2002.

[10] Michael Huang, Jose Renau and Josep Torrellas, "*Energy-Efficient Hybrid Wakeup Logic*", Proceedings of ISLPED August 2002 Page(s): 196-201, Monterrey California, USA.

[11] Masahiro Goshima, Kengo Nishino, Yasuhiko Nakashima, Shin-ichiro Mori, Toshiaki Kitamura, Shinji Tomita, "*A high-Speed Dynamic Instructions Scheduling Scheme for Superscalar Processors*" Proceedings of 34th Annual International Symposium on Microarchitecture, 2001.

[12] James A. Farrell and Timothy C. Fisher, "*Issue Logic for a 600-Mhz Out-of-Order Execution Microprocessors*" IEEE Journal of Solid State Circuits Vol. 33, No. 5 , May 1998. Page(s): 707-712.

[13] Steven E. Raasch, Nathan L. Binkert and Steven K. Reinhardt, "*A Scalable Instruction Queue Design Using Dependence Chains*", Proceedings of 29th Annual of International Symposium on Computer Architecture, 2002 Page(s): 318-329.

[14] L. Villa, M. Zhang M. and K. Asanovic, "*Dynamic Zero Compression for Cache Energy Reduction,*" Micro-33, Dec. 2000.

[15] Vasily G. Moshnyaga, "*Energy Reduction in Queues and Stacks by adaptive Bit-width Compression*",. Proceedings of International Symposium on Low Power Electronics and Design, August 2001 Page(s): 22-27 Huntington Beach California, USA.

[16] Manoj Franklin, Gurindar S. Sohi "*The Expandable Split Window Paradigm for Exploiting Fine Grain Parallelism*", Proceedings of 19th Annual of International Symposium on Computer Architecture, 1992 Page(s): 58-67.

[17] D. Brooks, V. Tiwari, and M. Martonosi. "*Wattch: A framework for architectural-level power analysis and optimizations.* Proceedings of 27th Annual International Symposium on Computer Architecture, June 2000.

Power–Performance Trade-Offs
in Wide and Clustered VLIW Cores
for Numerical Codes

Miquel Pericàs, Eduard Ayguadé, Javier Zalamea,
Josep Llosa, and Mateo Valero

Departament d'Arquitectura de Computadors,
Universitat Politècnica de Catalunya (UPC),
Jordi Girona, 1-3. Mòdul D6 Campus Nord, 08034 Barcelona, Spain,
{mpericas,eduard,jzalamea,josepll,mateo}@ac.upc.es

Abstract. Instruction-Level Parallelism (*ILP*) is the main source of performance achievable in numerical applications. Architectural resources and program recurrences are the main limitations to the amount of ILP exploitable from loops, the most time-consuming part in numerical computations. In order to increase the issue rate, current designs use growing degrees of *resource replication* for memory ports and functional units. But the high costs in terms of power, area and clock cycle of this technique are making it less attractive.

Clustering is a popular technique used to decentralize the design of wide issue cores and enable them to meet the technology constraints in terms of cycle time, area and power. Another approach is using *wide* functional units. These techniques reduce the port requirements in the register file and the memory subsystem, but they have scheduling constraints which may reduce considerably the exploitable ILP.

This paper evaluates several VLIW designs that make use of both techniques, analyzing power, area and performance, using loops belonging to the Perfect Club benchmark. From this study we conclude that applying either clustering, widening or both on the same core yields very power–efficient configurations with little area requirements.

1   Introduction

Smaller feature sizes are fueling resource *replication* as a technique to increase exploitable ILP. But this approach has many costs. Resource imbalance and recurrences limit the exploitable ILP. Other issues with replication are its huge power consumption and the long wire delays.

The access to centralized structures in a VLIW core makes the design very sensitive to wire delays. Several approaches exist that allow designers to increase the number of resources without increasing the cycle time and with power consumptions below N^3 [10][1]. *Clustering*[16] and *Widening*[8] are two such techniques. The strategy is similar in both cases. We reduce the number of functional units connected to the register files and try to localize communications.

---

[1] N is the number of resources

A. Veidenbaum et al. (Eds.): ISHPC 2003, LNCS 2858, pp. 113–126, 2003.

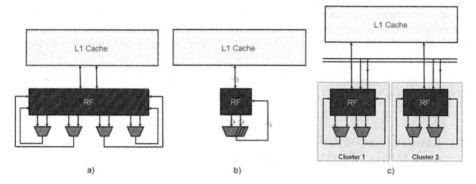

Fig. 1. (a) a centralized configuration, (b) a configuration with a wide–2 memory bus and a wide–4 FPU, and (c) a 2–clustered configuration. All configuration share the same amount of of arithmetic units and memory ports.

1.1 Widening

The width of the resources can be increased exploiting data parallelism at the functional unit level by using the *widening* technique [8]. Using this technique a single operation can be performed over multiple data. This technique exploits SIMD parallelism, similar to multimedia–like subword parallelism, but operating on full word–quantities.

Figure 1b) shows the effects of applying *widening*. As we see, this technique reduces the number of ports that access the register file, but the width of these ports increases with the widening degree. The same effect happens with the memory ports. Functional units in a wide architecture may execute both simple and wide operations. When simple operations are executed, the functional unit is used only partially and its computational bandwidth is not fully exploited.

The *widening* technique has been implemented only in limited quantities in commercial processors. For instance, the IBM POWER2 [13] applies widening to the memory ports. Vector processors like NEC's SX-3 [12], apply widening to the floating point units (FPUs).

1.2 Clustering

A different approach for dealing with wire delays is to partition the register file and functional units so that most of the communication is local and very few or no global components are used. This type of architecture is usually named *clustered* architecture. Fig. 1c) shows an example of such an architecture. In this case each cluster consists of several functional units connected to a local register file. Data sharing between clusters may result in additional execution cycles due to transfer operations. However, the reduced cycle time, area and power consumption compensate this overhead[16].

Clustered designs have been used in various commercial processors, such as the TI's TMS320C6x series [11], the HP Lx [3] and the Alpha EV6[6].

In this paper we evaluate VLIW processors combining both *clustering* and *widening* in the same core. Sect. 2 briefly tackles scheduling issues for our architectures. Hardware issues are discussed in Sect. 3, where an analysis of clock cycle and chip area is performed. Another important topic, power consumption, is dealt with in Sect. 4. Finally, performance numbers and conclusions are given in Sects. 5 and 6.

2   Modulo Scheduler for Clustered and Wide VLIW Architectures

The first step in our evaluation process has been to develop a tool to generate valid schedules for our VLIW cores. Our focus is on the loops that form the core of numerical applications. The scheduler we have developed is based on the MIRS (*Modulo Scheduling with Integrated Register Spilling*) algorithm [15] and on previous works on widening [8] and clustering [16].

Aggressive instruction scheduling for our loops is based on software pipelining. Modulo scheduling [9] is an algorithm for software pipelining in which each new iteration starts after a fixed number of cycles called the *initiation interval* (II).

2.1   The MIRS_C Scheduler

MIRS [15] is a modulo scheduler that aims at achieving the Minimum Initiation Interval (MII) with minimum register pressure. The pressure is reduced by making use of Hypernode Reduction Modulo Scheduling (HRMS) [7].

MIRS_C [16] is an extension of MIRS for clustered architectures. Scheduling on clustered architectures requires the algorithm to select specific clusters where operations will be executed and to insert data movement operations when the operants of an operation are not located in the same cluster.

2.2   Compaction of Wide Operations

To schedule one wide operation, a number of simple operations need to be compacted into a single wide operation. This process requires to unroll the loop and perform the following dependency analysis on the Data Dependence Graph (DDG):

- The nodes must represent the same operation in different iterations.
- They may not be dependent on each other.
- In case of load/store operations, addresses must be accessed with stride 1.

An example of compaction with widening degree 2 is shown in Fig. 2.

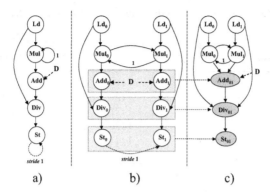

Fig. 2. (a) The original loop (DDG), (b) the DDG after applying unrolling, with compactible operations highlighted, and c) the DDG after applying wide–2 transformations

2.3   Integration of MIRS_C and Compaction Algorithms

Scheduling for a clustered architecture with wide units is performed via a two–step process in which the compaction algorithms are applied first and then the MIRS_C modulo scheduler. The final schedule contains both simple and wide instructions. At the ISA level, these are treated as independent instructions.

2.4   Preliminary Evaluation of Generated Schedules

At this point we perform a first evaluation of clustering and widening from the point of view of the scheduler. We have taken our workbench[2] and computed ΣII (the sum of the II for all the loops in the workbench) for a set of processor configurations using either clustering, widening or a combination of both. For all evaluations we have used architectures with equivalent amounts of hardware and storage capacity. In the results we distinguish between those loops which are fully compactible (ie. all operations can be compacted after unrolling), which account for 39.3% of the total number of loops, and those which are not (60.7%). Table 1 shows the results obtained in this test.

Several conclusions may be extracted from this table. The first three rows show that move operations inserted to communicate clusters almost do not affect the value of ΣII. The widening technique is more problematic. The value of ΣII depends a lot on the compactability of the loops. When loops have no compactible operations, wide functional units are used as simple units, which reduces the computation bandwidth of the architecture (compare rows 1, 4 and 6 of Table 1).

These values are significant from the point of view of code size. However, no definitive conclusion can be given without information on the processor clock

[2] The workbench consists of loops taken from the Perfect Club Benchmark. See section 5 for details.

Table 1. Impact of clustering and widening on ΣII

Clusters	Widening Degree	Registers per Cluster	Fully Compactible Loops (ΣII)	Not Fully Compactible Loops (ΣII)
1	1	64	1413	6409
2	1	32	1420	6441
4	1	16	1471	6514
1	2	32	1420	6815
2	2	16	1441	6912
1	4	16	1436	9456

cycle. Area and power consumption are also significant. These technological aspects may compensate the behavior in terms of ΣII. The rest of this paper is dedicated to the study of these three aspects. Results for the execution time and energy consumption are given in Sect. 5.

3   Clock Cycle and Area

3.1   Clock Cycle

While the previous section dealt with obtaining cycle counts, this section will concentrate in obtaining the clock cycle. Variations in the clock cycle will show up in the operation latencies and ultimately determine the execution time. It is thus important to develop a model to determine the frequency of a particular processor configuration.

All of the structures we are modeling (RF, DCACHE and FPUs) can be pipelined. However, to avoid complexity we decide not to pipeline the register file and thus determine the clock cycle as the access time of the register file.

Table 2 shows the frequency values for a set of 64–bit register files that correspond to a representative subset of the evaluated configurations in Sect. 5. These values have been obtained using a model for register files based on CACTI [14] with a technology parameter of 100nm.

Table 2. Processor frequencies for several configurations of the register file

Clusters	Widening Degree	Registers per Cluster	Register Width	Read Ports	Write Ports	Frequency
1	1	64	64	20	12	980 MHz
2	1	32	64	11	7	1880 MHz
4	1	16	64	6	4	2507 MHz
1	2	32	128	10	6	1961 MHz
2	2	16	128	6	4	2507 MHz
1	4	16	256	5	3	2618 MHz

The usage of the widening technique requires certain modifications on the structure of the register file so that we can access a wide data word in a single step. We have chosen to implement the register file as several replicated 64–bit register files that are accessed in parallel. This way no time penalty is paid.

3.2 Area

Miniaturization allows to pack more logic onto a die, but there is a limit. In order to design viable processor configurations this needs to be taken into account. We have estimated the area of our designs to analyze this constraint.

The area of the FPU has been estimated using the MIPS R10000 processor as reference. We estimate that at 100nm this FPU occupies an area of $1.92mm^2$. The total area that has to be accounted for the 8 FP units of our configurations is $0.1536cm^2$.

The data cache area has been estimated using the CACTI3 tool[14]. The simulated cache has a capacity of 32768 bytes and is direct-mapped with a block size of 32 bytes. We assume a single-banked cache with either 1, 2 or 4 read/write ports depending on the widening degree. The obtained areas of these three configurations are $0.123cm^2$ (4 ports, no widening), $0.0265cm^2$ (2 ports) and $0.0116cm^2$ (single ported).

The areas of the register files have been estimated using the same methodology as in [8].

Table 3 shows the areas of each component for the subset of configurations that have 64 registers along with the total area. Note how increasing widening and clustering degrees reduces the area. The widening technique requires a slightly smaller area because it also reduces the memory cache size and because wide–only configurations do not require inter–cluster communication ports.

The total areas of the configurations are small when compared to current microprocessors. The additional control logic needed in a VLIW processor is simple and occupies little on–chip area. The data suggests that no problems will appear due to die size constraints.

Table 3. Area costs for several configurations

Clusters	Wide	Registers	RF Area (cm^2)	FPU Area (cm^2)	DCACHE Area (cm^2)	Total Area (cm^2)
1	1	64	0.05537	0.1536	0.1229	0.3319
2	1	32	0.02084	0.1536	0.1229	0.2973
4	1	16	0.00643	0.1536	0.1229	0.2829
1	2	32	0.01627	0.1536	0.02648	0.1964
2	2	16	0.00643	0.1536	0.02648	0.1865
1	4	16	0.00398	0.1536	0.01155	0.1691

4   Energy and Power

Power consumption is a determinant factor for the success of any modern computing system. In our research we have tried to reduce power consumption introducing changes in the microarchitecture. In this section we explain how power consumption has been modeled.

Our power evaluation is based mainly on the Wattch power modeling infrastructure [2]. The following Wattch models have been used:

- Register files have been modeled as *Array Structures*.
- The power of the arithmetic units (FPUs) is obtained using the *Complex Logic Blocks* model.

The power consumption of the data cache has been obtained using the *CACTI3* tool. We choose this model because it is more recent and accurate than the Wattch–provided model.

For the purposes of this paper we have chosen to ignore the power consumption of smaller structures such as buses or control units. Clock distribution is very hard to model at high level and has also been left out.

4.1   Conditional Clocking

Most modern processors are designed with power consumption in mind. *Conditional clocking* is a technique that disables all or part of a hardware unit to reduce power consumption when it is not being used. In this work we have used the following conditional clocking style:

- When unit ports are only partially accessed, the power is scaled linearly with the port usage.
- Ports that are not accessed consume 10% of their maximum power.
- Unused FPUs consume 0% of their maximum power. This figure is not realistic, but as we will see, FPU power has little impact on the results.

5   Performance Evaluation

5.1   Methodology

In this section we detail the evaluation strategy that has been applied on 14 different processor configurations. All configurations have a total hardware of 8 floating point units and 4 memory buses, all 64 bits wide. The clustering and widening degrees range from 1 to 4, and the number of registers has been chosen so that configurations have mostly a storage capacity of 64–128 registers. The memory is assumed to be ideal. This means that memory accesses always hit in the first level of the cache and with the same latency.

For the evaluation we have constructed a workbench consisting of loops taken from the Perfect Club Benchmark [1]. We have only taken loops that are suitable for modulo scheduling and that have executed at least once during a sample execution. A total of 811 loops, accounting for 80% of the execution time of the benchmark, make up our workbench. Before scheduling, an unrolling degree of 4 has been applied on all loops. This allows us to schedule for all ranges of widening using the same unrolling degree.

A special terminology is used to name each of the fourteen configurations. From now one we will refer to each as $Cx.Wy.Rz$, where x is the number of clusters, y is the widening degree and z is the number of registers per cluster. The total register storage of these configurations can be obtained as the product of these three factors times 64 bits/register.

All clustered configurations have one extra read port and one extra write port in the register file to perform intercluster communication using *move* operations.

The latency of arithmetic operations depends on the FPU structure and the technology factor λ. It is thus the same for all configurations. The latency of the L1 cache memory has been obtained using the *CACTI3* tool [14]. For this structure, the access times depend on the widening degree that is being applied, because different widening degrees imply different port numbers. These latencies can be seen in Table 4 for a technology of $\lambda = 100nm$ [5]. Once the latency is known we can compute the number of clock cycles each operation spends in the execution state.

Table 4. Execution latencies at 100nm

Operation	Widening Degree	Execution Latency
Division and Modulo	–	7.56 ns
Square Root	–	11.304 ns
Other Arithmetic	–	2.52 ns
L1 Cache Load	1	1.423 ns
L1 Cache Load	2	0.862 ns
L1 Cache Load	4	0.726 ns

The methodology that has been applied to obtain the results is the following: First, all loops are scheduled for all processor configurations. Our scheduler determines the initiation interval and provides energy values for a single iteration of the loop. The total number of iterations is obtained from traces. Using this information we estimate the total execution cycles and the energy spent in each program by each configuration.

The following sections detail the results that have been obtained.

5.2 Execution Cycles

First of all we are interested in the number of execution cycles. We obtain an estimation by multiplying the initiation intervals (II) of all loops from all applications times the number of iterations and adding them. For this analysis we have used the maximum possible frequencies for each configuration. The resulting

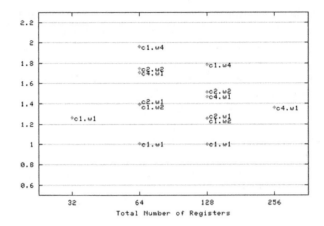

Fig. 3. Relative execution cycles

numbers are shown, relative to the base configuration (*C1.W1.R128*), in Fig. 3. To simplify the graphs, whenever possible, we have grouped configurations by their total number of registers. This allows to see how modifying the number of registers changes the results. As we saw in Sect. 2, clustering has little impact on the execution cycles. Thus, the principal observation here is that the execution cycles mainly increase due to the deeper pipelines. The fastest configurations are almost doubling the number of execution cycles.

Comparing configurations with similar frequencies allows to quantify the penalty due to the scheduling constraints of the clustering and widening techniques. The figure shows that the cycle count increase of widening is bigger than the clustering cost. The reason for this are the loops which are not fully compactible. This figure is just representing as cycle counts the results that we obtained in Table 1 so the same conclusions apply. At widening degree 4 this effect is most noticeable. Configuration *C1.W4.R16* executes up to 96% more cycles than the base configuration, while *C4.W1.R16* only executes 71% more cycles. Note that both configurations have similar clock frequencies (∼2.5GHz). Thus, the additional execution cycles indicate that on average widening reduces the amount of exploitable ILP more than clustering does.

5.3   Execution Time

Combining the number of execution cycles with the clock speed of the processor we obtain the execution time, our primary performance measurement. The results of this simple calculation are shown in Fig. 4, again relative to the base configuration. As in the previous figure, we group the configurations by the number of total register storage. The figure shows big gains in speed-up when going from centralized configurations (*c1.w1*) to clustered or wide architectures. Looking at the frequencies in Table 2 we see that the most aggressive configurations

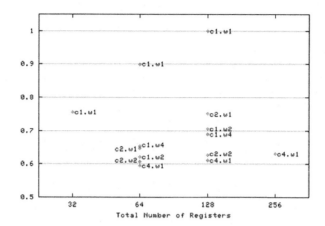

Fig. 4. Relative execution time

increase the clock speed in almost 200%. This compensates the increase in CPI discussed in the previous section. The best result is obtained with configuration *C4.W1.R16*. This configuration has a clock frequency that is 187% faster than the base configuration (873 MHz) and has only an execution cycle degradation of 71%. These results combined yield a speed-up of 1.68. Another configuration with very good results is *C2.W2.R16*. This configuration obtains a speed-up of 1.65 and it has a smaller area compared to configuration *C4.W1.R16*. It is interesting to observe how the best speed-ups are obtained with configurations that have only 16 registers per cluster. This shows that having high operating frequencies is more important than ILP when execution time is the main objective. However, parallelism should not be forgot. The *C1.W4.R16* configuration has the fastest clock of all, but due to low ILP extraction in non–compactible loops, the execution time is not among the best.

5.4   Energy Consumption

The energy consumption is an important measure that allows to determine whether a battery will provide enough energy to complete a certain task. This is very important in the embedded domain. The average energy per program is shown in Fig. 5 for each configuration. The contribution of each hardware structure is shown.

The figure shows big reductions in the energy consumption of the register file when using the clustering and widening techniques. In the most aggressive configurations, this reduction is over 90%. In the case of wide architectures, the smaller cache also yields big reductions in energy consumption (up to 57%). None of the techniques reduces the energy consumed by the floating point units. As the clustering and widening degrees increase, this energy becomes the predominant part of the total energy consumption Adding all three contributions we obtain a maximum 36% energy reduction for configuration *C1.W4.R16*

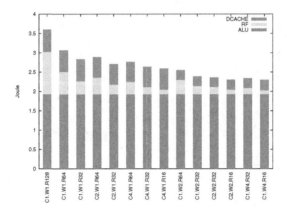

Fig. 5. Average energy per program

5.5   Power Consumption

If the design goal is to minimize the heat production of the system, looking at the energy consumption is not critical. Instead, we want to care about the power dissipation of the configuration. The bigger this quantity, the more heat is produced and the more electricity is consumed during a fixed operation time. The average power consumption per benchmark is shown in figure 6 for each configuration.

The figure basically confirms the common knowledge that fast configurations have higher power consumptions. This effect is more noticeable on configurations that rely on clustering, as no reduction in power consumption exists in the cache memories. Configuration $C4.W1.R16$ has the biggest power consumption, although the total energy is far below the total energy of the base configuration. Configurations with widening are less power-hungry as they benefit from cache–

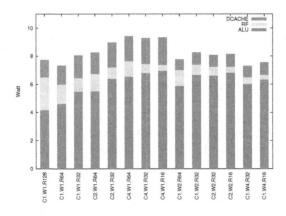

Fig. 6. Average power

energy reductions. Interestingly, configurations *C1.W4* have the smallest power consumptions, even though their execution times are far better than the base configuration.

Equalized Performance. Despite much research focusing on speed, there is not always a need to maximize performance. This happens, for example, in domains where it is only important whether an application is executed in a given time or not. Once the real–time specifications have been met there is no real advantage in making the processor faster. Instead, we will want to focus on power-efficiency as such devices are often portable and use batteries as power source. Following the guidelines of the last section we took the 14 configurations and underclocked them, adjusting latencies and frequency, until all execution times were identical to the *C1.W1.R128* configuration. Note that because now all configurations have the same execution times, the resulting figure is analogous to Fig. 5. We do not show it for this reason. The power values we obtained range from 4.91W (*C1.W4.R16*) to 7.79W (*C4.W1.R16*).

5.6   Energy–Delay

In the previous sections we have benchmarked the configurations from the point of view of performance and energy consumption. We see that clustering yields better performance but widening is less power–hungry. It would be interesting to have a global metric that takes into account both energy and performance. To quantify each of the configurations globally we have computed the energy–delay product. *EDP* is a metric that concentrates on high performance and low power simultaneously. The results are shown in Fig. 7 normalized to *C1.W1.R128*. From this figure we see that the configuration with the best power-performance relationship is *C2.W2.R16* (0.39). If we restrict ourselves to the configurations with a total storage of 128 registers the best configuration is *C2.W2.R32* (0.411). So, we confirm that configurations combining *clustering* and *widening* are very interesting structures with very good power–performance trade-offs.

6   Conclusions

In order to increase the performance of computers, aggressive processor configurations capable of executing more operations per second are required. Traditional techniques like resource replication or frequency scaling are very power hungry and increasingly inefficient. In this paper we have evaluated two techniques whose goal it is to obtain high performance while reducing power consumption. These two techniques are *clustering* and *widening*.

A total of 14 configurations, with clustering and widening degrees ranging from 1 to 4, have been evaluated. The results confirmed our expectations that clustering yields better execution time but widening has lower power consumption.

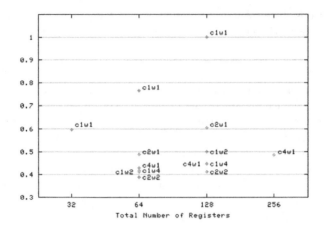

Fig. 7. Energy–Delay

To establish a global measurement, we computed the energy–delay for each of the configurations. This test yielded two conclusions. First, we saw that high degrees of clustering and widening, or combinations of both yield very efficient cores. Second, we found that the configurations with the smallest register files also tended to have better energy–delay products. In our tests, the configuration that obtained the best energy–delay value was *C2.W2.R16*.

The proposed techniques are well suited for the exploitation of Instruction–Level Parallelism (ILP) and moderate levels of Data–Level Parallelism (DLP). Many applications have little ILP and DLP to offer, taking more advantage from Thread–Level Parallelism (TLP). Such applications will scale much better using chip multiprocessors (CMP). This study has focused on numerical codes as found in scientific computing and engineering. These applications offer large quantities of ILP and DLP. They often contain also large TLP quantities to exploit, but for the current levels of ILP exploitation we believe that our approach is not limiting.

Acknowledgment

This work has been supported by Hewlett-Packard and by the Ministry of Science and Technology of Spain and the European Union (FEDER funds) under contract TIC 2001-0995-C02-0.

References

1. M. Berry, D. Chen, P. Koss and D. Kuck, *The Perfect Club Benchmarks: Effective Performance Evaluation of Supercomputers,* Technical Report 827, CSRD, Univ. of Illinois at Urbana-Champaign, Nov. 1988

2. D. Brooks, V. Tiwari and M .Martsoni, *Wattch: A Framework for Architectural-Level Power Analysis and Optimizations,* Int'l Symp. on Computer Architecture, 2000 (ISCA-00).
3. P. Faraboschi, G. Brown, G. Desoli and F. Homewood. *Lx: A technology platform for customizable VLIW embedded processing,* In Proc. 27th Annual Intl. Symp. on Computer Architecture, pages 203-213, June 2000.
4. L. Gwennap, *AltiVec Vectorizes PowerPC,* Microprocessor Report, vol. 12, no. 6, May 1998
5. M.S. Hrishikesh, N.P. Jouppi, K.I. Farkas, D. Burger, S.W. Keckler and P. Shivakumar, *The Optimal Logic Depth Per Pipeline Stage is 6 to 8 FO4 Inverter Delays,* In Proc. of the 29th Symp. on Comp. Arch (ISCA-02), May 2002.
6. R.E.Kessler, *The Alpha 21264 Microprocessor,* IEEE Micro, Volume 19 Issue: 2, Mar/Apr 1999
7. J. Llosa, M. Valero, E. Ayguadé and A. González. *Hypernode reduction modulo scheduling,* In Proc. of the 28th Annual Int. Symp. on Microarchitecture (MICRO-28),pages 350-360, November 1995.
8. D. Lòpez, J. Llosa, M. Valero and E. Ayguadé. *Cost–Conscious Strategies to Increase Performance of Numerical Programs on Aggressive VLIW Architectures,* IEEE Trans. on Comp., vol 50, no. 10, pp 1033–1051, October 2001.
9. B.R. Rau and C.D. Glaeser. *Some Scheduling Techniques and an Easily Schedulable Horizontal Architecture for High Performance Scientific Computing,* Proc. 14th Ann. Microprogramming Workshop, pp. 183-197, Oct. 1981
10. S. Rixner, W.J. Dally, B. Khailany, P. Mattson, U.J. Kapasi, J.D. Owens. *Register organization for media processing,* High-Performance Computer Architecture, 2000. HPCA-6. Proceedings. Sixth International Symposium on , 2000.
11. T.I.Inc. *TMS320C62x/67x CPU and Instruction Set Reference Guide,* 1998.
12. T. Watanabe, *The NEC SX-3 Supercomputer System.* Proc. ComCon91, pp. 303-308, 1991
13. S.W. White and S. Dhawan. *POWER2: Next Generation of the RISC System/6000 Family.* IBM J. Research and Development, vol. 38, no. 5, pp, 493-502, Sept. 1994
14. S.J.E. Wilton and N.P. Jouppi. *CACTI: An enhanced Cache Access and Cycle Time Model,* IEEE. J. Solid-State Circuits, vol 31, no. 5, pp. 677-688, May 1996.
15. J. Zalamea, J. Llosa, E. Ayguadé and M. Valero. *MIRS: Modulo Scheduling with integrated register spilling.* In Proc. of 14th Annual Workshop on Languages and Compilers for Parallel Computing (LCPC2001), August 2001.
16. J. Zalamea, J. Llosa, E. Ayguadé and M. Valero. *Modulo Scheduling with integrated register spilling for Clustered VLIW Architectures,* In Proc. 34th annual Int. Symp. on Microarch., December 2001.

Field Array Compression in Data Caches for Dynamically Allocated Recursive Data Structures

Masamichi Takagi[1] and Kei Hiraki[1]

Dept. of CS, Grad. School of Info. Science and Tech., Univ. of Tokyo,
{takagi-m, hiraki}@is.s.u-tokyo.ac.jp

Abstract. We introduce a software/hardware scheme called the Field Array Compression Technique (FACT) which reduces cache misses caused by recursive data structures. Using a data layout transformation, data with temporal affinity are gathered in contiguous memory, where recursive pointer and integer fields are compressed. As a result, one cache-block can capture a greater amount of data with temporal affinity, especially pointers, thereby improving the prefetching effect. In addition, the compression enlarges the effective cache capacity. On a suite of pointer-intensive programs, FACT achieves a 41.6% average reduction in memory stall time and a 37.4% average increase in speed.

1   Introduction

Non-numeric programs often use recursive data structures (RDS). For example, they are used to represent variable-length object-lists and trees for data repositories. Such programs using RDS make graphs and traverse them, however the traversal code often induces cache misses because: (1) there are too many nodes in the graphs to fit entirely in the caches, and (2) the data layout of the nodes in the caches is not efficient. One technique for reducing these misses is data prefetching [7, 8]. Another technique is data layout transformations, where data with temporal affinity are gathered in contiguous memory to improve the prefetch effect of a cache-block [3, 15]. Still another technique is enlargement of the cache capacity, however this has limitations caused by increased access time. Yet another technique is data compression in caches, which compresses data stored in the caches to enlarge their *effective* capacity [9, 11–13]. Not only does compression enlarge the effective cache capacity, it also increases the effective cache-block size. Therefore, applying it with a data layout transformation can produce a synergistic effect that further enhances the prefetch effect of a cache-block. This combined method can be complementary to data prefetching. We can compress data into 1/8 or less of its original size in some programs, but existing compression methods in data caches limit the compression ratio to 1/2, mainly because of hardware complexity in the cache structure. Therefore, we propose a method which achieves a compression ratio over 1/2 to make better use of the data layout transformation.

A. Veidenbaum et al. (Eds.): ISHPC 2003, LNCS 2858, pp. 127–145, 2003.

In this paper, we propose a compression method which we call the Field Array Compression Technique (FACT). FACT utilizes the combined method of a data layout transformation along with recursive pointer and integer field compression. It enhances the prefetch effect of a cache-block and enlarges the effective cache capacity, leading to a reduction in cache misses caused by RDS. FACT requires only slight modification to the conventional cache structure because it utilizes a novel data layout scheme for both the uncompressed data and the compressed data in memory; it also utilizes a novel form of addressing to reference the compressed data in the caches. As a result FACT surpasses the limits of existing compression methods, which exhibit a compression ratio of $1/2$. The remainder of this paper is organized as follows: Section 2 presents related works, and Sect. 3 explains FACT in detail. Section 4 describes our evaluation methodology. We present and discuss our results in Sect. 5 and give conclusions in Sect. 6.

2   Related Works

Several studies have proposed varying techniques for data compression in caches. Yang et al. proposed a hardware method [11], which compresses a single cache-block and puts the result into the primary cache. Larin et al. proposed another method [12], which uses Huffman coding. Lee et al. proposed yet another method [9], which puts the results into the secondary cache. These three techniques do not require source code modification. Assuming the compression ratio is $1/R$, these three methods must check R address-tags on accessing the compressed data in the caches because they use the address for the uncompressed data to point to the compressed data in the caches. These methods avoid adding significant hardware to the cache structure by limiting the compression ratio to $1/2$. FACT solves this problem by using a novel addressing scheme to point to the compressed data in the caches.

Zhang et al. proposed a hardware and software method for compressing dynamically allocated data structures (DADS) [13]. Their method allocates word-sized slots for the compressed data within the data structure. It finds pointer and integer fields which are compressible with a ratio of $1/2$ and produces a pair from them to put into the slot. Each slot must occupy one-word because of word alignment requirement; in addition, each slot must gather fields from a single instance. These conditions limit the compression ratio of this method. FACT solves these problems by using a data layout transformation, which isolates and groups compressible fields from different instances.

Truong et al. proposed a data layout transformation for DADS, which they call Instance Interleaving [3]. It modifies the source code of the program to gather data with temporal affinity in contiguous memory. This transformation enhances the prefetch performance of a cache-block. Compression after the transformation can improve the prefetch performance further.In addition, the transformation can isolate and group the compressible fields. Therefore we apply this method before compression.

Rabbah et al. automated the transformation using the compiler [15]. They proposed several techniques using compiler analysis to circumvent the problems caused by the transformation. We consider automating our compression processes as future work and we can utilize their method for the data layout transformation step.

3   Field Array Compression Technique

FACT aims to reduce cache misses caused by RDS through data layout transformation of the structures and compression of the structure fields. This has several positive effects. First, FACT transforms the data layout of the structure such that fields with temporal affinity are contiguous in memory. This transformation improves the prefetch performance of a cache-block. Second, FACT compresses recursive pointer and integer fields of the structure. This compression further enhances the prefetch performance by enlarging the effective cache-block size. It also enlarges the effective cache capacity.

FACT uses a combined hardware/software method. The steps are as follows: (1) We take profile-runs to inspect the runtime values of the recursive pointer and integer fields, and we locate fields which contain values that are often compressible (we call these compressible fields). These compressible fields are the targets of the compression. (2) By modifying the source code, we transform the data layout of the target structure to isolate and gather the compressible fields from different instances of the structure in the form of an array of fields. This modification step is done manually in the current implementation. (3) We replace the load/store instructions which access the target fields with special instructions, which also compress/decompress the data (we call these instructions cld/cst.) There are three types of cld/cst instructions corresponding to the compression targets and methods; we choose an appropriate type for each replacement. (4) At runtime, the cld/cst instructions perform compression/decompression using special hardware, as well as performing the normal load/store job.

We call this method the Field Array Compression Technique (FACT) because it utilizes a field array. The compressed data are handled in the same manner as the non-compressed data and both reside in the same cache. Aassuming a commonly-used two-level cache hierarchy, the compressed data reside in the primary data cache and the secondary unified cache. In the following, we describe details of the steps of FACT.

3.1   Compression of Structure Fields

We compress recursive pointer fields and integer fields because they often have exploitable redundancy.

Recursive Pointer Field Compression. While we use RDS for dynamic allocation of instances in different memory locations on demand, elaborate and fast

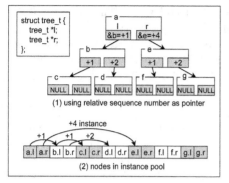

Fig. 1. Pointer compression

```
STDATA   | Data to be stored
BASE     | Base address
OFFSET   | Address offset
1/R      | Compression ratio
INCMP    | Codeword indicating incompressibility
```

```
/* operation of cst instruction */
cst(STDATA, OFFSET, BASE) {
    PA = physical address of (BASE + OFFSET)
    CA = PA/R
    CDATA = call compress_ptr(STDATA, BASE)
        or its family according to type of cst
    cache_write(CA, CDATA, C)
    if(CDATA == INCMP) {
        /* STDATA is incompressible */
        cache_write(PA, STDATA, N)
    }
}
```

Fig. 2. Compression algorithm of pointer fields

memory allocators create pools of instances for fast management of free objects and to exploit spatial locality [1]. For example, on an allocation request, the slab allocator [1] allocates a pool of instances if a free instance is not available; it then returns one instance from the pool. In addition, a graph created by RDS often consists of small sub-graphs each of which fits in a pool (e.g. binary tree). Therefore the distance between memory addresses of two structures connected by a pointer can often be small. In addition, one recursive pointer in the structure points to another instance of the same structure; moreover, the memory allocators using the instance pools can align the instances. Therefore we can replace the absolute address of a structure with a relative address in units of the structure size, which can be represented using a narrower bit-width than the absolute address. We abbreviate this method to **RPC**. The detailed steps of RPC are as follows: (i) We make a custom memory allocator for the structure, similar to [1], using instance pools as described above. (ii) We modify the source code to make it use the allocator. (iii) Using this layout, the cst instruction replaces the pointers at runtime with the relative sequence numbers in the pool. Figure 1 illustrates the compression. Assume we are constructing a balanced binary tree in depth-first order using RDS (1). When we use the memory allocator which manages the instance pool, the instances are arranged contiguously in memory (2). Therefore we can replace the pointers with relative sequence numbers in the pool (1).

Figure 2 shows the algorithm of RPC. The compression is done when writing the pointer field. Assume the head address of the structure which holds the pointer is BASE, the pointer to be stored is STDATA, and the compression ratio is $1/R$. Because the difference between the addresses of two neighboring instances is 8 bytes (pointer size) as a result of the data layout transformation described in Sect. 3.3, the relative sequence number in the pool is (STDATA-BASE)/8; we use this as the 64/R-bit codeword. A NULL pointer is represented by a special codeword. We use another special codeword to indicate incompressibility, and which is handled differently by the cld instruction if the difference is outside the range that can be expressed by a standard codeword. The address BASE can be obtained from the base-address of the cst instruction. Decompression is performed when reading pointer fields. Assume the address of the head of the structure which contains the pointer is BASE and the compressed pointer is LDDATA. The decompression calculates BASE+LDDATA×8, where BASE can be obtained from the base-address of the cld instruction.

Integer Field Compression. The integer fields often have exploitable redundancy. For example, there are fields which take only a small number of distinct values [10] or which use a narrower bit-width than is available [14]. FACT exploits these characteristics by utilizing two methods. In the first method, which we abbreviate to **DIC**, we locate 32-bit integer fields which take only a small number of distinct values over an entire run of the program, then compress them using fixed-length codewords and a static dictionary [11]. When constructing the N-entry dictionary, we gather statistics of the values accessed by all of the load/store instructions through the profile-run, and take the most frequently accessed N values as the dictionary. These values are passed to the hardware dictionary through a memory-mapped interface or a special register at the beginning of the program. In our evaluation, this overhead is not taken into account. The compression is done when writing the integer field. Assume the compression ratio is $1/R$. The cst instruction searches the dictionary, and if it finds the required data item, it uses the entry number as the $32/R$-bit codeword. If the data item is not found, the codeword indicating incompressibility is used. The decompression is done when reading the integer field. The cld instruction reads the dictionary with the compressed data. In the second method, which we abbreviate to **NIC**, we locate 32-bit integer fields which use a narrower bit-width than is available, and replace them with narrower bit-width integers. On compression, the cst instruction checks the bit-width of the data to be stored, and it omits the upper bits if it can. On decompression, the compressed data are sign-extended. While DIC includes NIC, DIC needs access to the hardware dictionary on decompression. Since a larger dictionary requires greater time, we use DIC when the dictionary contains less than or equal to 16 entries, otherwise we use NIC. That is, when we choose a compression ratio of $1/8$ or more in the entire program, we use DIC for the entire program; otherwise we use NIC.

3.2   Selection of Compression Target Field

In FACT, we locate the compressible fields of RDS in the program. We gather runtime statistics of the load/store instructions through profiling because the compressibility depends on the values in the fields that are determined dynamically. The profile-run takes different input parameters to the actual run. We collect the total access count of the memory instructions (A_{total}), the access count of fields for each static instruction (A_i) and the occurrence count of compressible data for each static instruction (O_i). Subsequently, we mark the static instructions that have A_i/A_{total} (access rate) greater than X and O_i/A_i (compressible rate) greater than Y. The data structures accessed by the marked instructions are the compression targets. We set $X = 0.1\%$ and $Y = 90\%$ empirically in the current implementation. We take multiple profile-runs that use different compression ratios. Taking three profile-runs with compression ratios of 1/4, 1/8, and 1/16, we select one compression ratio to be used based on the compressibility numbers shown. This selection is done manually and empirically in the current implementation.

3.3   Data Layout Transformation for Compression

We transform the data layout of RDS to make it suitable for compression. The transformation is the same as Instance Interleaving (I2) proposed by Truong et al. [3].

Isolation and Gathering of Compressible Fields through I2. FACT compresses recursive pointer and integer fields in RDS. Since the compression shifts the position of the data, accessing the compressed data in the caches requires a memory instruction to translate the address for the uncompressed data into the address which points to the compressed data in the caches. When we use the different address space in the caches for the compressed data, the translation can be done by shrinking the address for the uncompressed data by the compression ratio. Assuming the address of the instance is I and the compression ratio is $1/R$, the translated address is I/R. However, the processor must know the value of R which varies with structure types. To solve this problem, we transform the data layout of RDS to isolate and group the compressible fields away from the incompressible fields. Assume as an example compressing a structure which has a compressible pointer **n** and an incompressible integer **v**. Figure 3 illustrates the isolation. Since field **n** is compressible, we group all the **n** fields from the different instances of the structure and make them contiguous in memory. We fill one segment with **n** and the next segment with **v**, segregating them as arrays (B). Assume all the **n** fields are compressible at the compression ratio of 1/8, which is often the case. Then we can compress the entire array of pointers (C).

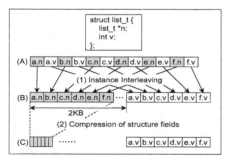

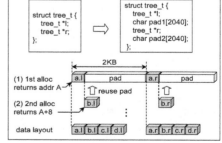

Fig. 3. Instance Interleaving (I2) and FACT

Fig. 4. I2 using a padded structure

In addition, the address translation becomes $I/8$, which can be done by a simple logical-shift operation.

Implementation of I2. We transform the data layout by modifying the source code. First, we modify the declaration of the structure to insert pads between fields. Figure 4 shows the declaration of example structure for binary tree before and after the modification. Assume the load/store instructions use addressing with a 64-bit base address register and a signed 16-bit immediate offset. Compilers set the base-address to the head of the structure when accessing a field with an offset of less than 32KB, and we use this property in the pointer compression. Because the insertion of pads makes the address of the n-th field (structure head address)+(pad size)×$(n − 1)$, we limit the pad size to 2KB so that the compiler can set the base address to the structure head when referencing the first 16 fields. Figure 4 illustrates the allocation steps using this padded structure. When allocating the structure for the first time, a padded structure is created, and the head address (assume it is A) is returned (1). On the next allocation request it returns the address $A+8$ (2) to reuse the pad with fields of the second instance. Second, we make a custom memory allocator to achieve this allocation, and we modify the memory allocation part of the source code to use it. The allocator is similar to that used in [3]. It allocates a memory block for the instance pool, which we call the arena as in [2]. Figure 5 illustrates the internal organization of one arena. The custom allocator takes the structure type ID as the argument and manages one arena type per structure type, which is similar to [1]. Since one arena can hold only a few hundred instances, additional arenas are allocated on demand. The allocator divides the arena by the compression ratio into compressed and uncompressed memory.

Problems of I2. The manual I2 implementation using padding violates C semantics. There are harmful cases: (1) copying an instance of the structure causes copying of other neighboring instances; (2) pointer arithmetic causes a problem because `sizeof` for the transformed data structure returns a value which does

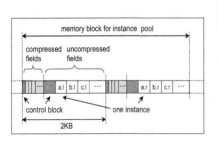

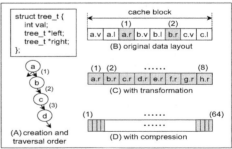

Fig. 5. Internal organization of one arena

Fig. 6. I2 and FACT can exploit temporal affinity

not match the offset to the neighboring instance; in addition, `sizeof` for one field does not match the offset to the neighboring field within one instance. However, (1) and (2) rarely occurs for programs using RDS. In cases where we find these problems in the current implementation, we simply stop applying the transformation. Alternatively, we can use a compiler to automate the transformation and circumvent these problems. Rabbah et al. studied this compiler implementation [15]. In the level the type information is available, their compiler can virtually pad target structures by modifying the address calculation function. In addition, it can modify memory allocation part of the program. Their technique modifies instance copy functions to circumvent case (1), and recognize and handle typical pointer arithmetic expressions in case (2) such as `p+c` and `(field_type*)((long)p+c)` (assume `p` is a pointer to a structure and `c` is a constant.). There is another problem: (3) the precompiled library cannot handle a transformed data structure. Rabbah et al. suggested that they need to access the entire library source code, or they need to transform an instance back to the original form before they pass it to a library function and transform it again after the function call.

3.4   Exploiting Temporal Affinity through I2 and FACT

We can exploit temporal affinity between fields through I2 and FACT. Figure 6 illustrates an example. Consider the binary tree in the program `treeadd`, which is used in the evaluation. The figure shows the declaration of the structure. After declaration of `val`, the structure has a 4-byte pad to create an 8-byte alignment. `treeadd` creates a binary tree in depth-first order, which follows the `right` edges first (A). Therefore the nodes of the tree are also organized in depth-first order in memory (B). After making the tree, `treeadd` traverses it in the same order using a recursive call. Therefore two `right` pointers that are contiguous in memory have temporal affinity. Using a 64-byte cache-block, 1 cache-block can hold two `right` pointers with temporal affinity (B) because each instance requires 24 bytes without I2. With I2, it can hold 8 of these pointers (C). With compression at a ratio of 1/8, it can hold 64 of these pointers (D). In this way, I2 and FACT enhance the prefetch effect of a cache-block.

The above example is an ideal case. In general there are two main conditions that programs must meet so that I2 and FACT can exploit temporal affinity. The main data structures used in the programs using RDS are graph structures. These programs make graphs using instances of RDS as the graph nodes and then traverse these instances. We can see a group of these instances as a matrix where one instance corresponds to one row and the fields with the same name correspond to one column. This matrix is shown in Fig. 7. While traditional data layout exploits temporal affinity between fields in the row direction, I2 and FACT exploit temporal affinity between fields in the column direction. Therefore, the first condition is: (**C1**) there exists greater temporal affinity between fields in the column direction (e.g. recursive pointers of instances) than between fields in the traditional row direction. In addition, we need high utilization in this direction. Therefore, the second condition is: (**C2**) in a short period, a single

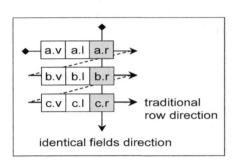

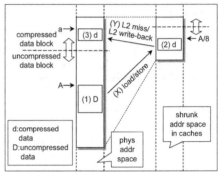

Fig. 7. Instances seen as a matrix and two directions in which we can exploit temporal affinity

Fig. 8. Address translations in FACT

traverse or multiple traverses utilize many nodes which are close in memory. When a program does not meet either condition, the traditional data layout is often more effective and I2 might degrade the performance.

3.5   Address Translation for Compressed Data

Because we attempt to compress data which changes dynamically, we find it is not always compressible. When we find incompressible data, space for storing the uncompressed data is required. There are two major allocation strategies for handling this situation. The first allocates space for only the compressed data initially, and when incompressible data is encountered, additional space is allocated for the uncompressed data [13]. The second allocates space for both the compressed and uncompressed data initially. Since the first approach makes the address relationships between the two kinds of data complex, FACT utilizes the second approach. While the second strategy still requires an address translation from the uncompressed data to the compressed data, we can calculate it using an affine transformation with the following steps: FACT uses a custom allocator, which allocates memory blocks and divides each into two for the compressed data and the uncompressed data. When using a compression ratio of 1/8, it divides each block into the compressed data block and the uncompressed data block with a $1:8$ ratio. This layout also provides the compressed data with spatial locality, as the compressed block is a reduced mirror image of the uncompressed block.

Consider the compressed data d and their physical address a, the uncompressed data D and their physical address A, and the compression ratio of $1/R$. When we use a to point to d in the caches, the compressed blocks only occupy $1/(R+1)$ of the area of the cache. On the other hand, when we use A to point to D in the caches, the uncompressed blocks occupy $R/(R+1)$ of the area. Therefore we prepare another address space for the compressed data in the caches and use A/R to point to d. We call the new address space the shrunk address space. We need only to shift A to get A/R, and we add a 1-bit tag to the caches to distinguish the address spaces. In the case of the write-back to and the fetch from main memory, since we need a to point to d, we translate A/R into a. This translation can be done by calculation, which although requires additional latency, slows execution time by no more than 1% in all of the evaluated programs. Figure 8 illustrates the translation steps. Assume we compress data at physical address $A(1)$ at the compression ratio of $1/8$. The compressed data are stored at the physical address $a(3)$. Then cld/cst instructions access the compressed data in the caches using the address $A/8(X)(2)$. When the compressed data need to be written-back to or fetched from main memory, we translate address $A/8$ into $a(Y)(3)$.

3.6   Deployment and Action of cld/cst Instructions

FACT replaces the load/store instructions that access the compression target fields with cld/cst instructions. They perform the compress/decompress operations as well as the normal load/store tasks at runtime. Since we use three types of compression we have three types of cld/cst instructions. We choose for each replacement one type depending on the target field type and the compression method chosen. We use only one integer compression method in the entire program as described in Sect. 3.1.

Figure 9 shows the operation of the cst instruction, and Fig. 10 shows its operation in the cache. In the figures, we assume a cache hierarchy of one level for simplicity. We also assume a physically tagged, physically indexed cache. The cst instruction checks whether its data are compressible, and if they are, it compresses them and puts them in the cache after the address translation which shrinks the address by the compression ratio. When cst encounters incompressible data, it stores the codeword indicating incompressibility, and then it stores

```
STDATA  Data to be stored
BASE    Base address
OFFSET  Address offset
1/R     Compression ratio
INCMP   Codeword indicating incompressibility

/* operation of cst instruction */
cst(STDATA, OFFSET, BASE) {
  PA = physical address of (BASE + OFFSET)
  CDATA = call compress_ptr(STDATA, BASE)
    or its family according to type of cst
  CA = PA/R
  cache_write(CA, CDATA, C)
  if(CDATA == INCMP) {
    /* STDATA is incompressible */
    cache_write(PA, STDATA, N)
  }
}
```

```
ADDR  Address in cache
DATA  Data to be stored
FLAG  C:compressed data
      N:uncompressed data

/* operation of cst instruction in cache */
cache_write(ADDR, DATA, FLAG) {
  if(cache-miss on {ADDR, FLAG}) {
    if(FLAG == C) {
      PA = calculate address of compressed data
        in main memory from ADDR
      access main memory with address PA
        and cache-fill
    } else {
      access main memory with address ADDR
        and cache-fill
    }
  }
  store DATA using {ADDR, FLAG}
}
```

Fig. 9. Operation of cst instruction

Fig. 10. Operation of cst inst. in the cache

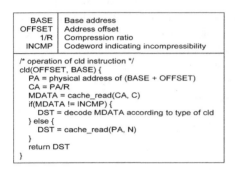

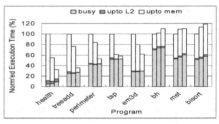

Fig. 12. Execution time of the programs with I2 and FACT. In each group, each bar shows from the left, execution time of the baseline configuration, with I2, and with FACT, respectively

Fig. 11. Operation of `cld` instruction

the uncompressed data to the address before translation. When the `cst` instruction misses the compressed data in all the caches, main memory is accessed after the second address translation, which translates the address of the compressed data in the caches to the address of the compressed data in main memory. Figure 11 shows the operation of the `cld` instruction. When the `cld` instruction accesses compressed data in the caches, it accesses the caches after the address translation and decompresses it. The translation shrinks the address by the compression ratio. When `cld` fetches compressed data from the cache and finds they are incompressible, it accesses the uncompressed data using the address before the translation. When `cld` misses the compressed data in the caches, it accesses main memory in a similar way to the `cst` instruction. Since the compressed and uncompressed memory can be inconsistent, we must replace all load/store instructions accessing the compression targets with `cld`/`cst` instructions so that the compressed memory is checked first.

4   Evaluation Methodology

We assume the architecture employing FACT uses a superscalar 64-bit microprocessor. For integer instructions, the pipeline consists of the following seven stages: instruction cache access 1, instruction cache access 2, decode/rename, schedule (SCH), register-read (REG), execution (EXE), and write-back/commit (WB). For load/store instructions, four stages follow REG stage: address-generation (AGN), TLB-access (TLB)/data cache access 1 (DC1), data cache access 2 (DC2), and WB. Therefore the load-to-use latency is 3 cycles. We assume the `cst` instruction can perform the compression calculation in its AGN and TLB stages; therefore no additional latency is required. We assume the decompression performed by the `cld` instruction requires one additional cycle to the load-to-use latency. When `cst` and `cld` instructions handle incompressible data, they must access both the compressed data and the uncompressed data. In this case, the pipeline stages for the first attempt to access the compressed data are wasted.

Table 1. Simulation parameters for processor and memory hierarchy

Fetch	fetch up to 8 insts, 32-entry inst. queue
Branch pred.	16K-entry GSHARE, 256-entry 4-way BTB, 16-entry RAS
Decode/Issue	decode/issue up to 8 insts, 128-entry inst. window
Exec. unit	4 INT, 4 LD/ST, 2 other INT, 2 FADD, 2 FMUL, 2 other FLOAT, 64-entry load/store queue, 16-entry write buf, 16 MSHRs, oracle resolution of load–store addr. dependency
Retire	retire up to 8 insts, 256-entry reorder buffer
L1 cache	inst: 32KB, 2-way, 64B block size data: 32KB, 2-way, 64B block size, 3-cycle load-to-use latency
L2 cache	256KB, 4-way, 64B block size, 13-cycle load-to-use latency, on-chip
Main mem.	200-cycle load-to-use latency
TLB	256-ent, 4-way inst TLB and data TLB, pipelined, 50-cycle h/w miss handler

Therefore we assume the penalty in this case is at least 4 cycles for `cst` (SCH, REG, data compression 1, and data compression 2) and 6 cycles for `cld` (SCH, REG, AGN, DC1, DC2, and decompression). When `cld/cst` instructions access the compressed data, they shrink their addresses. We assume this is done within their AGN or TLB stage; therefore no additional latency is required. We developed an execution-driven, cycle-accurate software simulator of a processor to evaluate FACT. Contentions on the caches and buses are modeled. Table 1 shows its parameters. We set the instruction latency and the issue rate to be the same as the Alpha 21264 [5].

We used 8 programs from the Olden benchmark [4], health (hea), treeadd (tre), perimeter (per), tsp, em3d, bh, mst, and bisort (bis), because they use RDS and they have a high rate of stall cycles caused by cache misses. Table 2 summarizes thier characteristics. All programs were compiled using Compaq C Compiler version 6.2-504 on Linux Alpha, using optimization option "-O4". We implemented a custom allocator using the instance pool technique, and modified the source codes of the programs to use it when evaluating the baseline configuration. This allocator is similar to [1, 2]. We did so to distinguish any improvement in execution speed caused by FACT from improvements resulting from the use of the instance pool technique, since applying this allocator speeds

Table 2. Program used in the evalutaion: input parameters for profile-run and evaluation-run, max. memory dynamically allocated, instruction count, The 4th and 5th columns show the numbers for the baseline configuration

Name	Input param. for profile-run	Input param. for evaluation	Max dyn. mem	Inst. count
hea	lev 5, time 50	lev 5, time 300	2.58MB	69.5M
tre	4K nodes	1M nodes	25.3MB	89.2M
per	128×128 img	16K×16K img[1]	19.0MB	159M
tsp	256 cities	64K cities	7.43MB	504M
em3d	1K nodes, 3D	32K nodes, 3D	12.4MB	213M
bh	256 bodies	4K bodies[2]	.909MB	565M
mst	256 nodes	1024 nodes	27.5MB	312M
bis	4K integers	256K integers	6.35MB	736M

---

[1] We modified **perimeter** to use a 16K×16K image instead of 4K×4K.

[2] The iteration number of **bh** was modified from 10 to 4.

up the programs. Similarly, when evaluating the configuration with I2, we apply a custom allocator which implements the data layout transformation but does not allocate memory blocks for the compressed data.

5   Results and Discussions

Compressibility of Fields. FACT uses three types of compression; RPC, NIC, and DIC. For each method, we show the dynamic memory accesses of the compression target fields (A_{target}) normalized to the total dynamic accesses (access rate), and the dynamic accesses of compressible data normalized to A_{target} (success rate). We use compression ratios of 1/4, 1/8, and 1/16. In these cases, 64-bit pointers (32-bit integers) are compressed into 16, 8, and 4 bits (8, 4, and 2 bits) respectively. Table 3 summarizes the results. Note that since the input parameters for the profile-run and the evaluation-run are different, success rates under 90% exist. The main data structures used in the programs are graph structures. With respect to pointer compression, tre, per, em3d, and tsp exhibit high success rates. This is because they organize the nodes in memory in a similar order as the traversal. In these programs we can compress many pointers into single bytes. On the other hand, bh, bis, hea, and mst exhibit low success rates, because they organize the nodes in a considerably different order to the traversal order, or because they change the graph structure quite frequently. With respect to integer compression, because tre, per, tsp, em3d, bh, and bis have narrow bit-width integer fields or enumeration type fields, they exhibit high success rates. Among them, tre, per and bis also exhibit high access rates. In the following, we use a compression ratio of 1/8 for tre, per, em3d, and tsp, and of 1/4 for hea, mst, bh, and bis.

Table 3. Access rate and compression success rate for compression target fields. Left table shows numbers for RPC; Middle for NIC; Right for DIC

Prog.	Access(%)			Success(%)			Access(%)			Success(%)			Access(%)			Success(%)		
ratio	1/4	1/8	1/16	1/4	1/8	1/16	1/4	1/8	1/16	1/4	1/8	1/16	1/4	1/8	1/16	1/4	1/8	1/16
hea	31.1	1.45	1.45	94.6	76.8	76.5	24.2	1.51	.677	83.7	88.6	93.7	24.3	24.1	.827	46.9	18.9	90.3
tre	11.6	11.6	11.5	100	98.9	96.5	5.80	5.78	5.77	100	100	100	5.80	5.78	5.77	100	100	100
per	17.6	17.5	17.6	99.8	95.9	85.6	12.4	12.4	4.33	100	100	83.1	12.4	12.4	12.4	100	100	91.0
tsp	10.2	10.2	10.2	100	96.0	67.1	.107	.107	.107	99.2	87.5	50.0	.107	.107	.107	50.0	50.0	50.0
em3d	.487	.487	.487	100	99.6	99.6	1.54	1.54	.650	100	99.1	68.8	1.54	1.54	1.14	100	100	72.6
bh	1.56	1.56	.320	88.2	51.3	52.2	.0111	.0111	.0111	100	100	100	.0176	.0111	.0111	100	100	100
mst	5.32	5.32	0	100	28.7	0	0	0	0	0	0	0	6.08	0	0	29.8	0	0
bis	43.0	41.2	41.0	90.8	65.6	59.2	27.8	0	0	100	0	0	27.8	0	0	100	0	0

Effect of Each Compression Method. FACT uses three types of compression, therefore we show the individual effect of each, and the effect of combinations of the three methods. The left side of Fig. 13 shows the results of applying each compression method alone. Each component in the bar shows from the bottom, busy cycles (busy) other than stall cycles waiting for memory data because

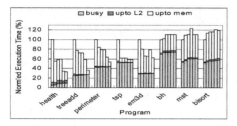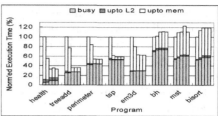

Fig. 13. Execution time of the programs. In each group, each bar shows from the left; Left graph: execution time of the baseline configuration, with I2, with NIC, with DIC, with RPC, and with FACT, respectively; Right graph: execution time of the baseline configuration, with I2, with NIC and RPC, with DIC and RPC, with FACT, respectively

of cache misses, stall cycles caused by accesses to the secondary cache (upto L2), and stall cycles caused by accesses to main memory (upto mem). All bars are normalized to the baseline configuration. First we compare NIC/DIC with RPC. In hea, tre, per, and tsp, RPC is more effective. This is because the critical path which follows the pointers does not depend on integer fields. On the other hand, in em3d, NIC/DIC is more effective. This is because the critical path depends on the integer fields, and there are more compressible integer fields than compressible pointer fields. Second we compare NIC with DIC. In hea and mst, NIC shows more speedup than DIC. This is because values not seen in the profile-run are used in the evaluation run. In other programs, the two methods exhibit almost the same performance. In bh, mst, and bis, no compression method reduces memory stall cycles. This will be explained in Sect. 5. The right side of Fig. 13 shows the results using a combination of pointer field compression and integer field compression. This combination leads to greater performance than either method alone in all programs except bh, mst, and bis. In addition, FACT is the best performing method in all programs except bh, mst, and bis.

Execution Time Results of FACT. Figure 12 compares execution times of programs using I2 and FACT. As the figure shows, FACT reduces the stall cycles waiting for memory data by 41.6% on average, while I2 alone reduces them by 23.0% on average. hea, tre, per, tsp, and em3d meet both conditions (C1) and (C2) in Sect. 3.4. Therefore, both I2 and FACT can reduce the memory stall cycles of these programs, with the latter reducing them more. FACT reduces most of the memory stall cycles which I2 leaves in tre and per because the whole structure can be compressed into 1/8 of its original size. bh builds an oct-tree and traverse it in depth-first order, but it skips the traversal of child nodes often. In addition, nodes are organized in a random order in memory. mst manages hashes using singly linked-lists and traverses multiple lists. The length of node-chain which one traversal accesses is two on average. In addition, nodes of each list are distributed in memory. bis changes the graph structure frequently. Therefore these programs do not satisfy condition (C2) and thus I2 is less efficient than the traditional data layout. In this case, (T1) I2 increases the memory stall cycles; (T2) in addition, it increases the TLB misses, which results in the increase in the busy cycles. These effects can be seen for these programs. These programs have many incompressible fields. A processor using FACT must access uncompressed data along with compressed data when it finds these fields. In this case, (S1) this additional access increases the busy cycles; (S2) it can increase the cache conflicts and cache misses, which results in the increase in the memory stall cycles. These effects further slow down these programs. Note that (S2) accounts for the increase in the busy cycles in hea because hea also has many incompressible fields.

Decompression Latency Sensitivity. We compare the execution time of I2 while varying the cycles taken by the compression operation of a cst instruction (L_c) and those taken by the decompression operation of a cld instruction (L_d). Note that up to this point, we have assumed that $(L_c, L_d) = (2, 1)$. In addition, note that two cycles of L_c are hidden by the address-generation stage and the TLB-access stage of the cst instruction, but no cycle of L_d is hidden. The increase in the execution time of the cases $(L_c, L_d) = (8, 4), (16, 8)$ over the case $(L_c, L_d) = (2, 1)$ is 3.95% and 11.4% on average, respectively. FACT is relatively insensitive up to the case $(L_c, L_d) = (8, 4)$ because an out-of-order issue window can tolerate additional latency of a few cycles.

Prefetch Effect vs. Capacity Effect. Both I2 and FACT reduce cache misses in two distinct ways: they increase the prefetch effect of a cache-block and they increase the reuse count of a cache-block. To show the breakdown of these effects, we use two cache-access events, which correspond to the two effects. We call them **spatial-hit** and **temporal-hit** and define them as follows: In the software simulator, we maintain access footprints for each cache-block residing in the caches. Every cache access leaves a footprint on the position of the accessed word. The footprints of a cache-block are reset on a cache-fill into a state which indicates that only the one word that caused the fill is accessed. When the same data block resides in both the primary and the secondary caches, cache accesses to the block in the primary cache also leave footprints on the cache-block in the secondary cache. We define the cache-hit on a word without a footprint as a spatial-hit and on a word with a footprint as a temporal-hit. Details of the relationship between temporal-hits and spatial-hits and the effects of I2/FACT are as follows: As is described in Sect. 3.4, I2 enhances the prefetch effect of a cache-block and increases the utilization ratio of a cache-block. The former reduces cache-misses and increases spatial-hits (we call this effect the prefetch effect). The latter reduces the number of cache-blocks used in a certain period of time, which increases temporal-hits. FACT enhances the prefetch effect further by compression, which increases spatial-hits (large-cache-block effect). In addition, compression reduces the number of cache-blocks used, which increases temporal-hits (compression effect).

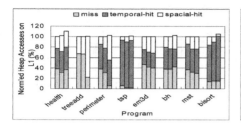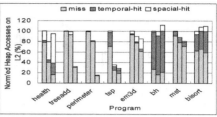

Fig. 14. Breakdown of accesses to the data caches, which refers to heap data. In each group, each bar shows from the left, the breakdown of accesses of the baseline configuration, with I2, and with FACT, respectively. Left: primary cache. Right: secondary cache

We observe cache accesses to the heap because I2 and FACT handle heap data. Figure 14 shows the results. First we compare I2 with the baseline. In all programs except bis, misses decrease and spatial-hits increase in either or both cache levels. In bh, temporal-hits also increase in the secondary cache. In summary, I2 mainly increases the number of spatial-hits, therefore its main effect is the prefetch effect. Second we compare FACT with I2 with respect to the compression of FACT. In hea, tre, per, em3d, and bis, misses decrease and spatial-hits increase in either or both cache levels. In hea, per, tsp, em3d, mst, and bis, temporal-hits increase in either cache levels. The spatial-hits increase have the larger impact. In summary, the compression of FACT exhibits both the large-cache-block effect and the compression effect, with the large-cache-block effect having the greater impact.

FACT Software-Only Implemention. FACT introduces new instructions and special hardware for compression and decompression operations to reduce the overhead associated with these operations. To depict the amount of this overhead, we implement FACT by modification of the source code only (we call this FACT-S) and compare FACT-S with FACT. We first apply I2 and then replace definitions and usages of the compression target fields with custom inline compression and decompression functions to implement FACT-S. In typical cases, the compression function requires 23, 49, and 15 instructions for RPC, DIC, and NIC, respectively. The decompression function requires 19, 23, and 13 instructions. Figure 15 shows the FACT-S execution time. We can see the software compression/decompression overhead as the increase in busy cycles. FACT-S is slower than I2 in all programs except tre because the overhead offsets the reduction in memory stall time. This observation confirms that additional instructions and hardware for FACT reduce the overhead effectively.

FACT with Cache-Conscious Data Layout Techniques. FACT exhibits poor performance in bh, mst, and bis because they do not meet condition (C2) in Sect. 3.4. There are two reasons that bh does not meet the condition: (B1) bh

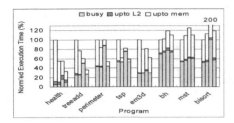

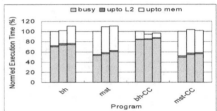

Fig. 15. Execution time of the programs with software FACT (FACT-S). In each group, each bar shows execution time of the baseline configuration, with I2, with FACT-S, and with FACT, respectively

Fig. 16. Execution time of the programs with CC, I2, and FACT. The "bh" and "mst" groups are a subset of Fig. 12. In the "bh-CC" and "mst-CC" groups, each bar shows execution time of the baseline configuration with CC, with CC+I2, and with CC+FACT, respectively

skips graph nodes, and (B2) its nodes are organized in memory in a random order. There are two reasons for mst: (M1) mst traverses a small number of nodes at a time, and (M2) the nodes in each list are distributed in memory. There is one reason for bis: (S1) bis changes the graph structure quite frequently. However, we can organize these nodes in memory in a similar order to the traversal order using cache-conscious data layout techniques (CC) [6] for those programs which do not modify the graph structures quite frequently. Using CC, we can correct properties (B2) and (M2) and make FACT exploit more affinity in these programs. Therefore we apply these techniques and show how the execution time with FACT changes. For bh, we applied ccmorph, which inserts a function in the source code, to reorganize the graph at runtime. We reorganize the graph in depth-first order in memory because depth-first order is the typical order of traversal. We applied ccmalloc for mst; it replaces the memory allocator and attempts to allocate a graph node in the same cache-block as the "parent" node. The "parent" node is specified through an argument. We select the node next to the allocating node in the list as the "parent" node.

Figure 16 shows the results: the "bh" and "mst" groups are a subset of Fig. 12. The "bh-CC" and "mst-CC" groups show the results with CC normalized to the baseline configuration running programs with CC. Using CC, FACT can reduce the memory stall time for bh and achieve speed beyond the baseline. In addition, FACT shows almost no degradation in the execution speed for mst.

6   Summary and Future Directions

We proposed the Field Array Compression Technique (FACT), which reduces cache misses caused by recursive data structures. Using a data layout transformation, FACT gathers data with temporal affinity in contiguous memory, which enhances the prefetch effect of a cache-block. It enlarges the effective cache-block size and enhances the prefetch effect further from the compression of recursive pointer and integer fields. It also enlarges the effective cache capacity. Through software simulation, we showed that FACT yields a 41.6% reduction of stall cycles waiting for memory data on average.

This paper has four main contributions. (1) FACT achieves a compression ratio of 1/8 and over. This ratio exceeds 1/2, which is the limit of existing compression methods. This ratio is achieved because of the data layout transformation and the novel addressing of the compressed data in the caches. (2) We are able to compress many recursive pointer fields into 8 bits, which is achieved partly because of grouping the pointers. (3) We represent the notion of a split memory space, where we allocate one byte of compressed memory for every 8 bytes of uncompressed memory. Each uncompressed element is represented in the compressed space with a codeword placeholder. This provides compressed data with spatial locality. Additionally, the addresses of the compressed elements and the uncompressed elements have an affine relationship. (4) We represent the notion of an address space for compressed data in the caches. The address space for uncompressed data is shrunk to be used as the address space for compressed data in the caches. This simplifies the address translation from the address for the uncompressed data to the address which points to the compressed data in the caches, and avoids cache conflicts.

Future work will entail building a framework to automate FACT using a compiler.

References

1. J. Bonwick. The slab allocator: An object-caching kernel memory allocator. *Proc. USENIX Conference*, pp. 87–98, Jun. 1994.
2. D. A. Barret and B. G. Zorn. Using lifetime prediction to improve memory allocation performance. *Proc. PLDI*, 28(6):187–196, Jun. 1993.
3. D. N. Truong, F. Bodin, and A. Seznec. Improving cache behavior of dynamically allocated data structures. *Proc. PACT*, pp. 322–329, Oct. 1998.
4. A. Rogers et al.Supporting dynamic data structures on distributed memory machines. *ACM TOPLAS*, 17(2):233–263, Mar. 1995.
5. Compaq Computer Corp. Alpha 21264 Microprocessor Hardware Reference Manual. Jul. 1999.
6. T. M. Chilimbi et al.Cache-conscious structure layout. *Proc. PLDI*, 1999.
7. C-K. Luk and T. C. Mowry. Compiler based prefetching for recursive data structures. *Proc. ASPLOS*, pp. 222–233, Oct. 1996.
8. A. Roth, A. Moshovos, and G. S. Sohi. Dependence based prefetching for linked data structures. *Proc. ASPLOS*, pp. 115–126, Oct. 1998.
9. J. Lee et al.An on-chip cache compression technique to reduce decompression overhead and design complexity. *Journal of Systems Architecture*, Vol. 46, pp. 1365–1382, 2000.
10. Y. Zhang, J. Yang, and R. Gupta. Frequent value locality and value-centric data cache design. *Proc. ASPLOS*, pp. 150–159, Nov. 2000.
11. J. Yang, Y. Zhang, and R. Gupta. Frequent value compression in data caches. *Proc. MICRO*, pp. 258–265, Dec. 2000.

12. S. Y. Larin. Exploiting program redundancy to improve performance, cost and power consumption in embedded systems. Ph.D. Thesis, ECE Dept., North Carolina State Univ., 2000.
13. Y. Zhang et al.Data compression transformations for dynamically allocated data structures. *Int. Conf. on Compiler Construction*, LNCS 2304, Springer Verlag, pp. 14–28, Apr. 2002.
14. D. Brooks et al.Dynamically exploiting narrow width operands to improve processor power and performance. *Proc. HPCA*, pp. 13–22, Jan. 1999.
15. R. M. Rabbah and K. V. Palem. Data remapping for design space optimization of embedded memory systems. *ACM TECS*, 2(2):186–218 May 2003.

FIBER: A Generalized Framework for Auto-tuning Software

Takahiro Katagiri[1,2], Kenji Kise[1,2], Hiroaki Honda[1], and Toshitsugu Yuba[1]

[1] Department of Network Science, Graduate School of Information Systems,
The University of Electro-Communications
[2] PRESTO,
Japan Science and Technology Corporation(JST),
1-5-1 Choufu-gaoka, Choufu-shi, Tokyo 182-8585, Japan,
Phone: +81 424-43-5642, FAX: +81 424-43-5644,
{katagiri,kis,honda,yuba}@is.uec.ac.jp

Abstract. This paper proposes a new software architecture framework, named FIBER, to generalize auto-tuning facilities and obtain highly accurate estimated parameters. The FIBER framework also provides a loop unrolling function, needing code generation and parameter registration processes, to support code development by library developers. FIBER has three kinds of parameter optimization layers—installation, before execution-invocation, and run-time. An eigensolver parameter to apply the FIBER framework is described and evaluated in three kinds of parallel computers; the HITACHI SR8000/MPP, Fujitsu VPP800/63, and Pentium4 PC cluster. Evaluation indicated a 28.7% speed increase in the computation kernel of the eigensolver with application of the new optimization layer of before execution-invocation.
Keywords: Auto-Tuning; Parameter optimization; Numerical library; Performance modeling; Eigensolver;

1 Introduction

Tuning work on parallel computers and other complicated machine environments is time-consuming, so an automated adjustment facility for parameters is needed. Moreover, library arguments should be reduced to make the interface easier to use and a facility is needed to maintain high performance in all computer environments. To solve these problems, many packages of SATF (Software with Auto-Tuning Facility) have been developed.

There are two kinds of paradigms for SATF. The first paradigm is known as computer system software. Examples of this software for tuning computer system parameters, like I/O buffer sizes, include Active Harmony [9] and AutoPilot [7].

The second paradigm is known as numerical library. PHiPAC [2], ATLAS and AEOS (Automated Empirical Optimization of Software) [1, 10], and FFTW [3] can automatically tune the performance parameters of their routines when they are installed, while in ILIB [4, 5], the facility of installation and run-time optimization is implemented.

A. Veidenbaum et al. (Eds.): ISHPC 2003, LNCS 2858, pp. 146–159, 2003.

For formalization of an auto-tuning facility, Naono and Yamamoto formulated the installation optimization known as SIMPL [6], an auto-tuning software framework for parallel numerical processing.

Although many facilities have been proposed, these conventional facilities of SATF have a limitation from the viewpoint of general applicability because the auto-tuning facility uses dedicated methods defined in each package. An example of this limitation is that several general numerical libraries contain direct solvers, iterative solvers, dense solvers, and sparse solvers. There is no software framework to adapt SATF to these solvers.

This paper proposes a new and general software framework for SATF called FIBER (Framework of Installation, Before Execution-invocation, and Run-time optimization layers) [8], a framework containing three types of optimization layers to solve these problems.

The remainder of this paper is organized as follows: in Section 2 we explain components and definitions for auto tuning in FIBER, in Section 3 we show an adaption and an evaluation for three types of optimization layers in FIBER, and in Section 4 we conclude with a summary of our research.

2   Components and Definitions of Auto-Tuning in FIBER

2.1   Components of FIBER

FIBER supports the following:

- **Code Development Support Functions:** This function performs automatic code generation to guarantee auto-tuning, parameterization, and its registration by specifying an instruction from library developers. The instruction is specified to library or sub-routine interfaces, or other parts of the program, by using a dedicated language.
- **Parameter Optimization Functions:** These functions perform the optimizations of specified parameters in the PTL (Parameter Tuning Layer). There are three kinds of timing for the optimizations (See Figure 2.)

Figure 1 shows software components in FIBER. The parameters in Figure 1 are added in the library, sub-routines, or other parts of the program described by library developers and a language is used to specify the parameters. The PTL optimizes parameters to minimize a function. Even the parameters in computer system libraries, such as MPI (Message Passing Interface), can be specified if the interface is open to users. PTL in FIBER, thus, can access system parameters. There are three kinds of optimization layers to optimize the specified parameters in PTL : IOL (Installation Optimization Layer), BEOL (Before Execution-invocation Optimization Layer), and ROL (Run-time Optimization Layer). These layers can access limited parameters to optimize. For instance, IOL-tuned parameters can be accessed in BEOL and ROL. BEOL-tuned parameters, however, can only be accessed by ROL. The main reason for this limited access is the necessity of obtaining highly estimated parameters in lower optimization layers.

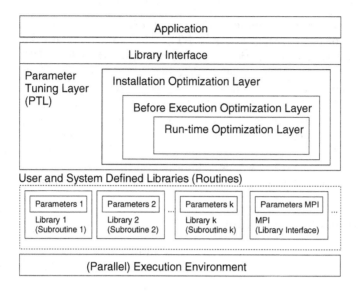

Fig. 1. Software components of FIBER.

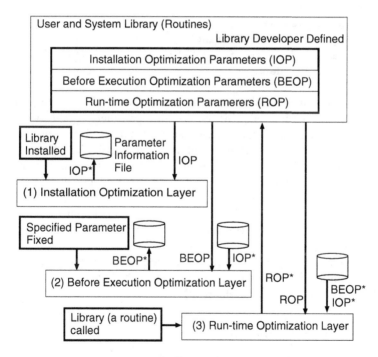

Fig. 2. Process view in library users for the three types of optimization layers in FIBER.

Figure 2 is a snapshot of the optimization process in library users applying the three types of optimization layers in PTL. The parameters are separated into IOP (Installation Optimization Parameters) accessed in IOL, BEOP (Before Execution-invocation Optimization Parameters) accessed in BEOL, and ROP (Run-time Optimization Parameters) accessed in ROL. The optimization procedure consists of: (1) IOL : When software installation is performed, (2) BEOL: After the special parameters are specified, such as problem sizes, by library users, and (3) ROL: When the target library, sub-routine, or other parts of the program are called or executed. These optimized parameters in the procedures are stored in a file, named "parameter information file". Lower level optimization layers can access the parameters stored in this file to perform their optimization.

2.2 Objective of Auto-tuning

This section presents the objective of the auto-tuning facility for FIBER. To clarify the facility, a library interface for parallel eigenvalue computation is given below.

[**Example 1**] A conventional parallel numerical library interface:

```
call PEigVecCal(
         A, x, lambda, n,                      ...(i)
         nprocs, myid, iDistInd,               ...(ii)
         imv, iud, ihit, icomm, kbi, kort, ...(iii)
         MAXITER, deps                         ...(iv)   )
```

The arguments in Example 1 are called as (i) basic information parameters, (ii) parallel control parameters, (iii) performance parameters, and (iv) algorithm parameters. For example, the dimension sizes of matrix A are specified in the parameters as (i). Data distribution information is stored in the parameters of (ii). Unrolling depths or block sizes are specified in the parameters of (iii). Maximum iterative numbers are defined in the parameters of (iv). Generally speaking, these arguments can be removed to design a better library interface for (ii), and to analyze numerical characteristics for (iv). The parameters of (iii), however, cannot be removed in conventional frameworks without the auto-tuning facility.

The aim of SATF is to maintain performance and by removing the parameters from (iii). Using SATF, the interface can be simplified as:

```
call PEigVecCal(A, x, lambda, n)
```

2.3 Definition of Auto-tuning

The auto-tuning facility is described as follows:[1]

---

[1] Similar definition was done by Naono and Yamamoto [6] In their definition, performance parameters are defined as CP (Critical Parameter). They defined two kinds of CP parameters: UCP (Users' Critical Parameter) and ICP (Internal Critical Param-

Let the parameter set for all library arguments be AP. The parameter set of (i) the basic information parameters as defined in Example 1 is defined here as BP, while the other is defined here as OP, where $OP \equiv AP/BP$.

Thus :

$$AP = BP \cup OP, \tag{1}$$

Where $BP \cap OP = \phi$.

The AP and BP sets are defined as:

Definition AP (All Parameters): The set of parameters for input, output, and performance is defined as AP for all interfaces in the library, sub-routine, or other parts of the program. □

Definition BP (Basic Parameters): The set of parameters for basic information in AP, such as matrix sizes for input or output matrices, and target machine environments information, such as the number of processors (PEs), is defined as BP. These parameters are specified by library users before the target libraries, sub-routines, or other parts of the program run. □

The following assumption is made:

Assumption 1: The execution time of the target library, sub-routine, or other parts of the program can be estimated by using the function F, which is derived from the set of AP. □

By Assumption 1, the execution time t for the target is obtained as

$$t = F(m), \tag{2}$$

where $m \subseteq AP$.

Let the set of PP, where $PP \subseteq OP$ be the set of (iii) performance parameters as defined in Example 1. The PP is defined here as follows.

Definition PP (Performance Parameters): The set of parameters in OP, which can affect the whole performance of the library, sub-routine, or other parts of the program is defined here as PP when the set of BP is fixed. Library users do not need to specify the parameters in PP. The performance of the library, sub-routine, and other parts of the program is controlled by the parameters. □

Let the set of OPP be $OPP \equiv OP/PP$.

Consequently,

$$OP = PP \cup OPP, \tag{3}$$

where $PP \cap OPP = \phi$.

For the OPP, the following assumption is made.

Assumption 2: The parameters in OPP do not affect the other parameters in AP, except for the parameters in BP. □

---

eter). UCPs are parameters specified in library arguments by users, and ICPs are parameters which do not appear library interface. ICPs are defined in the internal library.

Now the (i) of BP is fixed as $\bar{l} \subseteq BP$, and by using Assumption 1 and Assumption 2, the execution time t of the target is estimated as:

$$t = F(\bar{l}, g), \tag{4}$$

where $g \subseteq PP$.

The auto-tuning facility is defined as:

Definition of the Auto-tuning Facility: The auto-tuning facility is defined as the optimization procedure to minimize the function of F for execution time, when the set of BP is fixed.

Also, it is possible to define the facility as the following optimization problem: Finding the set of $g \subseteq PP$ in the condition of $\bar{l} \subseteq BP$, minimizing the function of F:

$$\min_g F(\bar{l}, g) = \min_g \hat{F}(g). \quad \square \tag{5}$$

In reality, there are many parameter optimization problems, such as minimization of fee for computer use, and the size of memory spaces used. The definition for F should be enhanced by taking into account these problems. The function of F is the target function to be minimized in the optimization process. Then, the function of F is called as a **cost definition function**

2.4   Definition of Auto-tuning in FIBER

In FIBER, the set of performance parameters PP is separated into three parameters:

$$PP = IOP \cup BEOP \cup ROP. \tag{6}$$

The parameters of IOP, $BEOP$, and ROP are defined as optimization parameters for installation, before execution-invocation, and run-time, respectively.

Definition of FIBER's Auto-tuning Facility: FIBER's auto-tuning facility is defined as an optimization procedure to estimate the parameter set PP fixing a part of BP in IOL, BEOL, and ROL.

FIBER's auto-tuning facility is a procedure to minimize the cost definition function F, which is defined as the user's programs and instructions in the three kinds of layers, to estimate the parameter set PP when a part of the parameter set BP is fixed in each layer. $\square$

The definition of auto-tuning in FIBER indicates that FIBER's installation optimization is an estimation procedure for PP when a part of BP, which is affected by machine environments, is fixed. FIBER's before execution-invocation optimization can be explained as an estimation procedure for PP when a part of BP, which is determined by the user's knowledge of the target process, is fixed by using optimized parameters in the installation layer. For this reason, the parameters estimated by the before execution-invocation layer have higher accuracy than parameters estimated by the installation layer. For the parameters of BP, which can not be fixed in the two optimization layers, FIBER's run-time optimization layers determine their values to estimate PP at run-time.

3   Adaptation and Evaluation
for FIBER Auto-tuning Facility

3.1   Specification of Performance Parameters by Library Developers

In the FIBER framework, library developers can implement detailed instructions to specify the performance parameters of PP and target areas of auto-tuning in their programs. Hereafter, the target area is called as *tuning region*. The typical instruction operator, named as *unrolling instruction operator* (unroll), is shown in the below.

Example of Unrolling Instruction Operator. In the source program, the line of !ABCLib$ is regarded as FIBER's instruction. The following code shows an example of an unrolling instruction operator.

```
!ABCLib$ install unroll (j) region start
!ABCLib$ varied (j) from 1 to 16
!ABCLib$ fitting polynomial 5 sampled (1-4,8,16)
 do j=0, local_length_y-1
    tmpu1 = u_x(j)
    tmpr1 = mu * tmpu1 - y_k(j)
    do i=0, local_length_x-1
       A(i_x+i, i_y+j) = A(i_x+i, i_y+j)
          + u_y(i)*tmpr1 - x_k(i)*tmpu1
    enddo
 enddo
!ABCLib$ install unroll (j) region end
```

The instruction operator of install can specify the auto-tuning timing of installation optimization. Similarly, the timings of before execution-invocation optimization (static) and run-time optimization (dynamic) can be specified.

The above code shows that the loop unrolled codes for j-loop, which adapts unrolling to the tuning region of region start – region end, are automatically generated. The depth of the loop unrolling is also automatically parameterized as PP.

The instruction operator specifies detailed functions for the target instruction operator and is called as the sub-instruction operator. The sub-instruction operator of varied defines the defined area of target variables. In this example, the area is {1,..,16}. For the cost definition function, the types of them can be specified by the sub-instruction operator fitting. In this example, a 5-th order linear polynomial function is specified. The sub-instruction operator of sampled is for definition of sampling points to estimate the cost definition function. This example of sampling is {1-4,8,16}.

3.2   Installation Optimization Layer (IOL)

The auto-tuning facility of FIBER is adapted to an eigensolver in this section. This is implemented using the Householder-bisection-inverse iteration method for computing all eigenvalues and eigenvectors in dense real symmetric matrices.

The cost definition function as the execution time for the solver is defined. For target parallel computers, the HITACHI SR8000/MPP at the Information Technology Center, The University of Tokyo is used[2].

Let the interface of the target library in the Householder-bisection-inverse iteration method be the same interface as `PEigVecCal`, as shown in Example 1. The main arguments of this library are:

- $PP \equiv$ { imv, iud, icomm, ihit, kbi, kort }
- $BP \equiv$ { n, nprocs }

The main definition area in this example is:

- iud $\equiv$ { 1,2,...,16 } : Unrolling depth for the outer loop of an updating process in the Householder tridiagonalization (a double nested loop, BLAS2).

A set of the parameters which optimize in IOL is:

- $IOP \equiv$ { imv, iud, kbi, ihit }.

How to Estimate Parameters in IOL. The IOL-parameters are estimated when the target computer systems, such as computer hardware architecture or compilers are fixed. This is because these parameters can be affected by the number of registers, the size of caches, the feature of vector processing factors, and other computer hardware characteristics.

The parameters are determined in the following way. First, the parameters in BP are fixed, and several points for execution time at the target process are sampled (referred to as sampled data). The cost definition function is then determined by using the sampled data.

[**Example 2**] Execute the installation optimization of the parameter of iud in Example 1.

Let the execution time be approximated by a linear pronominal formula. The sampling points to estimate the best parameter for iud are { 1,2,3,4,8,16 }. For the parameters of BP, the PE value is 8 and the parameters of n are sampled as { 200, 400, 800, 2000, 4000, 8000 }. The code of tuning region, cost definition function, and sampling points are the same as the example of an unrolling instruction operator in Section 3.1.

Table 1 shows the execution time on the HITACHI SR8000/MPP with the above sampled points.

---

[2] The HITACHI SR8000/MPP nodes have 8 PEs. The theoretical maximum performance of each node is 14.4 GFLOPS. Each node has 16 GB memory, and interconnection topology is a three dimensional hyper-cube. Its theoretical throughput is 1.6 Gbytes/s for one-way, and 3.2 Gbytes/s for two-way. In this example, the HITACHI Optimized Fortran90 V01-04 compiler specified option of *-opt=4 -parallel=0* was used. For the communication library, the HITACHI optimized MPI was used.

Table 1. HITACHI SR8000/MPP execution time 8PE. [sec.]

n\iud	1	2	3	4	8	16
200	.0628	.0628	.0629	.0623	.0621	.0625
400	.1817	.1784	.1763	.1745	.1723	.1719
800	.7379	.6896	.6638	.6550	.6369	.6309
2000	7.535	6.741	6.333	6.240	6.013	5.846
4000	54.06	48.05	44.85	44.36	42.89	41.19
8000	413.2	366.5	349.2	344.1	327.6	315.5

In this experiment, the basic parameter of **n** is fixed to estimate the function $f_n(iud)$ for **iud**, where $f_n(iud)$ is a k-th order polynomial function of $f_n(iud) = a_1 \cdot iud^k + a_2 \cdot iud^{k-1} + a_3 \cdot iud^{k-2} + \cdots + a_k \cdot iud + a_{k+1}$. By using an appropriate optimization method, the coefficients of $a_1, .., a_{k+1}$ can be determined.

The average of relative errors were investigated with polynomial functions from 0-th order to 5-th order. The result indicated that the 5-th order polynomial function was the best in this case. The 5-th order polynomial function, therefore, is chosen as the cost deification function in this example.

Figure 3 shows the overall of estimated function $f_n(iud)$ by using the estimated coefficients.

For this optimization, the sampled problem sizes of **n** are not always specified by library users at run-time. The estimated parameters in Figure 3, hence, cannot be used in any of the cases. If library users specify a different problem size against the sampling dimensions at run-time, FIBER estimates the appropriate values

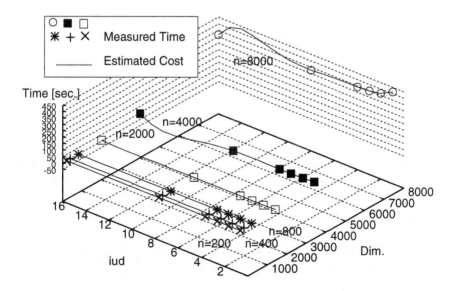

Fig. 3. The estimation of optimized parameter of **iud**. (fixed problem sizes)

by using the estimated coefficients in Figure 3 to calculate the estimation costs for the defined area of iud.

First, the costs for fixing the parameter iud can be calculated by varying the problem sizes of n as { 200, 400, 800, 2000, 4000, 8000 }. As a result, the all estimation costs in the iud defined area of { 1,2,...,16 } for the problem sizes n of { 200, 400, 800, 2000, 4000,8000 } can be obtained.

Second, these estimation costs can be regarded as new sampling points. By using the new sampling points, the function $f_{iud}(n)$ for the number of problems can be estimated. The least square method to estimate the function $f_{iud}(n)$ is also chosen. Let the $f_{iud}(n)$ also be approximated by a 5-th order polynomial function of $f_{iud}(n) = \bar{a}_1 \cdot n^5 + \bar{a}_2 \cdot n^4 + \bar{a}_3 \cdot n^3 + \bar{a}_4 \cdot n^2 + \bar{a}_5 \cdot n + \bar{a}_6$.

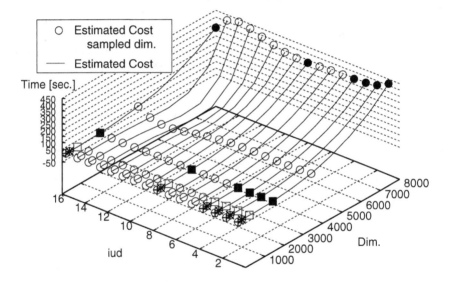

Fig. 4. The estimation of the optimized parameter of **iud**. (A case where the problem sizes are not fixed until run-time.)

Figure 4 shows the coefficients varied from the sampled number of n as { 200, 400, 800, 2000, 4000, 8000}. Figure 4 indicates that the function $f_{iud}(n)$ for n is determined in the entire definition area { 1,2,...,16 } of iud. The parameters of iud to minimize the cost function can be determined by substituting for n at run-time.

This is the parameter optimization procedure of FIBER's IOL.

Experiments on the Effect of IOL. An effect of IOL for the parameters of iud by using several kinds of machine environments is also evaluated.

- HITACHI SR8000/MPP
  - System configuration and compiler: Explained in Section 3.2.

- Fujitsu VPP800/63
  - System configuration: This machine is a vector-parallel style super-computer. The Fujitsu VPP800/63 at the Academic Center for Computing and Media Studies, Kyoto University is used. The total number of nodes for the VPP800 is 63. The theoretical maximum performance of each node is 8 GFLOPS for vector processing, and 1 GOPS for scalar processing. Each node has 8 GB memory, and inter-connection topology is a cross bar. Its theoretical throughput is 3.2 Gbytes/s. For the communication library, the Fujitsu optimized MPI was used.
  - Compiler : The Fujitsu optimized UXP/V Fortran/VPP V20L20 compiler specified option of -O5 -X9 was used.
- PC Cluster
  - System configuration: As a node of a PC cluster, the Intel Pentium4 2.0GHz is used. The number of PEs for the PC cluster is 4, and each node has 1GB (Direct RDRAM/ECC 256MB*4) memory. The system hardware board is the ASUSTek P4T-E+A (Socket478). The network card is the Intel EtherExpressPro100+. The Linux 2.4.9-34 and MPICH 1.2.1 are used as the operating system and communication library.
  - Compiler : The PGI Fortran90 4.0-2 compiler specified option of -fast was used.

[**Experiment 1**] Evaluate estimated errors in IOL for the parameter of iud.

Table 2 shows the estimated errors and relative errors to the best parameter with the sampled points of iud. The estimated errors in Table 2 were calculated by the sum of power from subtracting the measured time and the estimated time for the sampled points. The relative errors in Table 2 were calculated by using the execution time for estimated parameters and the execution time for best parameters. Table 2 indicates the enormity of relative errors in PC clusters are huge compared to super computers. This is because of the fluctuation for the execution time, observed in the PC cluster. The estimation of parameters is more sensitive compared to that of super computers.

3.3   BEOL (Before Execution-Invocation Optimization Layer)

In FIBER, the optimization of BEOL is performed when the parameters in BP are specified by the library user before executing the target process. This section explains several adoptions of this layer.

[**Situation 1**] In Example 1, the library users know the number of processors (=8 PEs), and matrix sizes (=8192) for the eigenvalue computation. ($n \equiv 8192$, nprocs $\equiv 8$) In Situation 1, the parameters to optimize in BEOL are

- $BEOP \equiv \{$ imv, iud, ihit, kbi $\}$.

Please note the IOL uses the totally estimated parameters with sampling points by library developer. In BEOP, however, library users directly specify the real sampling points to inform the FIBER optimization system, even though

Table 2. Estimated and relative errors for the parameters of `iud` on the sampled points with 5-th order polynomial cost definition function in IOL. One second is the unit for execution time.

(a) HITACHI SR8000/MPP (1Node, 8PEs)

Sampled Dim.	Estimated Parameter	Execution Time ET	Estimated Error	Best Parameter	Execution Time BT	Relative Error $(ET - BT)/BT * 100$
200	14	5.905E-2	8.749E-18	6	5.867E-2	0.64 %
400	14	0.1665	1.410E-16	14	0.1665	0%
800	14	0.6198	1.167E-15	14	0.6198	0%
2000	14	5.833	9.125E-14	16	5.824	0.15 %
4000	14	41.22	7.251E-12	15	41.00	0.54 %
8000	13	314.6	4.362E-10	15	314.4	0.04 %

(b) Fujitsu VPP800/63 (8PEs)

Sampled Dim.	Estimated Parameter	Execution Time ET	Estimated Error	Best Parameter	Execution Time BT	Relative Error $(ET - BT)/BT * 100$
200	7	3.073E-2	2.283E-18	2	3.058E-2	0.49 %
400	7	6.558E-2	5.277E-17	5	6.530E-2	0.44 %
800	7	0.1521	1.456E-16	10	0.1515	0.40 %
2000	5	0.6647	2.175E-15	4	0.6644	0.04 %
4000	6	3.418	2.414E-14	2	3.203	6.7 %
8000	7	23.06	1.412E-12	4	22.40	2.9 %

(c) PC Cluster (4PEs)

Sampled Dim.	Estimated Parameter	Execution Time ET	Estimated Error	Best Parameter	Execution Time BT	Relative Error $(ET - BT)/BT * 100$
200	7	0.2786	1.391E-16	13	0.2345	18.8 %
400	6	2.149	3.079E-15	4	0.6739	218 %
800	6	5.603	6.102E-14	14	2.7176	106 %
2000	6	20.38	1.533E-12	2	15.89	28.3 %
4000	6	106.5	6.107E-11	2	88.96	19.7 %
8000	2	583.2	1.901E-9	2	583.2	0 %

BEOP has the same PP parameters with respect to that of IOL. This is because, library users know the real number of PP, such as problem size of n. The accuracy of parameter estimated in BEOP, hence, is better than that of IOP. BEOP can be used in processes which need highly estimated parameters.

Experiments One: The Effect of BEOL. In this section, FIBER's BEOL will be evaluated on the three kinds of parallel computers.

[**Experiment 2**] Evaluate the effect of BEOL in Situation 1 where library users know the real problem sizes to execute. They are 123, 1234, and 9012 in this experiment.

Table 3 shows the execution time of FIBER's IOL-estimated parameters with the sampling points of `iud` (Estimated P1), the execution time of FIBER's IOL-estimated parameters with all definition area points (Estimated P2), and the execution time of FIBER's BEOL-estimated parameters (Best P). The results of Table 3 indicated that (1) modifying sampling points makes for better estimated accuracy and (2) FIBER's BEOL has 0.5% – 28.7% effectiveness compared to FIBER's IOL.

Table 3. Effect of optimization for the parameter of iud on the sampling points in BEOL. One second is the unit for execution time.

(a) HITACHI SR8000/MPP (1Node, 8PEs)

Specified Dim.	Estimated P1 (ExeT.$ET1$)	Estimated P2 (ExeT.$ET2$)	Best P (BEOL Optimized, ExeT.BT)	Rel.Err1 $(ET1 - BT)$ $/BT * 100$	Rel.Err2 $(ET2 - BT)$ $/BT * 100$
123	14 (0.0333)	11 (0.0341)	6 (0.0333)	0.00 %	2.4 %
1234	14 (1.668)	16 (1.662)	16 (1.662)	0.36 %	0 %
9012	16 (440.6)	12 (447.0)	16 (440.6)	0 %	1.4 %

(b) Fujitsu VPP800/63 (8PEs)

Specified Dim.	Estimated P1 (ExeT.$ET1$)	Estimated P2 (ExeT.$ET2$)	Best P (BEOL Optimized, ExeT.BT)	Rel.Err1 $(ET1 - BT)$ $/BT * 100$	Rel.Err2 $(ET2 - BT)$ $/BT * 100$
123	7 (0.0183)	1 (0.0182)	10 (0.0181)	1.1 %	0.5 %
1234	6 (0.2870)	6 (0.2870)	4 (0.2847)	0.8 %	0.8 %
9012	14 (34.67)	16 (34.29)	4 (32.03)	8.2 %	8.2 %

(c) PC Cluster (4PEs)

Specified Dim.	Estimated P1 (ExeT.$ET1$)	Estimated P2 (ExeT.$ET2$)	Best P (BEOL Optimized, ExeT.BT)	Rel.Err1 $(ET1 - BT)$ $/BT * 100$	Rel.Err2 $(ET2 - BT)$ $/BT * 100$
123	14 (0.1286)	4 (0.1285)	10 (0.1269)	1.3 %	1.2 %
1234	6 (7.838)	5 (6.283)	10 (6.090)	28.7 %	3.1 %
9012	6 (973.6)	1 (867.0)	2 (845.6)	15.1 %	2.5 %

4 Conclusion

This paper proposed a new framework for auto-tuning software, named FIBER. FIBER consists of three kinds of parameter optimization layers to generalize auto-tuning facilities and to improve parameter accuracy. The originality of the FIBER framework lies in its innovative optimization layer, the BEOL (Before Execution-invocation Optimization Layer). The experiment for BEOL effects indicated that the speed increased 28.7% compared to conventional optimization layers of installation in a computation kernel for eigensolver.

The key feature of the FIBER framework is how to determine the cost definition function of F according to the characteristics of libraries, sub-routines, or other parts of the program. Moreover, to extend the adaptation of auto-tuning and to obtain high quality estimated parameters, a more sophisticated method is needed. Evaluation of the cost definition function and extending the adaptation of FIBER are important future work.

A parallel eigensolver, named ABCLibDRSSED, has been developed which contains a part of FIBER's auto-tuning facility. The source code and manual for the alpha version are available at http://www.abc-lib.org/ The ABCLibScript language will be developed to support code generation, parameterization, and its registration for auto-tuning facilities based on FIBER's concept, as shown in Section 3.1.

Acknowledgments

The authors would like to thank Associate Professor Reiji Suda at the University of Tokyo for giving us useful comments and discussions for FIBER. This study was partially supported by PRESTO, Japan Science and Technology cooperation (JST).

References

1. ATLAS Project, available at http://www.netlib.org/atlas/index.html
2. J. Bilmes, K. Asanović, C.-W. Chin, and J. Demmel. Optimizing matrix multiply using PHiPAC: a portable, high-performance, ANSI C coding methodology. *Proceedings of International Conference on Supercomputing 97*, pages 340–347, 1997.
3. M. Frigo. A fast Fourier transform compiler. In *Proceedings of the 1999 ACM SIGPLAN Conference on Programming Language Design and Implementation*, pages 169–180, Atlanta, Georgia, May 1999.
4. T. Katagiri, H. Kuroda, K. Ohsawa, M. Kudoh, and Y. Kanada. Impact of auto-tuning facilities for parallel numerical library. *IPSJ Transaction on High Performance Computing Systems*, 42(SIG 12 (HPS 4)):60–76, 2001.
5. H. Kuroda, T. Katagiri, and Y. Kanada. Knowledge discovery in auto-tuning parallel numerical library. *Progress in Discovery Science, Final Report of the Japanese Discovery Science Project, Lecture Notes in Computer Science*, 2281:628–639, 2002.
6. K. Naono and Y. Yamamoto. A framework for development of the library for massively parallel processors with auto-tuning function and the single memory interface. *IPSJ SIG Notes*, (2001-HPC-87):25–30, 2001.
7. R. L. Ribler, H. Simitci, and D. A. Reed. The AutoPilot performance-directed adaptive control system. *Future Generation Computer Systems, special issue (Performance Data Mining)*, 18(1):175–187, 2001.
8. K. Takahiro, K. Kise, H. Honda, and T. Yuba. FIBER: A framework of installation, before execution-invocation, and run-time optimization layers for auto-tuning software. *IS Technical Report, Graduate School of Information Systems, The University of Electro-Communications*, UEC-IS-2003-3, May 2003.
9. C. Tapus, I.-H. Chung, and J. K. Hollingsworth. Active Harmony : Towards automated performance tuning. In *Proceedings of High Performance Networking and Computing (SC2002)*, Baltimore, USA, November 2003.
10. R. Whaley, A. Petitet, and J. J. Dongarra. Automated empirical optimizations of software and the ATLAS project. *Parallel Computing*, 27:3–35, 2001.

Evaluating Heuristic Scheduling Algorithms for High Performance Parallel Processing

Lars Lundberg, Magnus Broberg, and Kamilla Klonowska

Department of Software Engineering and Computer Science,
Blekinge Institute of Technology, S-372 25 Ronneby, Sweden,
Lars.Lundberg@bth.se

Abstract. Most cluster systems used in high performance computing do not allow process relocation at run-time. Finding an allocation that results in minimal completion time is NP-hard and (non-optimal) heuristic algorithms have to be used. One major drawback with heuristics is that we do not know if the result is close to optimal or not. Here, we present a method for finding an upper bound on the minimal completion time for a given program. The bound helps the user to determine when it is worth-while to continue the heuristic search for better allocations. Based on some parameters derived from the program, as well as some parameters describing the hardware platform, the method produces the minimal completion time bound. A practical demonstration of the method is presented using a tool that produces the bound.

1 Introduction

Parallel processing is often used in high performance computing. Finding an efficient allocation of processes to processors can be difficult. In most cases, it is impossible to (efficiently) move a process from one computer to another in a cluster or distributed system, and static allocation of processes to processors is thus unavoidable. Finding an allocation of processes to processors that results in minimal completion time is NP-hard [5], and heuristic algorithms have to be used. We do not know if the result from a heuristic algorithm is near or far from optimum, i.e. we do not know if it is worth-while to continue the heuristic search for better allocations.

Here we present a method for finding an upper bound on the optimal (i.e. minimal) completion time of a certain program. The method can be used as an indicator for the completion time that a good heuristic ought to obtain, making it possible to decide if it is worth-while to continue the heuristic search for better allocations. The performance bound is optimally tight given the available information about the parallel program and the target multiprocessor.

The result presented here is an extension of previous work [13][14]. The main difference compared to previous result is that we now take network communication time and program granularity into consideration. A practical demonstration of the method is included at the end of the paper.

A. Veidenbaum et al. (Eds.): ISHPC 2003, pp. 160–173, 2003.

2 Definitions and Main Result

A parallel program consists of n processes. The execution of a process is controlled by Wait(Event) and Activate(Event). Event couples a certain Activate to a certain Wait. When a process executes an Activate on an event, we say that the event has occurred. It takes some time for the event to travel from one processor to another. That time is the synchronization latency t. If a process executes a Wait on an event which has not yet occurred, that process becomes blocked until another process executes an Activate on the same event and the time t has elapsed. If both processes are on the same processor, it takes zero time for the event to travel. Each process is represented as a list of segments separated by a Wait or an Activate. These lists are independent of the way processes are scheduled. All processes are created at the start of the execution. Processes may be initially blocked by a Wait. The minimal completion time for a program P, using a system with k processors, a latency t and a specific allocation A of processes to processors, is $T(P, k, t, A)$. The minimal completion time for P, using k processors and latency t, is $T(P, k, t) = min_A T(P, k, t, A)$.

The upper part of Fig. 1 shows a program P with three processes (P1, P2, and P3). Processing for x time units is represented by Work(x). The lower part shows a graphical representation of P and two schedules resulting in minimum completion time for a system with three ($T(P, 3, t)$) and two processors ($T(P, 2, t)$). Each program has well defined start and end points, i.e., there is some code that is executed first and some other code that is executed last. The local schedule, i.e., the order in which processes allocated to the same processor are scheduled affects the completion time. We assume optimal local scheduling when calculating $T(P, k, t)$.

For each program P with n processes there is a parallel profile vector V of length n. Entry i in V contains the fraction of the completion time during which there are i active processes, using a schedule with one process per processor and no synchronization latency. The completion time for P, using a schedule with one process per processor and no synchronization latency is denoted $T(P, n, 0)$. $T(P, n, 0)$ is easy to calculate. In Fig. 1, $T(P, n, 0) = 3$. During time unit one there are two active processes (P1 and P3), during time unit two there are three active processes (P1, P2, and P3), and during time unit three there is one active process (P2), i.e. $V = (1/3, 1/3, 1/3)$. Different programs may yield the same parallel profile V.

By adding the work time of all processes in P, we obtain the total work time of the program. The number of synchronization signals in program P divided by the total work time for P is the granularity, z. In Fig. 1, $z = 5/6$.

Finding an allocation of processes to processors which results in minimal completion time is NP-hard. However, we will calculate a function $p(n, k, t, z, V)$ such that for any program P with n processes, granularity z, and a parallel profile vector V: $T(P, k, t) \leq p(n, k, t, z, V)T(P, n, 0)$. For all programs P with n processes, granularity z, and a parallel profile vector V: $p(n, k, t, z, V) = max_P(T(P, k, t)/T(P, n, 0))$.

```
process P1              process P2              process P3
begin                   begin                   begin
  Wait(Event_1);          Activate(Event_1);      Wait(Event_2);
  Work(1);                Activate(Event_2);      Work(2);
  Activate(Event_3);      Wait(Event_3);          Activate(Event_5);
  Work(1);                Work(2);              end P3;
  Activate(Event_4);      Wait(Event_4);
end P1;                   Wait(Event_5);
                        end P2;
```

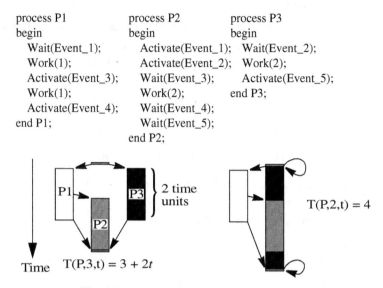

Fig. 1. Program P with synchronization signals.

3 Splitting a Program into a Thick and a Thin Part

We first multiply each time unit x, $x > 0$ of each process in program P by a factor y, $y > 1$, thus prolonging the segment with Δx and get program P'. P' is then transformed into P'' such that each time unit $x + \Delta x$ is prolonged with Δx. Obviously, $T(P'', n, 0) - T(P', n, 0) = T(P', n, 0) - T(P, n, 0) = \Delta x\, T(P,n,0)/x$.

The situation with $t > 0$ is more complicated, since we do not prolong t. Let ΔL denote the difference in length between P and P'. If we denote the length of program P with L, the length of P' will be $L' = L + \Delta L$. In the same way we create P'' with a difference between P' and P'' called $\Delta L'$. The length of P'' will then be the length of $L'' = (L + \Delta L) + \Delta L'$

For the moment we assume that there is one process per processor. We will relax this restriction later. The *critical path* is the longest path from the start to the end of the program following the synchronizations. In the case when two (or more) paths are the longest, the path with the minimum number of synchronizations is the critical path. Let $arr(P)$ be a number of synchronizations (arrows) in the critical path in P, and $arr(P')$ be the number of arrows in the critical path in P'.

When we prolong a program the critical path may change, i.e. when path two is longer than path one, and path two has less execution (and thus more synchronizations) than path one. During the prolongation, path one grows faster than path two.

Lemma 1: $arr(P) \geq arr(P')$.

Proof: Suppose that $arr(P) = x, (x \geq 0)$ and that there is another path in P that has more than x arrows. When we prolong the processes, the length of the path with more arrows (and thus less execution) increases slower than the critical path.

Theorem 1: $\Delta L \leq \Delta L'$.

Proof: Let $E_1 = y_1 + arr(E_1)t$ be the length of path one, where y_1 is the sum of the segments in path one, $arr(E_1)$ is the number of arrows with the communication cost t. Let $E_2 = y_2 + arr(E_2)t$ be the corresponding length for path two. Further, let $y_1 > y_2$ and path two be the critical path, i.e. $E_1 < E_2$. Then let

$$E_1' = y_1 \frac{x + \Delta x}{x} + arr(E_1)t \quad \text{and} \quad E_2' = y_2 \frac{x + \Delta x}{x} + arr(E_2)t . \quad \text{Let also}$$

$$E_1'' = y_1 \frac{x + 2\Delta x}{x} + arr(E_1)t \text{ and } E_2'' = y_2 \frac{x + 2\Delta x}{x} + arr(E_2)t . \text{ There are three}$$

possible alternatives:

- $E_1' < E_2'$ and $E_1'' < E_2''$. In this case $\Delta L = \Delta L'$.
- $E_1' < E_2'$ and $E_1'' \geq E_2''$. In this case $\Delta L < \Delta L'$.
- $E_1' \geq E_2'$ and $E_1'' > E_2''$. In this case $\Delta L < \Delta L'$.

Theorem 2: $2L' \leq L + L''$.

Proof: $2L' = L + \Delta L + L + \Delta L \leq L + \Delta L + L + \Delta L' = L + L + \Delta L + \Delta L' = L + L''$.

We now look at the case with more than one process per processor. Let P, P' and P'' be such programs; P, P' and P'' are identical except that each work-time in P'' is twice the corresponding work-time in P', and all work-times are zero in P. Consider an execution of P'' using allocation A. Let Q'' be a program where we have merged all processes executing on the same processor, using allocation A, into one process. Let Q' be the program which is identical to Q'' with the exception that each work-time is divided by two. Let Q be the program which is identical to Q'' and Q' except that all work-times are zero. From Theorem 2 we know that $2T(Q', k, t, A) \leq T(Q, k, t, A) + T(Q'', k, t, A)$. From the definition of Q'' we know that $T(P'', k, t, A) = T(Q'', k, t, A)$. Since the optimal local schedule may not be the same for P' and P'' we know that $T(P', k, t, A) \leq T(Q', k, t, A)$, since Q' by definition is created with optimal local scheduling of P''.

Consider a program R. The number of processes in R and P are equal, and R also has zero work-time. The number of synchronizations between processes i and j in R is twice the number of synchronizations between i and j in P. All synchronizations in R must be executed in sequence. It is possible to form a sequence, since there is an even number of synchronizations between any pair of processes.

Sequential execution of synchronizations clearly represents the worst case, and local scheduling does not affect the completion time of a sequential program, i.e. $2T(P, k, t, A) \le T(R, k, t, A)$ and $2T(Q, k, t, A) \le T(R, k, t, A)$.   Consequently, $4T(P', k, t, A) \le 4T(Q', k, t, A) \le 2T(Q'', k, t, A) + 2T(Q, k, t, A) \le 2T(P'', k, t, A) + T(R, k, t, A)$.

We now take an arbitrary program P'. First we create m copies of P', where $m = 2^x$ ($x \ge 2$). We then combine the m copies in groups of four and transform the four copies of P' to two programs P'' and one program R. From the discussion above we know that $4T(P', k, t, A) \le 2T(P'', k, t, A) + T(R, k, t, A)$ for any allocation A, i.e. $4T(P', k, t) \le 2T(P'', k, t) + T(R, k, t)$.

$4T(P', n, 0) = 2T(P'', n, 0) + T(R, n, 0)$, and n, V and z are invariant in this transformation. We now end up with 2^{x-1} programs P''. We combine these P'' programs in groups of four and use the same technique and end up with 2^{x-2} programs P''' (with twice the execution time compared to P'') and one program R. We repeat this technique until there are only two very thick programs. By selecting a large enough m, we can neglect the synchronization times in these two copies. This is illustrated in Fig. 2, where $m = 4$. Note that the execution time using k processors may increase due to this transformation, whereas $T(P, n, 0)$ is unaffected.

Thus, we are able to transform a program into a thin part consisting only of synchronizations, and a thick part consisting of all the execution time.

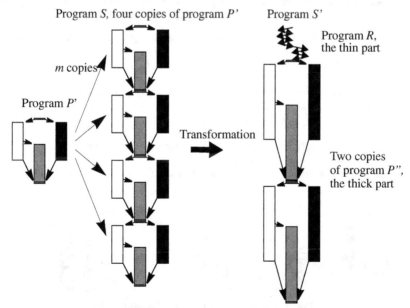

Fig. 2. The transformation of m copies of the program P into a thick and a thin part.

4 The Thick Part

The thick part has the nice property that we can assume $t = 0$, since selecting a large enough m in Section 3 will cause the synchronization time in the thick part to be negligible since virtually all synchronizations will be in the thin part. We are thus also free to introduce a constant number of new synchronizations. More elaborate proofs concerning the transformations in this section can be found in [9] and [10].

We consider $T(P, n, 0)$. This execution is partitioned into m equally sized time slots such that process synchronizations always occur at the end of a time slot. In order to obtain Q we add new synchronizations at the end of each time slot such that no processing done in slot r, $(1 < r \leq m)$, can be done unless all processing in slot $r - 1$, has been completed. The net effect of this is that a barrier synchronization is introduced at the end of each time slot. Fig. 3 shows how a program P is transformed into a new program Q by this technique.

We obtain the same parallel profile vector for P and Q. The synchronizations in Q form a superset of the ones in P, i.e. $T(P, k, 0) \leq T(Q, k, 0)$ (in the thick part we assume that $t = 0$). Consequently, $T(P, k, 0)/T(P, n, 0) \leq T(Q, k, 0)/T(Q, n, 0)$.

We introduce an equivalent representation of Q (see Fig. 3), where each process is represented as a binary vector of length m, where m is the number of time slots in Q, i.e. a parallel program is represented as n binary vectors. We assume unit time slot length, i.e. $T(Q, n, 0) = m$. However, $T(Q, k, 0) \geq m$, because if the number of active processes exceeds the number of processors on some processor during some time slot, the execution of that time slot will take more than one time unit.

With P_{nv} we denote the set of all programs with n processes and a parallel profile V. From the definition of V, we know that if $Q \in P_{nv}$, then $T(P, n, 0)$ must be a multiple of a certain minimal value m_v, i.e. $V = (x_1/m_v, x_2/m_v, ..., x_n/m_v)$ where $T(P, 1, 0) = xm_v$, for some positive integer x. Programs in P_{nv} for which $T(Q, n, 0) = m_v$ are referred to as minimal programs. For instance, the program in Fig. 1 is a minimal program for the parallel profile vector $V = (1/3, 1/3, 1/3)$.

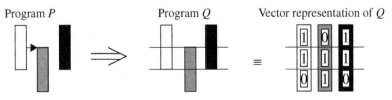

Program P Program Q Vector representation of Q

Fig. 3. The transformation of program P.

We create $n!$ copies of Q. The vectors in each copy are reordered such that each copy corresponds to one of the $n!$ possible permutations. Vector v $(1 \leq v \leq n)$ in copy c $(1 \leq c \leq n!)$ is concatenated with vector v in copy $c + 1$, forming a new program Q' with n vectors of length $n!m$. The execution time from slot $1 + (c - 1)m$ to cm cannot be less than $T(Q, k, 0)$, using k processors, i.e. $T(Q', k, 0)$ cannot be less than $n!T(Q, k, 0)$. We thus know that $T(Q, k, 0)/T(Q, n, 0) = T(Q, k, 0)/m = n!T(Q, k, 0)/(n!m) \leq T(Q', k, 0)/T(Q', n, 0)$. We obtain the same V for Q' and Q.

The n vectors in Q' can be considered as columns in a $n!m \times n$ matrix. The rows in Q' can be reordered into n groups, where all rows in the same group contain the same number of ones (some groups may be empty).

Due to the definition of minimal programs, we know that all minimal programs Q result in the same program Q_m', i.e. all programs Q' in P_{nv} can be mapped onto the same Q_m'. We have now found the worst possible program Q_m' $(Q_m' \in P_{nv})$.

We will now show how to calculate the thick part for any allocation of processes to processors. The allocation is specified using an ordered set. The number of elements in an ordered set A is a, i.e. $|A| = a$. The $i{:}th$ element of the ordered set is denoted a_i, thus $A = (a_1, ..., a_a)$.

The function $f(A, q, l)$ calculates the execution time of a program with n processes and $\binom{n}{q}$ rows when q processes are executing simultaneously using allocation A, compared to executing the program using one processor per process. The function f is recursively defined and divides the allocation A into smaller sets, where each set is handled during one (recursive) call of the function.

$f(A, q, l)$ uses q for the number of ones in a row. The l is the maximum number of ones in any previous set in the recursion. Note that we have the parameters n and k implicit in the vector, $n = \sum_{i=1}^{a} a_i$ and $k = a$. From the start $l = 0$. We have $w = a_1$, $d = |\{a_x, ...\}|$ $(1 \leq x \leq a$ and $a_x = a_1)$ and $b = n - wd$.

If $b = 0$ then $f(A, q, l) = \sum_{l_1 = max(0, \lceil q/d \rceil)}^{min(q, w)} \pi(d, w, q, l_1)max(l_1, l)$, otherwise: $f(A, q, l) =$

$$\sum_{l_1 = max(0, \lceil \frac{q-b}{d} \rceil)}^{min(q, w)} \sum_{i = max(l_1, q-b)}^{min(l_1 d, q)} \pi(d, w, i, l_1)f(A - \{a_1, ..., a_d\}, q - i, max(l_1, l)).$$

$\pi(k, w, q, l)$ [9] [10] is a help function to f and denotes the number of permutations of q ones in kw slots, which are divided into k sets with w slots in each, such that the set with maximum number of ones has exactly l ones.

The binary matrix can be divided into n parts such that the number of ones in each row is the same for each part. The relative proportion between the different

parts is determined by V; $s(A, V) = \sum_{q = 1}^{n} v_q f(A, q, 0)/\binom{n}{q}$.

5   The Thin Part

From Section 3 we know that the thin part of a program for which the ratio $T(P, k, t)/T(P, n, 0)$ is maximized consists of a sequence of synchronizations, i.e. program R in Section 3. From Section 4 we know that the identity of the processes allocated to a certain processor does not affect the execution time of the thick part. In the thin part we would like to allocate processes that communicate frequently to the same processor.

Let y_i be a vector of length i-1. Entry j in this vector indicates the number of synchronizations between processes i and j. Consider an allocation A such that the number of processes allocated to processor x is a_x. The optimal way of binding processes to processors under these condition is a binding that maximizes the number of synchronizations within the same processor.

Consider also a copy of the thin part of the program where we have swapped the communication frequency such that process j now has the same communication frequency as process i had previously and process i has the same communication frequency as process j had previously. It is clear that the minimum execution time of the copy is the same as the minimum execution time of the original version. The mapping of processes to processors that achieves this execution time may however, not be the same. For instance, if we have two processors and an allocation $A = (2,1)$, we obtain minimum completion time for the original version when P1 and P3 are allocated to the same processor, whereas the minimum for the copy is obtained when P1 and P2 share the same processor.

If we concatenate the original (which we call P) and the copy (which we call Q) we get a new program P' such that $T(P', k, t, A) \geq T(P, k, t, A) + T(Q, k, t, A)$ for any allocation A (all thin programs take zero execution time using a system with one process per processor and no communication delay). By generalizing this argument we obtain the kind of permutation as we had for the thick part.

These transformations show that the worst case for the thin part occurs when all synchronization signals are sent from all n processes $n - 1$ times - each time to a different process. All possible synchronization signals for n processes equals $n(n - 1)$ and all possible synchronization signals between the processes allocated to processor i equals $a_i(a_i - 1)$. Because some processes are executed on the same processor, the communication cost for them equals zero. That means that the number of synchronization signals in the worst-case of program (regarding communication cost) is equal to $n(n-1) - \sum_{i=1}^{k} a_i(a_i - 1)$; the parameters n and k implicit in the vectors,

$n = v = \sum_{i=1}^{a} a_i$ and $k = a$.

If $\quad\quad\quad n = 1 \quad\quad\quad$ then $\quad\quad\quad r(A, t, V) = 0, \quad\quad\quad$ otherwise

$$r(A, t, V) = \frac{n(n-1) - \sum_{i=1}^{k} a_i(a_i - 1)}{n(n-1)} \cdot t \sum_{q=1}^{n} (qv_q).$$

6   Combining the Thick and Thin Parts

Combining the thick and thin parts is done by adding $s(A, V)$ and $r(A, t, V)$, weighted by the granularity z. Thus, we end up with a function $p(n, k, t, z, V) = min_A(s(A, V) + zr(A, t, V))$. N.B. A implicitly includes the parameters $n = \sum_{i=1}^{a} a_i$ and $k = a = |A|$. We use the granularity z as a weight between the thick and thin parts. In programs with high granularity (large z) the thin part has larger impact than in programs with low granularity (small z).

As previously shown the thick section is optimal when evenly distributed over the processors [9][10], but the thin section is optimal when all processes reside on the same processor. The algorithm for finding the minimum allocation A is based on the knowledge about the optimal allocation for the thick and thin part respectively.

The basic idea is to create *classes* of allocations. An allocation class consists of three parts:

- A common allocation for the thick and thin parts, i.e. the number of processes on the first processors. The allocation must be decreasing, i.e. the number of processes on processor i must be greater than or equal to the number of processes on processor $i + 1$.
- The allocations for the remaining processors for the thick part, the allocations are evenly distributed. The largest number of processes on a processor must be equal to or less than the smallest number of assigned processes for any processor in the common assignment.
- The allocations for the remaining processors for the thin part, the allocations make use of as few of the remaining processors as possible and with as many processes as possible on each processor. The largest number of processes on a processor must be equal to or less than the smallest number of assigned processes for any processor in the common assignment.

For finding an optimal allocation we use a classical branch-and-bound algorithm [1]. The allocations can be further divided, by choosing the class that gave the minimum value and add one processor in the common allocation. All the subclasses will have a higher (or equal) value than the previous class. The classes are organized as a tree structure and the minimum is now calculated over the *leaves* in the tree and a value closer to the optimal is given. By repeatedly selecting the leaf with minimum value and create its subclasses we will reach a situation where the minimum leaf no longer has any subclasses, then we have found the optimal allocation.

7 Method Demonstration

Previously, we have developed a tool for monitoring multithreaded programs on a uni-processor workstation and then, by simulation, predict the execution of the program on a multiprocessor with any number of processors [2][3]. We have modified this tool to extract the program parameters n, V and z.

We demonstrate the described technique by bounding the speed-up of a parallel C-program for generating all prime numbers smaller than or equal to X. The algorithm generates the primes numbers in increasing order starting with prime number 2 and ending with the largest prime number smaller than or equal to X. A number Y is a prime number if it is not divisible by any prime number smaller than Y. Such divisibility tests are carried out by filter threads; each filter thread has a prime number assigned to it. In the front of the chain there is a number generator feeding the chain with numbers. When a filter receives a number it will test it for divisibility with the filter's prime number. If the number is divisible it will be thrown away. If the number is not divisible it will be sent to the next filter. In the case that there is no next filter, a new filter is created with the number as the new filter's prime number.

The number generator stops by generating 397, which is the 78th prime number. This means the program contains 79 threads (including the number generator). The program's speed-up has been evaluated using three networks; *fast* network with $t = 0$, an *ordinary* network with a synchronization time of four microseconds, and a *slow* network with eight microseconds latency. The value of the ordinary network has been defined by the remote memory latency found in three distributed shared memory machines: SGI's Origin 2000 [8], Sequent's NUMA-Q [12], and Sun's WildFire [6]. The remote memory latency varies between 1.3 to 2.5 micro seconds and a synchronization must at least include two memory accesses. The slow network is simply twice the ordinary network. The results are compared to the predictions of our tool when applying the corresponding cost for synchronization. The allocation algorithm used in the simulated case is simple. The first n/k processes (process one is the number generator, process two is the first filter, and so on) are allocated to the first processor, the next n/k processes to the second processor and so on. In the case when n is not divisible with k, the first n modulus k processors are allocated $\lceil n/k \rceil$ processes and the remaining processors are allocated $\lfloor n/k \rfloor$ processors. This is a common way to allocate chain-structured programs [4].

As can be seen in Fig. 4 (zero latency) the simple allocation scheme performs worse than function p, and we can thus conclude that we for sure are able to find a better allocation. In Fig. 5 (latency is four micro seconds) the simple allocation scheme performs worse than function p for less than eight processors, and we can thus conclude that we for sure are able to find a better allocation when we have less than eight processors. In the case with eight processors or more we can not conclude if there is a better allocation or not. However, it is important to know that the fact that we are above the bound does not guarantee that the simple allocation is the optimal. In Fig. 6 ($t = 8\mu s$) the simple allocation scheme performs worse than function p for four (or less) processors, and we can thus conclude that we for sure are able to find a better allocation when we have less than five processors. In the case with five processors or more we can not conclude if there is a better allocation or not.

As indicated earlier, the function p gives a bound. The intended usage of this bound is to indicate that there is a better allocation strategy if the current allocation strategy gives results below the bound, as in the case in Fig. 4 (the prime program with no latency). We now look at other allocation strategies and tries a round robin strategy that assigns (on a multiprocessor with k processors) thread number i to processor (i mod k)+1. In Fig. 7 we see that the round robin strategy gives better result than the bound. However, a better result than the bound does not automatically mean that we have found an optimal result.

In order to illustrate the benefits of our branch-and-bound algorithm we consider the prime number program on 16 processors. In total we can find 6,158,681 unique allocations of the 79 processes on 16 processors. The number of allocations that needs to be evaluated using our approach is very small, roughly only one allocation out of 80,000 needs to be evaluated. The time to calculate the optimal allocation for the prime number program ($k = 16, t = 8\mu s$) is five minutes on a 300MHz Ultra 10. Without the allocation classes and the branch-and-bound technique the same calculation would require two and a half year.

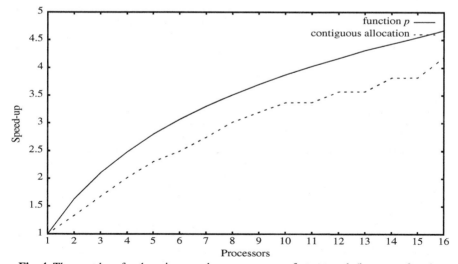

Fig. 4. The speed-up for the prime number program on a fast network (latency $= 0$ μs).

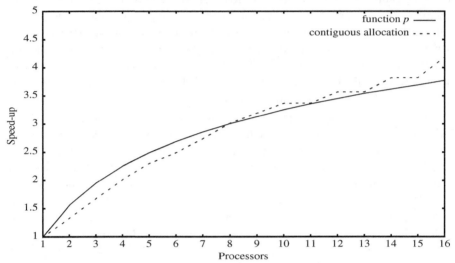

Fig. 5. The speed-up for the prime number program on an ordinary network (latency = 4 μ s).

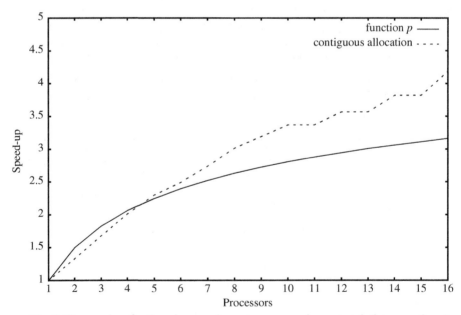

Fig. 6. The speed-up for the prime number program on a slow network (latency = 8 μ s).

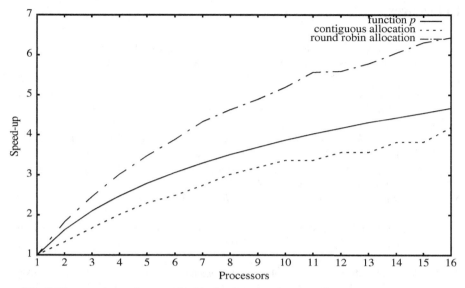

Fig. 7. The speed-up using round robin for the prime number program (latency = 0 μ s).

8 Conclusion

Clusters do not usually allow run-time process migration, and it is essential to find an efficient allocation of processes to processors. Finding the optimal allocation is NP-hard and we have to resort to heuristic algorithms. One problem with heuristic algorithms is that we do not know if the result is close to optimal or if it is worthwhile to continue the search for better allocations.

In this paper we show how to analytically calculate a bound on the minimal completion time for a given parallel program. The program is specified by the parameters n (number of processes), V (parallel profile vector), and z (synchronization frequency); different programs may yield the same n, V, and z. The hardware is specified by the parameters k (number of processors) and t (synchronization latency). The bound on the minimal completion time is optimally tight. This means that there are programs for which the minimal completion time is equal to the bound. The parameters z and t give a more realistic model compared to previous work [13][14]. The bound makes it possible to determine when it is worth-while to continue the heuristic search. Our method includes a branch-and-bound algorithm that has been shown to reduce the search space to only 0.0004% for finding the optimal bound.

The method has been implemented in a practical tool. The tool is able to automatically obtain the values of n, V, and z for a program and calculate the minimal completion bound with the given hardware parameters. Based on this tool we have demonstrated the usability of the method. One feature of the tool is that it can be totally operated on a single-processor workstation.

References

1. J. Blazewicz, K. Ecker, E. Pesch, G. Schmidt, and J. Weglarz, *Scheduling Computer and Manufacturing Processes*, Springer-Verlag, ISBN 3-540-61496-6, 1996.
2. M. Broberg, L. Lundberg, and H. Grahn, *Visualization and Performance Prediction of Multithreaded Solaris Programs by Tracing Kernel Threads*, in Proc. of the 13th International Parallel Processing Symposium, pp. 407-413, 1999.
3. M. Broberg, L. Lundberg, and H. Grahn, *VPPB - A Visualization and Performance Prediction Tool for Multithreaded Solaris Programs*, in Proceedings of the 12th International Parallel Processing Symposium, pp. 770-776, 1998.
4. S. H. Bokhari, *Partitioning Problems in Parallel, Pipelined, and Distributed Computing*, IEEE Transactions on Computers, Vol. 37, No. 1, 1988.
5. M. Garey and D. Johnson, *Computers and Intractability*, W. H. Freeman and Company, 1979.
6. E. Hagersten, and M. Koster, *WildFire: A Scalable Path for SMPs*, in Proc. of the Fifth International Symposium on High Performance Computer Architecture, pp. 172-181, 1999.
7. S. Kleiman, D. Shah, and B. Smaalders, *Programming with threads*, Prentice Hall, 1996.
8. J. Laudon and D. Lenoski, *The SGI Origin: A ccNUMA Highly Scalable Server*, in Proc. of the 24th Annual International Symposium on Computer Architecture (ISCA), pp. 241-251, 1997.
9. H. Lennerstad and L. Lundberg, *An optimal Execution Time Estimate of Static versus Dynamic Allocation in Multiprocessor Systems*, SIAM Journal of Computing, Vol. 24, No. 4, pp. 751-764, 1995.
10. H. Lennerstad and L. Lundberg, *Optimal Combinatorial Functions Comparing Multiprocess Allocation Performance in Multiprocessor Systems*, SIAM Journal of Computing, Vol. 29, No. 6, pp. 1816-1838, 2000.
11. H. Lennerstad and L. Lundberg, *Optimal Worst Case Formulas Comparing Cache Memory Associativity*, SIAM Journal of Computing, Vol. 30, No. 3, pp. 872-905, 2000.
12. T. Lovett and R. Clapp, *STiNG: A cc-NUMA Computer System for the Commercial Marketplace*, in Proc. of the 23rd Annual International Symposium on Computer Architecture (ISCA), pp. 308-317, 1996.
13. L. Lundberg, H. Lennerstad, *Using Recorded Values for Bounding the Minimum Completion Time in Multiprocessors*, IEEE Transactions on Parallel and Distributed Systems, Vol. 9, No. 4, pp. 346-358, 1998.
14. L. Lundberg, H. Lennerstad, *An Optimal Upper Bound on the Minimal Completion Time in Distributed Supercomputing*, in Proc. of the 1994 ACM International Conference on Supercomputing, pp. 196-203, 1994.

Pursuing Laziness for Efficient Implementation of Modern Multithreaded Languages*

Seiji Umatani, Masahiro Yasugi, Tsuneyasu Komiya, and Taiichi Yuasa

Department of Communications and Computer Engineering,
Graduate School of Informatics,
Kyoto University,
Sakyo-ku Kyoto 606-8501, Japan,
{umatani,yasugi,komiya,yuasa}@kuis.kyoto-u.ac.jp

Abstract. Modern multithreaded languages are expected to support advanced features such as thread identification for Java-style locks and dynamically-scoped synchronization coupled with exception handling. However, supporting these features has been considered to degrade the effectiveness of existing efficient implementation techniques for fine-grained fork/join multithreaded languages, e.g., lazy task creation. This paper proposes efficient implementation techniques for an extended Java language OPA with the above advanced features. Our portable implementation in C achieves good performance by pursuing 'laziness' not only for task creation but also stealable continuation creation, thread ID allocation, and synchronizer creation.

1 Introduction

There are many multithreaded programming languages for parallel processing. These languages have simple constructs to dynamically create threads and allow them to cooperate with each other. We are designing and implementing a programming language OPA (Object-oriented language for PArallel processing)[1–4], which is an extended Java language with simple but powerful multithread constructs. OPA provides fork/join multithread constructs for structuring divide-and-conquer style parallel processing. Furthermore, OPA employs the object-oriented paradigm to express other types of synchronization, e.g., communication (synchronization) through shared objects and mutually exclusive method invocation, and it supports exception handling extended for multithread constructs. The details of OPA will be explained in Sect. 2.

Multithread constructs and their implementation should satisfy the following properties. First, they must be efficient. In other words, we must reduce the overhead of thread creation and synchronization. We would like to write fine-grained multithreaded programs using them. So, if they incur unacceptable overhead, most programmers would not use them and, instead, would create coarse-grained threads and manage fine-grained work objects explicitly. For this

---

* This research is supported in part by the 21st century COE program.

A. Veidenbaum et al. (Eds.): ISHPC 2003, LNCS 2858, pp. 174–188, 2003.

reason, we propose efficient implementation (compilation) techniques for OPA. Next, the language system should be portable. To achieve this, the OPA compiler generates standard C code. In contrast, some other systems exploit assembly-level techniques for efficiency. Finally, they must provide sufficient expressiveness. Some other languages achieve efficiency by limiting their expressiveness. For example, the fork/join constructs of Cilk[5, 6] employ 'lexical-scope', and Cilk does not provide other types of synchronization. Cilk's constructs are simple but not flexible enough to write various irregular programs.

In this paper, we propose three techniques which reduce fork/join style constructs' overhead. Their common key concept is *laziness*. Laziness means that we delay certain operations until their results become truly necessary. First, allocating activation frames in heap can be lazily performed. In the case of implementing fine-grained multithreaded languages in C, since each processor has only a single (or limited number of) stack(s) for a number of threads, frames for multiple threads are generally allocated in heap. For the purpose of efficiency, we allocate a frame at first in stack, and it remains there until it prevents other thread's execution. Second, a *task* can be lazily created. In this paper, a task is (a data structure of) a schedulable active entity that does not correspond to a single (language-level) thread. A blocked thread becomes a task to release the processor and the stack it uses. Also, for the purpose of load balancing, each thread may become a task to move around the processors. Our approach decreases the number of actually created tasks while it keeps good load balancing. Furthermore, we delay some other operations related to thread creation and, as a result, we can make the cost of thread creation close to zero. Third, data structures for synchronization can be lazily made. In OPA, such a data structure is necessary because join synchronizers are dynamically-scoped[3] and each thread may be blocked.

In this paper, we compare our efficient implementation of OPA with that of the Cilk language, which is a parallel extension of the C language. Cilk has fork/join style constructs, and its system achieves good dynamic load balancing on shared-memory multiprocessors. Since Cilk does not provide other types of synchronization such as communication (synchronization) through shared objects and mutually exclusive method execution as in OPA, the implementation techniques of Cilk cannot directly be applied for OPA. However, by pursuing laziness, our OPA implementation obtains better performance than the Cilk implementation even with the richer expressiveness of OPA.

The rest of this paper is organized as follows. Section 2 overviews the OPA language. Section 3 describes our previous implementation of OPA. In Sect. 4, we propose lazy normalization schemes for the implementation of OPA. By 'lazy normalization', we mean that we create a normal and heavy version of an entity (e.g., a heap-allocated frame) from a temporary and lightweight version of the entity (e.g., a stack-allocated frame) only when the normal one is truly necessary. Section 5 presents related work and discusses expressiveness, efficiency, and portability. Section 6 shows benchmark results of some programs written in OPA and Cilk. Finally, in Sect. 7, we present concluding remarks.

2   Overview of the OPA Language

As mentioned earlier, OPA has fork/join constructs to express structured divide-and-conquer parallel algorithms in a straightforward manner. These constructs are very popular and also adopted in many parallel languages[6–8]. (their detailed syntax and semantics are rich in variety, and will be discussed later in Sect. 5.)

We introduce join statements, syntactically denoted by "join *statement*," to join a number of threads; by "join *statement*," the *statement* are executed with the current thread and new threads created during the execution of the *statement* are joined with the completion of the join statement. Par statements, syntactically denoted by "par *statement*," are used to fork new threads; by "par *statement*," a thread is forked to execute the *statement*. For instance, the statement:

```
join {par o.m(); par o2.m2(); o3.m3(); };
```

forks two threads to execute o.m() and o2.m2() respectively and executes o3.m3() with the current thread then waits for the completion of the forked threads (see Fig. 1). Furthermore, if a forked thread forks another new thread, the new thread will also be joined. For example, if the definition of m, which is executed by the first forked thread, is as follows:

```
m(){par o4.m4(); o5.m5(); };
```

the thread to execute o4.m4() is also joined at the end of the join statement.

When a thread is forked, its join-destination is automatically determined as the inner most join statement within which the thread is forked. In summary, OPA has fork/join constructs where join targets (i.e., synchronizers) have dynamic scope. If you are interested in more details, see [3].

Figure 2 shows a simple Fibonacci program written in OPA. par fib(n-1) at line 5 means that its method invocation is executed concurrently by a newly created thread. Note that fib is a usual method; that is, we can also call it sequentially (line 6). This is desirable because a parent thread can participate in the computation and the number of thread creation can be reduced. In this program, the parent uses a join label, the variant of join synchronizers, for

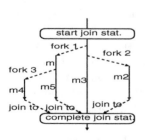

Fig. 1. Fork/join parallelism.

```
1 class Fib{
2   private static int fib(int n){
3     if(n < 2) return n;
4     else{
5       int x = par fib(n-1);
6       int y = fib(n-2);
7     join:
8       return x+y;
9     }
10  }
11  public static void main(String[] args){
12    int r = fib(36);
13  }
14 }
```

Fig. 2. fib code in OPA language.

safely using the value returned by its child. This `join` label is required before the statement `return x+y` to avoid the anomaly that would occur if x is summed to y before it is computed.

Fork/join constructs are simple and powerful enough to express divide-and-conquer programs. However, we may often want to describe more complex and irregular programs which enable threads to access an object exclusively or to notify another thread that a certain event has occurred. Although we do not describe details so much, OPA provides these features in an object-oriented style[2]. (Similar functions appear in COOL[9], but its implementation is fairly straightforward and incurs considerable overhead.) At least, what we want to keep in mind is that threads in an irregular program using these features may be blocked in general situations and this fact makes the language system more complicated than other languages' systems which only provide fork/join constructs.

3   Implementation Schemes for OPA

We implemented an OPA system on shared-memory parallel computers[4]. The OPA system consists of a compiler from OPA to C and a runtime system in C.

In this section, we describe our previous OPA implementation. In particular, there are three issues related to the paper: (1) activation frame allocation, (2) thread creation and scheduling, and (3) nested fork/join management. In the next section, we will show lazy normalization schemes to improve them.

3.1   Activation Frame Allocation

Each processor has only one (or a limited number of) C execution stack(s) (we call it simply 'stack' unless specified) to manage a number of threads. Thus, naïve implementations tend to allocate activation frames for threads in heap, resulting in lower performance than stack allocation. Many research projects have addressed this problem[6, 7, 10–12].

Hybrid Execution Model[12] and the OPA system adopt almost the same technique for frame allocation. They use two kinds of frames (i.e., stack-allocated frames and heap-allocated ones) and two versions of C code (i.e., fast/slow versions). Each processor provisionally executes threads with its stack using a sequential call of the fast version C function. Since a method can invoke a method sequentially or concurrently in OPA, a thread consists of several activation frames. When a thread that is running in stack is blocked for some reason (e.g., for mutual exclusion), for each stack frame that is a part of the thread an activation frame is allocated in heap and its state (*continuation*) is saved in it. Because heap frames must keep the caller-callee relations (implicit on stack), a callee's frame should point to the caller frame, and so one thread is represented as a list of frames in heap. To restart a thread, slow version C code restores the state from heap frames.

Fast version C code for `fib` generated by the OPA compiler is shown in Fig. 3. Lines 18–24 corresponds to a sequential method invocation. A method invocation

```
1  int f__fib(private_env *pr, int n) {
2    f_frame *callee_fr; thread_t *nt;
3    f_frame *nfr; int x, y; f_frame *x_pms;
4    if(n < 2) return n;
5    else{
6      /* enter a join block */
7      frame *njf = ALLOC_JF(pr);
8      njf->bjf = pr->jf; njf->bjw = pr->jw;
9      pr->jf = njf; pr->jw = njf->w;
10     /* create a new thread */
11     nt = ALLOC_OBJ(pr, sizeof(thread_t));
12     nt->jf = pr->jf; nt->jw = SPLIT_JW(pr, pr->jw);
13     nt->cont = nfr = MAKE_CONT(pr, c__fib);
14     // save continuation (including n-1)
15     x_pms = MAKE_CONT(pr, join_to); nfr->caller_fr = x_pms;
16     enqueue(pr, nt);
17     /* sequential call */
18     y = f__fib(pr, n-2);
19     if((y==SUSPEND) && (callee_fr=pr->callee_fr)){
20       f_frame *fr = MAKE_CONT(pr, c__fib);
21       // save continuation
22       callee_fr->caller_fr = fr;
23       pr->callee_fr = fr; return SUSPEND;
24     }
25     WAIT_FOR_ZERO(pr->jf);
26     pr->jf = pr->jf->bjf; pr->jw = pr->jf->bjw;
27     x = x_pms->ret.i;
28     return x+y;
29   }
30 }
```

Fig. 3. Compiled C code for `fib` (pseudo).

in OPA is compiled to a C function call and additional code fragments. (To avoid confusion, we refer to invocations in OPA as method invocations and C calls as function calls.) After returning to the caller of a function call, it is checked if the callee has been suspended (line 19). In such a case, since the caller belongs to the same thread as the callee, the caller also saves its own continuation (lines 20–21), links the callee to itself (line 22), and informs its own caller that it has been suspended (line 23).

A callee must return a pointer to its own heap frame (if exists, othewise 0) besides the method's return value. In Hybrid Execution Model, it is returned as a return value of a C function call and the method's return value is stored in another place allocated by the caller. To reduce the overhead, the OPA system returns the method's return value as a return value of the C function call, and the special value SUSPEND (that is selected from rarely used valued, e.g., -5) indicates the suspension of the callee. In such a case, the system checks further if a pointer to the callee frame is stored in a fixed place (pr->caller_fr).

3.2   Thread Creation and Scheduling

The OPA runtime manages each thread with a meta *thread object* (that is invisible to programmers). A thread object includes the following fields: (a) continuation and (b) join block information. The continuation field stores a pointer to the corresponding list of heap-allocated frames if there exists. The join block information fields are described later. In addition, the OPA system uses thread objects for thread identification; for example, it enables Java's **synchronized** methods which allows a thread holding a lock to acquire the same lock more than one times.

The previous OPA implementation processes thread creation in the following manner (Fig. 3 lines 11–16):

1. creates a thread object (line 11),
2. performs operations related to join frame information (line 12), (described later)
3. saves a continuation for starting from the beginning of the forked method's body (lines 13–14).
4. appends the frame join_to to process join synchronization after completion of the new thread (line 15). The frame is also used as a placeholder. It has a field defined as an union type so that it can hold any type of value (e.g., ret.i for int),
5. enqueues the thread object to the processor's local task queue (line 16).

A task queue is a doubly-ended queue (deque) and thread objects are enqueued at its tail. When a processor's stack becomes empty, it takes a task from the local task queue's tail and executes it. The reason why a task queue is realized as a deque is that other idle processors may steal a task that is expected to be largest from its head, thus enabling dynamic load balancing. Note that every thread has been converted to a task (i.e., a representation that can be passed around processors) before enqueued.

3.3 Nested Fork/Join Management

Finally, we explain management of nested fork/join constructs. When exiting from a join block, of course, the parent thread needs to wait for the completion of all the child threads created within the join block. The number of child threads is, in general, known only at run-time. Thus, the OPA system prepares a counter for each join block which holds the (maximum) number of child threads that has not yet joined to the join synchronizer.

OPA synchronizers are implemented as *join frames*. Join frames are used with thread objects, as shown in Fig. 4. Each join frame keeps track of the maximum

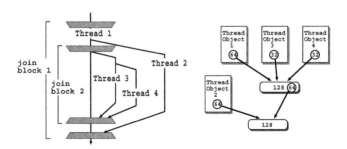

Fig. 4. Structure of join frames with thread objects.

number of threads those have not yet joined in its counter, and counter value 0 means that the entire synchronization has completed. That is, we adopt weighted reference counting[13]. Each thread keeps a weighted reference (i.e., a pointer and its weight) to a join frame in the thread object. The benefit of this scheme is that creating a new child thread only needs to split the parent weight between the parent and the child, and no more operation and synchronization. Each join frame has a pointer to the one-level outer join frame to keep nested structures of join blocks and it also provides a place for suspending a parent thread.

In Fig. 3, a parent enters a join block (lines 7–9), splits a weight for newly created thread (line 12), and exits from the join block (lines 25–26). At line 27, it gets a return value from the placeholder after synchronization has completed.

In summary, our previous implementation schemes of OPA incur unacceptable overhead for fine-grained multithreaded programs.

4   Lazy Normalization Schemes

In the implementation schemes described in the previous section, some operations can be done lazily by changing thread scheduling strategy. In Sect. 4.1, we propose a scheme to perform thread object creation lazily, and in Sect. 4.2 to perform join frame allocation lazily.

4.1   Lazy Task Creation

As described previously, the conventional OPA system creates a task, i.e., a full-fledged thread object, at the time of thread creation. In other words, every thread creation is realized as 'eager' task creation.

Lazy Task Creation (LTC)[14] is used for an efficient Multilisp[15] implementation originally. In brief, the basic idea is as follows:

1. At thread creation, the processor starts a child thread's execution prior to the parent. The parent's continuation is (automatically) saved in stack.
2. An idle processor steals a task by extracting (a continuation of) a thread from the bottom of the processor's stack and making a task from it.

We call the processor that steals a task *thief*, and the one that is stolen *victim*. The advantages of LTC are (a) saving the parent's continuation takes no more cost than a sequential function call. (b) It enables dynamic load balancing by allowing a thief to extract a thread from the victim's stack.

Now, consider that we may realize these two processes in C. First, saving a parent's continuation is accomplished by a normal C function call, so we need no more consideration. Second, we encounter a problem that we cannot extract a continuation from a C stack in a portable manner. In contrast, the Multilisp system manipulates the stack at assembly level. For example, the Cilk implementation, realized at C level, saves a parent's continuation into its heap-allocated frame in advance. In this way, a thief can take it without manipulating victim's stack although it incurs a certain amount of overhead on every thread creation.

In order to solve this problem, we adopt a message-passing implementation of LTC[16]. Its feature is that a thief does not access another processor's local data including its stack. The thief simply sends a request message for a task to a randomly selected victim and waits for the response. A victim, when it notices the request, extracts a continuation from its own stack, makes a task from it, and sends it back to the thief. To achieve this, the OPA system can exploit the lazy heap frame allocation mechanism (Fig. 6). In order to extract a continuation from the bottom of a C stack, the system temporarily (internally) suspends all continuations (threads) above it.

We describe the details with Fig. 5. It is fast version C code for `fib` with laziness and a thread is created in lines 13–26. A function call at line 14 corresponds to a child thread's execution. A suspension check at line 15 is the same as the conventional one. That is, while extracting the bottom continuation, all functions on stack act as if all threads were blocked. The difference appears at line 17, it checks a task request. Since the non-zero value (the thief processor ID) means that this processor is requested to extract the bottom continuation,

```
 1 int f__fib(private_env *pr, int n) {
 2   f_frame *callee_fr; thread_t **ltq_ptr = pr->ltq_tail;
 3   int x, y; f_frame *x_pms = NULL;
 4   if(n < 2) return n;
 5   else{
 6     /* enter a join block */
 7     pr->js_top++;
 8     /* polling here */
 9     if(pr->thief_req){
10       // suspension code here
11     }
12     /* create a new thread */
13     pr->ltq_tail = ltq_ptr+1;
14     x = f__fib(pr, n-1);
15     if((x==SUSPEND) && (callee_fr=pr->callee_fr)){
16       f_frame *x_pms = MAKE_CONT(pr, join_to); callee_fr->caller_fr = x_pms;
17       if(pr->thief_req){
18         f_frame *fr = MAKE_CONT(pr, c__fib);
19         (*ltq_ptr)->jf = *(pr->js_top); (*ltq_ptr)->jw = SPLIT_JW(pr, pr->jw);
20         (*ltq_ptr)->cont = fr;
21         // save continuation
22         SUSPEND_IN_JOIN_BLOCK(pr);
23         pr->callee_fr = fr; return SUSPEND;
24       }
25     }
26     pr->ltq_tail = ltq_ptr;
27     /* sequential call */
28     y = f__fib(pr, n-2);
29     if((y==SUSPEND) && (callee_fr=pr->callee_fr)){
30       f_frame *fr = MAKE_CONT(pr, c__fib);
31       // save continuation
32       callee_fr->caller_fr = fr;
33       SUSPEND_IN_JOIN_BLOCK(pr);
34       pr->callee_fr = fr; return SUSPEND;
35     }
36     /* exit join block */
37     if(*(pr->js_top)){
38       WAIT_FOR_ZERO(pr->jf);
39       if(x_pms) x = x_pms->ret.i;
40     }
41     pr->js_top--;
42     return x+y;
43   }
44 }
```

Fig. 5. Compiled C code for `fib` with laziness (pseudo).

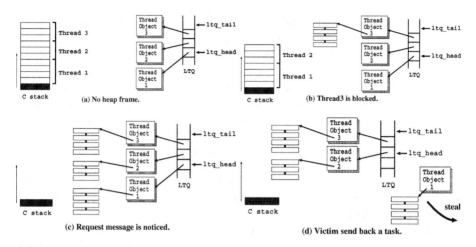

Fig. 6. Each thread in victim's stack is converted to a task during a task steal.

it starts suspending the parent thread (lines 18–23). ltq_ptr points to the corresponding thread object in the task queue. The way of splitting the weight differs from that of conventional code, and we explain this in the next subsection. (We use polling to check a request, and an efficient polling technique is found in [17].)

After these operations, the bottom continuation has been converted into a task at the task queue's head and can be sent to the thief. The disadvantage is that it converts all the continuations on stack into tasks, not only the bottom continuation. However, LTC assumes well-balanced divide-and-conquer programs, and during the execution of such programs, we expect for only a few times of stealing to happen. Also, even in unbalanced programs, we avoid the overhead as follows: the victim does not resume all tasks in the task queue with stack frames (i.e., completely recover the C stack) to restart, but the victim resumes a task taken from its tail with slow version code, and uses the remaining tasks for the later request without the conversion overhead.

So far, we present how we can extract the bottom continuation from C stack. However, for now, only two of five operations of task creation (listed in the Sect. 3.2), saving continuation and queuing, have we done lazily. In order to bring the cost of thread creation as close to that of sequential call as possible, we need adequate modifications to the remaining three operations.

The important point is that a child thread is called from a parent as normal function call. If a child can continue its execution until the end without blocking (i.e., on stack), (a) the thread's return value can be passed through the function's return value (not through placeholder), and (b) since the child thread always complete its execution before the parent thread reaches the end of join block, there is no need to split weight between them. In Fig. 5, at line 14, thread's return value is set to x directly. Only when a child thread has blocked, a frame for synchronization (and a placeholder) is appended (line 16), and only in such a case the parent thread examines a pointer to the placeholder and get a return value (line 39). (Note that weight has been split at the point of blocking (e.g., after polling) and set into the child's thread object.)

In this way, we can notice that we do not use the contents of thread objects which is created at each thread creation at all (at least, in current implementation), and so that they are only for thread ID. It means that a thread object needs to be unique only while the corresponding thread lives in stack. Then, it is redundant to allocate a thread object at each thread creation in the same place of task queue for many times. We decide not to free the thread object at thread's termination, and reuse it for the next thread creation by keeping it in task queue. Thread objects are allocated at initialization of the task queue and when a thread leaves stack, since it brings its thread object together, a substitute one is allocated. When a processor resumes a task, it frees the thread object that has been already in task queue's head.

Ultimately, in the case of no suspension, code for a thread creation has reduced to lines 13–15, and 26. In other words, the overhead as compared with sequential call is only a pair of increment/decrement of the task queue pointer and a check for suspension.

4.2   Lazy Join Frame Allocation

In the previous subsection, there is a lazy normalization scheme in which we do not have to split weight at all times. In this section, we extend this laziness to the process of entering/exiting a join block.

As described before, it can be said that there is an implicit synchronization between a parent and a child thread on stack. Then, as long as all the threads that synchronize to a certain join block execute in the same stack, there is no need to use a counter for this join block. In such a case, the corresponding join frame exists only for maintain the depth of nested join blocks (recall that they have a pointer field to the outer join frame). Alternatively, instead of allocating join frames, we prepare a stack per processor to maintain the depth of nested join blocks. The reason why we use a stack (we call it *join stack*), not a simple depth counter, is that we want to allocate a join frame and store it into the join stack when the above condition is not satisfied.

The fork/join structure in Fig. 4 is improved to Fig. 7. At initial state, the join stack is empty. After entering join blocks for several times, it only needs to increment the top pointer (Fig. 5 line 7) and the state becomes to Fig. 7 (a) (Be sure that there is one join frame at the bottom, and this is what resumed task has already had.) In the case of exiting a join block without making the join frame, it passes through the check at line 37, and only decrements the top pointer (line 41). If certain thread blocks and join stack's top has no pointer to a join frame, it allocates a new join frame, set the pointer to stack top, and split the weight (Fig. 7 (b).) When extracting a continuation for steal, all the join frames are allocated and split weights for all the threads (line 19). In addition, functions that executed in its local join block need to link a pointer between join frames to remain nested join block structure in heap (lines 22, 33). The final state of join block structure looks like (c).

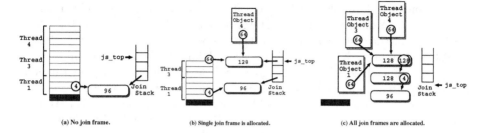

Fig. 7. Lazy join frame allocation.

5   Related Work

There are many multithreaded languages or multithreading frameworks that realize low cost thread creation and/or synchronization with automatic load balancing. We classify these languages/frameworks roughly into two categories. One class is for those that support only restricted parallelism. The other class is for those that supports arbitrary parallelism.

WorkCrews[8] is a model for controlling fork/join parallelism in an efficient and portable manner. When it creates a new thread, it creates stealable entity, (i.e., task) and continues the parent's execution. If the task is not yet stolen when the parent reaches its join point, the parent calls it sequentially. If the task is stolen, the parent thread blocks while waiting for the stolen task's join. Note that, in this model, once a parent thread calls a child thread sequentially, it is impossible to switch context to the parent thread even if, for example, the child thread blocks. So, this model can only be applicable to well-structured fork/join parallelism.

Lazy RPC[7] is an implementation scheme of parallel function call in a C compatible manner. It uses a similar technique as WorkCrews, and so the same restriction on parallel call.

FJTask[18, 19] is a Java's library for fine-grained fork/join multithreading and its technique is similar to Lazy RPC.

Since WorkCrews, Lazy RPC and FJTask employ restricted (well-structured) fork/join parallelism, each task can be started with the stack which is already used by some other tasks; thus the cost of task creation is comparable to that of object allocation.

Cilk[5, 6] is an extended C language with fork/join constructs and, like OPA, its implementation is a compiler from Cilk to C (and runtime). Its implementation technique is also based on LTC-like work steal. However, in several points, it differs from OPA. First, a join construct is lexically-scoped, and does not support other types of synchronization. These simplify the management of child threads. Second, its base language is C, so it does not provide exception handling. Third, it does not have a `synchronized` construct, that is, there is no need to manage thread identity. Fourth, for work steal, Cilk saves a parent thread's continuation in a heap-allocated frame at every thread creation. Indolent Closure Creation[20] is a variant of Cilk implementation and it employs a polling method similar to OPA for LTC. A different point from OPA is that a victim reconstructs the whole stack from all the tasks except the stolen one before continuing its execution.

As compared with Cilk's lexically-scoped join-destination, we think that OPA's dynamically-scoped one is more applicable. This is because a thread's join-destination can be determined as dynamically as a function's caller (return-destination) is determined, and as additionally as an exception handler (throw-destination) is determined. Furthermore, in Cilk, functions that fork some other threads should be called concurrently to avoid poor load balancing, since the sequential call prevents the caller from proceeding without the completion of all threads forked in the callee.

The lexically-scoped join-destination makes the Cilk implementation simple and efficient with the programming restriction. By contrast, OPA realizes a dynamically-scoped join-destination in an efficient manner using laziness, so it does not impose any restriction on programmers. In addition, it is possible to write 'lexically-scoped' style programs in OPA and the compiler can confirm that these programs conform to lexically-scoped style.

LTC[14] (and message passing LTC[16]) is an efficient implementation technique for Multilisp. Multilisp provides dynamic thread creation and general synchronization through future (and implicit touch) constructs, but, it does not have fork/join constructs and so programmers may need some skill to write correct programs. Stack manipulation for work steal is implemented in assembly level, then limited portability. Employing a polling method for LTC is originally proposed in message passing LTC.

StackThreads/MP[10] is also a stack-based approach for multithreading. It enables one processor to use other processor's stack-allocated frames for work steal. It enables general synchronization without heap-allocated frames. To realize this, it only works on shared-memory multiprocessors, and is implemented by assembly level manipulation of stack/frame pointers.

As compared with these languages, OPA has benefits of both categories: simple fork/join constructs, high portability, and general synchronization. In addition, it supports other advanced features like exception handling, synchronized method, and so on.

6   Performance

In this section, we evaluate the performance of our OPA implementation compared with Cilk 5.3.2[5], which is known as a good implementation of fine-grained multithreaded languages on multiprocessors. The configurations of the shared-memory parallel computer for this measurements are shown in Table 1.

We ported two Cilk benchmark programs, fib.cilk and matmul.cilk, that come with the Cilk distribution into OPA. Fig. 6 are measurement results of two programs on the SMP. Both OPA and Cilk systems show almost ideal speedups. This means both systems can efficiently distribute workloads among processors, while OPA system achieves better absolute performance than the Cilk system.

Table 1. Computer settings.

Machine	Sun Ultra Enterprise 10000 (Starfire)
CPU	Ultra SPARC II 250MHz, 1MB L2 cache
Main Memory	10GB
Num of CPUs	64
Compiler	gcc version 3.0.3 (with -O3 option)

Table 2. Performance results on 1 processor.

		fib(38)	matmul(450) (sec)
C		12.8	17.8
OPA		30.3	22.3
Cilk		50.9	24.4

Table 2 shows execution times of OPA programs, Cilk programs and sequential C programs on an uniprocessor of the SMP. For comparison with Cilk's paper [6], these measurements are taken on a Sun Enterprise 3000 with 167MHz UltraSPARC-I processors. In both benchmark programs, OPA system achieved better performance than Cilk. These results mean that our OPA implementation incurs less overhead for thread creation and synchronization than Cilk. In Cilk, the relative execution time of fib to C is 3.97. In the paper [6], this number is 3.63; the overhead of heap frame allocation is about 1.0 and that of the THE protocol is 1.3. The THE protocol is a mostly lock-free protocol to resolve the race condition that arises when a thief tries to steal the same frame that its victim is attempting to pop. In OPA, the relative execution time of fib to C is about 2.36, and the breakdown of OPA's serial overhead for fib is shown in Fig. 8. The total execution time is smaller than that of Cilk, primarily because the OPA system lazily performs heap frame allocation and the OPA system employs a polling method to resolve the race condition between a thief and its victim; that is, the OPA system only incurs the overhead of suspension check for each method (thread) call, that is about 0.28, and the overhead of polling, that is about 0.13, rather than the overhead of heap frame allocation (stealable continuation creation) plus the overhead of the THE protocol. StackThreads/MP uses the same technique as the OPA system, but it only exhibits almost comparable performance to the Cilk system. This seems to be because the language of Stack-

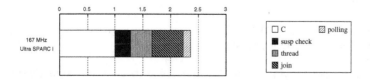

Fig. 8. Breakdown of overhead for fib on 1 processor.

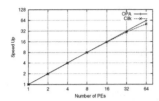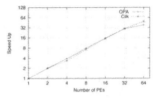

Fig. 9. Speedups of OPA and Cilk.

Threads/MP does not directly permit a thread to return a value or to join to the parent thread; those operation must explicitly be expressed and performed with additional overhead.

In practice, the above evaluation needs to be fixed because of the richer expressiveness of OPA. First, the second recursive call of `fib` can be expressed as a sequential call in OPA, reducing a thread creation cost. Second, supporting advanced features such as thread identification for Java-style locks, dynamically-scoped synchronization, thread suspension requires additional overhead. More specifically, the overhead of task queue manipulation is about 0.39 and that of join stack manipulation and counter check for synchronization is about 0.56. Even with these additional overhead for the richer expressiveness than Cilk, our implementation of OPA incurs smaller overhead than Cilk by pursuing 'laziness'.

7   Conclusions

In this paper, we proposed efficient and portable implementation techniques for OPA's fork/join constructs. OPA supports several advanced features such as mutual exclusion, Java's **synchronized** method and dynamically-scoped synchronization coupled with exception handling.

Supporting these features has been considered to degrade the effectiveness of existing efficient implementation techniques for fine-grained fork/join multithreaded languages, e.g., lazy task creation. Our implementation techniques pursued 'laziness' for several operations such as stealable continuation creation, thread object allocation and join frame allocation.

We compared the OPA implementation with Cilk. We confirmed that the performance of OPA programs exceeded that of Cilk programs, which indicates the effectiveness of our techniques.

References

1. Yasugi, M., Taki, K.: OPA: An object-oriented language for parallel processing — its design and implementation —. IPSJ SIG Notes 96-PRO-8(SWoPP'96) **96** (1996) 157–162 (in Japanese).
2. Yasugi, M., Eguchi, S., Taki, K.: Eliminating bottlenecks on parallel systems using adaptive objects. In: Proc. of International Conference on Parallel Architectures and Compilation Techniques, Paris, France. (1998) 80–87

3. Yasugi, M.: Hierarchically structured synchronization and exception handling in parallel languages using dynamic scope. In: Proc. of International Workshop on Parallel and Distributed Computing for Symbolic and Irregular Applications. (1999)

4. Yasugi, M., Umatani, S., Kamada, T., Tabata, Y., Ito, T., Komiya, T., Yuasa, T.: Code generation techniques for an object-oriented parallel language OPA. IPSJ Transactions on Programming **42** (2001) 1–13 (in Japanese).

5. Supercomputing Technologies Group: Cilk 5.3.2 Reference Manual. Massachusetts Institute of Technology, Laboratory for Computer Science, Cambridge, Massachusetts, USA. (2001)

6. Frigo, M., Leiserson, C.E., Randall, K.H.: The implementation of the Cilk-5 multithreaded language. ACM SIGPLAN Notices (PLDI'98) **33** (1998) 212–223

7. Feeley, M.: Lazy remote procedure call and its implementation in a parallel variant of C. In: Proceedings of International Workshop on Parallel Symbolic Languages and Systems. Number 1068 in LNCS, Springer-Verlag (1995) 3–21

8. Vandevoorde., M.T., Roberts, E.S.: WorkCrews: An abstraction for controlling parallelism. International Journal of Parallel Programming **17** (1988) 347–366

9. Chandra, R., Gupta, A., Hennessy, J.L.: COOL. In Wilson, G.V., Lu, P., eds.: Parallel Programming Using C++. The MIT Press (1996)

10. Taura, K., Tabata, K., Yonezawa, A.: StackThreads/MP: Integrating futures into calling standards. In: Proceedings of ACM SIGPLAN Symposium on Principles & Practice of Parallel Programming (PPoPP). (1999) 60–71

11. Wagner, D.B., Calder, B.G.: Leapfrogging: A portable technique for implementing efficient futures. In: Proceedings of Principles and Practice of Parallel Programming (PPoPP'93). (1993) 208–217

12. Plevyak, J., Karamcheti, V., Zhang, X., Chien, A.A.: A hybrid execution model for fine-grained languages on distributed memory multicomputers. In: Proceedings of the 1995 conference on Supercomputing (CD-ROM), ACM Press (1995) 41

13. Bevan, D.I.: Distributed garbage collection using reference counting. In: PARLE: Parallel Architectures and Languages Europe. Number 259 in LNCS, Springer-Verlag (1987) 176–187

14. Mohr, E., Kranz, D.A., Halstead, Jr., R.H.: Lazy task creation: A technique for increasing the granularity of parallel programs. IEEE Transactions on Parallel and Distributed Systems **2** (1991) 264–280

15. Halstead, Jr., R.H.: Multilisp: a language for concurrent symbolic computation. ACM Transactions on Programming Languages and Systems (TOPLAS) **7** (1985) 501–538

16. Feeley, M.: A message passing implementation of lazy task creation. In: Proceedings of International Workshop on Parallel Symbolic Computing: Languages, Systems, and Applications. Number 748 in LNCS, Springer-Verlag (1993) 94–107

17. Feeley, M.: Polling efficiently on stock hardware. In: Proc. of Conference on Functional Programming Languages and Computer Architecture. (1993) 179–190

18. Lea, D.: A Java fork/join framework. In: Proceedings of the ACM 2000 conference on Java Grande, ACM Press (2000) 36–43

19. Lea, D.: Concurrent Programming in Java: Design Principles and Patterns. Second edn. Addison Wesley (1999)

20. Strumpen, V.: Indolent closure creation. Technical Report MIT-LCS-TM-580, MIT (1998)

SPEC HPG Benchmarks for Large Systems

Matthias S. Müller[1], Kumaran Kalyanasundaram[2], Greg Gaertner[3],
Wesley Jones[4], Rudolf Eigenmann[5], Ron Lieberman[6],
Matthijs van Waveren[7], and Brian Whitney[8]

[1] High Performance Computing Center Stuttgart (HLRS),
Allmandring 30, 70550 Stuttgart, Germany,
Phone: +49 711 685 8038, Fax: +49 711 678 7626,
mueller@hlrs.de
[2] SPEC HPG Chair, Silicon Graphics, Inc.,
kumaran@sgi.com
[3] SPEC HPG Vice Chair, Hewlett-Packard Corporation,
Greg.Gaertner@hp.com
[4] National Renewable Energy Lab,
Wesley_Jones@nrel.gov
[5] Purdue University,
eigenman@ecn.purdue.edu
[6] Hewlett-Packard Corporation
lieb@rsn.hp.com
[7] SPEC HPG Secretary, Fujitsu Systems Europe Ltd.,
waveren@fujitsu.fr
[8] Sun Microsystems,
Brian.Whitney@sun.com

Abstract. Performance characteristics of application programs on large-scale systems are often significantly different from those on smaller systems. In this paper, we discuss such characteristics of the benchmark suites maintained by SPEC's High-Performance Group (HPG). The Standard Performance Evaluation Corporation's (SPEC) High-Performance Group (HPG) has developed a set of benchmarks for measuring performance of large-scale systems using both OpenMP and MPI parallel programming paradigms. Currently, SPEC HPG has two lines of benchmark suites: SPEC OMP and SPEC HPC2002. SPEC OMP uses the OpenMP API and includes benchmark suites intended for measuring performance of modern shared memory parallel systems. SPEC HPC2002 is based on both OpenMP and MPI, and thus it is suitable for distributed memory systems, shared memory systems, and hybrid systems. SPEC HPC2002 contains benchmarks from three popular application areas, Chemistry, Seismic, and Weather Forecasting. Each of the three benchmarks in HPC2002 has small and medium data sets in order to satisfy the need for benchmarking a wide range of high-performance systems. We present our experiences regarding the scalability of the benchmark suites. We also analyze published results of these benchmark suites based on application program behavior and systems' architectural features.

Key words: SPEC OMP, SPEC HPC, Benchmark, High-Performance Computing, Performance Evaluation, OpenMP, MPI

A. Veidenbaum et al. (Eds.): ISHPC 2003, LNCS 2858, pp. 189–201, 2003.

1   Introduction

SPEC (The Standard Performance Evaluation Corporation) is an organization for creating industry-standard benchmarks to measure various aspects of modern computer system performance. SPEC's High-Performance Group (SPEC HPG) is a workgroup aimed at benchmarking high-performance computer systems. In June of 2001, SPEC HPG released the first of the SPEC OMP benchmark suites, SPEC OMPM2001. This suite consists of a set of OpenMP-based[8, 9] application programs. The data sets of the SPEC OMPM2001 suite (also referred to as the medium suite) are derived from state-of-the-art computation on modern medium-scale (4- to 16-way) shared memory parallel systems. Aslot et al.[1] have presented the benchmark suite. Aslot et al.[2] and Iwashita et al.[6] have described performance characteristics of the benchmark suite. The second, large suite, SPEC OMPL2001, focusing on 32-way and larger systems, was released in May 2002. SPEC OMPL2001 shares most of the application code base with SPEC OMPM2001, but the code and the data sets have been improved and made larger to achieve better scaling and also to reflect the class of computation regularly performed on such large systems[10]. However, both suites are limited to shared memory platforms. The largest system where results have been reported is a 128-way system.

SPEC HPC2002 is the latest release of the HPC benchmark suite. It is suitable for shared and distributed memory machines or clusters of shared memory nodes. SPEC HPC applications have been collected from among the largest, most realistic computational applications that are available for distribution by SPEC. In contrast to SPEC OMP, they are not restricted to any particular programming model or system architecture. Both shared-memory and message passing methods are supported. All codes of the current SPEC HPC2002 suite are available in an MPI and an OpenMP programming model and they include two data set sizes. Performance characteristics of application programs on large-scale systems are often significantly different from those on smaller systems. In our previous paper[10] we have discussed the scaling of SPEC OMP benchmarks. In this paper, we characterize the performance behavior of large-scale systems (32-way and larger) using the SPEC OMP and HPC2002 benchmark suites. In Section 2, we provide a short description of the applications contained in the benchmarks. Section 3 analyzes the published results of SPEC OMPL2001 and SPEC HPC2002 on large systems, based on application program behavior and systems' architectural features. Section 4 concludes the paper.

2   Description of the Benchmarks

2.1   Overview of the SPEC OMPL2001 Benchmark

The SPEC OMPL2001 benchmark suite consists of 9 application programs, which represent the type of software used in scientific technical computing. The applications include modeling and simulation programs from the fields of chemistry, mechanical engineering, climate modeling, and physics. Of the 9 application programs, 7 are written in Fortran, and 2 are written in C. The benchmarks require a virtual address space of about 6.4 GB in a 16-processor run. The rationale for this size were to provide data sets significantly larger than those of the SPEC OMPM benchmarks, with a requirement for a 64-bit address space.

Descriptions of the 9 applications codes follow.

APPLU solves 5 coupled non-linear PDEs on a 3-dimensional logically structured grid, using the Symmetric Successive Over-Relaxation implicit time-marching scheme[4]. Its Fortran source code is 4000 lines long.

APSI is a lake environmental model, which predicts the concentration of pollutants. It solves the model for the mesoscale and synoptic variations of potential temperature, wind components, and for the mesoscale vertical velocity, pressure, and distribution of pollutants. Its Fortran source code is 7500 lines long.

MGRID is a simple multigrid solver, which computes a 3-dimensional potential field. Its Fortran source code is 500 lines long.

SWIM is a weather prediction model, which solves the shallow water equations using a finite difference method. Its Fortran source code is 400 lines long.

FMA3D is a crash simulation program. It simulates the inelastic, transient dynamic response of 3-dimensional solids and structures subjected to impulsively or suddenly applied loads. It uses an explicit finite element method[7] Its Fortran source code is 60000 lines long.

ART (Adaptive Resonance Theory) is a neural network, which is used to recognize objects in a thermal image[5]. The objects in the benchmark are a helicopter and an airplane. Its C source code is 1300 lines long.

GAFORT computes the global maximum fitness using a genetic algorithm. It starts with an initial population and then generates children who go through crossover, jump mutation, and creep mutation with certain probabilities. Its Fortran source code is 1500 lines long.

EQUAKE is an earthquake-modeling program. It simulates the propagation of elastic seismic waves in large, heterogeneous valleys in order to recover the time history of the ground motion everywhere in the valley due to a specific seismic event. It uses a finite element method on an unstructured mesh[3]. Its C source code is 1500 lines long.

WUPWISE (Wuppertal Wilson Fermion Solver) is a program in the field of lattice gauge theory. Lattice gauge theory is a discretization of quantum chromodynamics. Quark propagators are computed within a chromodynamic background field. The inhomogeneous lattice-Dirac equation is solved. Its Fortran source code is 2200 lines long.

2.2   Overview of the SPEC HPC2002 Benchmark Suite

SPEC HPC2002 is a benchmark suite based on high-performance computing (HPC) applications and the MPI and OpenMP standards for parallel processing. It is targeted at those who evaluate performance for HPC systems, including users, system vendors, software vendors, and researchers. It uses a set of actual applications to measure the performance of the computing system's processors, memory architecture, and operating system. SPEC HPC2002 improves upon and replaces the SPEC HPC96 benchmark suite. The SPEC HPC2002 suite comprises three benchmarks, each with a small- and medium-sized data set:

SPECenv (WRF) is based on the WRF weather model, a state-of-the-art, non-hydrostatic mesoscale weather model, see http://www.wrf-model.org. The WRF (Weather Research and Forecasting) Modeling System development project is a multi-year project being undertaken by several agencies. Members of the WRF Scientific Board include representatives from EPA, FAA, NASA, NCAR, NOAA, NRL, USAF and several universities. SPEC HPG integrated version 1.2.1 of the WRF weather model into the SPEC tools for building, running and verifying results. This means that the benchmark runs on more systems than WRF has officially been ported to. It is written in C and F90 and contains about 25000 lines of C and 145000 lines of F90 code. It can run in OpenMP, MPI or mixed MPI-OpenMP mode (hybrid). Fig. 1 shows the scalability for the different data sets and the OpenMP and MPI mode on a Sun Fire 6800. The medium data set shows better scalability. The best programming model will depend on the platform and data set. Here, OpenMP is better for the small data set and MPI for the large data set.

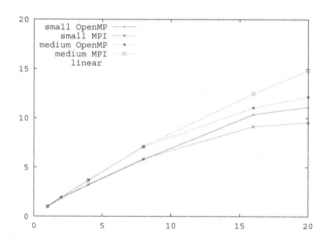

Fig. 1. Speedup of SPECenv for different data sets and programming models on a Sun Fire 6800 platform.

SPECseis (seis) was developed by ARCO beginning in 1995 to gain an accurate measure of performance of computing systems as it relates to the seismic processing industry for procurement of new computing resources. It consists of a modeling phase which generates synthetic seismic traces for any size of data set, with a flexibility in the geometry of shots and receivers, ground structures, varying lateral velocity, and many other options. A subsequent phase stacks the traces into common midpoint stacks. There are two imaging phases which produce the valuable output seismologists use to locate resources of oil. The first of the two imaging phases is a Fourier method which is very efficient but which does not take into account variations in the velocity profile. Yet, it is widely used and remains the basis of many methods for acoustic imaging. The second imaging technique is a much slower finite-difference method, which can handle variations in the lateral velocity. This technique is used in many seismic migration codes today. SPECseis contains about 25,000 lines of Fortran and C code. It can run in OpenMP or MPI mode. Fig. 2 shows the scalability for the different data sets and the OpenMP and MPI mode on a Sun Fire 6800. The medium data set shows better scalability. The best programming model will depend on the platform and data set. In this case OpenMP has better scalability.

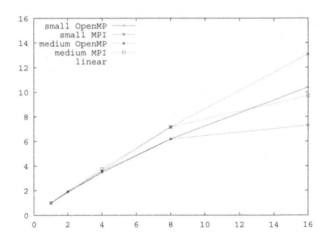

Fig. 2. Speedup of SPECseis for different data sets and programming models on a Sun Fire 6800 platform.

SPECchem (Gamess) is used to simulate molecules ab initio, at the quantum level, and optimize atomic positions. It is a research interest under the name of GAMESS at the Gordon Research Group of Iowa State University and is of interest to the pharmaceutical industry. Like SPECseis, SPECchem is often used to exhibit performance of high-performance systems among the computer vendors. Portions of SPECchem codes date back to 1984. It comes with many built-in functionalities, such as various field molecular wave-functions, certain energy corrections for some of the wave-functions, and simulation of several different phenomena. Depending on what wave-functions you choose, SPECchem has the option to output energy gradients of these functions, find saddle points of the potential energy, compute the vibrational frequencies and IR intensities, and more. SPECchem contains

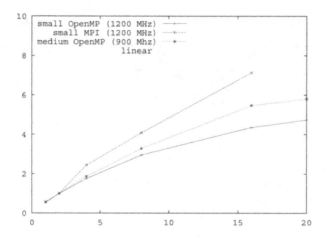

Fig. 3. Speedup of SPECchem for different data sets and programming models on a Sun Fire 6800 platform.

over 120,000 lines of code written in Fortran and C. It can run in OpenMP, MPI or mixed MPI-OpenMP mode (hybrid). Fig. 3 shows the scalability for the different data sets and the OpenMP and MPI mode on a Sun Fire 6800. The medium data set shows better scalability, despite the fact that this data set was measured on a machine with faster processors. In this case MPI has better scalability for the small data set.

3   Large System Performance of SPEC Benchmark Suites

Performance characteristics of application programs on large-scale systems are often significantly different from those on smaller systems. Figure 4 shows a scaling of Amdahl's speedup for 32 to 128 threads, normalized by the Amdahl's speedup of 16 threads.

Amdahl's speedup assumes perfect scaling of the parallel portion of the program. Actual programs and actual hardware have additional sources of overhead, which degrade the performance obtained on a real system relative to the upper bound given by Amdahl's law. Figure 5 show the scaling data for published benchmark results of SPEC OMPL2001. The numbers listed in the following figures have been obtained from the results published by SPEC as of May, 2003. For the latest results published by SPEC, see http://www.spec.org/hpg/omp. All results shown conform to Base Metrics reporting rules. For better presentation of the graph, we have normalized all results with the 32-processor results of the same type of system configuration. If the same system has faster and slower processor configurations, we used the scores with the faster processors. In order to make the graphs readable, we have selected the systems that provided at least 32- and 64-processor results.

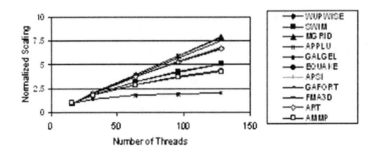

Fig. 4. Scaling of Amdahl's Speedup of OMPM2001 through 128 threads, normalized by the 16-thread speedup.

3.1 Scalability of OMP Benchmarks

Figure 5 shows the scalability of the SPEC OMPL benchmark applications. The benchmarks WUPWISE, SWIM, APSI, and GAFORT show good scalability up to 128 processors. In order for SWIM to scale well, the bandwidth to main memory needs to scale with the number of processors. To increase the scalability OMPL2001 SWIM has more parallel loops than the OMPM2001 version. In addition, some scalar computation is performed in parallel, in favor of improved locality. Compared to OMPM2001 APSI, OMPL2001 APSI has a larger trip count of 240 for the corresponding loop. OMPL2001 APSI also has an improved work array distribution scheme as well as an improved handling of parallel reduction operations.

The benchmark APPLU shows superlinear scaling on HP Superdome. In our previous paper, we also presented superlinear scaling for the SGI Origin 3800 and the Fujitsu PRIMEPOWER 2000[10]. This is due to a more efficient usage of the cache as more processors are used. The same effect is visible on the Sun Fire 15K. According to the cache sizes of these systems, the sweet spot of the aggregate cache amount is between 64MB and 96MB. In the OMPL2001 version, false sharing was reduced by moving one of the OpenMP DO directives from the outermost loop to the second-level loop. The benchmarks EQUAKE, MGRID, and ART show good scaling up to 64 processors, but poor scaling for larger numbers of processors. HP Superdome and SGI Origin 3800 scaled less on EQUAKE. MGRID and EQUAKE are sparse matrix calculations, which do not scale well to large numbers of processors. In order to gain more scalability in the OMPL2001 version, we exploited more parallelism in EQUAKE, resulting in better scaling on HP Superdome and SGI Origin 3800. OMPL2001 ART calls malloc() more efficiently than the OMPM2001 version. This change reduces contention on malloc(), and thus improved the scalability of ART. Larger data set in OMPL2001 helped the scaling of MGRID.

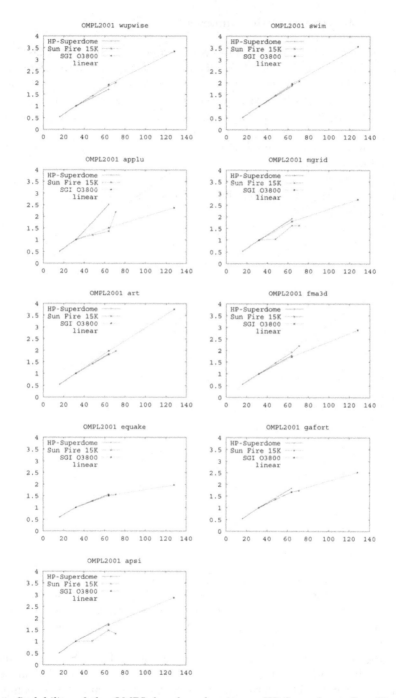

Fig. 5. Scalability of the OMPL benchmark suite on HP-Superdome, Sun Fire 15K and SGI O3800.

3.2   Scalability of HPC Benchmarks

As of May 2003 nineteen results have been submitted for SPEC HPC2002. Eight
for the medium and eleven for the small data set. Results on up to 128 processes-
threads have been submitted. In this section we focus on the medium data set
results on an IBM SP with 128 Power-3 CPUs running at 375MHz and an SGI
3800 with 128 R14000A CPUs at 600 MHz. In addition we show the results of
a Sun Fire 6800 and 15K with UltraSparc III CPUs at 900MHz or 1200MHz
(indicated in the graphs), and a 64 CPU Sun Fire 880 cluster with 8way SMPs
connected by Myrinet.

Fig. 6 shows the scalability of SPECenv for the medium data set. All results
use the MPI model of execution. The benchmark shows good scalability up to
128 processors.

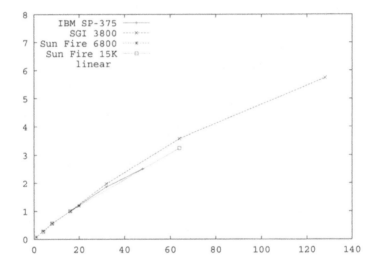

Fig. 6. Speedup of SPECenv for the medium data set.

Fig. 7 shows the scalability of SPECseis for the medium data set. The scaling
behavior depends strongly on the programming model and platform. While it
shows almost perfect scaling (81% efficiency) on 16 processors, the efficiency on
a Sun Fire 15K is much lower. From this preliminary results no general scaling
behavior of this application can be deduced.

For the SPECchem benchmark only results of IBM and Sun are available
for the medium data set. For both submitted IBM results the MPI model of
execution was used. The efficiency of the 32 processor run is 82% compared
to the 16 processor run on the IBM. Using OpenMP the Sun Fire 15K shows a
perfect, almost superlinear scaling from 16 to 32 processors, but for 64 processors
the efficiency is only 57% compared to the 16 processor run. Better efficiency is

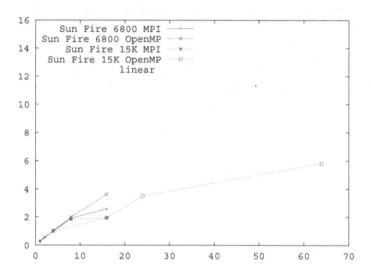

Fig. 7. Speedup of SPECseis for the medium data set.

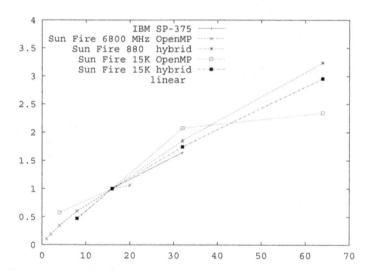

Fig. 8. Speedup of SPECchem for the medium data set.

achieved with the hybrid execution model, where 81% is reached with 64 CPUs on the Sun Fire 880 cluster.

Although two of the HPC2002 benchmarks can be used in hybrid mode all of the submitted results are limited to either pure OpenMP or MPI mode. Tab. 1 shows the potential benefit of an hybrid execution of SPECenv. Similar benefit

Table 1. Scaling of different execution models of SPECenv for the small and medium data set. Please note the different base points.

MPI x OMP	16x1	1x16	8x2	8x1	4x1	1x4	2x2	1x1
4way Itanium , small data set					1.72	1.39	1.98	1
Sun Fire 6800, medium data set	1.76	1.56	1.83	1				

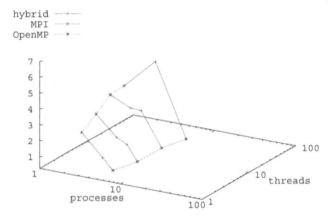

Fig. 9. Speedup of SPECchem for MPI, OpenMP and hybrid execution on a Sun Fire 15K.

is visible for SPECchem in Fig. 9, since the MPI and OpenMP parallelism is on a different level of granularity.

4   Conclusion and Future Work

In this paper we have analyzed the performance characteristics of published results of the SPEC OMPL2001 and HPC2002 benchmark suites. We have found that many of the benchmark programs scale well up to 128 processors. We also have demonstrated the impact of the choice of the execution model. The results show that the best choice of MPI, OpenMP or hybrid depends on the used hardware architecture as well as on the program and the data sets. Although HPC2002 is not limited to shared memory platforms, there are no results of larger machines available, so far. We attribute this to the relative recent release of HPC2002 and expect it to change in the near future.

The trends of the SPEC HPC2002 codes indicate clear limits of scalability. We conclude that, even given sizeable data sets, large-scale, realistic applications do not exhibit the near-ideal speedups that some of the smaller benchmarks suggest. While this is an expected finding for many readers, demonstrating the evidence is an important result of the SPEC HPC2002 development. The fact that SPEC benchmarks and reports are fully disclosed will allow both scientists and engineers identify the causes that limit performance and develop remedies.

Current efforts of SPEC/HPG are the update of SPECchem and the general improvement regarding the portability of the benchmark suites. This includes the provision of pre-compiled tools for new platforms. SPEC/HPG is open to adopt new benchmark programs. A good candidate program would represent a type of computation that is regularly performed on high-performance computers. Among others, SPEC/HPG currently examines SPEC CPU2004 for codes that satisfy this requirement.

Acknowledgments

The authors would like to thank all of those who participated in the benchmark development, especially the authors of the application programs and the contributors of the data sets. Benchmark development would have been impossible without their hard work. Special thank go to Hideki Saito for his contributions and valuable comments.

References

[1] Vishal Aslot, Max Domeika, Rudolf Eigenmann, Greg Gaertner, Wesley B. Jones, and Bodo Parady. SPEComp: a new benchmark suite for measuring parallel computer performance. In *WOMPAT'01: Workshop on OpenMP Applications and Tools*, volume 2104 of *LNCS*, pages 1–10. Springer, July 2001.

[2] Vishal Aslot and Rudolf Eigenmann. Performance characteristics of the SPEC OMP2001 benchmarks. In *3rd European Workshop on OpenMP, EWOMP'01*, Barcelona, Spain, September 2001.

[3] Hesheng Bao, Jacobo Bielak, Omar Ghattas, Loukas F. Kallivokas, David R. O'Hallaron, Jonathan R. Shewchuk, and Jifeng Xu. Large-scale simulation of elastic wave propagation in heterogeneous media on parallel computers. *Computer Methods in Applied Mechanics and Engineering*, 152(1-2):85–102, January 1998.

[4] E. Barszcz, R. Fatoohi, V. Venkatkrishnan, and S. Weeratunga. Solution of regular sparse triangular systems on vector and distributed-memory multiprocessors. Technical Report Rept. No: RNR-93-007, NASA Ames Research Center, 1993.

[5] M.J. Domeika, C.W. Roberson, E.W. Page, and G.A. Tagliarini. Adaptive resonance theory 2 neural network approach to star field recognition. In Steven K. Rogers and Dennis W. Ruck, editors, *Applications and Science of Artificial Neural Networks II*, volume 2760 of *Proc. SPIE*, pages 589–596. International Society for Optical Engineering, 1996.

[6] Hidetoshi Iwashita, Eiji Yamanaka, Naoki Sueyasu, Matthijs van Waveren, and Kenichi Miura. The SPEC OMP 2001 benchmark on the Fujitsu PRIMEPOWER system. In *3rd European Workshop on OpenMP, EWOMP'01*, Barcelona, Spain, September 2001.

[7] S.W. Key and C. C. Hoff. An improved constant membrane and bending stress shell element for explicit transient dynamics. *Computer Methods in Applied Mechanics and Engineering*, 124:33–47, 1995.

[8] OpenMP Architecture Review Board. *OpenMP C and C++ Application Programming Interface Version 2.0, March 2002.*
    http://www.openmp.org/specs/mp-documents/cspec20.pdf.

[9] OpenMP Architecture Review Board. *OpenMP Fortran Application Programming Interface Version 2.0, November 2000.*
    http://www.openmp.org/specs/mp-documents/fspec20.pdf.

[10] Hideki Saito, Greg Gaertner, Wesley Jones, Rudolf Eigenmann, Hidetoshi Iwashita, Ron Lieberman, Matthijs van Waveren, and Brian Whitney. Large system performance of SPEC OMP benchmark suites. *Int'l Journal of Parallel Programming*, 31(3), June 2003.

Distribution-Insensitive Parallel External Sorting on PC Clusters*

Minsoo Jeon and Dongseung Kim

Department of Electrical Engineering, Korea University
Seoul, 136-701, Korea
{msjeon, dkim}@classic.korea.ac.kr
Tel. no.: +822 3290 3232,   Fax. no.: +822 928 8909

Abstract.   There have been many parallel external sorting algorithms reported such as NOW-Sort, SPsort, and hill sort, etc. They are for sorting large-scale data stored in the disk, but they differ in the speed, throughput, and cost-effectiveness. Mostly they deal with data that are *uniformly* distributed in their value range. Few research results have been yet reported for parallel external sort for data with *arbitrary distribution*. In this paper, we present two distribution-insensitive parallel external sorting algorithms that use sampling technique and histogram counts to achieve even distribution of data among processors, which eventually contribute to achieve superb performance. Experimental results on a cluster of Linux workstations show up to 63% reduction in the execution time compared to previous NOW-sort.

1   Introduction

The capacity of digital data storage worldwide has doubled every nine months for at least a decade, at twice the rate predicted by Moore's Law [12] of the growth of computing power during the same period [5], thus, more data are generated as the information era continues [14]. Processing and storing data will take longer time than before. *External sort* orders large-scale data stored in the disk. Since the size of data is so large compared to the capacity of the main memory, a processor can process only a fraction of the data at a time by bringing them into main memory from the disk, sorting, then storing back onto the disk to make room for the next data. Hence, the sorting usually demands multiple iterations of data retrieval from the disk, ordering computation in main memory, and writing-back to the disk. The time for pure sorting computation grows in general as a function of $O(N)$ or $O(N \log N)$ as the input size of N increases. Dealing with data stored in disk memory additionally needs a great

---

* This research was supported by KOSEF Grant (no. R01-2001-000341-0).

A. Veidenbaum et al. (Eds.): ISHPC 2003, pp. 202–213, 2003.

amount of time since disk memory is very slow compared to the speed of recent CPUs. So, keeping the frequency of disk I/Os as small as possible is very important to achieve high performance.

Many parallel/distributed sorting algorithms have been devised to shorten the execution time. They can be classified into either *merge-based* sorts or *partition-based* sorts. Most merging based algorithms such as *parallel binary merge sort* [16] and *external hillsort* [17] are not fast algorithms because they have to access disk many times in reading and writing data during log P iterations of merging, where P is the number of processors used in the sorting. Similarly, neither are those bitonic sort [3], odd-even merge sort [3, 7] since they need $O(\log^2 P)$ iterations of *merge split* operations [1]. Although they are flexible and keep even load among processors, they are slow due to $O(N)$ data accesses in each merge split operation.

Partition-based external sorts such as *sample sort* [10] and *NOW-sort* [2] execute in two phases regardless of the data amount: key partitioning into P processors using P-1 pivots, then local sort in each processor. The first phase roughly relocates all data in the order of the indexes of processors by classifying and exchanging keys among processors. Then, all keys in a processor P_i are less than or equal to any key stored in P_j if $i < j$ ($i, j = 0, 1, ..., P$-1). The second phase needs only local sorting computation without interprocessor communication. These sorts are generally fast compared to merge based sort. However, if there is imbalance in the number of data among processors, the completion time is lengthened by the heaviest loaded processor (having the maximum data amount). Hence, finding a set of proper *pivots* leading to even partitioning is extremely important to make the best use of the computing power of all processors. These sorts have superior performance due to a small number of disk accesses, but their performance is *sensitive* to the input distribution.

NOW-Sort is known as a fast and efficient external sorting method implemented on a network of workstations with a large scale of disk data (from dozens of gigabytes to terabytes). It uses partial radix sort and bubble sort algorithms, communication primitives called active message, and fast disk I/O algorithms. It is fastest external sorting algorithm in that the number of accesses to hard disk is *2R 2W* (2 read accesses, 2 write accesses) *per key*, which is the fewest disk-accesses among algorithms mentioned above [4, 13]. The detailed algorithm is described in section 2.

Various dialects of NOW-sorting algorithm were devised to achieve high speed, high throughput, or cost-effectiveness. SPsort [18] is a large-scale SMP version of NOW-Sort implemented on IBM RS/6000 SP with 1952 processors, and has a record of sorting a terabyte of data in 1,057 seconds. Datamation, MinuteSort, PennySort, TerabyteSort benchmarks [19] are respectively to achieve a maximum throughput, greatest size of data sorted in a minute, cost-effectiveness, and a minimum time for sorting a fixed number of data. Most of them did not concern about the distribution of the data, and they usually included only *uniform* distribution for simplicity [19]. Few research results have been yet reported for *arbitrary* distribution.

In this paper, we develop a distribution-insensitive NOW-Sort algorithm together with sample sort to solve the problem. They employ sampling technique and histogram counts for the even distribution of data among all processors. Conditions to get the best performance are investigated, and experimental results on PC cluster are reported.

The rest of this paper is organized as follows. In section 2 and section 3 NOW-Sort and our algorithms are described, respectively. In section 4 the analysis and discussion on the experimental results of the proposed algorithms are given. Conclusion is given in the last section.

2   Partition Based Sort: NOW-Sort

NOW-Sort is a distributed external sorting scheme, implemented on a network of workstations (NOWs) that are interconnected via fast ethernet or Myrinet. The sorting consists of two phases: a data redistribution phase and an individual local sort phase. Detail description is given below.

In the first phase, each processor (node) classifies its data into P groups according to their key values, and it sends them to the corresponding nodes. In the process, keys are classified and replaced according to the order of processor indexes: P_0 contains the smallest keys, P_1 the next smallest, and so on. If the keys are integers, classification is accomplished by bucket sort using a fixed number of *most significant bits* (MSBs) of their binary representation. With uniform distribution a prefix of $\log_2 P$ MSBs is sufficient. It determines where the key should be sent and stored for further processing. Thus, P_0 will be in charge of the smallest keys whose $\log P$ MSBs are $0...00$ in the binary representation, P_1 the next smallest keys with the prefix of $0...01$, ... , and P_{P-1} the biggest keys with the prefix of $1...11$. To reduce the communication overhead, keys are sent to other processors only after the classified data are filled up in temporary buffers. While processors receive keys from others, they again bucket-sort the incoming keys before storing them into disk to reduce the time for final local sort. The bucket sort uses the next few MSBs after the first $\log_2 P$ bits of each key. Incoming keys are ordered according to the indexes of the buckets, and each bucket will include keys that do not exceed the main memory capacity. The buckets are stored as separate files according to their indexes for future references.

Although keys have been ordered with respect to the order of processors at the end of the phase 1, they are not yet sorted within each processor. As soon as all processors finish the redistribution, the second phase begins: each node repeats the following: load a file of keys to main memory from disk, sort it, and write back to disk. In-place sort is possible since files are loaded and stored with same names and sizes. If all files are processed, the sorting completes.

To shorten the execution time in the external sort, reducing the number of disk accesses per key is critical. In the first phase of NOW sort, each key is read once from the disk, sent to a certain node, and written back into the disk there. In the local sort phase there is also one disk-read and one disk-write operations per key. Thus, there are 2R2W (2 read accesses, 2 write accesses) disk accesses per key, no matter how many data there are and how many processors the sort includes. 2R2W is the fewest disk accesses of external sort algorithms known so far [4, 13].

In the benchmarking reports such as Datamation and MinuteSort, they do not include all kinds of data distribution. To make the sort simple and fast, they only use keys with *uniform* distribution in developing a best sorting algorithm. In case of non-uniform distribution processors may have uneven work load after redistribution, thus, the execution time may increase considerably. The following section describes our solutions for the problem that are insensitive to the data distribution.

3   Distribution-Insensitive External Sort

3.1   Sample Sort

In this section we modify NOW-Sort algorithm in order to use sampling techniques. Sample sort is a two-stage parallel sorting algorithm capable of ordering keys with *arbitrary* distribution. Keys initially stored evenly throughout the nodes are partitioned using *splitters* and relocated according to their values so that P_0 will have the smallest keys, P_1 the next smallest, ..., P_P the largest. Thus, no more key exchanges among nodes are needed in the later stage. Now each node sorts its keys locally, then, the whole keys are ordered among processors and within the processors as well. More explanation follows.

The splitters (or pivots) are obtained by *sampling*. Each node samples a fixed number of keys, and then they are collected and sorted in a master processor. P-1 splitters are collected from them having equal interval in the sorted list, and they are broadcasted to all nodes. Now, all nodes read their own keys, classify them into P groups using the pivots, and transmit them to P nodes according to the indexes of the groups. At the same time, each node receives keys from others. After the exchange, each node individually sorts its own keys. In this parallel computation, if the load of each node is roughly same, all nodes can finish the job at about the same time, which gives the best performance in the sample sort. If not, the performance is aggravated by the most heavily loaded processor, in which case the time should be longer than the balanced case.

Sample sort is a very excellent sorting scheme when the distribution of keys is not uniform, and especially when the information is not known ahead. Depending on how many samples are extracted from the input, the quality of the resultant partitions of the keys varies. Although the use of many samples produces a good result, it consumes computation time significantly. On the other hand, too few samples may give coarse data distribution. In section 4, three different sampling rates are experimented and compared.

3.2   Histogram-Based Parallel External Sort

This algorithm is similar to the sample sort in the aspect that it classifies and distributes using P-1 splitters, and then local sort follows. It differs from the sample sort in the way to find the splitters: it employs histogram instead of sampling [6, 9]. The algorithm executes in two phases too: histogram computation and bucket sort phase (phase 1) and data distribution and post-processing phase (phase 2). In phase 1, each of the P nodes reads data from disk, computes local histogram to get the information about the distribution of its keys. While a node computes histogram using a fixed number of MSBs of the keys, it also performs bucket sort and places data into disk files bucket by bucket. Although keys have been ordered with respect to the order of buckets, they have not been sorted yet within each bucket. Histogram data in each processor are sent to a master processor (for example, P_0), and then the global histogram is computed by summing up them. P-1 pivots are selected from the boundary values of the histogram bins that divide keys evenly to all processors. The pivots are broadcasted to all P nodes from the master. In phase 2, each node reads a fixed number of data from disk, classifies the data into P buffers in memory, and sends each buffered data to the corresponding node. While it transmits its data to other nodes, it also receives data from other nodes. Incoming data are gathered, sorted, and stored into disk bucket by bucket. This process continues until all disk data are consumed. This sorting demands 2R2W accesses per key, too. The detailed algorithm is described below.

Let N, P denote the number of total keys and the number of processors, respectively. P_0 is the master node that calculates the global histogram and finds splitters. N/P keys are initially stored in the disk of each node.

Phase 1: Histogram computation and bucket sort phase
(S1) Each node iterates the following computation:
   1. Each node reads a fixed number of keys from disk.
   2. Each node calculates local histogram. While a node computes histogram, it also performs bucket sort such that each key is assigned to the corresponding histogram buffer where the key belongs. It stores (appends) the buffered data into disk files according to the indexes whenever they are full.
(S2) Each node sends local histogram information to the master node (P_0). It later receives splitters from P_0.

The master node performs the following computation additionally:
1. It collects local histograms and calculates the global histogram.

2. It computes the accumulated sums of the histogram counts and selects P-1 splitters: The first splitter is chosen to the histogram boundary to include the $\frac{N}{P}$ smallest keys in the range, the next splitter to include the next $\frac{N}{P}$ smallest keys, and so on. Notice that splitters derived from very fine-grain histogram will produce quite even partitioning, however, moderate histogram suffices for good load balancing in practice.

Phase 2: Data redistribution and post-processing phase
(S3) Each node reads data from disk bucket by bucket, partitions them into P-1 bins by the splitters, and sends them to the corresponding nodes. At the same time it gathers, sorts, and stores into disk the incoming data from other nodes bucket by bucket. The process continues until all data in the disk are read out and there are no more incoming data.

An example is given in Fig. 1 to illustrate the pivot selection for two processors (P=2) with N=40 keys stored in each processor. The master node gathers histograms (Fig. 1(a)), merges them, and computes the accumulated sums (Fig. 1(b)). The pivot (in this case only one pivot is needed) is the upper boundary value of fourth histogram bin with the accumulated sum of 41, which is closest to the even partitioning point of 40. Now, those keys in the shaded bins are exchanged with the other processor. P_0 and P_1 then have 41 & 39 keys after redistribution, respectively.

The complexity of the algorithm will now be estimated. In phase 1, step S1.1 requires the time to read all data from local disk (T_{read}). S1.2 requires the time to calculate histogram ($T_{comp.hist}$), to bucket-sort (T_{bucket}), and to write data to disk (T_{write}). S2 requires the communication time to send and receive histogram information and splitters ($T_{comm.hist}$). In phase 2, S3 requires the time to read data from disk bucket by bucket (T_{read}), the time to classify and send to the corresponding nodes (T_{comm}), the time to sort the incoming data (T_{sort}), and the time to write the sorted data onto disk (T_{write}). The expected time can be estimated as below:

$$T_{total} = (T_{read} + T_{comp.hist} + T_{bucket} + T_{write} + T_{comm.hist}) + (T_{read} + T_{comm} + T_{sort} + T_{write})$$

It is simplified as

$$T_{total} = 2T_{read} + 2T_{write} + T_{comm} + T_{sort} + T_{etc} \qquad (1)$$

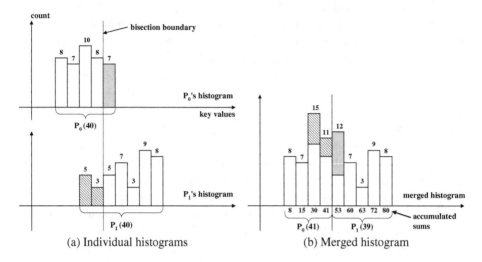

Fig. 1. Example of pivot selection.

where $T_{etc} = T_{comp.hist} + T_{comm.hist} + T_{bucket}$ Let B_r, B_w, B_c, B_s, and B_b be the disk read bandwidth, disk write bandwidth, network bandwidth, the number of data that can be sorted by radix sort per unit time, and the number of data that can be processed by bucket sort in unit time, respectively. They are all constants in general. By substituting the parameters, Eq. 1 is rewritten as follows:

$$T_{total} = \frac{2N}{P}(\frac{1}{B_r}+\frac{1}{B_w})+(\frac{P-1}{P})\cdot\frac{2N}{P}\cdot\frac{1}{B_c}+\frac{N}{P}\cdot\frac{1}{B_s}+\frac{N}{P}\cdot\frac{1}{B_b}$$

$$= \frac{N}{P}\cdot\{\frac{2}{B_r}+\frac{2}{B_w}+(\frac{P-1}{P})\cdot\frac{2}{B_c}+\frac{1}{B_s}+\frac{1}{B_b}\} \tag{2}$$

$$\approx O(\frac{N}{P})$$

In Eq. 2 since $\{\frac{2}{B_r}+\frac{2}{B_w}+\cdots+\frac{1}{B_b}\}$ is a constant with many processors, the running time of the algorithm is approximately $O(\frac{N}{P})$.

4   Experiments and Discussion

The algorithms are implemented on a PC cluster. It consists of 16 PCs with 933 MHz Pentium III CPUs interconnected by a Myrinet switch. Each PC runs under linux with 256MB RAM and 20GB hard disk. The programs are written in C language with MPI communication library (MPICH-GM). Experiments are performed on the cluster of up to $P=16$ nodes, and the size of input data ranges from $N=256MB$ to $N=1GB$ per node. Input keys are 32-bit integers synthetically generated with three distribution functions (*uniform*, *gauss*, and *stagger*). More distribution will be included in the near future.

Fig. 2, 3, and 4 show the execution times of the algorithms for uniform, gaussian, and staggered distribution. *Sample, histogram*, and *NOW* in the figures represent sample sort, histogram-based sort, and generic NOW-Sort, respectively. For uniform and staggered distribution as observed in Fig. 2 and 4, generic NOW-Sort outperforms the two distribution-insensitive sorting schemes since it does not have any overhead to reflect data distribution characteristics during the sort. However, it delivers quite inferior performance for gaussian distribution when more than two processors are used (see Fig. 3). *Optimum* sample sort and histogram-based sort have about an order of magnitude speedup in the case, up to 63% reduction in the execution time compared to NOW-sort, and the histogram-based algorithm has a shorter execution time by 13.5% to 20% than sample sort, as observed in Fig. 5. The performance improvement comes from the way to classify keys. While the histogram-based algorithm classifies keys using prefix of each key, with the complexity of $O(N)$, the sampling adopts binary search for the classification with the complexity of $O(N \log P)$. The job consumes up to 20% of the overall execution time. In all cases the histogram-based algorithm has better performance than the sample sort.

Three different sample counts of $2P(P-1)$, $\log N$, and $\sqrt{N}$ of the total N data, are chosen for *regular* [15], *mid-range* [8], and *optimum* degrees [11], respectively. Fig. 5 and 6 show the respective execution times and the degree of load imbalance of the sorting. Load distribution is well balanced with $\sqrt{N}$ sampling rate among them. Load balancing of histogram-based sort is as good as that of sample sort with $\sqrt{N}$ sampling rate. The term *optimum* comes from the fact that it produces the best result in our experiments, hinted by quicksort and quickselect achieving the shortest execution time when the sample count is $\sqrt{N}$ [11]. Regular sample judges with insufficient information, thus, the load allocation is very poor, whereas mid-range and optimum maintain well balanced load among processors. We can easily tell the reason from Fig. 7 which analyzes the execution times of the heaviest loaded processors: The regular sampling has the greatest computation time, the most data communication, and the longest disk access time, all due to having the largest number of data to process.

Load balancing in the two parallel sorting methods is maintained fairly well when histogramming and $\sqrt{N}$ samples are used. In case of gaussian distribution, the maximum deviation of load from the perfectly even distribution for all three distributions is lesser than 1% as observed in Fig. 6(b). The load imbalance is also minimum in the histogram-based algorithm for other distribution functions.

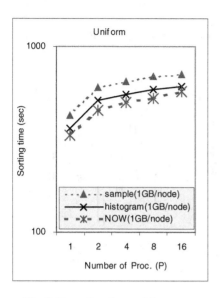

Fig. 2. Execution times of three sorts (sample, histogram, and NOW) for *uniform* distribution.

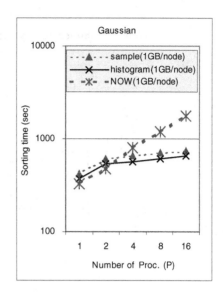

Fig. 3. Execution times of three sorts (sample, histogram, and NOW) for *gaussin* distribution.

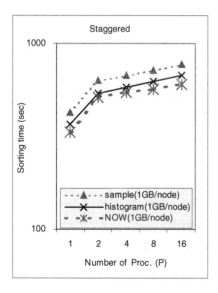

Fig. 4. Execution times of three sorts (sample, histogram, and NOW) for *staggered* distribution.

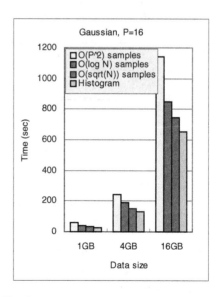

Fig. 5. Execution times of histogram- based sort and sample sort for *gaussian* distribution. Sample sort uses three different sampling rates.

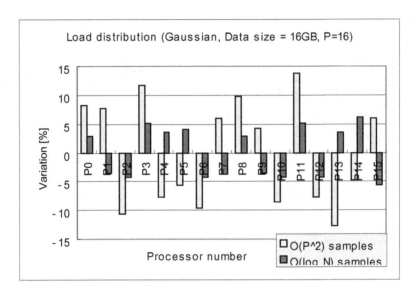

(a) Sample sort with different sample counts.

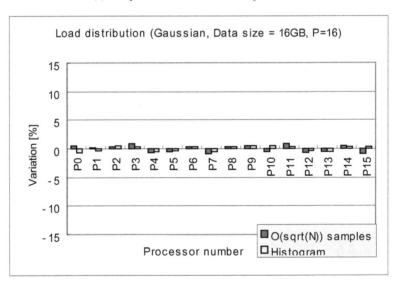

(b) Optimal sample sort and histogram-based sort.

Fig. 6. Degree of load imbalance in participating processors for *gaussian* distribution.

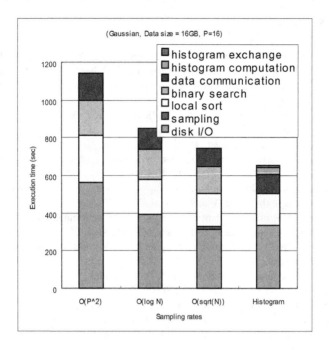

Fig. 7. Analysis of execution time for *gaussian* distribution.

5   Conclusion

In this paper we report enhanced external sorting algorithms that can be applied to data with arbitrary distribution. Sample sorting is used, and histogram-based NOW-Sort is developed. Both algorithms partition and redistribute data so that all processors have even load. They differ in the way to select the splitters for the distribution. The algorithms require 2R2W disk accesses regardless of the size of data and the number of processors. Thus, they are as fast as NOW-Sort, and the execution time reduces to an order of magnitude with respect to NOW-Sort for *non-uniform* data.

   The difficulty in the algorithms is to find a proper number of histogram bins, and the size of buckets in the redistribution process, both of which should be chosen to maximally use the main memory. We are developing a better method to classify data in the first phase of the sample sort, and extending the experiments to include other distribution functions.

References

1. S. G. Akl: The design and analysis of parallel algorithms, Chap. 4, Prentice Hall (1989)
2. A. C. Arpaci-Desseau, R. H. Arpaci-Desseau, D. E. Culler, J. M. Hellerstein, and D. A. Patterson: High-performance sorting on networks of workstations. ACM SIGMOD '97, Tucson, Arizona (1997)
3. K. Batcher: Sorting networks and their applications. Proc. AFIPS Spring Joint Computer Conference 32, Reston, VA (1968) 307-314
4. A. A. Dusseau, R. A. Dusseau, D. E. Culler, J. M. Hellerstein and D. A. Patterson: Searching for the sorting record: experiences in tuning NOW-Sort. Proc. SIGMETRICS Symp. Parallel and Distributed Tools (1998) 124-133
5. U. Fayyad and R. Uthurusamy: Evolving data mining into solutions for insights. Communications of the ACM, Vol. 45, No. 8 (2002) 29-31
6. M. Jeon and D. Kim: Parallel merge sort with load balancing. Int'l Journal of Parallel Programming, Kluwer Academic Publishers, Vol. 31, No.1 (2003) 21-33
7. D. E. Knuth: The Art of Computer Programming, Volume III: Sorting and Searching, Addison-Wesley (1973)
8. J.-S. Lee, M. Jeon, and D. Kim: Partial sort. Proc. Parallel Processing System, Vol. 13, No. 1, Korea Information Science Society (2002) 3-10
9. S.-J. Lee, M. Jeon, D. Kim, and A. Sohn: Partitioned parallel radix sort. Journal of Parallel and Distributed Computing, Academic Press, Vol. 62 (2002) 656-668
10. X. Li, et. al.: A practical external sort for shared disk MPPs. Proc. Supercomputing '93 (1993) 666-675.
11. C. C. Mcgeoch and J. D. Tygar: Optimal sampling strategies for quicksort. Random Structures and Algorithms, Vol. 7 (1995) 287-300
12. G. E. Moore: Cramming more components onto integrated circuits. Electronics, Vol. 38, No. 8 (1965)
13. F. Popovici, J. Bent, B. Forney, A. A. Dusseau, R. A. Dusseau: Datamation 2001: A Sorting Odyssey. In Sort Benchmark Home Page.
14. J. Porter: Disk trend 1998 report. http://www.disktrend.com/pdf/portrpkg.pdf
15. R. Raman: Random sampling techniques in parallel computation. Proc. IPPS/SPDP Workshops (1998) 351-360
16. D. Taniar and J. W. Rahayu: Sorting in parallel database systems. Proc. High Performance Computing in the Asia Pacific Region, 2000: The Fourth Int'l Conf. and Exhibition Vol. 2 (2000) 830-835
17. L. M. Wegner, J. I. Teuhola: The external heapsort. IEEE Trans. Software Engineering, Vol. 15, No. 7 (1989) 917-925
18. J. Wyllie: SPsort: How to sort a terabyte quickly. Technical Report, IBM Almaden Lab., http://www.almaden.ibm.com/cs/gpfs-spsort.html (1999)
19. http://research.microsoft.com/barc/SortBenchmark, In Sort Benchmark Home Page.

Distributed Genetic Algorithm for Inference of Biological Scale-Free Network Structure

Daisuke Tominaga, Katsutoshi Takahashi, and Yutaka Akiyama

Computational Biology Research Center,
National Institute of Advanced Industrial Science and Technology,
Aomi 2-43, Koto, Tokyo, 135-0064 Japan,
{tomianga,sltaka,akiyama}@cbrc.jp, http://www.cbrc.jp/

Abstract. We propose an optimization algorith based on parallelized genetic algorithm (GA) for inference biological scale-free network. The optimization task is to infer a structure of biochemical network only from time-series data of each biochemical element. This is a inverse problem which can not be solved analytically, and only heulistic searches such as GA, simulated annealing, etc. are practically effective. We applied our algorithm for several case studies to prove its effectiveness. The results shows high parallelization efficiency of our GA based algorithm and importance of scale-free property.

1 Introduction

The minimum unit of life is a cell. To infer internal mechanisms of a cell is considered that to reveal the principle of life. There are very large numbers of chemical compounds in a cell, and interactions among them form many complex networks, such as metabolic systems, gene regulatory networks, protein interaction networks, etc. Schemes of such networks are essence of life and developing a inference method for biochemical networks is very important.

Recent experimental methods of molecular biology make possible to obtain a huge amount of data from living cells directly. For example, DNA microarray is one of such novel method and can measure expression levels of thousands of genes (mRNA) at once, and interactions between genes (regulatory relationships) can be estimated from changes of expression levels in different conditions or in time. In the case, observed data are vectors contains thousands of real numbers. Estimated interaction networks also contains thousands of elements (genes). Precise physical models of networks such as simultaneous differential equations generally based on the mass-action low or michaeris-menten's low of enzyme reactions are not realistic or practical for biochemical systems because its may be very stiff or can not be solved for large system. Although bionary network model (directed graph or connection matrix) are simple and scalable, and used for biochemical networks in many case, these distributed models can not deal with continuous values represent system components and continuous time. For such networks, simple differential equation models ignoring details of physical phenomena are effective to understand and represent schemes of networks.

A. Veidenbaum et al. (Eds.): ISHPC 2003, LNCS 2858, pp. 214–221, 2003.

To infer a network structure from observed time series data, we developed numerical optimization algorithm for canonical differential equation representing a biochemical network. The equation we introduced is the S-system [6, 8]. It has simple form and easily be treated by computer algorithm because it has canonical form. Our algorithm based on Genetic Algorithm (GA) [2] optimizes a S-system model to fit to observed time series data of biochemical elements such as expression levels of genes, amount of proteins or concentration of metabolites.

The problem, fitting a model to observed data, is inverse problem [4, 7]j Many possible interaction schemes (network structure) will show very similar behavior and it is impossible to choice which one is the best model. We introduced and compared two method to our algorithm, one is simple threshold method and the other is assumption of scale-free property. However, it has not been proved that all biological network have scale-free property, some scale-free networks found in microorganisms[5]. Typical type of biological networks is expected which has scale-free property and developing inference algorithm for scale-free network is important for biochemical studies and data analysis. We estimated effectiveness of our algorithm with two link truncation method for the inverse problem in case studies.

2   Method

The S-system [6, 8] is a simultaneous differential equation having canonical form which was developed to represent biochemical network systems, Each variable in a S-system model correspond to each component in a network. Its canonical form are defined by a set of real value parameters. By solving a S-system numerically (numerical integration), time series of each variables are obtained. Then, the task of the optimization is finding a set of parameters defining a S-system whose behavior matches to given (experimentally observed) time series data [4, 7].

2.1   S-system

The S-system is defined as follows:

$$\frac{dX_i}{dt} = \alpha_i \prod_{j=1}^{n} X_j^{g_{ij}} - \beta_i \prod_{j=1}^{n} X_j^{h_{ij}} \tag{1}$$

where X_i is a variable which represents network component such as expression level of mRNA or amount of protein, and so on, t is the time. n is the number of components in a network to be modeled. In an ordinary way, it is the number of genes or proteins in a target system. right-hand side of the equation contains two terms, the first term represents total increasing effect of all other variables $X_j (j \neq i)$ to X_i, the second is total decreasing effect by all other variables. α_i and β_i are rate constants having positive or zero value. g_{ij} represents a influence of X_j to increasing processes of X_i, h_{ij} represents a influence of X_j to decreasing processes of X_i. If g_{ij} is positive, X_j accelerating synthesis processes of X_i. If it is zero, it has no effect to synthesis processes. If it is negative, it suppress synthesis process of X_i. This has different meaning in the case of h_{ij} having positive value. Positive h_{ij} means decomposition of X_i is accelerated by X_j. Negative h_{ij} represents X_j suppress decomposition processes of X_i. If both g_{ij} and h_{ij} have zero value, it represents no direct interaction from X_j to X_i.

Signs and values of exponential parameters g_{ij} and h_{ij} represent links in a network structure, such as gene regulations. Defining parameter values of the S-system is equal to estimate a network structure of target system. Although large numbers of different networks (S-system models having different sets of parameters) show similar behavior, realistic structures as a biochemical systems are fewer. By introducing some hypothesis, the inverse problem become easier than original very hard problem. We introduce assume that biochemical systems are Scale-free network [1] in which the number of components having k links in a network is decreasing proportionally to $k^{-\gamma}$ [5]. The number of links (the number of g_{ij} and h_{ij} which have non-zero value) is very limited by this assumption. It is reported that $\gamma = 2$ in typical biochemical systems.

The S-system is numerically solved very fast and precisely by using Taylor expansion in log space[3].

2.2 Cost Function

The task is find a S-system model which minimizes a difference between given time-course data and simulation result of the model. The cost function to be minimized is defined as follows:

$$f = \sum_{i=1}^{n} \sum_{t=1}^{T} \frac{X_i^t - x_i^t}{x_i^t} \tag{2}$$

where n is the number of variables, i is suffix of variables, t is suffix of sampling points, X_i^t is simulated value of a variables from a model to be evaluated by the function and x_i^t is observed value of a variables. If a model (S-system differential equation) can not be solved nemerically in some region of the domain, the function is not be defined in such region. The function (eq. 2) is not continuous.

2.3 Genetic Algorithm

In this study, cost function to be optimized is difference between behavior of a model and given time series data, and the optimization is search a model (a set of S-system parameter) to minimize the difference. This is a inverse problem [7].

The genetic algorithm (GA) is one possible approach for the inverse problem. It can optimize highly non linear function in high dimensional real space. The number of parameters to be estimated to fit given time series data is $2n(n+1)$, and it is dimension of search space.

Design of a Individual. An individual in GA context is a candidate for optimized solution. In this case, it is a one model of S-system or a set of $2n(n+1)$

real value parameters (real vector in $2n(n+1)$ dimensional space) which define a model. This means that the optimization (search) is an inverse problem which is done in $2n(n+1)$ dimentional space. A group of individual called population in GA context applied GA operations and will reach to optimal point step by step.

GA Operations. GA mimic evolution of life to optimize a function. It contains three operators which to be applied to group of for final solution individuals, evaluation, crossover and mutation. Evaluation is to score each individual. In many case and this study, it is calculation of cost function. Crossover operation generate new individual from 'parent' individuals. Mutation operations changes parts of individuals randomly. We introduce SPX (simplex crossover) operation which do crossover and mutation at once. SPX generates new child individual from $D+1$ parents (D is dimension of search space) and keeps variance and arithmetic mean of parent individual group. It can avoid convergence in early stage in the optimization and make possible global search over local minima.

The optimization steps in GA are (1) make P individuals randomly. Each individual contains a set of S-system parameters, (2) evaluate all individuals, (3) make $P-1$ individuals by SPX, (4) replace $P-1$ individuals in parent group except the best individual by new $P-1$ new child individuals, and then (5) return to step (2). Just after evaluation, if sufficiently good individual is found, optimization loop is terminate. Output of the algorithm is the last good individual.

3   Case Study

We applied our algorithm to a case study to check effectiveness of our algorithm and scale-free assumption. 10 sets of time series data were prepared and given to our algorithm. One set of its are shown in Fig. 1. This set contain 105 sampling points (21 time points *times* 5 components) and are calculated from a S-system

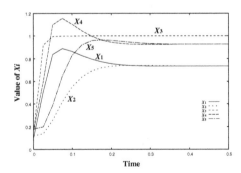

Fig. 1. Time-series data made by calculation from Table 1. Initial values are: $X_1 = 0.1$, $X_2 = 0.12$, $X_3 = 0.14$, $X_4 = 0.16$, $X_5 = 0.18$.

Table 1. A set of S-system parameters which represents a network shown in Fig. 2.

i	α_i	g_{i1}	g_{i2}	g_{i3}	g_{i4}	g_{i5}	β_i	h_{i1}	h_{i2}	h_{i3}	h_{i4}	h_{i5}
1	5.0	0.0	0.0	1.0	0.0	-1.0	10.0	2.0	0.0	0.0	0.0	0.0
2	10.0	2.0	0.0	0.0	0.0	0.0	10.0	0.0	2.0	0.0	0.0	0.0
3	10.0	0.0	-1.0	0.0	0.0	0.0	10.0	0.0	-1.0	2.0	0.0	0.0
4	8.0	0.0	0.0	2.0	0.0	-1.0	10.0	0.0	0.0	0.0	2.0	0.0
5	10.0	0.0	0.0	0.0	2.0	0.0	10.0	0.0	0.0	0.0	0.0	2.0

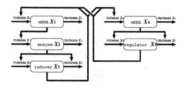

Fig. 2. A model of a typical gene expression scheme

model which is defined by a set of parameters shown in Table 1. This model represents a typical scheme of gene regulatory network shown in Fig. 2. other nine sets were calculated from same S-system model by changing initial values of components. All sets of initial values are shown in Table 2.

We compared three methods which are (1) Simple GA introducing only SPX, (2) Simple GA introducing only SPX and parameter threshold and, (3) Simple GA introducing only SPX and Scale-free assumption. In the method (2), after SPX and before evaluation, all g_{ij} and h_{ij} in all individuals are checked, and reset a value of parameter to zero when the value of parameter is less than a given threshold. We set the threshold to 0.2. In the method (3), we set γ to 2.0.

Other optimizing conditions were: GA population was 50, search range for α_i was $[0, 10]$, search range for g_{ij} was $[-3, 3]$. When a average of squared error between an individual and given data at all sampling points was less than 0.05, or GA loop iteration reached 20000 times, the optimization was terminated.

Table 2. 10 sets of initial values to prepare 10 sets of time-series data.

X_1	X_2	X_3	X_4	X_5
0.1	0.12	0.14	0.16	0.18
1.0	0.12	0.14	0.16	1.0
0.1	1.0	0.14	0.16	0.18
0.1	0.12	1.0	0.16	0.18
0.1	0.12	0.14	1.0	0.18
0.1	0.12	0.14	0.16	1.0
1.0	1.0	0.14	0.16	0.18
0.1	1.0	1.0	0.16	0.18
0.1	0.12	1.0	1.0	0.18
0.1	0.12	0.14	1.0	1.0

Table 3. Comparison of three algorithms. a) A number of runs of GA optimization, b) Conversion ratio (terminated GA runs before maximum generation), c) Average number of generations, d) False positive, e) False negative

algorithm	a	b	c	d	e	d + e
1	64	0.438	11064	4	2	6
2	48	0.0	20000	10	1	11
3	64	0.297	14078	3	0	3

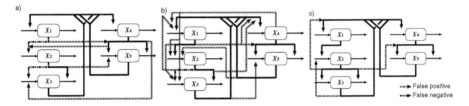

Fig. 3. Network structures obtained by a) algorithm 1 (normal method), b) algorithm 2 (threshold cutting), c) algorithm 3 (scale-free). Dashed lines represent false positive link which do not exist in original model but found in obtained model. Dotted lines are false negative links which are in original model and not found in obtained model. While false negative links must be avoided because it means the algorithm can not find links which originally exists, false positive links are acceptable.

Only $alpha_i$ and g_{ij} were optimized. The total number of optimized parameter (dimension of search space) is 60.

The results are shown in Table 3. In each result of three methods, we analyzed frequency of non-zero parameters and defined structure of a network. Obtained three structures shown in Fig. 3. All obtained structure were compared with original structure shown in figure fig:original. False-positive links which is contained obtained structure but not included in original structure, and False-negative which is not contained obtained structure but originally exist were counted and shown in Table 3. Both of false-positive and false-negative are minimized by scale-free assumption. Especially, false-negative link is eliminated. Successful convergence were more than simple GA method, and average number of GA iteration ('generation') were less than threshold method. Scale-free assumption make optimization faster. In the case of threshold method, sufficiently good individual was not obtained in 48 runs.

4 Conclusion

4.1 Optimization

Under the assumption of scale-free property, our optimization algorithm can find all links in original model. Since false-negative links must be avoided in real situations in which researchers are estimating whether interaction is or not between components in target systems, scale-free property of biochemical systems are very significant as shown in table 3. Simple cut off method by threshold (method (2) in table 3) has worse effect than Simple GA without any link trancation. The reason why threshold method has negative effecet may be the threshold value is fixed. Low threshold value allows wide variety of network structure can exisits in inidividual group, and high threshold value is hard restriction because more links will be eliminated. Changing threshold value like simulated annealing may have better effecet.

Generally, most significant advantage of GA is that it can search all over the searching domain and can avoid trapping local minima, however, GA do local search because wide part of the domain is covered by regions in which the cost function can not be calculated and the value of the function defined as zero. In these regions S-system models are very stiff or collapsed. This is a defect of the S-system because it has power-low form and easily become stiff and causes floating point overflow. Especially networks containing many links are stiff. Our algorithm introducing scale-free property may avoid these regions.

We are going to improve our algorithm to be able to applicable to larger and realistic biochemical systems. Scale-free property is just a statistical property and not considering topology of networks. Biological networks have some typical types of topology (feedback loop, cascade, etc.) and time series data may reflect it. Regulation order among network components will be estimated roughly by fitting functions such as Gaussian or Sigmoid to given time series data. By this rough estimation prior to GA optimization, a number of links to be estimated will be limited, invalid stiff region in the domain can be reduced, and scalability of our algorithm will be improved.

The island model or other types of distributed GA have possibility to improve optimization time by communications. The possibility depend significantly on profile of the function to be optimized and is unpredictable in our case study bacause the function is not continuous. S-system has a power-low form and easy to become stiff. It can be solved in very limitted region in definition domain. In the other region in the domain, it can not be solved and the cost function (eq. 2)

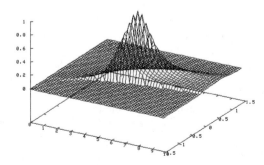

Fig. 4. Profile of the function to be optimized in case studies. X and y axises are $alpha_1$ and g_{11} respectively. Other S-system parameters are fixed as shown in Talbe 1 and the cost function is solved and plotted.

value is can not be defined (Fig. 4). This undefined region make the task hard. Traditional gradient method can not be applied and only heulistic searches are effective. GA is one of population search methods and has better performance than other sequential search such as simulated annealing.

4.2   Parallel Computation

Genetic Algorithm we introduced in this study is a variation of distributed Genetic Algorithm in which large population is devided into sub groups and each group is optimized independently. Generally good individuals will be exchanged among sub groups in distributed GA, however, our GA is mostly simplified with no exchange. In thish study, multiple runs of GA was done by parallel computing using MPI and SCore on Linux PC cluster. GA processes had no communication each other after initialization throughout optimization. Results of each process are written in NFS mounted file system. Then performance of the optimization as parallel computing is almost ideal efficiency. Elapse time of optimization equals to the latest optimization process among multiple runs. If a variance of CPU times of each optimization processes is large, elapse time will be long. In this case study, the variance of CPU times is very small. It is a great adavantage of our GA having no communications while optimization.

References

1. Reka Albert and Albert-Laszlo Barabasi. Power-law distribution of the world wide web. *Science*, 287(2115a), 2001.
2. D. E. Goldberg. *Genetic Algorithms in Search, Optimization, and Machine learning*. Addiso-Wesley, Reading, M.A., USA, 1989.
3. D. H. Irvine and M. A. Savageau. Efficient solution of nonlinear ordinary differential equations expressed in s-system canonical form. *SIAM Journal of Numerical Analysis*, 27:704–735, 1990.
4. Shinichi Kikuchi, Daisuke Tominaga, Masanori Arita, Katsutoshi Takahashi, and Masaru Tomita. Dynamic modeling of genetic networks using genetic algorithm and s-system. *Bioinformatcis*, 19(5):643–650, 2003.
5. J. Podani, Z.N. Oltvai, H. Jeong, B.Tombor, A.-L. Barabasi, and E. Szathmary. Comparable system-level organization of archaea and eukaryotes. *Nature Genetics*, 29:54–56, 2001.
6. Michael A. Savageau. *Biochemical System Analysis. A Study of Function and Design in Molecular Biology*. Addiso-Wesley, Reading, M.A., USA, 1976.
7. Daisuke Tominaga, Nobuto Koga, and Masahiro Okamoto. Efficient numerical optimization algorithm based on genetic algorithm for inverse problem. *Proc. of Genetic and Evolutionary Computation Conference (GECCO 2000)*, pages 251–258, 2000.
8. Everhard O. Voit. *Computational Analysis of Biochemical Systems*. Cambridge University Press, 2000.

Is Cook's Theorem Correct for DNA-Based Computing?

Towards Solving NP-complete Problems on a DNA-Based Supercomputer Model

Weng-Long Chang[1], Minyi Guo[2], and Jesse Wu[3]

[1] Department of Information Management,
Southern Taiwan University of Technology,
Tainan County, Taiwan,
changwl@csie.ncku.edu.tw
[2] Department of Computer Software,
The University of Aizu,
Aizu-Wakamatsu City, Fukushima 965-8580, Japan,
minyi@u-aizu.ac.jp
[3] Department of Mathematics,
National kaohsiung Normal University,
Kaohsiung 802, Taiwan,
jwu@nknucc.nknu.edu.tw

Abstract. Cook's Theorem [4, 5] is that if one algorithm for an NP-complete problem will be developed, then other problems will be solved by means of reduction to that problem. Cook's Theorem has been demonstrated to be right in a general *digital electronic* computer. In this paper, we propose a DNA algorithm for solving the *vertex-cover problem*. It is demonstrated that if the size of a reduced NP-complete problem is equal to or less than that of the vertex-cover problem, then the proposed algorithm can be directly used for solving the reduced NP-complete problem and Cook's Theorem is correct on DNA-based computing. Otherwise, Cook's Theorem is incorrect on DNA-based computing and a new DNA algorithm should be developed from the characteristic of NP-complete problems.

Index Terms – Biological Computing, Molecular Computing, DNA-based Computing, NP-complete Problem.

1 Introduction

Nowadays, producing roughly 10^{18} DNA strands that fit in a test tube is possible through advances in molecular biology [1]. Those 10^{18} DNA strands can be employed for representing 10^{18} bit information. Basic biological operations can be applied to simultaneously operate 10^{18} bit information. This is to say that there are 10^{18} data processors to be executed in parallel. Hence, it is very clear that biological computing can provide very huge parallelism for dealing with the problem in real world.

A. Veidenbaum et al. (Eds.): ISHPC 2003, LNCS 2858, pp. 222–233, 2003.

Adleman wrote the first paper in which it was demonstrated that DNA (*DeoxyriboNucleic Acid*) strands could be applied for figuring out solutions to an instance of the NP-complete Hamiltonian path problem (HPP) [2]. Lipton wrote the second paper in which it was shown that the Adleman techniques could also be used to solving the NP-complete satisfiability (SAT) problem (the first NP-complete problem) [3]. Adleman and his co-authors proposed *sticker* for enhancing the Adleman-Lipton model [8].

In this paper, we use *sticker* to constructing solution space of DNA library sequences for the *vertex-cover problem*. Simultaneously, we also apply DNA operations in the Adleman-Lipton model to develop a DNA algorithm. The main result of the proposed DNA algorithm shows that the vertex-cover problem is resolved with biological operations in the Adleman-Lipton model from solution space of sticker. Furthermore, if the size of a reduced NP-complete problem is equal to or less than that of the vertex-cover problem, then the proposed algorithm can be directly used for solving the reduced NP-complete problem.

2   DNA Model of Computation

In subsection 2.1, a summary of DNA structure and the Adleman-Lipton model is described in detail.

2.1   The Adleman-Lipton Model

A DNA (*DeoxyriboNucleic Acid*) is a *molecule* that plays the main role in DNA based computing [9]. In the biochemical world of large and small *molecules, polymers*, and *monomers*, DNA is a polymer, which is strung together from monomers called *deoxyriboNucleotides*. The monomers used for the construction of DNA are deoxyribonucleotides, which each deoxyribonucleotide contains three components: a *sugar*, a *phosphate* group, and a *nitrogenous* base. This sugar has five carbon atoms - for the sake of reference there is a fixed numbering of them. Because the base also has carbons, to avoid confusion the carbons of the sugar are numbered from 1' to 5' (rather than from 1 to 5). The phosphate group is attached to the 5' carbon, and the base is attached to the 1' carbon. Within the sugar structure there is a hydroxyl group attached to the 3' carbon.

Distinct nucleotides are detected only with their bases, which come in two sorts: purines and pyrimidines [1, 9]. Purines include *adenine* and *guanine*, abbreviated A and G. Pyrimidines contain *cytosine* and *thymine*, abbreviated C and T. Because nucleotides are only distinguished from their bases, they are simply represented as A, G, C, or T nucleotides, depending upon the sort of base that they have. The structure of a nucleotide is illustrated (in a very simplified way) in Fig. 1. In Fig. 1, **B** is one of the four possible bases (A, G, C, or T), **P** is the phosphate group, and the rest (the "stick") is the sugar base (with its carbons enumerated 1' through 5').

In the Adleman-Lipton model [2, 3], *splints* were used to correspond to the edges of a particular graph the paths of which represented all possible binary

Fig. 1. A schematic representation of a nucleotide.

numbers. A s it stands, their construction indiscriminately builds all splints that lead to a complete graph. This is to say that hybridization has higher probabilities of errors. Hence, Adleman et al. [8] proposed the sticker-based model, which was an abstract model of molecular computing based on DNAs with a random access memory and a new form of encoding the information, to enhance the Adleman-Lipton model.

The DNA operations in the Adleman-Lipton model are described below [2, 3, 6, 7]. These operations will be used for figuring out solutions of the vertex-cover problem. The DNA operations in the Adleman-Lipton model are described below [2, 3, 6, 7]. These operations will be used for figuring out solutions of the vertex-cover problem.

The Adleman-Lipton Model

A (test) tube is a set of molecules of DNA (i.e. a multi-set of finite strings over the alphabet $\{A, C, G, T\}$). Given a tube, one can perform the following operations:

1. *Extract*. Given a tube P and a short single strand of DNA, S, produce two tubes $+(P, S)$ and $-(P, S)$, where $+(P, S)$ is all of the molecules of DNA in P which contain the strand S as a sub-strand and $-(P, S)$ is all of the molecules of DNA in P which do not contain the short strand S.
2. *Merge*. Given tubes P_1 and P_2, yield $\cup(P_1, P_2)$, where $\cup(P_1, P_2) = P_1 \cup P_2$. This operation is to pour two tubes into one, with no change of the individual strands.
3. *Detect*. Given a tube P, say 'yes' if P includes at least one DNA molecule, and say 'no' if it contains none.
4. *Discard*. Given a tube P, the operation will discard the tube P.
5. *Read*. Given a tube P, the operation is used to describe a single molecule, which is contained in the tube P. Even if P contains many different molecules each encoding a different set of bases, the operation can give an explicit description of exactly one of them.

3   Using Sticker for Solving the Vertex-Cover Problem in the Adleman-Lipton Model

In subsection 3.1, the vertex-cover problem is described. Applying sticker to constructing solution space of DNA sequences for the vertex-cover problem is introduced in subsection 3.2. In subsection 3.3, one DNA algorithm is proposed to resolving the vertex-cover problem. In subsection 3.4, the complexity of the proposed algorithm is offered. In subsection 3.5, the range of application to famous Cook's theorem is described in molecular computing.

3.1 Definition of the Vertex-Cover Problem

Assume that G is a graph and $G = (V, E)$, where V is a set of vertices in G and E is a set of edges in G. Also suppose that V is $\{v_1, \ldots, v_n\}$ and E is $\{(v_a, v_b)| v_a$ and v_b are, respectively, vertices in $V\}$. Assume that $|V|$ is the number of vertex in V and $|E|$ is the number of edge in E. Also suppose that $|V|$ is equal to n and $|E|$ is equal to m.

Mathematically, a *vertex cover* of a graph G is a subset $V^1 \subseteq V$ of vertices such that for each edge (v_a, v_b) in E, at lease one of v_a and v_b belongs to V^1 [4, 5]. The vertex-cover problem is to find a minimum-size vertex cover from G. The problem has been shown to be a NP-complete problem [5].

The graph in Fig. 2 denotes such a problem. In Fig. 2, the graph G contains three vertices and two edges. The minimum-size vertex cover for G is $\{v_1\}$. Hence, the size of the vertex-cover problem in Fig. 2 is one. It is indicated from [5] that finding a minimum-size vertex cover is a NP-complete problem, so it can be formulated as a search problem.

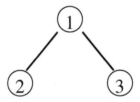

Fig. 2. the graph G of our problem.

3.2 Using Sticker for Constructing Solution Space of DNA Sequence for the Vertex Cover Problem

The first step in the Adleman-Lipton model is to yield solution space of DNA sequences for those problems solved. Next, basic biological operations are used to remove illegal solution and find legal solution from solution space. Thus, the first step of solving the vertex-cover problem is to generate a test tube, which includes all of the possible vertex covers. Assume that an n-digit binary number corresponds to each possible vertex cover to any n-vertex graph, G. Also suppose that V^1 is a vertex cover for G. If the i-th bit in an n-digit binary number is set to 1, then it represents that the corresponding vertex is in V^1. If the i-th bit in an n-digit binary number is set to 0, then it represents that the corresponding vertex is out of V^1.

By this way, all of the possible vertex covers in G are transformed into an ensemble of all n-digit binary numbers. Hence, with the way above, Table 1 denotes the solution space for the graph in Fig. 2. The binary number 000 in Table 1 represents that the corresponding vertex cover is empty. The binary numbers 001, 010 and 011 in Table 1 represent that those corresponding vertex

Table 1. The solution space for the graph in Fig 2.

3-digit binary number	The corresponding vertex cover
000	$\emptyset$
001	$\{v_1\}$
010	$\{v_2\}$
011	$\{v_2, v_1\}$
100	$\{v_3\}$
101	$\{v_3, v_1\}$
110	$\{v_3, v_2\}$
111	$\{v_3, v_2, v_1\}$

covers are $\{v_1\}$, $\{v_2\}$ and $\{v_2, v_1\}$, respectively. The binary numbers 100, 101 and 110 in Table 1 represent that those corresponding vertex covers, subsequently, are $\{v_3\}$, $\{v_3, v_1\}$ and $\{v_3, v_2\}$. The binary number 111 in Table 1 represents that the corresponding vertex cover is $\{v_3, v_2, v_1\}$. Though there are eight 3-digit binary numbers for representing eight possible vertex covers in Table 1, not every 3-digit binary number corresponds to a *legal* vertex cover. Hence, in next subsection, basic biological operations are used to develop an algorithm for removing illegal vertex covers and finding legal vertex covers.

To implement this way, assume that an unsigned integer X is represented by a binary number $x_n, x_{n-1}, \ldots, x_1$, where the value of x_i is 1 or 0 for $1 \leq i \leq n$. The integer X contains 2^n kinds of possible values. Each possible value represents a vertex cover for any n-vertex graph, G. Hence, it is very obvious that an unsigned integer X forms 2^n possible vertex cover. A bit x_i in an unsigned integer X represents the i-th vertex in G. If the i-th vertex is in a vertex cover, then the value of x_i is set to 1. If the i-th vertex is out of a vertex cover, then the value of x^i is set to 0.

To represent all possible vertex covers for the vertex-cover problem, *sticker* [8, 14] is used to construct solution space for that problem solved. For every bit, x_i, two distinct 15 base value sequences are designed. One represents the value 1 and another represents the value 0 for x_i. For the sake of convenience of presentation, assume that x_i^1 denotes the value of x_i to be 1 and x_i^0 defines the value of x_i to be 0. Each of the 2^n possible vertex covers is represented by a library sequence of 15*n bases consisting of the concatenation of one value sequence for each bit. DNA molecules with library sequences are termed library strands and a combinatorial pool containing library strands is termed a library. The probes used for separating the library strands have sequences complementary to the value sequences.

The Adleman program [14] is modified for generating those DNA sequences to satisfy the constraints above. For example, for representing the three vertices in the graph in Fig. 2, the DNA sequences generated are: $x_1^0 = $ **AAAACTCACC CTCCT**, $x_2^0 = $ **TCTAATATAATTACT**, $x_3^0 = $ **ATTCTAACTCTACCT**, $x_1^1 = $ **TTTCAATAACACCTC**, $x_2^1 = $ **ATTCACTTCTTTAAT** and $x_3^1 = $ **AACATACCCCTAATC**. Therefore, for every possible vertex cover to the graph in Fig. 2, the corresponding library strand is synthesized by employing a mix-and-split combinatorial synthesis technique [15]. Similarly, for any n-vertex graph, all of the library strands for representing every possible vertex cover could be also synthesized with the same technique.

3.3 The DNA Algorithm for Solving the Vertex Cover Problem

The following DNA algorithm is proposed to solve the vertex cover problem.

Algorithm 1 *Solving the vertex cover problem.*

(1) *Input (T_0), where the tube T_0 includes solution space of DNA sequences to encode all of the possible vertex covers for any n-vertex graph, G, with those techniques mentioned in subsection 3.2.*

(2) *For $k = 1$ to m, where m is the number of edges in G.*
  Assume that e_k is (v_i, v_j), where e_k is one edge in G and v_i and v_j are vertices in G. Also suppose that bits x_i and x_j, respectively, represent v_i and v_j.
  (a) $\theta^1 = +(T_0, x_i{}^1)$ and $\theta = -(T_0, x_i{}^1)$.
  (b) $\theta^2 = +(\theta, x_j{}^1)$ and $\theta^3 = -(\theta, x_j{}^1)$.
  (c) $T_0 = \cup(\theta^1, \theta^2)$.
  EndFor

(3) *For $i = 0$ to $n - 1$*
  For $j = i$ down to 0
  (a) $T_{j+1}{}^{ON} = +(T_j, x_{i+1}{}^1)$ and $T_j = -(T_j, x_{i+1}{}^1)$.
  (b) $T_{j+1} = \cup(T_{j+1}, T_{j+1}{}^{ON})$.
  EndFor
  EndFor

(4) *For $k = 1$ to n*
  (a) If (detect (T_k) = 'yes') then
  (b) Read (T_k) and terminate the algorithm.
  EndIf

EndFor

Theorem 31 *From those steps in Algorithm 1, the vertex cover problem for any n-vertex graph, G, can be solved.*

Proof. Omitted.

3.4 The Complexity of the Proposed DNA Algorithm

The following theorems describe time complexity of Algorithm 1, volume complexity of solution space in Algorithm 1, the number of the tube used in Algorithm 1 and the longest library strand in solution space in Algorithm 1.

Theorem 32 *The vertex-cover problem for any undirected n-vertex graph G with m edges can be solved with $O(n^2)$ biological operations in the Adleman-Lipton model, where n is the number of vertices in G and m is at most equal to $(n * (n - 1)/2)$.*

Proof. Omitted.

Theorem 33 *The vertex-cover problem for any undirected n-vertices graph G with m edges can be solved with sticker to construct $O(2^n)$ strands in the Adleman-Lipton model, where n is the number of vertices in G.*

Proof. Refer to Theorem 3-2.

Theorem 34 *The vertex-cover problem for any undirected n-vertices graph G with m edges can be solved with $O(n)$ tubes in the Adleman-Lipton model, where n is the number of vertices in G.*

Proof. Refer to Theorem 3-2.

Theorem 35 *The vertex-cover problem for any undirected n-vertices graph G with m edges can be solved with the longest library strand, $O(15 * n)$, in the Adleman-Lipton model, where n is the number of vertices in G.*

Proof. Refer to Theorem 3-2.

3.5   Range of Application to Cook's Theorem in DNA Computing

Cook's Theorem [4, 5] is that if one algorithm for one NP-complete problem will be developed, then other problems will be solved by means of reduction to that problem. Cook's Theorem has been demonstrated to be right in a general digital electronic computer. Assume that a collection C is $\{c_1, c_2, \cdots, c_m\}$ of clauses on a finite set U of variables, $\{u_1, u_2, \cdots, u_n\}$, such that $—c_x—$ is equal to 3 for $1 \leq x \leq m$. The 3-satisfiability problem (3-SAT) is to find whether there is a truth assignment for U that satisfies all of the clauses in C. The simple structure for the 3-SAT problem makes it one of the most widely used problems for other NP-completeness results [4]. The following theorems are used to describe the range of application for Cook's Theorem in molecular computing

Theorem 36 *Assume that any other NP-complete problems can be reduced to the vertex-cover problem with a polynomial time algorithm in a general electronic computer. If the size of a reduced NP-complete problem is not equal to or less than that of the vertex-cover problem, then Cook's Theorem is uncorrected in molecular computing.*

Proof. We transform the 3-SAT problem to the vertex-cover problem with a polynomial time algorithm [4]. Suppose that U is $\{u_1, u_2, \cdots, u_n\}$ and C is $\{c_1, c_2, \cdots, c_m\}$. U and C are any instance for the 3-SAT problem. We construct a graph $G = (V, E)$ and a positive integer $K \leq |V|$ such that G has a vertex cover of size K or less if and only if C is satisfiable.

For each variable u_i in U, there is a truth-setting component $T_i = (V_i, E_i)$, with $V_i = \{u_i, u_i{}^1\}$ and $E_i = \{\{u_i, u_i{}^1\}\}$, that is, two vertices joined by a single edge. Note that any vertex cover will have to contain at least one of u_i and

$u_i{}^1$ in order to cover the single edge in E_i. For each clause c_j in C, there is a satisfaction testing component $S_j = (V_j{}^1, E_j{}^1)$, consisting of three vertices and three edges joining them to form a triangle:

$$V_j{}^1 = \{a_j[j], a_2[j], a_3[j]\}$$

$$E_j{}^1 = \{\{a_1[j], a_2[j]\}, \{\{a_1[j], a_3[j]\}, \{\{a_2[j], a_3[j]\}\}.$$

Note that any vertex cover will have to contain at least two vertices from $V_j{}^1$ in order to cover the edges in $E_j{}^1$.

The only part of the construction that depends on which literals occur in which clauses is the collection of communication edges. These are best viewed from the vantage point of the satisfaction testing components. For each clause c_j in C, assume that the three literals in c_j is denoted as x_j, y_j, and z_j. Then the communication edges emanating from S_j are given by:

$$E_j{}^2 = \{\{a_1[j], x_j\}, \{a_2[j], y_j\}, \{a_3[j], z_j\}\}.$$

The construction of our instance to the vertex-cover problem is completed by setting $K = n + 2 * m$ and $G = (V, E)$, where

$$V = (\bigcup_{i=1}^{n} V_i) \cup (\bigcup_{j=1}^{m} V_j^1)$$

and

$$E = (\bigcup_{i=1}^{n} E_i) \cup (\bigcup_{j=1}^{m} E_j^1) \cup (\bigcup_{j=1}^{m} V_E^2).$$

Therefore, the number of vertex and the number of edge in G are, respectively, $(2*n+3*m)$ and $(n+6*m)$. Algorithm 1 is used to determine the vertex-cover problem for G with $2^{2*n+3*m}$ DNA strands. Because the limit of DNA strands is 10^{21}, n is equal to or less than 15 and m is also equal to or less than 15. That is to say that Algorithm 1 at most solves the 3-SAT problem with 15 variables and 15 clauses. However, a general digital electronic computer can be applied to directly resolve the 3-SAT problem with 15 variables and 15 clauses. Hence, it is at once inferred that if the size of a reduced NP-complete problem is not equal to or less than that of the vertex-cover problem, then Cook's Theorem is uncorrected in molecular computing.

From Theorem 33-36, if the size of a reduced NP-complete problem is equal to or less than that of the vertex-cover problem, then Algorithm 1 can be directly used for solving the reduced NP-complete problem. Otherwise, a new DNA algorithm should be developed according to the characteristic of NP-complete problems.

4   Experimental Results of Simulated DNA Computing

We finished the modification of the Adleman program [14] in a PC with one Pentium(R) 4 and 128 MB main memory. Our operating system is Window 98 and the compiler is C++ Builder 6.0. This modified program is applied to generate DNA sequences for solving the vertex-cover problem. Because the source code of the two functions *srand48()* and *drand48()* was not found in the *original* Adleman program, we use the standard function *srand()* in C++ builder 6.0 to replace the function *srand48()* and added the source code to the function *drand48()*. We also added subroutines to the Adleman program for simulating biological operations in the Adleman-Lipton model in Section 2. We add subroutines to the Adleman program to simulate Algorithm 1 in subsection 3.3.

The Adleman program is used for constructing each 15-base DNA sequence for each bit of the library. For each bit, the program is applied for generating two 15-base random sequences (for 1 and 0) and checking to see if the library strands satisfy the seven constraints in subsection 3.2 with the new DNA sequences added. If the constraints are satisfied, the new DNA sequences are greedily accepted. If the constraints are not satisfied then mutations are introduced one by one into the new block until either: (A) the constraints are satisfied and the new DNA sequences are then accepted or (B) a threshold for the number of mutations is exceeded and the program has failed and so it exits, printing the sequence found so far. If n-bits that satisfy the constraints are found then the program has succeeded and it outputs these sequences.

Consider the graph in Fig. 2. The graph includes three vertices: v_1, v_2 and v_3. DNA sequences generated by the Adleman program modified were shown in Table 2. This program, respectively, took one mutation, one mutation and ten mutations to make new DNA sequences for v_1, v_2 and v_3. With the nearest neighbor parameters, the Adleman program was used to calculate the enthalpy, entropy, and free energy for the binding of each probe to its corresponding region on a library strand. The energy was shown in Table 3. Only G really matters to the energy of each bit. For example, the delta G for the probe binding a '1' in the first bit is thus estimated to be 24.3 kcal/mol and the delta G for the probe binding a '0' is estimated to be 27.5 kcal/mol.

Table 2. Sequences chosen to represent the vertices in the graph in Fig. 2.

Vertex	5' → 3' DNA Sequence
$x_3{}^0$	ATTCTAACTCTACCT
$x_2{}^0$	TCTAATATAATTACT
$x_1{}^0$	AAAACTCACCCTCCT
$x_3{}^1$	AACATACCCCTAATC
$x_2{}^1$	ATTCACTTCTTTAAT
$x_1{}^1$	TTTCAATAACACCTC

The program simulated a mix-and-split combinatorial synthesis technique [15] to synthesize the library strand to every possible vertex cover. Those library strands are shown in Table 4, and represent eight possible vertex covers: $\emptyset, \{v_1\}, \{v_2\}, \{v_2, v_1\}, \{v_3\}, \{v_3, v_1\}, \{v_3, v_2\}$ and $\{v_3, v_2, v_1\}$, respectively. The program is also applied to figure out the average and standard deviation for

Table 3. The energy for the binding of each probe to its corresponding region on a library strand.

Vertex	Enthalpy energy (H)	Entropy energy (S)	Free energy (G)
$x_3{}^0$	105.2	277.1	22.4
$x_2{}^0$	104.8	283.7	19.9
$x_1{}^0$	113.7	288.7	27.5
$x_3{}^1$	112.6	291.2	25.6
$x_2{}^1$	107.8	283.5	23
$x_1{}^1$	105.6	271.6	24.3

Table 4. DNA sequences chosen represent all possible vertex covers.

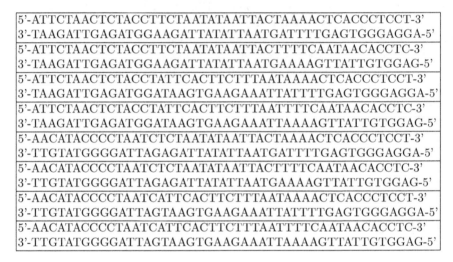

```
5'-ATTCTAACTCTACCTTCTAATATAATTACTAAAACTCACCCTCCT-3'
3'-TAAGATTGAGATGGAAGATTATATTAATGATTTTGAGTGGGAGGA-5'
5'-ATTCTAACTCTACCTTCTAATATAATTACTTTTCAATAACACCTC-3'
3'-TAAGATTGAGATGGAAGATTATATTAATGAAAAGTTATTGTGGAG-5'
5'-ATTCTAACTCTACCTATTCACTTCTTTAATAAAACTCACCCTCCT-3'
3'-TAAGATTGAGATGGATAAGTGAAGAAATTATTTTGAGTGGGAGGA-5'
5'-ATTCTAACTCTACCTATTCACTTCTTTAATTTTCAATAACACCTC-3'
3'-TAAGATTGAGATGGATAAGTGAAGAAATTAAAAGTTATTGTGGAG-5'
5'-AACATACCCCTAATCTCTAATATAATTACTAAAACTCACCCTCCT-3'
3'-TTGTATGGGGATTAGAGATTATATTAATGATTTTGAGTGGGAGGA-5'
5'-AACATACCCCTAATCTCTAATATAATTACTTTTCAATAACACCTC-3'
3'-TTGTATGGGGATTAGAGATTATATTAATGAAAAGTTATTGTGGAG-5'
5'-AACATACCCCTAATCATTCACTTCTTTAATAAAACTCACCCTCCT-3'
3'-TTGTATGGGGATTAGTAAGTGAAGAAATTATTTTGAGTGGGAGGA-5'
5'-AACATACCCCTAATCATTCACTTCTTTAATTTTCAATAACACCTC-3'
3'-TTGTATGGGGATTAGTAAGTGAAGAAATTAAAAGTTATTGTGGAG-5'
```

Table 5. The energy over all probe/library strand interactions.

	Enthalpy energy (H)	Entropy energy (S)	Free energy (G)
Average	108.283	282.633	23.7833
Standard deviation	3.58365	6.63867	2.41481

the enthalpy, entropy and free energy over all probe/library strand interactions. The energy is shown in Table 5. The standard deviation for delta G is small because this is partially enforced by the constraint that there are 4, 5, or 6 Gs (the seventh constraint in subsection 3.2) in the probe sequences.

The Adleman program is employed for computing the distribution of the types of potential mishybridizations. The distribution of the types of potential mishybridizations is the absolute frequency of a probe-strand match of length k from 0 to the bit length 15 (for DNA sequences) where probes are not supposed to match the strands. The distribution is, subsequently, 106, 152, 183, 215, 216, 225, 137, 94, 46, 13, 4, 1, 0, 0, 0 and 0. It is pointed out from the last four zeros that there are 0 occurrences where a probe matches a strand at 12, 13, 14, or 15 places. This shows that the third constraint in subsection 3.2 has been satisfied. Clearly, the number of matches peaks at 5 (225). That is to say that there are 225 occurrences where a probe matches a strand at 5 places.

5   Conclusions

Cook's Theorem is that if one algorithm for one NP-complete problem will be developed, then other problems will be solved by means of reduction to that problem. Cook's Theorem has been demonstrated to be right in a general digit electronic computer. From Theorem 33-36, if the size of a reduced NP-complete problem is equal to or less than that of the vertex-cover problem, then Cook's Theorem is right in molecular computing. Otherwise, Cook's Theorem is uncorrected in molecular computing and a new DNA algorithm should be developed from the characteristic of NP-complete problems.

Chang and Guo [11, 13] applied splints to constructing solution space of DNA sequence for solving the vertex-cover problem in the Adleman-Lipton. This causes that hybridization has higher probabilities for errors. Adleman and his co-authors [8] proposed *sticker* to decrease probabilities of errors to hybridization in the Adleman-Lipton. The main result of the proposed algorithms shows that the vertex cover problem is solved with biological operations in the Adleman-Lipton model from solution space of sticker. Furthermore, this work represents clear evidence for the ability of DNA based computing to solve NP-complete problems.

Currently, there still are lots of NP-complete problems not to be solved because it is very difficulty to basic biological operations for supporting mathematical operations. We are not sure whether molecular computing can be applied for dealing with every NP-complete problem. Therefore, in the future, our main work is to solve other NP-complete problem unsolved with the Adleman-Lipton model and the sticker model, or develop a new model.

References

1. R. R. Sinden. DNA Structure and Function. Academic Press, 1994.
2. L. Adleman. Molecular computation of solutions to combinatorial problems. Science, 266:1021-1024, Nov. 11, 1994.
3. R. J. Lipton. DNA solution of hard computational problems. Science, 268:542:545, 1995.
   A. Narayanan, and S. Zorbala. DNA algorithms for computing shortest paths. In Genetic Programming 1998: Proceedings of the Third Annual Conference, J. R. Koza et al. (Eds), 1998, pp. 718–724.
4. T. H. Cormen, C. E. Leiserson, and R. L. Rivest. Introduction to algorithms.
5. M. R. Garey, and D. S. Johnson. Computer and intractability. Freeman, San Fransico, CA, 1979.
6. D. Boneh, C. Dunworth, R. J. Lipton and J. Sgall. On the computational Power of DNA. In Discrete Applied Mathematics, Special Issue on Computational Molecular Biology, Vol. 71 (1996), pp. 79-94.

7. L. M. Adleman. On constructing a molecular computer. DNA Based Computers, Eds. R. Lipton and E. Baum, DIMACS: series in Discrete Mathematics and Theoretical Computer Science, American Mathematical Society. 1-21 (1996)

8. S. Roweis, E. Winfree, R. Burgoyne, N. V. Chelyapov, M. F. Goodman, Paul W. K. Rothemund and L. M. Adleman. Sticker Based Model for DNA Computation. 2nd annual workshop on DNA Computing, Princeton University. Eds. L. Landweber and E. Baum, DIMACS: series in Discrete Mathematics and Theoretical Computer Science, American Mathematical Society. 1-29 (1999).

9. G. Paun, G. Rozenberg and A. Salomaa. DNA Computing: New Computing Paradigms. Springer-Verlag, New York, 1998. ISBN: 3-540-64196-3.

10. W.-L. Chang and M. Guo. Solving the Dominating-set Problem in Adleman-Lipton's Model. The Third International Conference on Parallel and Distributed Computing, Applications and Technologies, Japan, 2002, pp. 167-172.

11. W.-L. Chang and M. Guo. Solving the Clique Problem and the Vertex Cover Problem in Adleman-Lipton's Model. IASTED International Conference, Networks, Parallel and Distributed Processing, and Applications, Japan, 2002, pp. 431-436.

12. W.-L. Chang, M. Guo. Solving NP-Complete Problem in the Adleman-Lipton Model. The Proceedings of 2002 International Conference on Computer and Information Technology, Japan, 2002, pp. 157-162.

13. W.-L. Chang and M. Guo. solving the 3-Dimensional Matching Problem and the Set Packing Problem in Adleman-Lipton's Mode. IASTED International Conference, Networks, Parallel and Distributed Processing, and Applications, Japan, 2002, pp. 455-460.

14. Ravinderjit S. Braich, Clifford Johnson, Paul W.K. Rothemund, Darryl Hwang, Nickolas Chelyapov and Leonard M. Adleman.Solution of a satisfiability problem on a gel-based DNA computer. Proceedings of the 6th International Conference on DNA Computation in the Springer-Verlag Lecture Notes in Computer Science series.

15. A. R. Cukras, Dirk Faulhammer, Richard J. Lipton, and Laura F. Landweber. "Chess games: A model for RNA-based computation". In Proceedings of the 4th DIMACS Meeting on DNA Based Computers, held at the University of Pennsylvania, June 16-19, 1998, pp. 27-37.

16. T. H. LaBean, E. Winfree and J.H. Reif. Experimental Progress in Computation by Self-Assembly of DNA Tilings. Theoretical Computer Science, Volume 54, pp. 123-140, (2000).

LES of Unstable Combustion
in a Gas Turbine Combustor

Jyunji Shinjo, Yasuhiro Mizobuchi, and Satoru Ogawa

CFD Technology Center, National Aerospace Laboratory of Japan
7-44-1 Jindaiji-higashimachi, Chofu, Tokyo 182-8522 JAPAN
Phone: +81-422-40-3316
FAX: +81-422-40-3328
shinjou@nal.go.jp

Abstract. Investigation of lean premixed combustion dynamics in a gas turbine swirl-stabilized combustor is numerically conducted based on Large Eddy Simulation (LES) methodology. Premixed flame modeling is based on the flamelet (G-equation) assumption. Unsteady flame behavior is successfully captured inside the combustor. The unstable flame motions are mainly governed by acoustic resonance, velocity field fluctuations and heat release fluctuations. They are in phase and coupled to sustain combustion instabilities. Future combustion control will lie in how to change these links. It is demonstrated that numerical methods based on LES are effective in comprehending real-scale combustion dynamics in a combustor.

Key Words: large-eddy simulation, premixed combustion, instabilities, gas turbine and swirl combustor.

1 Introduction

Regulations regarding pollutant emissions from combustor systems are getting more and more stringent worldwide due to environmental concerns. Conventionally, combustor systems have relied on diffusion combustion. Fuel and air are not mixed until they enter the combustor. This combustion method can be stable over a wide range of operation conditions. The flame temperature, however, is rather high because chemical reactions occur only around the stoichiometric conditions. This leads to large NOx emissions, because the amount of NOx formation is dependent on local flame temperature.

NOx reduction is strongly related to the local flame temperature. Premixed flames can be formed even the local fuel/air ratio (equivalence ratio) is in the lean or rich side. This means that the premixed flame temperature can be lower than the diffusion flame temperature. Thus, lean premixed combustion is a promising way to reduce NOx emissions from modern combustors. However, this combustion method is prone to combustion instabilities and this sometimes causes mechanical damage to the combustor system. A wide range of stable operation should be assured to use lean premixed combustion in industrial burners. Combustion control may be one solution

A. Veidenbaum et al. (Eds.): ISHPC 2003, pp. 234–244, 2003.

to this problem. Various control methods are now being investigated worldwide and understanding flame dynamics is important.

Flames in practical devices are usually used in turbulent regimes. Turbulent combustion can be characterized by complicated phenomena such as turbulence, heat conduction, diffusion, chemical reactions and so on. Mechanisms of unstable combustion are not yet fully understood due to the complexity. Experimental measurements are sometimes difficult because the temperature in a combustor is high. Optical measurement methods are useful in combustion research, but data sets are usually two-dimensional.

Recent progress in computational fluid dynamics (CFD) and computer hardware has enabled us to conduct combustion simulations. Direct numerical simulations (DNS) are now possible with complicated chemical reactions. Detailed flame structures can be investigated, but the computational domain should remain small.

Practical-scale turbulent simulations must incorporate turbulent models. When dealing with unsteady phenomena like combustion instabilities, large-eddy simulation (LES) is the most appropriate choice. LES is based on the spatially filtered Navier-Stokes equations. Grid-scale phenomena are directly solved on a numerical mesh and subgrid-scale phenomena are modeled. Flames, especially premixed flames, are usually thin and subgrid modeling is important. In LES, total integration time becomes much longer compared to Reynolds-averaged Navier-Stokes simulations (RANS) because unsteady phenomena must be observed for a longer time period. But recent and future computer performance improvement is expected to make LES simulations of combustion systems easier to conduct.

In the present study, we conduct numerical simulations of unsteady flame behavior in a gas turbine swirl-stabilized combustor to investigate the mechanism of unsteady combustion and demonstrate the effectiveness of numerical simulation in combustion LES. This will finally lead to combustion control in the near future.

2 Numerical Implementation

The target combustor is a swirl-stabilized gas turbine combustor, which is usually used in modern gas turbine systems. The key concept of flame holding is to make a recirculation zone in the center region of the combustor and push the burned gas upstream. The combustion chamber length is L=0.3m and the diameter D=0.15m in this research. The combustor has an inlet section of inner diameter of D_{in}=0.3D and outer diameter D_{out}=0.6D. The exit is contracted to 0.5D.

Methane and air are premixed and injected into the combustor. The global equivalence ratio is set around 0.55-0.6. The incoming flow is swirled to the swirl number of about 0.8 in the inlet section. Swirl is numerically given to the flow by adding a hypothetical body force, not by solving swirler vanes directly. The unburned gas enters the combustion chamber with sudden expansion.

Some part of the burned gas is re-circulated upstream and the rest of burned gas goes out through the contracted exit where the flow is accelerated. The pressure and temperature are set at 1 atm and 400-700K, respectively.

The governing equations of the system are the three-dimensional Navier-Stokes equations. Compressibility must be included because pressure wave propagation inside the combustor plays an important role in determining flame dynamics.

$$\frac{\partial \rho}{\partial t} + \frac{\partial \rho u_j}{\partial x_j} = 0$$

$$\frac{\partial \rho u_i}{\partial t} + \frac{\partial \left(\rho u_i u_j + \delta_{ij} p \right)}{\partial x_j} = \frac{\partial \tau_{ij}}{\partial x_j} \qquad (1)$$

$$\frac{\partial E}{\partial t} + \frac{\partial (E + p) u_j}{\partial x_j} = \frac{\partial \left(\tau_{ij} u_i + q_j \right)}{\partial x_j}$$

LES techniques are used to capture unsteady turbulent phenomena in the combustor. In LES, large-scale coherent structures are directly solved by numerical mesh and small-scale (subgrid-scale) eddies are modeled based on the similarity law of turbulence. Spatial filtering operation decomposes the governing equations into the two scales. For example, the filtered momentum equation is

$$\frac{\partial \overline{\rho} \tilde{u}_i}{\partial t} + \frac{\partial \overline{\rho} \tilde{u}_i \tilde{u}_j}{\partial x_j} = -\frac{\partial \overline{p} \delta_{ij}}{\partial x_j} + \frac{\partial \overline{\tau}_{ij}}{\partial x_j} - \frac{\partial \tau_{ij}^{sgs}}{\partial x_j} \qquad (2)$$

where the bar denotes averaging and the tilde Favre averaging. The last term is an unresolved subgrid term and modeled using the dynamic procedure as

$$\tau_{ij}^{sgs} = \overline{\rho} \left(\widetilde{u_i u_j} - \tilde{u}_i \tilde{u}_j \right) = -2\overline{\rho} \left(C_s \Delta \right)^2 \left| \overline{S}_{ij} \right| \overline{S}_{ij} \qquad (3)$$

where C_s is the dynamic coefficient, Δ the filter scale and $\overline{S}_{ij}$ the resolved strain rate tensor.

Premixed flames are thin when the flow turbulence intensity is not extremely high, which is the case in gas turbine combustors. In an LES grid system, grid resolution is usually not fine enough to resolve the internal structures of flame, such as temperature layer and reaction zone. Thus a premixed flame model should be added to the above equations.

One simple way to model a premixed flame is to assume that the flame thickness is infinitely small and the flame front is a discontinuity surface between burned (G=1) and unburned (G=0) gases. This approach is called the flamelet (G-equation) approach and commonly used in premixed combustion simulations. One of the advantages of using this model is that complicated chemical reactions can be separated from the flow calculation. Chemical reactions are solved in advance and the data are stored in a table. At each time step, the computational code refers the table to include the heat release effect. This strategy can save computational time and memory. However, near the lean limit, phenomena such as extinction or re-ignition, in which the balance between chemical reactions and diffusion transport is important, cannot be simulated. The present conditions are slightly above the lean flammability limit, so using the G-equation approach is justified.

The motion of flame front $G=G_0$ is described by the balance between the convection velocity and the propagation speed of flame called the laminar burning velocity s_L.

$$\frac{\partial \rho G}{\partial t} + \frac{\partial \rho G u_j}{x_j} = \rho s_L |\nabla G| \qquad (4)$$

The filtered G-equation yields

$$\frac{\partial \bar{\rho} \tilde{G}}{\partial t} + \frac{\partial \bar{\rho} \tilde{G} \tilde{u}_j}{x_j} = -\frac{\partial}{\partial x_j}\left[\bar{\rho}\left(\widetilde{u_j G} - \tilde{u}_j \tilde{G} \right) \right] + \rho s_L \widetilde{|\nabla G|} \qquad (5)$$

The last term is an unresolved term and includes turbulent effects like flame front wrinkling. This flame propagation term is modeled as

$$\rho s_L \widetilde{|\nabla G|} = \bar{\rho} \tilde{s}_T |\nabla \tilde{G}| \qquad (6)$$

where s_T is called the turbulent burning velocity. Here, it is empirically given by

$$\frac{\tilde{s}_T}{s_L} = 1 + C\left(\frac{u'}{s_L}\right)^n \qquad (7)$$

where u' is the subgrid turbulent intensity. It means that turbulent wrinkling increases the apparent flame speed. Details of modeling are given for example in [3,4].

Numerical methods are based on the finite-volume discretization. The convective terms are constructed by Roe's upwind numerical flux formulation. The time integration method is two-stage Runge-Kutta integration. Walls are treated as non-slip adiabatic walls. Incoming and outgoing boundary conditions are basically given as non-reflecting conditions. Perfectly non-reflecting conditions, however, do not sustain the global mass flow rate. Here, partially non-reflecting conditions are imposed based on Poinsot's method [5] to set the time-averaged mass flow rate at a fixed value.

The computational domain covers the inlet section, combustor and exit section. The computational domain is decomposed into twelve sub-domains. One processor is assigned to each sub-domain, and the domains exchange boundary information with each other. Figure 1 shows the numerical grid system used in this computation. The number of grid point for the swirl-stabilized combustor is about 1 million. The computer system used is NAL CENSS system. The peak performance of the CENSS system is about 9 TFLOPS and the total memory size is about 3.7 TB. The number of time step of integration is about 2 million steps and the time interval of one step is 0.01 microseconds.

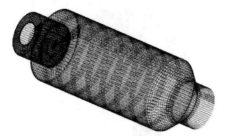

Fig. 1. Schematic of grid system.

3 Results and Discussion

3.1 Flame Dynamics

First, results of the equivalence ratio of 0.55 and the initial temperature of 700K are shown.

Figure 2 shows the time-averaged flame shape defined by the G value. By the swirling motion given to the inlet unburned gas, the flame is stabilized behind the dump plane. The flame shape indicates the surface where the local flow velocity and the turbulent burning velocity balance. In the center region, a recirculation zone is formed and the hot burned gas is moving upstream to hold the flame here. This can be also confirmed by seeing the averaged velocity field.

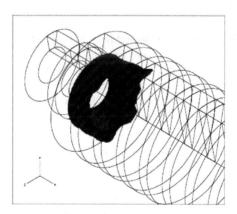

Fig. 2. Averaged flame shape.

Figure 3 shows the time-averaged velocity field around the dump plane. A central recirculation zone is formed along the centerline. Recirculation zones are also created in corner regions due to sudden expansion. This type of flowfield is typically observed in swirl-stabilized combustors like this. Also in Figure 3, shown a typical result obtained by experiment. The swirl number is almost the same as that of the present calculation, although there are some differences in detailed combustor configuration. They are qualitatively in good agreement.

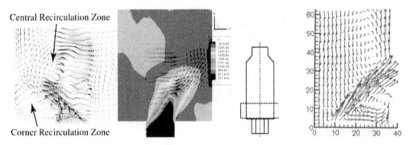

Fig. 3. Averaged velocity field and temperature field around the dump plane (left) and typical experimental result (right).

The time-averaged flowfield shown above seems rather simple. But instantaneous flowfield structures are more complicated and they change unsteadily. From here, these unsteady structures are analyzed.

By observing the combustor pressure history, one can usually find several resonant frequencies inside the combustor. These frequencies are determined from the combustor length and the burned gas temperature. The basic mode is usually the quarter-wave mode, in which a pressure node exists at the combustor exit and the pressure fluctuation amplitude gets largest at the combustor dump plane. The frequency can be estimated from the combustor length and the speed of sound of burned gas.

$$f_{1/4} = \frac{a}{4L} \approx 700Hz \tag{8}$$

The temporal pressure history obtained from the present calculation is shown for a certain period in figure 4. This pressure trace is measured at a point on the combustor dump plane. The FFT result of the pressure trace is also shown in this figure. The basic frequency is around 740 Hz. The numerical and analytical frequencies are close and this means the present simulation well produces acoustic resonance in the combustor. Higher modes are also observed as harmonics.

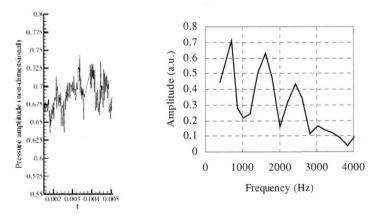

Fig. 4. Pressure history (left) and FFT result (right).

The time-averaged spatial distribution of pressure amplitude also indicates the existence of quarter-wave mode. Figure 5 shows the calculated time-averaged pressure fluctuation amplitude. The amplitude is the largest at the combustor dump plane and the smallest near the combustor exit.

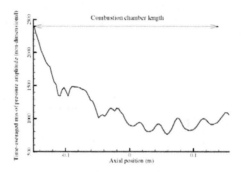

Fig. 5. Time-averaged spatial distribution of pressure fluctuation amplitude.

Local flame shape is determined by the balance between convective and flame velocities. Inside the combustor, pressure waves are propagating and this causes local velocity field fluctuations. The inlet unburned gas velocity at the dump plane is changing temporally, leading to periodic vortex shedding from the dump plane. Figure 6 shows an example sequence of coherent vortical structures and flame shape. Vortical structures are identified by the vorticity magnitude and the flame surfaces are identified the G value.

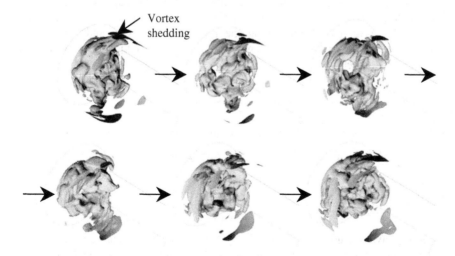

Fig. 6. Vortex/flame interaction: time interval between each shot is about 0.28 ms.

Figure 6 corresponds to nearly one acoustic period of the basic quarter-wave mode. The vortex shedding frequency in this case is determined mostly by acoustic resonance in the combustor. Figure 7 shows a time history of the vorticity magnitude at a point near the dump plane. The vorticity magnitude gets periodically stronger and this frequency corresponds to the basic quarter-wave mode. Thus, in the present case, vortex shedding is correlated to acoustic resonance in the combustor. Shear layer instability may also play a role in vortex forming. The relationship between acoustic shedding and shear layer instability is still not clear and should be analyzed in the future.

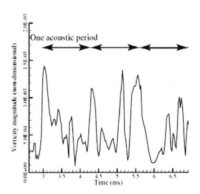

Fig. 7. Trace of vorticity magnitude at a point near the dump plane.

The flame shape is locally distorted when vortical structures are present in the vicinity. This vortex/flame interaction changes the total flame surface area and flame position. Because heat release by combustion occurs at the flame front, fluctuations in flame shape and location are directly related to heat release fluctuations.

The Rayleigh index is a criterion of combustion system stability. The global Rayleigh index is given by integrating the product of pressure and heat release fluctuations over time and space dimensions. In the present model, the local heat release rate is estimated by

$$\dot{q} \propto \rho_u s_T \Delta h A_f / V \tag{9}$$

where ρ_u is unburned density, Δh heat produced by combustion, A_f local flame surface area and V local control volume. The Raleigh index is

$$R = \frac{1}{V}\frac{1}{T}\int_{V,T} p'(\mathbf{x},t)\dot{q}'(\mathbf{x},t)dtdV \tag{10}$$

If the global Rayleigh index is positive, the system is unstable by heat release.

In the present case, contours of the local Rayleigh index integrated over time around the dump plane are shown in figure 8.

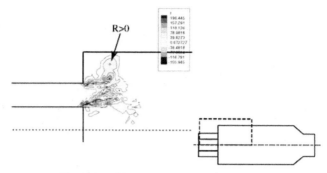

Fig. 8. Local Rayleigh index distribution.

In some regions the local Rayleigh index is negative, but in most areas, the index is positive. The global Rayleigh index integrated over space is 2.498 (non-dimensional) and positive. Thus heat release and pressure fluctuations are coupled to make the system unstable.

From these results, it can be said that combustion instabilities are driven by interactions between acoustic pressure oscillations, velocity fluctuations leading to vortex shedding, and heat release fluctuations. These three factors are in phase and driving the unstable combustion behavior in the combustor. Attenuation of combustion instabilities depends on how to weaken or change these links. This requires a detailed understanding of effects of various parameters, and we need much more calculation cases for this purpose. Below, a preliminary comparison is made between other cases in different conditions.

3.2 Effect of Inlet Temperature and Equivalence Ratio

Some experimental results indicate that preheating of unburned gas makes combustion more stable [6] to some extent. Comparing pressure traces of case of preheated (700K) and less preheated (400K) (case 2), the mean amplitude of the latter case becomes larger. This trend is the same as that of experimental observations.

Furthermore, if the global equivalence ratio is increased slightly to 0.6 (case 3), the amplitude becomes again larger. Figure 9 shows the relative pressure amplitude of the three cases.

Heat release and thermal expansion ratio may be part of the reason. Thermal expansion ratio of preheated case is 1970/700=2.8 while the ratio for the initial temperature of 400K is 1740/400=4.4. And the heat release rate is higher when the equivalence ratio is higher. But these factors are non-linearly inter-connected, so further investigation is needed for quantitative evaluation.

And also as a future plan, we need to examine the effect of equivalence ratio fluctuations. In real combustor systems, perfectly constant equivalence ratio premixing is almost impossible because fuel and air injection rates can be affected by pressure waves from the combustor. This causes fluctuations in local equivalence ratio, thus local heat release and is expected to make the system unstable. We will soon conduct simulations including this effect.

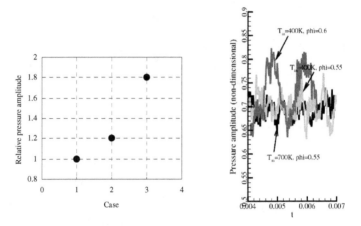

Fig. 9. Parametric variation of pressure amplitude.

These results have indicated that the present LES method can be applied to understanding flame dynamics in a combustor. In terms of computational resources, this LES method does not require large memory storage, but the total integration time should be large enough to capture acoustic resonance modes. And parametric calculations are necessary to design a combustor system. As the time step interval is determined by the minimum grid size, it now takes about a month to conduct one case. Increasing the number of CPUs in parallel computation may reduce the computational turnaround time and that will make this kind of calculation much easier to conduct.

4 Concluding Remarks

LES calculations of unstable combustion behavior have been conducted. The present method succeeded in reproducing acoustic resonance in the combustor, and pressure fluctuations causes velocity field fluctuations. The flame position and shape are affected by the local flow field. The Rayleigh index of the system is positive indicating that these factors are in phase to sustain the unstable combustion behavior. The oscillation amplitude can change according to flow field conditions, such as the equivalence ratio and flow temperature. Combustion control may lie in how to change the links between these factors. Further investigation will be conducted to demonstrate the effectiveness of combustion control in gas turbine applications.

References

1. Taupin, B., Vauchelles, D., Cabot, G. and Boukhalfa, A., "Experimental study of lean premixed turbulent combustion", 11th International Symposium on Applications of Laser Techniques to Fluid Mechanics, 2002

2. Stone, C. and Menon, S., "Swirl control of combustion instabilities in a gas turbine combustor", Proceedings of the Combustion Institute, 2002
3. Poinsot, T. and Veynante, D., "Theoretical and numerical combustion", Edwards, 2001
4. Peters, N., "Turbulent combustion", Cambridge University Press, 2000
5. Poinsot, T. J., and Lele, S. K., "Boundary Conditions for Direct Simulations of Compressible Viscous Flows", Journal of Computational Physics, 101, pp-104-129, 1992
6. Lefebvre, A. H., "Gas turbine combustion", Edwards, 1998

Grid Computing Supporting System on ITBL Project

Kenji Higuchi[1], Toshiyuki Imamura [2], Yoshio Suzuki[1], Futoshi Shimizu[1],
Masahiko Machida[1], Takayuki Otani[1], Yukihiro Hasegawa[1],
Nobuhiro Yamagishi[1], Kazuyuki Kimura[1], Tetsuo Aoyagi[1],
Norihiro Nakajima[1], Masahiro Fukuda [1], and Genki Yagawa[3]

[1] Center for Promotion of Computational Science and Engineering,
Japan Atomic Energy Research Institute,
6-9-3 Higashi-Ueno, Taito-ku, Tokyo 110-0015, Japan
{higuchi, suzukiy, shimizu, mac,
otani, hasegawa, yama, ka-kimu,
aoyagi, nakajima, Fukuda}@koma.jaeri.go.jp
http://www2.tokai.jaeri.go.jp/ccse/english/index.html
[2] Department of Computer Science, The University of Electro-Communications,
1-5-1, Chohugaoka, Chohu, Tokyo 182-8585, Japan
imamura@ im.uec.ac.jp
http://www.uec.ac.jp/index-e.html
[3] School of Engineering, The University of Tokyo,
7-3-1, Hongo, Bunkyo-ku, Tokyo 113-8656, Japan
yagawa@ q.t.u-tokyo.ac.jp
http://www. t.u-tokyo.ac.jp /index-e.html

Abstract. Prototype of the middleware for Grid project promoted by national institutes in Japan has been developed. Key technologies that are indispensable for construction of virtual organization were already implemented onto the prototype of middleware and examined in practical computer/network system from a view point of availability. In addition several kinds of scientific applications are being executed on the prototype system. It seems that successful result in the implementation of those technologies such as security infrastructure, component programming and collaborative visualization in practical computer/network systems means significant progress in Science Grid in Japan.

1 Introduction

Recently the progress of computers and networks makes possible to perform research work efficiently by sharing computer resources including experimental data over geographically separated laboratories even in different organizations. As a supporting tool kits/system for the advanced research style, the notion of GRID [1] has been proposed. Based on the notion, many projects have been started to implement the grid system from the middle of 90's (for example, see [2], [3], [4]).

A. Veidenbaum et al. (Eds.): ISHPC 2003, pp. 245–257, 2003.
© Springer-Verlag Berlin Heidelberg 2003

In Japan also, one of Grid projects, namely R&D for construction of a virtual research environment, ITBL (Information Technology Based Laboratory) [5] has started since 2001 aiming to connect computational resources including database at mainly national laboratories and universities in Japan. The project is now conducted under the collaboration of six organizations of Ministry of Education, Culture, Sports, Science and Technology (MEXT); Japan Atomic Energy Research Institute (JAERI), RIKEN, National Institute for Materials Science (NIMS), National Aerospace Laboratory of Japan (NAL), National Research Institute for Earth Science and Disaster Prevention (NIED), and Japan Science and Technology Corporation (JST). JAERI constructs a software infrastructure consisting of an authentication mechanism and a basic tool kit for supporting the development of various meta-applications. RIKEN provides a secure network infrastructure based on the VPN (Virtual Private Network) technology. As for applications, the followings are planned to be developed by all members of the projects; Integrated aircraft simulation system by NAL, full cell simulation system by RIKEN, three dimensional full-scale earthquake simulation system by NIED, materials design simulation system by NIMS and JST, and regional environmental simulation system by JAERI.

In the ITBL project, JAERI is constructing a software infrastructure named by 'ITBL Middleware' and has developed a prototype by which an experiment in public was successfully executed on third ITBL symposium held on February 17th in Tokyo. In the experiment, coupled simulation of heat conduction and fluid dynamics on different kinds of supercomputers was executed. A user at NAL site stated a job onto computers at JAERI and RIKEN sites and shared simulation image with a user at JAERI site. Some kinds of technology indispensable for Grid computing have been demonstrated in the experiment, as follows. i) Security infrastructure; It is necessary to securely access computational resources such as supercomputer of each organization through firewalls installed for each local area network. It is also necessary to realize Single-Sign-On to supercomputers. ii) Component programming on different kinds of supercomputers; This technology is necessary to realize facility to organically joint program and data files for job execution on different kinds of supercomputers, namely task mapping and communication library. iii) Collaborative visualization; This technology is necessary to realize facility for sharing image as simulation results among users.

Based on the experiment, several kinds of scientific applications are now being executed on the prototype system. We believe that it is a large and important step for Science Grid in Japan that the above technologies indispensable for grid computing have been successfully installed onto a software infrastructure such as ITBL Middleware developed in a national project and demonstrated over the practical supercomputer and network systems.

2   Design Philosophy of ITBL Middleware

ITBL Middleware was designed so as to realize; i) selection of a light-loaded computer from available ones, ii) very large or complicated simulation that can not be executed on a single computer due to resource restrictions, or can not be executed efficiently due to their complexity, iii) execution of a task which consists of programs and data on distributed resources, for example, processing experimental data on a supercomputer which is apart from the equipment, iv) sharing information with cooperative research group members in different laboratories.

For the purpose, we basically provide a component programming environment and global data space, called community, on a security system. Components are programs written in existing language and considered as the ingredient of a meta-application. The component programming environment is a tool kit which supports the development of components on distributed computers, the composition of them into a meta-application, and the selection of computing resources on which programs are executed, as described below. Community is a federated data repository which realizes the transparent access to the distributed data for members, while excludes the access from non-members. Security system provides a Single-Sign-On mechanism based on PKI (Public Key Infrastructure) X.509 certificates. It establishes secure and easy access to resources in multi sites.

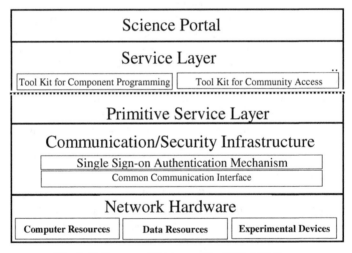

Fig. 1. Software architecture of the ITBL Middleware.

The above concept has been implemented in the form of enhancement of the functions developed in STA[6] that is programming environment JAERI has developed since 1995 aiming to support 'Seamless Thinking Aid' in program development/execution. The system already has various kinds of tools which support network computing including the component programming. The developing costs of the ITBL infrastructure were considerably reduced by using the STA system as the basis of the infrastructure.

The software architecture of the ITBL Middleware is shown in Fig. 1. User's applications are constructed on science portal. Two kinds of tool kits are provided on the service layer. The former supports component programming consisting of the tools called by TME, Stampi, Patras and so on. The latter supports community access exchanging image/data among researchers. It is necessary to provide various kinds of primitive tools such as process manager, task scheduler, and resource monitor on the primitive service layer. A single sign-on authentication mechanism and common communication interface independent of the underlining protocols is provided on the infrastructure layer.

3   Implementation of Grid Technologies onto ITBL Middleware

Outline of key technologies used in the construction of ITBL Middleware, that is, security infrastructure, component programming and collaborative visualization are described in this section.

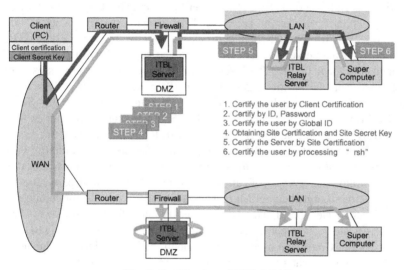

Fig. 2. Certification of ITBL Middleware.

3.1   Security Infrastructure

In the construction of the Grid system it is necessary to realize securely access to the computational resources such as supercomputer of each organization and Single-Sign-On to supercomputers resulting in organic linkage of computational resources. ITBL Middleware is designed supposing that computational resources are protected by firewall installed for each local area network. Thus only https was employed as communication protocol between sites in terms of commonly used one because it is required that the system commonly works under the various kinds of security policy. As for certification, ITBL Middleware employs public key-based security mechanism. This mechanism is implemented on ITBL server, which is established at each site. The server consists of three workstations where security data and communication facility are distributed on. In other words, security is enhanced by physical separation of data-file and -accessibilities to those servers as follows. i) ITBL Front Server: Communication from/to other site to/from own site goes through this server firstly. User certification, access control by global ID and so on are made on this server. ii) ITBL Relay Server: Intermediate process between user process and computers in use works on this server. iii) ITBL Data Server: Data for certification or user management is stored in this server. ITBL Front Server executes certification of user/server by using data on this server, as described below.

Public Key-Based Security Mechanism in ITBL Middleware: Security mechanism made in ITBL Middleware is shown in Fig. 2. Certification in ITBL system is made by public-key infrastructure with X.509 certificate, where encryption keys are used. In other words, all kinds of information are encoded by secret key and decoded by public key in ITBL system. When a user opens ITBL portal from his terminal, user's client certificate is certified on ITBL Front Server by using ITBL CA certificate (Step 1). Process on ITBL Front Server access to ITBL Data server by using ID and password (Step 2). ITBL Front Server certifies the user by referring list of ITBL global ID on ITBL Data Server. If user certification was successfully made, global ID is transformed into local ID on ITBL Front Server. After that user processes work with transformed local ID on the target computer (Step 3). ITBL Front Server gets site certificate and site's secret key used at next step (Step 4). ITBL Relay Server certifies ITBL Front Server by site certificate (Step 5). Target supercomputer certifies the user/server by using rsh commands issued by ITBL Relay Server (Step 6).

Management of certificates in ITBL is made by ITBL Certification Authority (CA) using certificates issued by Japan Certification Services Inc. (JCSI). Certificates used in ITBL are as follows. Server certificate: Certificate issued by third party, used to certify identification of organization that manage a site and to encode communication by SSL. Site certificate: Certificate used to certify identification of site. Client certificate: Certificate used to certify identification of user and issued by ITBL CA. ITBL CA certificate: Certificate used to certify identification of ITBL CA including public key. CA root certificate: Certificate used to certify identification of ITBL CA signed by third party.

3.2   Component Programming on Different Kinds of Supercomputers

Component programming is a programming style which tries to construct an application by combining components through well defined interface. In ITBL Middleware, components are existing sequential programs with I/O function, or MPI based parallel programs. The grid application is constructed according to the following three steps. In the first step, a user develops component programs on computers. After that, a user will combine component programs by considering dependencies among them (step 2). When executing the application, a user will map component programs onto computing resources (step 3). It should be noted that the step 2 and step 3 should be separated, because the target computer should be selected just when the component programs are executed due to the dynamic nature of computer loads.

For integration of component programs on distributed computing resources, we provide two kinds of methods. i) To execute an MPI-based parallel program on a parallel computer cluster. ii) To connect input and output of independent programs. By using these two methods, a user can construct a complicated application on a distributed environment. In order to support the first mechanism, we provide a communication library for a heterogeneous parallel computer cluster, called Stampi[7]. The second mechanism is supported by a task mapping editor TME [8].

MPI-Based Communication Library Stampi: Stampi was developed to remove a barrier of heterogeneity in communication in distributed computing, and to provide all communication functionality in the MPI semantics. In the development we focused on utilizing applications developed in the MPI2 standard with minimal modification. Stampi realizes distributed computing in which a virtual supercomputer comprises any parallel computer connected via LAN or WAN. The minimum configuration of Stampi is illustrated in Fig. 3. Stampi library is linked to the user's parallel application, the vendor's supplied MPI library and message routers bridging the gap between two user applications on the different machines. Here we assume that parallel machine A and B are managed in interactive and batch(NQS) mode respectively. In addition to the basic configuration shown in Fig.3, it is assumed that NQS commands (qsub, qstat, etc.) are available on only the frontend node of machine B. All nodes in a cabinet including the frontend are IP-reachable on the private address and separated in global Ips.

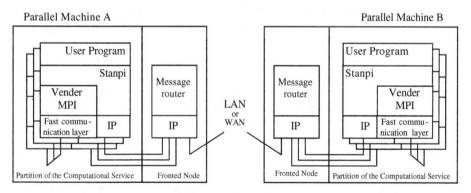

Fig. 3. Concept of Stampi.

In addition to the communication mechanism described above, we should introduce common rules to share the distributed heterogeneous resources in order to establish real distribute computing. They involve to remove or hide the difference in various version of software or handling of data and application caused by heterogeneity. In the above mentioned configurations, we considered communication layer, segmenting in private addresses, execution mode(interactive and batch), handling of remote processes, compile and link (commands, options etc.), data formats and so on.

In order to realize distributed executables without modification to the MPI applications, Stampi uses a profiling interface shift in its implementation. Therefore users only need to recompile their codes before they run the application. Stampi manages hosts and the local rank internally by Stampi-communicator, and when it detects a communication, it chooses the best way of inter- or intra- communication mechanism. In the current implementation, Stampi uses a TCP-socket for inter-communication and the vendor supplied communication mechanism for intra-communication. Introducing such a hierarchal mechanism lighten the disadvantages in a common usage of communication. In the case of communication for other sites Stampi uses a socket supplied by ITBL Middleware resulting in encapsulation by https between ITBL Front Servers.

As for common data representation that is also considerably most important in a heterogeneous environment, Stampi adopted external32 format defined in the MPI2. Stampi hooks a converting procedure on the subroutines, MPI_Pack and MPI_Unpack, also in the external communication layer.

As for other term, especially the handling of remote processes is significant, and it was realized by introducing a router process and new API. Router process was employed to resolve the problem that direct external communication is not available usually in the case of supercomputers. As for API, MPI2 was basically employed for dynamic process management. Although optimization of routing and control of load balance are indispensable for distributed computing, they are located on the upper layer of Stampi.

TME (Task Mapping Editor): TME that supports visual programming has architecture of four structures, GUI, control, communication, and adaptor layers. i) The GUI layer faces user and some components for design and monitoring are facilitated. ii) The control layer comprises five components, Proxy Server (PS), Tool Manager (TM), File Manager (FM), Exec Manager (EM) and Re-source Manager (RM), and they perform the communication control between distribut-ed machines, the management of any components on TME, handling of distributed files, executing and monitoring the application on local/remote machines, monitoring the state of batch systems and the load of interactive nodes, respectively. iii) The communication layer relays the RPC-based requests from the control layer to the adapter layer deployed on distributed machines via PSs. iv) The adaptors play a role of an outlet between RPC called from EM and the existing applications. The features of TME corresponding to four trend issues shown above are as follows.

Visual design of a distributed application is one of features of TME that supports to draw specifications of distributed application graphically. On the TME console, any re-sources, for example, data files, programs and devices, are expressed as icons with higher abstraction on the TME naming space. By linking program icons and data icons, users can specify data dependencies and execution order. Thus, TME requires users little knowledge of distributed architecture and environment, and they can handle the resources as on a single computer. Parallelism and concurrency can also be taken advantage of the nature of the data-flow diagram. Consequently, TME supports the intuitive understanding for the structure of distributed applications.

TME has automatic job submission facility. Users can choose computers on which they submit jobs interactively. According to the execution order of programs and required data files as defined on the TME console, the RM and EM assign and invoke programs, and transfer data onto specified machines. Complex applications that comprise multiple executables and a lot of data transfer can be encapsulated and automated on the TME system.

As for job monitoring and scheduling, EM watches the state of the application and it sends a report to the TME control system, thus user can monitor the application virtually on real-time. On the monitoring console, any resources and dependency are represented as icons and lines as well as the design console. Thus, it helps detecting structural bottlenecks. EM and RM also coordinate semi-automatic resources mapping. When user registers his or her scheduling policy and priority, the user can take advantage of efficient allocation and reservation of the resources. This function optimizes the economical restriction under the circumstance where accounting is a severe problem.

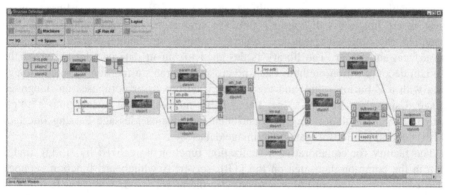

Fig. 4. Data-flow of the genome sequence analysis system and a snapshot of a visualization tool.

Flexible integration framework distinguishes our system from others. On the TME environment, various kinds of built-in components are integrated. Main built-in components are 'blocking IO', 'pipelining IO', 'tee (duplicate the output stream)', 'container (sequential list of input/output files)', 'text browser', 'GUI layout-composer', 'simple data-viewer', and so forth. This mechanism is realized by the development framework of the STA. The applet developed by a user is also registered as a new built-in component and it can be shared among other users. In addition, the GUI layout composer separates the design phase and the execution phase from the task of an application user. It means that TME facilitates a common frame or a collaborative space between the developer and the application user.

An example of TME in bioinformatics is shown in Fig. 4. One of typical analysis methods is to search the specified genome pattern from huge amount of DNA data. However, it is realized by the combination of a lot of applications and databases. On the TME console, such a complex procedure can be defined as the figure and the user may simplify the whole process. The integration of the applications and database may further reduce the burden of bioinformatics analyses.

3.3 Collaborative Visualization

As for visualization facility of ITBL, user can use Patras/ITBL when he wants to execute real-time or collaborative visualization for steering simulation parameters, or AVS/ITBL when he wants to execute off line visualization for detailed examination of his work [9].

Patras/ITBL Visualization System: ITBL Middleware has facility of real-time visualization, that is, the simultaneous execution of a simulation program and visualization of results, as well as visualization of extremely large data. The Patras/ITBL server-client remote visualization system carries out the visualization process on each processor of a supercomputer (server) and displays images onto a user terminal (client). The user can track and steer simulation parameters and control the simulation procedure. To actually use the Patras/ITBL visualization library, users must insert function call routines into the simulation program. Patras/ITBL can visualize the following entities: object, contour, streamline, vector, grid line, isosurface, and tracer. The library is designed to run in data-parallel based on the domain decomposition method. Each processor generates a fraction of a total image data with a Z-buffer value, and then a specified processor composes an integrated image. Using this Z-buffer value, Patras/ITBL merges partial images into a single image at each simulation time step and then implements Message Passing Interface (MPI) communications to gather the image data.

The facility for collaborative visualization function is realized by storing image data in the common data area on the ITBL server. A whiteboard function supports such collaborative visualization sessions. Furthermore, we developed a function to display the simulation progress in real time for coupled multi physics numerical simulations executed simultaneously on two remote hosts. A Stampi MPI-based library enables communications between different architectures in a heterogeneous environment.

In a collaborative session of ITBL, users should be distinguished as a master client or multiple slave clients. Only a master client can change the visual simulation parameters and control the simulation program; the slave clients can only monitor the displayed image. The collaboration function lets a user change a master client to a slave client and make one of the slave clients the new master client. A collaboration servlet processes the collaboration function. When a master client executes Patras/ITBL, a graphic server sends image data to the ITBL server at every constant time interval; the ITBL server sends the image data to a user terminal. Slave clients only obtain image data from a graphic server if a meaningful change exists in the image data from a previous image. The image update flag, which takes a value of true or false, distinguishes whether the image data is new. Storing image data in the common data area on the ITBL server aids in this collaborative visualization.

Besides visualization of the simulation results, ITBL user can use whiteboard facility. When users switch their display to a whiteboard window, the whiteboard operation buttons appear. A user can use a mouse to draw freehand figures in the viewing window. The drawings are then displayed on every participant's terminals.

AVS/ITBL Visualization System: Besides real-time processing described above, a facility for post-processing visualization is available in ITBL Middleware, which is realized by the AVS/ITBL visualization system. The facility enables remote post-processing visualization of any data stored on any supercomputer located in the ITBL network. Registered users can graphically access both systems through Web browsers. The remote visualization function based on a Web browser lets users visualize their simulation results on their own terminals. Here, users can select from a showImage or showPolygon entity. With showImage, an image is displayed in 2D (GIF data); with showPolygon, it's displayed in 3D (GFA data). In both cases, users can rotate, resize, and translate a graphical image. To use this function, users must prepare an AVS visualization network file using the system's common gateway interface (CGI) relay and image data generation modules. The CGI relay module links with the visualization parameter of an arbitrary visualization module in the AVS visualization network and enables referencing and updating of that visualization parameter. When a CGI relay module is introduced, the AVS/Express network editor displays the visualization-parameter definition screen. Data on this screen defines variable names and arrays, HTML tags, and so on. Next, a visualization parameter input interface is automatically generated by the interface generation filter according to the definition of the CGI relay module and is displayed in the right frame of a Web browser.

4   Current Status of Installation of ITBL Server/Middleware

ITBL Server and ITBL Middleware have been already installed into practical computer/network systems of national institutes and universities including six members in Japan as follows. Firstly, ITBL servers have been established at JAERI, RIKEN, NAL, Japan Advanced Institute of Science and Technology (JAIST) and Institute for Materials Research(IMR) of Tohoku University.   Thus, they are registered as ITBL site. Especially JAERI consists of three sites because three sets of ITBL Server were installed into three establishments geographically separated. As for ITBL middleware, it has been installed onto the following computers at JAERI: ITBL computer (Fujitsu PrimePower2000, 512CPUs, 1TFlops), Compaq Alpha Server SC/ES40(2056CPUs, 1.5TFlops), SGI Origin3800(768CPUs, 768GFlops), Fujitsu VPP5000(64CPUs, 614GFlops), Hitachi SR8000(8CPUs×20nodes, 240GFlops), NEC SX-6i(1CPU, 8GFlops) and so on. Scientific applications constructed on the above systems are as follows: Atomistic Simulation, Laser Simulation, Analyze Protein Structural Information Databases, Numerical Environmental System and so on. As for Other institutes, ITBL middleware has been installed onto Hitachi SR2201(8CPUs×4nodes, 9.6GFlops) at Kyoto University, Fujitsu VPP5000(2CPUs,

19.2GFlops) at JAIST, Hitachi SR8000(8CPUs×64nodes, 921.6GFlops) at IMR of Tohoku University. ITBL Middleware is now being installed the following systems, besides the computers shown above: NEC SX-6 (12CPUs, 96GFlops), IBM pSeries690(16CPUs, 83.2GFlops) and SGI Onyx300 of JAERI, Numerical Window Tunnel 3rd(Fujitsu Primepower2500, 9.3TFlops) of NAL, Cray SV1ex(32GFlops) of NIED and so on.

5   Scientific Applications on Prototype System

Some scientific applications are being executed on the prototype system described above by mainly member of Center for promotion of Computational Science and Engineering (CCSE) of JAERI.

Superconducting Device Simulation: In CCSE of JAERI, superconducting device simulations are intensively performed. Superconducting device has quite unique and useful characters in contrast to semiconductor devices. Study for some kinds of device functions in several complex configurations has been effectively made by using ITBL middleware on supercomputers sharing a visualized simulation results from geographically separated sites in real time. This enables an interactive and rapid collaboration between experiments, engineers, and computer scientists.

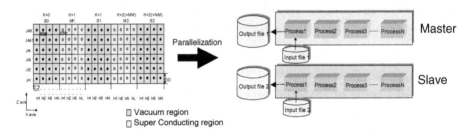

Fig. 5. Outline of the superconducting device simulation.

In addition, coupled simulation is being made on the ITBL system. In the simulation superconducting-device is divided into two regions, of which physical attributes are exchanged at every time step. In the production run, physical phenomena are simulated changing the mesh size of two regions independently. As for vacuum region, maxwell equation is solved by linear-linear partial differential equation. Though calculation is not suitable to vector processing, computational cost is low compared with the rest. As for super-conducting region, Klein-gordon equation is solved by non-linear partial differential equation. Calculation is suitable to vector processing because of trigonometric functions frequently used in the simulation. In addition computational cost is high compared with the rest. In order to reduce the simulation time and effectively use the computational resources, coupled simulation by vector and scalar processor should be employed. Figure 5 shows the outline of the simulation.

Atomistic Simulation: 'An atomistic simulation on ITBL' being performed in CCSE of JAERI is an example of TME application, which would demonstrate a usefulness of the container module. A molecular-dynamics (MD) simulation is one way to study defect cluster formation and its interaction with dislocations in irradiated metals. Because the simulation often requires longer CPU-time over the utilization rule, the whole calculation is divided into a series of chunks. A pseudo-iteration structure of chunks is defined on TME, where series of input- and output- files are set in input- and output- containers, respectively. Parallel computation of each MD simulation is executed by a portable parallel program called Parallel Molecular Dynamics Stencil (PMDS). The TME scheduler dynamically assigns these calculations to the free or less busy computer with the consequence that a high throughput computing is realized. Figure 6 shows illustration of the simulation.

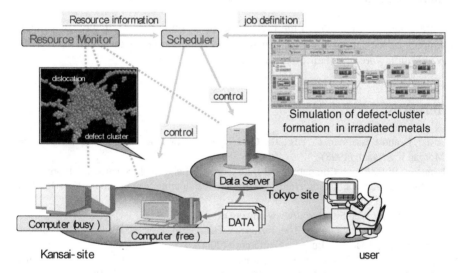

Fig. 6. Illustration of the atomistic simulation.

6 Concluding Remarks

A prototype of Grid middleware has been developed in the ITBL project. The middleware is now being evaluated resulting in improvement of facilities such as security mechanism. The evaluation and improvement are being made mainly over four establishments geographically separated, Tokyo, Tokai, Naka and Kansai in JAERI, which are connected through a high speed backbone network SuperSINET. In addition, we have adapted the software infrastructure to the practical computer/network systems of other institutes of ITBL project and constructed Grid applications. For instance, coupled simulation of heat and fluid analyses has been demonstrated. In the demonstration, calculation result of heat analysis on SMP machine at JAERI site and succeeding calculation result of fluid dynamics on vector processor at RIKEN site were simultaneously visualized to the users at JAERI and NAL sites.

A prototype of the middleware has been constructed in ITBL project as a research product at first stage. The project is ready to move to next stage. In other words, the software infrastructure is being sophisticated and prevailing over national institutes and universities in Japan. Thus various kinds of scientific applications on organizations of MEXT are being constructed on the prototype system. After the evaluation and improvement through the scientific applications, the software and hardware infrastructure developed in this project will be offered for practical use to scientific researchers in Japan.

Acknowledgments

We appreciate the members of the ITBL task force committee. They gave us valuable comments on designing the infrastructure and cooperation in installations of ITBL Server/Middleware into their practical systems.

References

1. Foster, I., and Kesselman, C., eds.: The Grid: Blueprint for a New Computing Infrastructure, Morgan Kaufmann (1998)
2. Johnston, W.E.: Using Computing and Data Grids for Large-scale Science and Engineering, the International Journal of High Performance Computing Applications, Vol. 15, No.3 (2001), pp.223-242
3. Unicore project Web Site: http://www.unicore.de
4. Segal, B.: Grid Computing: The European Data Grid Project, IEEE Nuclear Science Symposium and Medical Imaging Conference (2000)
5. Yamaguchi, Y., and Takemiya, H.: Software Infrastructure of Virtual Research Environment ITBL, Journal of Japan Numerical Fluid Dynamics, Vol.9, No.3, PP83-88 (2001)(in Japanese)
6. Takemiya, H., Imamura, T., Koide, H., Higuchi, K., Tsujita, Y., Yamagishi, N., Matsuda, K., Ueno, H., Hasegawa, Y., Kimura, T., Kitabata, H., Mochizuki, Y., and Hirayama, T.: Software Environment for Local Area Metacomputing, SNA2000 Proceedings, (2000)
7. Imamura, Toshiyuki, Tsujita, Y., Koide, H., and Takemiya, H.: An Architecture of Stampi: MPI Library on a Cluster of Parallel Computers, J.Dongarra, P.Kacsuk, N.Podhorzski(Eds), Recent Advances in Parallel Virtual Machine and Message Passing Interface, LNSC 1908, pp.200-207, Springer, Sep. 2000
8. Imamura, T., Takemiya, H., Hironobu Yamagishi, N., and Hasegawa, Y.: TME - a Distributed resource handling tool, the International Conference on Scientific and Engineering Computation, IC-SEC2002, December 3rd-5th, Singapore
9. Suzuki, Y., Sai, K., Matsumoto, N., and Hazama, O. : Visualization Systems on the Information-Technology-Based Laboratory, IEEE Computer Graphics and Applications, Vol. 23, No. 2(2003)

A Visual Resource Integration Environment for Distributed Applications on the ITBL System

Toshiyuki Imamura[1,2], Nobuhiro Yamagishi[2], Hiroshi Takemiya[3], Yukihiro Hasegawa[2], Kenji Higuchi[2], and Norihiro Nakajima[2]

[1] Department of Computer Science,
the University of Electro-Communications,
1-5-1 Chofugaoka, Chofu, Tokyo 182-8585, Japan,
`imamura@im.uec.ac.jp`
[2] Center for Promotion of Computational Science and Engineering,
Japan Atomic Energy Research Institute,
6-9-3 Higashi-Ueno, Taitoh-ku, Tokyo 110-0015, Japan,
{yama,hasegawa,higuchi,nakajima}@koma.jaeri.go.jp
[3] Hitachi East Japan Solutions, Ltd.,
2-16-10 Honcho, Aoba-ku, Sendai, Miyagi 980-0014, Japan

Abstract. TME, the Task Mapping Editor, has been developed for handling distributed resources and supporting the design of distributed applications on the ITBL. On the TME window, a user can design a workflow diagram, just like a drawing tool, of the distributed applications. All resources are represented as icons on the TME naming space and the data-dependency is defined by a directed arrow linking the icons. Furthermore, it is equipped with an important mechanism allowing integration and sharing of user-defined applets among the users who belong to a specific community. TME provides users a higher-level view of schematizing the structure of applications on the Grid-like environment as well as on the ITBL system.

1 Introduction

The progress of distributed processing is permeating successively with the rapid growth of Grid [1] and the development of a grid middleware like the globus toolkit [2]. On the Grid environment, geographically scattered resources, machines, huge and countable databases and experiments can be linked organically and uniformly with higher abstraction. Globus toolkit and other existing grid middleware support primitive functions to couple several components, however, it is believed difficult for a beginner to quickly master the functionalities. On the other hand, component programming is thought as one of prospective frameworks. Simplification of the interfaces, handling of any resources, and construction of the coupled services have become important issues for the next step of the Grid.

A. Veidenbaum et al. (Eds.): ISHPC 2003, LNCS 2858, pp. 258–268, 2003.

Generally, it is believed that the visual support that a GUI offers is more efficient than a script. GUI provides not only intuitive understanding but also the design of the combination of multiple components. In addition, it enables users to detect the structural bottleneck and errors via real-time monitoring of execution circumstance. There exist several projects implementing a GUI-based steering and monitoring on the Grid environment. WebFlow [3] is one of the pioneers in visual designing tools for distributed computing. GridMapper [4] is also a GUI tool, among several, which focuses on the monitoring of geographical network. The UNICORE project [5] intends to connect multiple computer centers and support illustration of the dependencies among data and jobs in order to realize a meta-computing facility. Triana [6] is also a graphical programming environment adopted as a test bed application in the GridLab project [7]. From the trends witnessed in the related works, common features, visual design, automatic execution, monitoring, and flexible integration framework can be considered as the key technologies for increasing the usability of the Grid.

This paper covers the concepts of Task Mapping Editor developed at the Japan Atomic Energy Research Institute (JAERI) and its implementation on the ITBL systems [8], also known as one of Grid environments initiated within the e-Japan strategy.

2   Task Mapping Editor

The Task Mapping Editor, hereafter 'TME', was originally developed as an execution environment in the STA (Seamless Thinking Aid) basic software project [9]. It was intended to facilitate a framework in distributed computing focused on a local area network and to support a user-friendly graphical interface (see Figure 1). The STA basic system is composed of three elements : development environments, communication infrastructures, and TME.

The main objectivity of TME was to provide application users as well as developer with a highly developed aspect to the structure of the application. At present, the STA project has shifted to become part of the ITBL(IT-Based Laboratory) project, one of e-Japan projects advanced among research institutes under the administration of the Ministry of Education, Culture, Sports, Science and Technology (MEXT). The TME has been restructured by using both security and communication infrastructures equipped on the ITBL. In the ITBL project, TME is prioritized as a core environment where users design and perform meta-applications. Due to the revisions made to run on the ITBL system, it is now possible to couple the computer resources located apart in various institutes with the help of the authentication mechanism introduced in the ITBL system. Such a feasible extension to its usability makes TME a prominent Grid software or toolset.

This section describes the software architecture of TME on the ITBL.

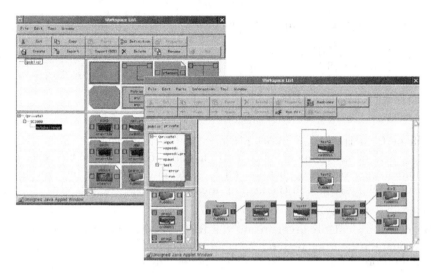

Fig. 1. Screenshot of the TME main windows, the windows display an example of the dataflow design. On the left window, all repositories are represented as icons with a thumbnail and a bitmap image of computers, and built-in functions are listed on the upper canvas. On the right window, user defines a distributed application.

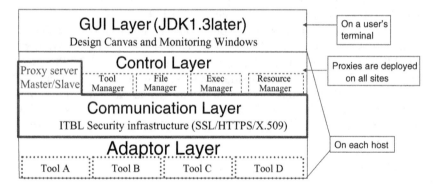

Fig. 2. Software architecture of TME, assumed that TME is built on the ITBL system.

2.1   Software Architecture of TME

The basic architecture of TME layers four structures, GUI, control, communication, and adapter layers as depicted in Figure 2.

The GUI layer is the only layer seen by the users. It plays important roles in relaying users' intentions to the system and vice versa. The visual design and monitoring functions have been implemented with Java language. Basically, the TME executes on any platform, where JDK1.3 or JRE1.3 later is installed, and uniform user interface is supplied.

The control layer is comprised of following five subcomponents:

1. Proxy Server (PS),
2. Tool Manager (TM),
3. File Manager (FM),
4. Execution Manager (EM) and,
5. Resource Manager (RM),

which perform the communication control between distributed machines, the management of any components on TME, handling of distributed files, execution and monitoring of the application on local/remote machines, monitoring the state of batch systems and the load of interactive nodes, respectively.

The communication layer relays the RPC-based requests from the control to the adapter layer deployed on distributed machines via Proxy Servers. Due to the limitation of the Java programming model, any communication other than a web server is normally not allowed. However, the introduced Proxy server running on the web server passes any messages to the backend servers. Hence, the control layer can connect any adapter layer on any computers seamlessly. Figure 3 illustrates a typical software configuration of the TME system on the ITBL.

For all connections between TME client and ITBL servers, and servers and proxy servers, authentication is carried out by checking mutually the X.509-based certificates as seen in Figure 3. It is also worth mentioning that all connections are established on the SSL (Secured Socket Layer) or the HTTPS (Secured HTTP). Thus, all accesses from outside to a server within ITBL are secured. The authorized connections ranging among multiple sites are demanded an approval from all participants for security purposes and policies on other sites. On the real test bed of the ITBL sytem, execution of TME across multiple institutes have been made possible under this security framework [8].

Other communication library for coupling multiple applications running on several machines, and on several sites, is the Stampi [10] developed by JAERI. The implementation of Stampi according to the MPI-2 specification enables users to invoke multiple MPI applications on multiple sites dynamically and to establish a single virtual computer system on the ITBL. The remote and distributed I/O function, called MPI-I/O, is also supported [11]. Both dynamic process management and remote-parallel file handling are very helpful in building a meta application. By using the Stampi library, TME enables users to develop a Grid-enabled MPI application and execute it on the ITBL facilities. Normal connection is established via ITBL relayers at all the sites. When power-intensive computing and high-throughput communication are required, VPN (Virtual Private Network) connection can be chosen as the dedicated network.

In the case of the STA version, Nexus 4.1.0 developed at Argonne National Laboratory was used in the communication infrastructure. Since the ITBL version is developed basically as an upper-compatible of the STA version, much more details of it can be found in literatures on the STA version (see *e.g.* [9]).

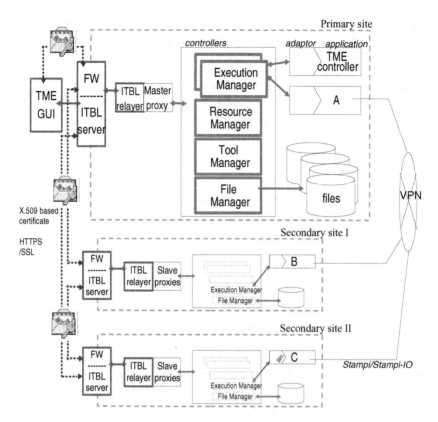

Fig. 3. Typical configuration of TME components on a multi-institutional distributed environment. Each site has its own firewall, and the ITBL front end server is connected on DMZ (DeMilitarized Zone) of a subnet of the FW. All connections among different sites are established via HTTPS or SSL, and user is authenticated by X.509 based security mechanism.

The adapters play a role of an outlet between EM and the existing applications. For example, an adapter interprets machine-dependent options, data formats, and so forth.

Rest of this section describes the features of the TME corresponding to the four issues in trend shown in the previous section.

2.2   Visual Design of a Distributed Application

TME is an editor that supports a component programming of distributed application graphically as shown in Figure 1. On the TME console, any component, for example, data files, programs and devices, are expressed as icons with higher abstraction on the TME naming space.

Since taking the burden off the users is the primal motivation of the development of the TME, the user interface is designed in a simple fashion. For example, both input and output ports are represented as small buttons found on the left and right sides of the icons respectively. When a user connects an output port and an input port, one has to only pick up both buttons representing the ports on the icons, and select an I/O connection operation button from the TME menu bars. If a user wants to cancel the last action and delete the icons from the canvas, undo operation button and delete operation on the bar will guide them like a drawing tool. All icons placed on the TME system have several properties, for example the number of input and output data files, working directory, default options, and so on. These properties can be designated dynamically and the TME will guide through the setups of the core items using dialog windows at run-time.

When a user designs a distributed application on the TME by linking program and data icons, the user can specify the dependencies of data and the execution order simultaneously. This implies that a starting operation of each program and data transfers inspire according to the dependency like a Petri-net model, and the pre and post processing can be presented. Furthermore, a burden of these procedures can be reduced dramatically through the automation of the procedures by the TME system as will be presented in the proceding passage. Thus, TME requires from the users least amount of knowledge in distributed architecture and environment. The users can handle the resources as on a single computer. Parallelism and concurrency can also be taken advantage of the nature of the data-flow diagram. Consequently, TME supports the intuitive understanding for the structure of distributed applications.

2.3   Supporting Automatic Job Submission

Users can choose computers on which they submit jobs interactively. According to the execution order of programs and required data files as defined on the TME console, the RM and EM assign and invoke programs, and transfer data onto the specified machines. Complex applications that comprise multiple executables and a lot of data transfer can be encapsulated and automated on the TME system.

Because the development is carried out to conserve local management policies and alliance, the ITBL system has no global queuing system, and only plays a role of an agent to submit jobs to local queuing systems. Although one can easily imagine that introducing a global queuing system will improve the total performance and load average, ITBL solves the scheduling problem by introducing a user specific scheduling script and resource monitoring systems described in the next paragraph.

2.4   Job Monitoring and Scheduling

EM watches the state of the corresponding application and sends a report to the TME control system, allowing for a user to monitor the application virtually in real-time. On the monitoring window, all resources and dependencies are represented by icons and lines as on the design console. In addition, EM and the TME controller daemon collect every status of components, for example the start time, chosen queue class, generated output file and time stamps, and so forth. Thus, such functions help detect structural bottlenecks.

EM and RM also coordinate semi-automatic resources mapping. When a user registers one's scheduling policy and priority, the user can take an advantage of efficient allocation and reservation of the resources. TME supports two types of scheduling or resource selection mechanism: filtering/ordering and script setting. The filtering excludes computers and queue classes, which do not satisfy user conditions, from the list of registered resources. On the other hand, the ordering prioritizes the candidates, and then TME selects the best suited machine and queue with the highest priority rank after evaluating all the ordering conditions.

TME also allows for customization of the registered scheduling policy by the users. Scheduling is represented by a Perl script as depicted in Figure 4, and TME supplies several APIs for inquiring the status of the machines and queue systems. More detailed control becomes possible by specifying the user scheduling recipes. For example, if the execution times on every platform can be estimated, exact throughput calculated in the priority evaluation saves the total cost. This function optimizes the economical control where accounting is a severe problem.

2.5   Flexible Integration Framework

On the TME environment, various kinds of built-in components are integrated. The main built-in components are the blocking IO, pipelining IO, tee (duplicate the stream), container (sequential list of files and directories), text browser, GUI layout-composer, simple data-viewer, and so forth. This mechanism is realized by the development framework of the STA. The applet developed by a user is also registered as a new built-in component which can be shared among other users. Therefore, sharing of scenarios and parameters of the numerical experiments is to be expected on the TME, which provides a virtual collaborative environment.

In addition, the GUI layout composer separates the design and execution phases from the application end user tasks. What this means is that TME facilitates a common frame or a collaborative space between the developer and the application user.

3   TME Applications

We have designed several applications by using the TME. In this section, three applications are presented as examples.

1. The radioactive source estimation system requires high response capability, speed and accuracy [12]. The core simulations should be carried out on as many computational servers and on most sites available in order to minimize the calculation time when an emergent accident should arise. This application is designed on TME as shown in Figure 5 (left). Users can easily specify spawning slave processes over distributed sites by linking master and slave icons.

```
sub order_cpu {
    my $ref = shift(@_);
    my @server = @$ref;
    my @result;
    my $schedule;
    $schedule = info::get_schedule();
    if(!defined $schedule) {
        return @server;
    }
    @server = sort sort_order @server;
    for(my $i = 0; $i <= $#server; $i++) {
        $server = $server[$i]->{server};
        $queue = $server[$i]->{queue};
        for($p = $schedule; $p; $p = $p->{next}) {
            if(($server eq $p->{name}) &&
               ($queue eq $p->{allocation})) {
                $num = get_count($p->{peno});
                $server[$i]->{data} = $num;
                last;
            }
        }
    }
    @result = ordering(0, @server);

    return @result;
}
```

Fig. 4. Example of a scheduling script, the function returns a list of available queue classes which is sorted by a specific ordering policy and CPU priority.

2. The data analysis for nuclear experiments is also computationally intensive. In this case, huge amounts of data is recorded to the DB server at every observation. The physicists analyze these data to study specific phenomena. This work was also formulated on the TME as depicted in Figure 5 (right).
3. In the bioinformatics field, one of typical analysis methods is to search for a specific genome pattern from the huge amount of DNA data. This is usually realized by combining multiple applications and databases. On the TME console, such a complex procedure can be defined as Figure 6 and the user may simplify the whole process. The integration of the applications and databases may further reduce the burden of carrying out bioinformatics analysis.

4   Discussions and Current Status

Cooperation with the Grid middleware is also one of the greatest interests for developers and users. Currently, TME uses the nexus library developed by ANL, the predecessor of globus-nexus in the globus toolkit (GTK) [2]. Since in GTK

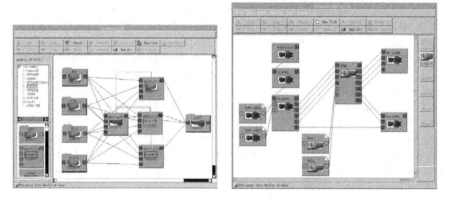

Fig. 5. Dataflow of the radioactive source estimation system (left) and the data analysis for nuclear experiments

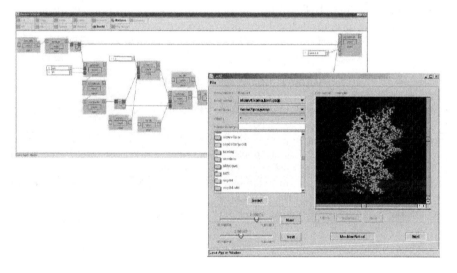

Fig. 6. Dataflow of the genome sequence analysis system and a snapshot of a visualization tool which is customized by a bioinformatics user.

version 2 or later, the communication and security frameworks are shifting to that based on OGSA, the future release of TME will adopt a globus-based communication infrastructure and collaborate with many Grid services.

The main feature of TME is the description by dataflow; however, the control-flow mechanism has been introduced in the latest version of TME for the expert developers. This extension provides the structure of conditional branch and loop, and enables the development of more advanced applications. Authors believe that this can contribute to the development of an advanced PSE (Problem Solving Environment) on a distributed environment, which is one of the ultimate goals of Grid computing.

For definition of the distributed application, TME adopts a subset of XML. The XML is implemented for describing universal documents; however, it is also a powerful tool for defining a distributed application. Using the XML format suggests that other tools like an emulator, a debugger, and a scheduler share the TME components. In the related works, such topics as the definition of the Grid services or Grid applications by the XML format have been discussed. In the FHRG project at Fraunhofer [13], the Grid Application Description Language (GADL) based on the XML is adopted for defining all Grid resources on their sites. WSDL (Web Service Description Language) [14], WPDL (Workflow Process Definition Language), etc are currently proposed as a standardized specification. The authors recognize this extension to be a significant issue for the next stage of TME.

The current status of the TME is as follows. As described previously, TME GUI is implemented as Java applets, and has been made available on a web browser such as Netscape Navigator 6 or later and Microsoft Internet Explorer 5.0 or later, on which JRE 1.3 or 1.4 has been plugged in. TME main adapter and control layer programs has been ported to ten kinds of parallel computers, several kinds of WS serves, and WS/PC clusters as shown in Table 1.

Table 1. List of the TME platforms

Vector Parallel Computers	Fujitsu VPP300, VPP5000, NEC SX-4, SX-5, SX-6, Cray T90
Scalar Parallel Computers	Fujitsu PrimePower, AP3000, Hitachi SR2201, SR8000, IBM RS6000/SP, pSeriese, Cray T3E, Intel Paragon, SGI Origin2000, Origin3000, Compaq Alpha server
WS Servers	Fujitsu GP8000, HP HP9000, NEC TX-7, Sun Enterprise, SGI Onyx
PC cluster	Linux, FreeBSD

5   Concluding Remarks

TME supports a component programming and handling of the computational resources distributed over multiple sites. It has a higher-level view of schematizing the structure which helps to intuitively understand the applications. The automatic submission and monitoring can improve the efficiency of the jobs. In addition, the framework, which makes possible the integration and co-operation of various user-defined functions, suggests the potential of the TME as collaboration environment and PSE. In the near future, there are plans to improve the scalability and reliability of TME and through such developments, the authors would like to contribute to the advancement of Grid computing.

Finally, the authors would like to thank Dr. Kei Yura and Prof. Dr. Hironobu Go for their support in the construction of bioinformatics applications and databases, and Dr. Osamu Hazama for his helpful comments.

References

1. Foster, I. and Kesselman C. eds., The Grid: Blueprint for a Future Computing Infrastructure, Morgan Kaufmann Publishers (1999) and activities of GGF in http://www.globalgridforum.org/
2. The Globus Project, http://www.globus.org/
3. Bhatia, D., et al. WebFlow - a visual programming paradigm for Web/Java based coarse grain distributed computing. Concurrency — Practice and Experience, Vol. 9, No. 6. Wiley (1997) 555–577
4. Allcock, W. et al., GridMapper: A Tool for Visualizing the Behaviour of Large-Scale Distributed Systems," In Proceedings of the eleventh IEEE International Symposium on High Performance Distributed Computing, HPDC-11, IEEE Computer Society (2002) 179–187
5. Erwin, D.W, UNICORE — a Grid computing environment, Concurrency and Computation: Practice and Experience Vol. 14, No. 13–15. Wiley (2002) 1395–1410
6. The Triana Project, http://www.trianacode.org/
7. Allen, G., et al., Enabling Applications on the Grid: A GridLab Overview, International Journal of High Performance Computing Applications, Special issue on Grid Computing: Infrastructure and Applications, SAGE Publications (to appear August 2003)
8. The ITBL project, http://www.itbl.jp/
9. Takemiya, H., et al., Software Environment for Local Area Metacomputing, In Proceedings of the fourth International Conference Supercomputing and Nuclear Applications, SNA2000 (2000)
10. Imamura, T., et al., An Architecture of Stampi: MPI Library on a Cluster of Parallel Computers, In Dongarra, J., et al. (eds.): Recent Advances in Parallel Virtual Machine and Message Passing Interface, seventh European PVM/MPI Users' Group Meeting Balatonfüred, LNCS 1908. Springer-Verlag (2000) 200–207
11. Tsujita, Y., et al. Stampi-I/O: A Flexible Parallel-I/O Library for Heterogeneous Computing Environment, In Kranzlmüller, D., et al. (eds.): Recent Advances in Parallel Virtual Machine and Message Passing Interface, ninth European PVM/MPI User's Group Meeting Linz, LNCS 2474. Springer-Verlag (2002) 288–295
12. Kitabata, H. and Chino, M., Development of source term estimation method during nuclear emergency, In Proceedings of the International Conference on Mathematics and Computation, Reactor Physics and Environmental Analysis, M&C99 Vol. 2 (1999) 1691–1698
13. The Fraunhofer Research Grid Project, http://www.fhrg.fhg.de/
14. Web Services Description Language (WSDL) Version 1.2, http://www.w3.org/TR/wsdl12/

Development of Remote Visualization and Collaborative Visualization System in ITBL Grid Environment

Yoshio Suzuki and Nobuko Matsumoto

Center for Promotion of Computational Science and Engineering,
Japan Atomic Energy Research Institute
6-9-3 Higashi-Ueno, Taito-ku, Tokyo 110-0015, Japan
{suzukiy,matsumoto}@koma.jaeri.go.jp
http://www2.tokai.jaeri.go.jp/ccse/english/index.html

Abstract. The Information-Technology-Based Laboratory (ITBL) project aims to build a virtual research environment for sharing intellectual resources such as remote computers, programs and data in research institutions and for supporting cooperative studies among researchers throughout Japan. The ITBL infrastructure software forms the basis of its ITBL framework. The application visualization system (AVS)/ITBL is one of core visualization systems on the ITBL system infrastructure software and has been developed based on the existing commercial visualization tool AVS/Express. The AVS/ITBL visualization system enables remote postprocessing visualization of any data stored on any supercomputer located in the ITBL network through Web browsers. In addition, a collaborative visualization function, which refers to researchers in remote locations simultaneously accessing and monitoring the same visualized image, has been developed based on the AVS/ITBL. The global structure as well as the performance of the AVS/ITBL is presented.

1   Introduction

In recent years, the research and development for construction of the grid environment which aims to contribute to the development of science have been furthered globally [1]. Also in Japan, some researches related to the grid have been furthered. As one of them the project called Information-Technology-Based Laboratory (ITBL) has been advanced [2]. The goal of the ITBL project is to build a virtual research environment for sharing intellectual resources such as remote computers, large-scale experimental facilities and data bases in universities and institutes and for supporting cooperative studies among researchers throughout Japan. Six institutions have joined as initial project members: National Institute for Materials Science (NIMS), National Research Institute for Earth Science and Disaster Prevention (NIED), National Aerospace Laboratory of Japan (NAL), the Physical and Chemical Research Institute (RIKEN), Japan Science and Technology Corporation (JST) and Japan Atomic En-

A. Veidenbaum et al. (Eds.): ISHPC 2003, pp. 269–277, 2003.

ergy Research Institute (JAERI). Now, some other institutions are also participating in the project. In these institutions, application software has mainly been developed.

In the Center for Promotion of Computational Science and Engineering (CCSE) of JAERI, the ITBL system infrastructure software has been developed. In this development, construction of various tools which can efficiently perform simulation research has been advanced. Almost all tools can be used easily through a Web browser. As one of these tools, the application visualization system (AVS)/ITBL has been constructed on the ITBL system infrastructure software [3]. The related work is also described in reference [3]. Here, by examining what visualization function is useful for simulation researchers in the grid environment, the following four functions have been developed:

1. Data sets located in remote locations can be directly read from a visualization tool on a local graphics server.
2. Results can be easily visualized using a Web browser at a local user terminal.
3. Visualization processes can be executed in command mode.
4. Visualized images are sharable among user terminals from remote locations.

Functions 1 to 3, which have already been described in our previous papers [3] and [4], are briefly introduced in section 2. Function 4 is presented in section 3. The performance is described in section 4. Finally in section 5, the concluding remarks and a brief summary will be given.

2   AVS/ITBL Visualization System

The AVS/ITBL visualization system has been developed based on the existing commercial visualization tool AVS/Express [5]. The AVS/ITBL visualization system reads and visualizes files stored in remote computers connected to the ITBL environment (Fig.1).

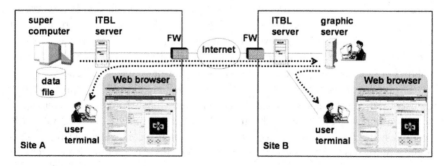

Fig. 1. Configuration of AVS/ITBL in the ITBL environment. A user in site B visualizes the data file stored on a supercomputer in site A using a graphics server. Other users in site A and site B visualize the file using a Web browser on a user terminal. All of them can access and monitor the same visualized image.

In the ITBL environment, a site is an independent operating unit composed of several computers. The user terminal is a PC or workstation with a Web browser equipped with a plug-in activated to run Java applets. Generally, a site has a firewall (FW), and ITBL can communicate through the FW to share resources over multiple sites. HTTPS enables a secure connection between the user terminal and ITBL servers.

2.1 Visualization with a Remote Module

AVS/Express is widely used for postprocessing visualization of scientific data but can read only data on a computer (graphics server) where AVS/Express is installed. Therefore, when a simulation is performed on a remote computer, it is necessary to transfer data from the remote computer to the graphics server. In AVS/ITBL, the function, which enables to read data on remote computers over a firewall, has been developed. The data file is temporarily copied from a remote computer to a local graphics server. This is realized using Starpc communication library application programming interface (API) which is a remote procedure-call-based communication library for parallel computer clusters [6] and is a communication tool of the ITBL system infrastructure software. The function is mounted as the AVS relay process (1st). The global structure and the execution procedure of this function are described in detail in the reference [3].

2.2 Visualization on a Web Browser

The function, which can display the visualized image on a web browser as well as controlling visualization parameters from a web browser, has been developed. Users can execute the visualization on a web browser specifying the network file (v file) and the data file which are used in AVS/Express. That is, if the network file and the data file of AVS/Express have already been created by a researcher, it is possible for other joint researchers to execute the visualization easily even if they are not well-versed in the usage of AVS/Express. Users can select from a showImage or show-Polygon entity. With showImage, an image is displayed in 2D (GIF file format data); with showPolygon, it's displayed in 3D (GFA file format data). GFA stands for Geometry Flipbook Animation and is a special format adopted by AVS/Express. In both cases, users can rotate, resize and translate graphic images, and make changes in visualization parameters interactively. In order to generate a GIF file or a GFA file, two modules: the image data generation module which incorporates the function of the geomcapture module of AVS/Express and the CGI relay module which generalizes visualization parameters on AVS/Express have been developed. Furthermore, AVS relay process (2nd) has been developed for communication control of the visualization parameter generalized by the CGI relay module. Starpc is used in transfer of the GIF file or the GFA file and in communication of visualization parameters.

The function is similar to AVS/Express Web Server Edition (WSE) [http://www.kgt.co.jp/english/product/avs/vizserver/]. The WSE is the system which unifies MAST and the AVS/Express. MAST has been developed by KGT Inc. and is the framework software which can easily build the Web system for using the various existing applications through a network. On the other hand, the function in AVS/ITBL is premised on the handling of large-scale data through SuperSINET. The global structure and the execution procedure are described in detail in references [3] and [4].

2.3 Visualization in Batch Processing Mode

The functions described previously are premised on the interactive use of the visualization operation. Next, the batch processing mode in which the visualization process including the remote module function is attained by execution in command form has been developed. Development of this function enables execution from the program execution support tool TME (Task Mapping Editor), which is one of the tools on the ITBL system infrastructure software, and enables performing a simulation and visualization collectively. To achieve this, the following items have been developed:

1. Users can save the visualization parameter file using an icon on the Web browser.
2. Users can execute AVS/ITBL as a command.
3. Users can designate the visualization parameter file as an augment of AVS/ITBL command.
4. Users can designate the data file as an augment of AVS/ITBL command.

AVS/ITBL in command form has been developed based on AVS relay process (2nd) described in section 2.2. AVS relay process (2nd) is a command itself can read the information of the data file and the visualization parameter file directly in the batch processing mode, while it acquires their information from servlet by communication of Starpc in the visualization function on a Web browser. The example of utilization to simulation research (numerical environment system which is one of the applications in the ITBL project) of the batch processing mode is described in reference [4].

3 Collaborative Visualization

The function described previously presumes that one researcher performs visualization operation. It is possible that some researchers belonging to the same research group are far apart from one another. In order for two or more such researchers to evaluate simulation results together through visualization operation, a new visualization function has been developed.

This function is realized by communication between visualization processes of AVS/ITBL which two or more users started. Researchers can refer to the same visualization image so that other processes read the visualization parameter which one certain visualization process outputted. The visualization process which outputs the shared visualization parameter is called "operation master", and the other visualization process is called "operation slave". Moreover, the visualization process which is started in advance of other visualization processes and manages the whole visualization collaboration is called "collaboration master".

As a method of exchanging a visualization parameter between visualization processes, the CGI relay module described in section 2 is used. The value of the visualization parameter defined in the CGI relay module can be exchanged between servlet and the visualization processes in AVS/Express. Thus, the expansion of the CGI relay module has been carried out and the visualization parameter of one visualization process is reflected in that of other visualization processes through the file. Fig. 2 shows the global structure of the collaborative visualization function on AVS/ITBL.

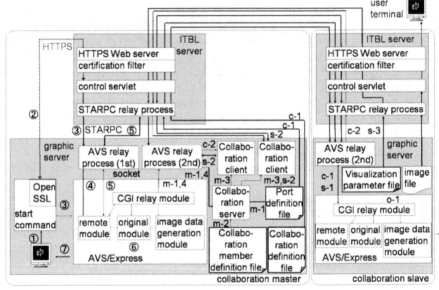

Fig. 2. Global structure of the collaborative visualization function on AVS/ITBL.

Execution procedures of the collaborative visualization function are as follows:
(a) A user starts AVS/ITBL on a graphics server. A user selects the network file incorporating the CGI relay module and the data file used for visualization.
(b) In AVS/ITBL started by procedure (a), clicking on the 'Collaboration start' icon on the panel of the CGI relay module starts the collaboration server and outputs a collaboration definition file. Then, the visualization process of AVS/ITBL started by the procedure (a) serves as a collaboration master.
(c) Other users who use the collaborative visualization function start AVS/ITBL from the web browser, and read the collaboration definition file.
(d) The same visualization result as the collaboration master can be displayed on the browser of other users by clicking the 'showImage' or the 'showPolygon' on the web browser started by the procedure (c).

Detailed processing procedures in each visualization process are as follows (signs in the text such as (m-1) and (c-1) correspond to each process in Fig. 2):
(1) Procedure of the collaboration master

(m-1) In order to start the collaborative visualization function, the following procedures are executed:

- Clicking the 'Collaboration start' icon starts the collaboration server and records the port number on a port definition file. This port is used to receive the demand from a collaborator.
- The CGI relay module creates the copy of a network file.
- The collaboration server outputs a collaboration definition file which includes the site and host name of the collaboration master, the directory of the port definition file, and the directory of the copy of the network file.

(m-2) A collaboration server receives the demand that a collaborator wishes to participate, and registers the site, host, and user name of the collaborator. When a collaboration member definition file exists, the demand from the user who is not registered in this file is refused.

(m-3) A collaboration server answers the site, host and user name of the current operation master to the inquiry from a collaborator.

(m-4) The collaboration master can change the operation master.

(2) Procedure of visualization processes other than the collaboration master

(c-1) Users other than the collaboration master start a visualization process by selecting the collaboration definition file. From the information on the collaboration definition file, the site and host name of the collaboration master, the path of the copy of the network file and the path of the port definition file are specified.

(c-2) Using the information acquired in the procedure (c-1), the collaborator is started at the host where the collaboration master exists. The collaborator connects with the collaboration server using the gained port number and registers the site, host and user name with the collaboration server.

(3) Procedure of an operation master

(o-1) The operation master exports the visualization parameter defined in the CGI relay module to the file with a fixed name.

(4) Procedure of an operation slave

(s-1) The operation slave requires the import of the visualization parameter file which the operation master exports.

(s-2) By the procedure (s-1), the operation slave asks the collaboration master the present operation master.

(s-3) The operation slave imports the visualization parameter which the operation master exports using the site, host and user name given in procedures (s-2) and (m-3). Thus, the same visualization image as the operation master is obtained.

4 Performance

This section describes the performance of the visualization function with a remote module and the visualization function on a Web browser.

4.1 Visualization with a Remote Module

By reading data on the remote computer Origin3800 using the local graphics server GP400S, the performance of the data input remote module on AVS/ITBL is evaluated. Here, GP400S and Origin3800 are currently installed at the CCSE/JAERI and at the Naka Fusion Research Institute (Naka FRI)/JAERI, respectively. It is about 130km between CCSE and Naka FRI, and the SuperSINET is installed as an infrastructure network. The SuperSINET, which National Institute of Informatics (NII) provides as a cooperation study, is the network through Japan at the speed of 10Gbps.

Table 1 shows the time required for the data input process for two files with different sizes. Measurements using the normal read module and the data input remote module are compared. The times in the parenthesis are the value descried in reference [3]. The performance is again evaluated in this paper, since the performance of Starpc has been improved. The times are expressed with the average value of three times measurements.

It is found that the remote-data read becomes about 19 times faster when times before and after the improvement of Starpc are compared. The case involving the data input remote module firstly copies the data file using the function of Starpc and then reads it at local. That is, the time required to 'remote-data read' is almost the same as the sum of times of 'file-management function' and 'local-data read'. As a result of measurement, however, 'remote-data read' requires a little more time. It is likely that the data input remote module needs time to recognize the end of the data transfer and to prepare for AVS/ITBL reading data. Moreover, it is found that the time difference between 'file-management function' and 'scp' is less than 1% when the file size is bigger (step1.inp). It means that the communication speed using the improved Starpc becomes the same level as that with ssh.

Table 1. Time of the data input process using the usual read and the data input remote modules. The wind.fld is from the AVS/Express samples. Step1.inp is from the simulation of blood flow coupled with a diseased blood vessel[8] which is performed in CCSE

File name	File size (kB)	Local-data read (s)	Remote-data read (s)	File-management function (s)	scp (s)
wind.fld	241	< 1	5	4	< 1
		(1)	(96)	(55)	(1)
step1.inp	11012	4	52	36	34
		(-)	(-)	(-)	(-)

Table 2. Time of the visualization process using a Web browser. In comparison, time of the visualization process using the graphics server is also shown

File name	File size (kB)	'visualize' click (s)	'showImage' click (s)	Visualization using graphics server (s)
wind.fld	241	10.81	1.55	9.45
step1.inp	11012	39.32	21.81	18.90

4.2 Visualization on a Web Browser

Next, the performance of the visualization function on a Web browser is evaluated. The results are shown in Table 2.

By clicking the 'visualize' icon after selecting the data file and the visualization parameter file on a Web browser, AVS/Express on the graphics server performs the visualization process getting the information in these files from the CGI relay module. These files are firstly transferred from the user terminal to the ITBL server and are further transferred to the CGI relay module through AVS relay process (2nd). After that, by clicking the 'showImage' on the browser, an image is displayed on the user terminal. That is, the sum of times for these two processes is needed to display an image. Comparing this time with that when AVS/Express is performed only on the graphics server, the time of the visualization process using a Web browser is not so different from that using only the graphics server when the file size is smaller (about 1.3 times longer for wind.fld), but the former is longer than the latter when the file size is larger (about 3.2 times longer for step1.inp). One reason is that much time is required in order to create image file using the image data generation module, when the file size is larger. Another one is that in the process that the information of the visualization data is passed to the CGI relay module, the data file is temporally copied to a work domain on a graphics server even if it is located on the local graphics server. The larger the file size is, the longer the time required to copy it is. This is because the same way for the remote data is adopted when reading the local data. It should be improved in near future.

4 Summary

In this paper, as one of the tools on the ITBL system infrastructure software, the research and development of the visualization function based on AVS/Express have been described by examining how visualization function is useful for simulation researchers in grid environment. Here, four functions as follows have mainly been described:

1. Visualization with a remote module.
2. Visualization on a Web browser.
3. Visualization in a batch processing mode.
4. Collaborative visualization.

Although there are some simulation researches which perform visualization using these functions such as the numerical environment system described previously and the blood-flow simulation in an aneurysm, the visualization at sufficiently high speed has not been achieved when the data size is larger. It is necessary to develop these functions to enable the visualization at high speed to such large-scale data.

Now, GP400S are mainly used as a visualization server. So far some parts of the functions have already been transplanted to Onyx300 and in the current year the transplant of all functions will be performed. Thereby, the improvement in an overall processing performance is expected since the hardware performance improves.

Acknowledgments

The authors gratefully acknowledge the support of Prof. Yagawa, Dr. Hirayama, Dr. Yamagiwa, Dr. Nakajima, Dr. Aoyagi, Dr. Higuchi, Mr. Maesako, Mr. Sai, Mr. Kimura, Mr. Yamagishi, Mr. Hasegawa of JAERI, Mr. Takemiya (Hitachi East Japan Solutions, Ltd.) and Mr. Fujisaki (Fujitsu Limited). We also thank Mr. Kanazawa, Mr. Suzuki, Mr. Teshima, Mr. Arakawa, Mr. Ishikura of Fujitsu Limited, Mr. Nakai, Mr. Sakamoto of KGT Inc., Mr. Maesaka, Mr. Sakamoto, Mr. Miyamoto, Mr. Kakegawa, Mr. Endo, Mr. Yamazaki, Mr. Nakamura and Mr. Otagiri of Fujitsu Nagano Systems Engineering Limited for their insightful advice on this development.

References

1. Foster, I. and C.Kesselman, Ed.: The GRID, Blue Print for a New Computing Infrastructure. Morgan Kaufmann (1998).
2. Yagawa, G. and Hirayama, T.: Japan IT-Based Laboratory. to be published in Proc. of Global Grid Forum 5 (2002)
3. Suzuki, Y., Sai, K., Matsumoto, N. and Hazama, O.: Visualization system on the Information-Technology-Based-Laboratory. IEEE Computer Graphics and Applications March/April (2003) pp.32-39
4. Suzuki, Y., et al.: Development of Multiple Job Execution and Visualization System on ITBL System Infrastructure Software and Its Utilization for Parametric Studies in Environmental Modeling, in Proceedings of the 3rd International Conference on Computational Science (Melbourne, Australia and St. Petersburg, Russia 2-4 June 2003, Proceedings, Part III), LNCS2659, (2003) pp.120-129
5. Upson, C. et al.: The Application Visualization System: A Computational Environment for Scientific Visualization. IEEE Computer Graphics and Applications, vol.9, no.4 (1989) pp.30-42
6. Takemiya, H. and Yamagishi, N.: Starpc: A Library for Communication among Tools on a Parallel Computer Cluster-User's and Developers Guide to Starpc. JAERI-Data/Code 2000-005, JAERI (2000)
7. Imamura, T. et al.: An Architecture of Stampi: MPI Library on a Cluster of Parallel Computers. Recent Advances in Parallel Virtual Machine and Message Passing Interface (LNCS 1908), Springer (2000) pp.200-207
8. Guo, Z., Hirayama, T. and Matsuzawa, T: Large-scale parallel simulation of blood flow coupled with a diseased blood vessel. Application of High-Performance Computers in Enginnering VII, Advances in High Performance Computing, Vol.7, WIT Press (2002)

Performance of Network Intrusion Detection Cluster System

Katsuhiro Watanabe[1], Nobuhiko Tsuruoka[1], and Ryutaro Himeno[1]

RIKEN, Hirosawa 2-1, Wako, Saitama 351-0198, Japan,
(kwata,tsuruoka,himeno)@riken.go.jp

Abstract. We develop a PC based new style NIDS called DNIDS. This system provides ubiquitous, high-performance, high accuracy and cost-effective monitoring network. In this paper, we show packet capturing performance of single PC NIDS and cluster PC NIDS. Finally, we show a concept of DNIDS.

1 Introduction

In a computer network management, Intrusion Detection System (IDS) has become to be general for detecting intrusions as Internet grows. The intrusion means that a person attempting to break into or misuse a system (e.g., stealing confidential data, sending spam mails using another person's email server). For the presents, almost IDS are Network IDS (NIDS [1]) which captures packets of traffic and infers if it's an intrusion or not. NIDS is widely used, however, enough attention hasn't given to the performance like the routing/switching devices [2].

An important factor of NIDS performance is packet capturing. All network devices have the limitations of capacity to receive packets per unit of time, the device fail to receive packets if traffic exceeds. This failure is called "packet drop". The packet drop impacts on performance and accuracy of NIDS, and therefore we understand the limitations of NIDS and study methods to improve the packet capturing. In section 3 we show packet capturing performance of ordinary PC, and effectiveness of several tuning methods. In section 5 we show NIDS cluster and its performance. Finally, we propose DNIDS which is combination of packet drop avoidance and another improve method.

2 IDS, NIDS, DNIDS

Administrator controls the system to provide service to users, and users accept it, however, all of users are not trusty and users do not always operate computer correctly. This is a very similar phenomenon of the natural human society. It is difficult that administrator checks if operations of users are right, in contrast with the policeman actually see a criminal action in the human society, the administrator thus needs tools to inspect users. IDS is general term for above kind of supporting system. IDS is classified into Host based Intrusion Detection

A. Veidenbaum et al. (Eds.): ISHPC 2003, LNCS 2858, pp. 278–287, 2003.

System (HIDS) and Network Intrusion Detection System (NIDS). The former is inspects users in one host (e.g., PC, WorkStation) using various system information, the later is inspects remote network user using network information. HIDS is precisely than NIDS, although, the sphere of HIDS is restricted. At present, the greater part of IDS is NIDS. Because, One NIDS is able to monitor network nodes at same time, this fact is important for an administrator or network manager who has not enough time to spend for inspection users. However, little attention has given to the performance and accuracy of NIDS, hence also little attention has given to improve method of NIDS. We propose Distributed Network Intrusion Detection System (DNIDS) which integrates the combination of HIDS, NIDS and other network connected system. Modules of the DNIDS cooperate through the EEXN to inspect large scale network entirely or high-speed network. To arrange the DNIDS, we must know basic performance of a NIDS for calculation necessary number of NIDS box, and we should know the method to improve performance for keeping the DNIDS building cost low. Packet capturing is first process of almost all NIDS. We pay attention to packet capturing performance, which is important factor of NIDS performance rather than other factors. In this paper, we make the packet capturing performance clear and try to improve it using several methods.

3 Packet Capturing Performance

In this section, we study the packet capturing performance of a PC and the improve method. A packet capturing is general method to monitoring a network. Normally, a network device receives only a packet whose destination address is the device, however, a network monitoring device receives all packets arrives the network interface. It is possible to monitor all packets using special mode, which is called "promiscuous mode", of Network Interface Card (NIC), and this type of monitoring is called "packet capting". We need an indicator of NIDS performance, and define "packet capturing performance" as following:

Definition 1. *A maximum number of received packets per unit time using promiscuous mode without a packet drop.*

We will measure capacity of various NIDS using this definition.

3.1 PCAP Library and Snort

Packet Capture library (PCAP library)[4] is an architecture independent packet capturing API which is widely implemented in many operating systems and used by various network monitoring programs. Snort[3] is free distributed NIDS software; it is application of PCAP library. We measured packet capturing performance using these popular packages PCAP and Snort.

3.2   Peroformance Test

The test environment is in Fig. 1.

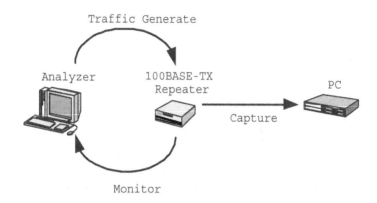

Fig. 1. single sensor test environment

The specification of PC is below:

- Motherboard: Supermicro Super P3TDLR+
- CPU: Pentium III 1.2GHz Single
- Chipset: ServerWork Serverset III LE
- Memory: PC133 DIMM 235MB x 2
- HDD: IBM Ultramaster DDYS-T36950M 36.7G
- SCSI: Adaptec AIC-7892 SCSO controller
- Network: Onboard Intel 82559 Ethernet controller

We measured the performance of each frame length traffic in 64byte length (=minimal Ethernet frame length), 256byte length and greater. In general, network device difficult to handle 64byte frame traffic, although capability of handling frames is important by two reasons:

- 64 byte frame traffic is ordinary in a real network (in Fig. 2).
- NIDS should resist 64 byte jamming frames by an attacker.

By our definition of packet capturing performance, packet capturing performance of the PC with this specification is 70Mbps at 64byte length, and 100Mbps at 256byte length or greater. This result is shown in Fig. 3.

The packet drop is occurred only in a range over the 70Mbps at 64bytes. There are two main reasons of the packet drop:

The traffic is in the range $[70Mbps, 90Mbps)$. The main reason of packet drop is that Linux kernel fails handling of the NIC's interrupt. In the case of 64 bytes, Ethernet NIC generates a maximum number of interrupts which is difficult to handle by the kernel in spite of low CPU usage.

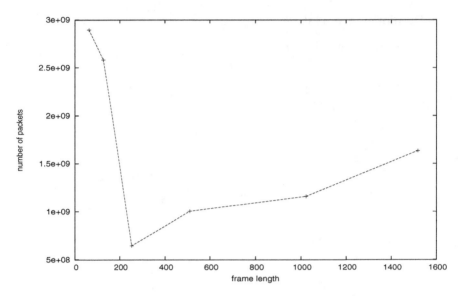

Fig. 2. frame length distribution on real network

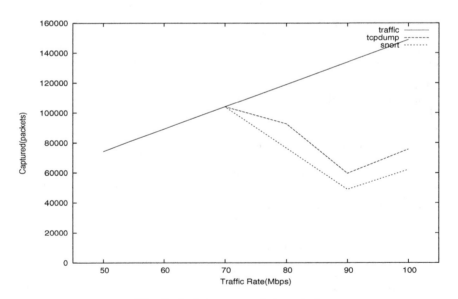

Fig. 3. single sensor packet capturing

The traffic is in the range $[90Mbps, 100Mbps)$. There is one more reason to drop packets. In this case, NIC itself drops packets and number of interrupt is decreased. As a result, the number of captured packets is increased in this range.

As the result, packet capturing performance of single PC is in the table below.

Table 1. A measurement result of single PC packet capturing performance.

frame length (bytes)	maximum number of capture (Mbps)	packet capturing performance (Mbps)
64byte	61.1	53.3
256byte	92.8	92.8
512byte	96.2	96.2

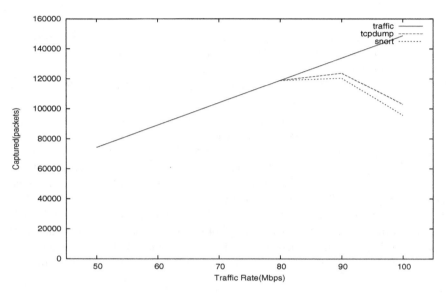

Fig. 4. device polling sensor packet capturing

3.3   Device Polling

We measured also about the packet capturing performance with Device Polling. Device Polling is an implementation technique of operating system kernel to reduce interrupt handling load. A kernel with device polling does active checking if a packet arrives in NICs instead of receiving NIC's interrupts passively. We show the result of performance test of a Device Polling PC in Fig. 4. As you see, Device Polling is highly effective to reduce the packet drop. However, this feature works on FreeBSD and particular devices.

4   NIDS Performance

In this section, we discuss reasons that NIDS should keep the packet drops minimum.

4.1   NIDS Mechanism and Packet Drop

Except few systems all NIDS has same mechanism, the procedure to inspection as follows:

Table 2. clustering NIDS performance

frame length (bytes)	maximum number of capture (Mbps)	packet capturing performance (Mbps)
64	304.8	152.4
256	463.8	371.0
512	481.2	385.0

1. Capture packets. Do de-fragmentation, construct UDP/TCP session streams, etc. It depends on the architecture of NIDS that how far reconstruction traffic information.
2. Consult a database of known attacking patterns to collate with previous extract information.

It is different from other systems like the amount of network traffic monitor, NIDS is not statistical system. It inspects nothing if one attack packet in millions ordinary packets was lost. Furthermore, there are many avoiding methods to NIDS's inspection. If the NIDS has no de-fragmentation function, an attacking pattern like a buffer over flow in fragments is pass by. NIDS without TCP session constructor misses to find an attacking pattern over packets. Of cause, a failure of de-fragmentation leads to fail of session extraction. The packet drop impacts on de-fragmentation, session extraction and other upper layer information extraction. To prevent NIDS avoiding, ought to prevent packet drop.

5   NIDS Cluster Performance

As previous sections, there are the limitations of single PC's packet capturing performance; its capacity is insufficient for the bandwidth of 1000BASE Ethernet. To improve the capacity of NIDS, we apply clustering-method to NIDS. (NIDS cluster is shown in [2]

5.1   Performance Test

To apply clustering-method for NIDS, the traffic of network must be shared with plural NIDS. There are two type of special device to share the network traffic, load balancer and network TAP.

Load balancer captures packets from target port and distributes it into monitoring ports. Network tap is more physical device; it transfers electric signals into multiple ports. We choose hardware accelerated load balancer for this purpose. The test environment is consists of one load balancer and five PCs, in Fig. 5.

We show result of cluster NIDS in Fig. 6, 7 and table below.

As result, the combination of load balancer and PC-NIDS make improved the packet capturing performance.

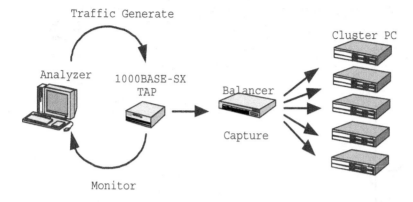

Fig. 5. cluster sensor packet capturing

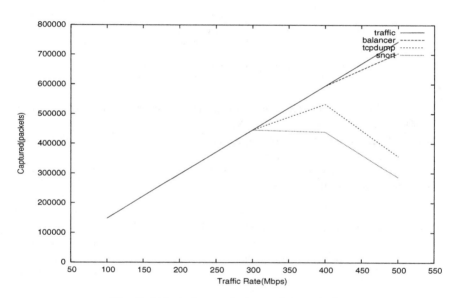

Fig. 6. 64 byte frame cluster packet capturing

6   DNIDS

We have developed a new IDS framework called Distributed Network Intrusion Detection System (DNIDS) which improves accuracy of the IDS. The DNIDS provides intrusion detection which covers a whole large system with gathering events from various systems and analyzing correlations among the events.

The DNIDS consists of an event information exchange network (EEXN), event analyzing modules and input modules which brings event information in the DNIDS from generic system logs. An explanation of the DNIDS process, which is along a flow of event information, is below:

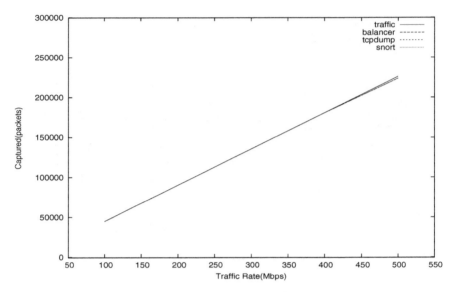

Fig. 7. 256 byte frame cluster packet capturing

- The input module transfers event information, which is gathered from syslog, SNMP or an IDS logs to the EEXN.
- The analyzing modules accept requested kind of event from the EEXN and send back a result onto the EEXN.
- The output modules notify for an administrator with E-MAIL, controls a firewall to prevent illegal traffic, etc.

IDS has a basic problem which informs false reports, and hence it is important task that improve accuracy of intrusion detection. We showed one method to improve accuracy of intrusion detection, however, there are other methods in addition to the packet capturing performance improvement. The DNIDS uses correlations of events identified by various systems, and thereby the DNIDS further improves the accuracy of IDS.

There are many caes that IDS possibly detects the intrusion or misuse with high accuracy using low accuracy events. We will give an example of case that a NIDS detects "FTP buffer overflow" (which is an attack to execute any process on a FTP process). The NIDS possibly detects "FTP buffer overflow", although the NIDS itself has no effective method to make sure whether "FTP buffer overflow" successes. However, the NIDS detects intrusion with high accuracy when the NIDS possible inquires the FTP server about FTP session logs and health of FTP process.

We need a new system to put the mechanism into practice, which system exchanges and analyzes event information among IDSs and non-IDSs. The system must input event information from various computer systems, analyzes on the system, and send the results. The system must recursively analyze event information and the result of the system.

We propose the DNIDS using following approaches:

- The DNIDS communicate with flexible format called "record" to unify various formats of system logs.
- The each module, which connects on the EEXN, only requests to the EEXN with a announce of necessary kinds of record. There is no need that the module searches other modules which output necessary records or ask to send result to the module.
- The each analyzing module does simple analyzes, which checks threshold of numbers, counting records, filtering matching record and etc., consequently a combination of simple logics makes flexible, complex analysis mechanism.

The analyzing logic, which is result of module combination, is robust because communications of the EEXN is based on kind of record. If the analyzing logic is based on directly connection of modules, it is changed by new connection, disconnection or process termination of module, the logic is consequently very unstable. This connection free logic has necessary property to operate the DNIDS in the real world.

The DNIDS does intrusion detection with fine-grained distributed processes; the DNIDS is very scalable accordingly. Many processes on many CPU parallel evaluate processes of the intrusion detection logic in the DNIDS at same time. A bottle-neck of the DNIDS scalability depends on scalability of the EEXN which has very scalable mechanism, and there is no restriction about CPU power. Consequently, the DNIDS is very scalable than other IDSs.

The Internet will have grown; number of network nodes in the world will have been exponentially increasing accordingly. The DNIDS will treat much number of devices, and will be very effective framework for intrusion detection.

7   Conclusion and Future Work

We discussed about the effect of packet drop and defined packet capturing performance as an indicator of NIDS quality. Next, we show packet capturing performance of ordinary PC and device polling tuned PC. Furthermore, we show NIDS cluster system and its packet capturing performance and effectiveness. Finally, we gave the insufficient points of current NIDS and introduced the DNIDS. The mechanism of the DNIDS is purly log analysing system. We thinks the DNIDS is appricable for other fields which needs large-scale log monitoring, and plan a fine-grained system monitoring for a large-scale network and HPC clustering system.

Acknowledgement

This work is a part of Information Technology Based Laboratory (ITBL) project, we thanks for this work is conducted at the Physical and Chemical Research Institute (RIKEN) , and support by the Ministry of Education, Culture, Education, Sports, Science and Technology (MEXT).

References

1. Biswanath Mukherjee, L. Todd Heberlein, and Karl N. Levitt. Network intrusion detection. *IEEE Network*, 8(3):26–41, 1994.
2. Christopher Kruegel, Fredrik Valeur, Giovanni Vigna, Richard Kemmerer. Stateful intrusion detection for high-speed networks. In *IEEE Symposium on Security and Privacy*. University California, 2002.
3. Martin Roesch. Snort - light weight intrusion detection for networks. In *Proceeding of LISA'99: 13th Systems Administrator Conference*, 1999. http://www.snort.org/docs/lisapaper.txt.
4. Steaven McCanne, Van Jacobson. The bsd packet filter: A new architecture for user-level packet capture. In *USENIX Winter*, Lawrence Berkley Laboratory, One Cyclotron Road, Berkley, CA 94720, 1993. Lawrence Berkley Laboratory.

Constructing a Virtual Laboratory on the Internet: The ITBL Portal

Yoshinari Fukui[1], Andrew Stubbings[2], Takashi Yamazaki[2], and Ryutaro Himeno[1]

[1] Computer and Information Division
The Institute of Physical and Chemical Research (RIKEN)
2-1,Hirosawa, Wako-shi, Saitama 351-0198, Japan
{yfukui, himeno}@riken.go.jp
[2] Grid Research Inc.
2-1-6, Sengen, Tsukuba, Ibaraki 305-0047, Japan
{ajs, tyama}@grid-research.com

Abstract. The benefits of computer simulation are effective over a very large range of subjects, but it is mainly universities, research institutes or large companies that can actually perform such large computations. There are also many organizations where computer simulation is not performed that can receive its benefits. By using the ITBL portal, we remove the obstacles of performing a computer simulation. In doing so, we want to be able to gain universal benefit for certain fields by using effective computer simulations.

1 Introduction

We will describe our vision for the ITBL portal. The benefits of computer simulation are effective over a very large range of subjects, but it is mainly universities, research institutes or large companies that can actually perform such large computations. There are also many organizations to which computer simulation is not performed that can receive the benefit of computer simulation.

There are several reasons for this state of affairs,

- Organizations are unfamiliar with the availability of a computer simulation for their subject problem.
- There are no users in the organization proficient in computer simulation.
- The organization does not have such large computer resources.

By using the ITBL portal, we remove the obstacles of performing a computer simulation. In doing so, we want to be able to gain universal benefit for certain fields by using effective computer simulations.

Also, in the field and type of industry to which the computer simulation has not been performed until now, it is thought that there are many areas where computer simulation can be very effective.

A. Veidenbaum et al. (Eds.): ISHPC 2003, pp. 288–297, 2003.

2   ITBL Portal and Computer Simulation Procedure

The process of executing a computer simulation is examined, and in doing so we consider the reasons why computer simulations are not performed. The procedure for solving an actual problem using a computer simulation is considered to be as follows (Fig. 1).

(1) Modeling
  - Modeling of a phenomenon
  - Mathematical modeling

(2) Creation of a program
  - Make a simulation model for the computer
  - Programming

(3) Execution of the calculation
  - Obtain the results (numerical value)

(4) Making the results easily recognizable by a human (visualization)

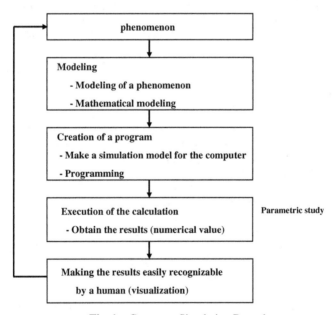

Fig. 1.   Computer Simulation Procedure.

In considering the above procedure we have prepared the following three functions that comprise the ITBL portal.

(a) Groupware function

(b) Calculation service function

(c) Application showroom

(3) and (4) of the computer simulation procedure can be solved using the calculation service function of (b).

For (2), those proficient in creating a computer program can use the function in (b) to solve their problem. For those who cannot create a program personally, the application showroom (c) can be used to run simulations using the applications already integrated with the ITBL portal. Although the groupware function of (a) is not a fundamental solution of (1), collaboration and planning with well-informed people allows a problem to be solved on a more personal level.

Furthermore, in order to overthrow the myth that "the availability of a computer simulation is not known", for instance, by performing a parametric study, that it is shown that the optimization and the improvements in the yield of a product are realizable. It is difficult to realize the optimization and the improvement in the yield of a product in a single simulation. To give an example, consider an auto accident of some kind. In the past there were many relatively poor computer simulations using a single compute engine to perform post-analysis and verification of a design when studying the accident. Effective computer simulations, such as optimization of a product and improvement in yield, will become possible by performing parametric study. To use a fashionable term, frontloading becomes possible. Using the fact that parametric study can be done, a breakthrough in design can be made and thus an improvement in quality of optimization and yield rate of the product becomes possible.

3   Parametric Study and Grid

To calculate a problem realistically, it is necessary to use parameter sets that number thousands to tens of thousands. For example, in order to complete a problem that has ten parameters using about 1000 calculations per parameter, it may actually only by possible to calculate two parameters because of the size of the problem. When the parameter count is increased to three, approximately 60,000 calculation runs are needed. In order to perform parametric studies, each parameter run must calculate for a short time.

Although other papers describe parametric studies in more detail, modeling is more important than an improvement in speed-up for coding. Thus, for each parameter set it is necessary to calculate many parametric runs. Many calculations corresponding to a parameter set are possible by "obvious parallel processing". This type of calculation is very suitable for the ITBL portal grid system. Although great efforts are required to create each parallel application, it is very easy to create a parallel computer calculation if an effective parameter set has been considered.

4   Implementation of the ITBL Portal

The ITBL Portal is an experimental system that gives people the opportunity to evaluate the benefits of computer simulation that were previously not available to them. The ITBL Portal system consists of three parts,

- Groupware part
- Computation service part
- Application showroom part using calculation service

Although the ITBL Portal logically consists of three parts, the application showroom part actually uses the calculation service part. Thus, the ITBL Portal consists of two subsystems, a groupware part and a computation service part. We carried out the construction of the portal system by creating a system using standard technology that does not depend on hardware availability (Fig. 2, Fig. 3).

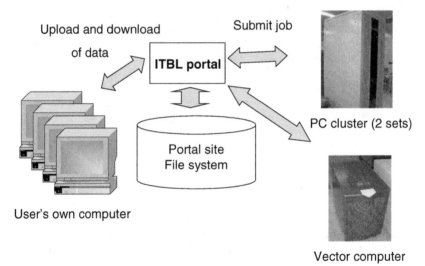

Fig. 2. System of ITBL portal.

4.1   Groupware Functions

Construction of the groupware part uses Java servlet technology as a base for the server side. Using a servlet, we remove the reliance on the operating system and hardware platform, thus the software is easily maintained. We use JavaServer Pages for HTML file generation, and 'tomcat' as the application server.

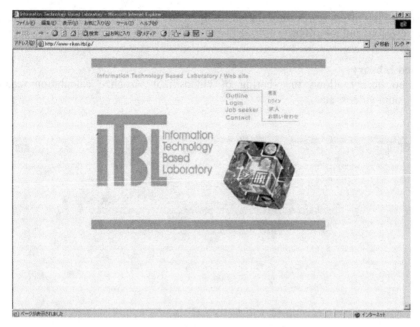

Fig.3. ITBL Portal initial screen.

- Group Function

A user can create a research group, and users of the ITB portal can then join that group. Within the group, messages can be transmitted and replies received from members of the group. It then becomes possible to carry out information sharing within the research group. Moreover, we can also send mail to other users not in the research group by using a mailing list.

A shared folder is created for each research group, enabling the transfer between members of files and data that cannot be easily transmitted using the messaging service. Furthermore, the schedule of members or events for the research group can be coordinated through the use of the event calendar function (Fig. 4).

- Link

Links can be made from the ITBL portal to the home page of an organization or an individual.

- Database

A researcher database provides information about each individual. This contains address and a telephone number, as well as work experience, institute information of the researcher, career, patents, research field, published papers, reward history, etc. The member has the choice as to whether his or her personal information is made public to other members or is kept private. A research subject database stores information about each individual research subject. A research resources database contains information about the resources available for research. A database for researcher job offers stores information for applicants looking for a research position, joint research, and job offers.

- Video Library
A video library allows the sharing of videos that visualize calculation results, symposium videos, etc.

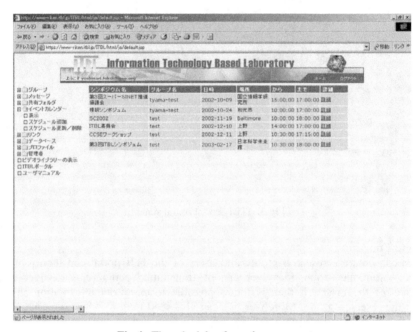

Fig.4. The schedule of members or events.

Fig. 5. Current ITBL portal computers (from left to right, ITBL portal sever, PC cluster, Vector computer SX6-i).

4.2 Function of Computation Service Part

The calculation service part is built using a Web interface, and interfaces with calculation job management. The GridPort Toolkit was used as a base for the ITBL Portal. The GridPort Toolkit is used for Web Portal construction in the Grid environment using Perl/CGI scripts to interface to the Globus Toolkit. However, GridPort provides base functionality and many portions of the Calculation Service are original specifications in the ITBL portal (Fig. 5).

The calculation job management section submits user jobs. We use the Globus Toolkit for job management, as this is a de-facto standard in the Grid environment. Using the Globus Toolkit it is easy to add or remove computer resources from the ITBL portal without rewriting large portions of the portal. The computer resources that can currently be accessed from the ITBL Portal are several PC clusters and a vector-type computer (Fig. 6).

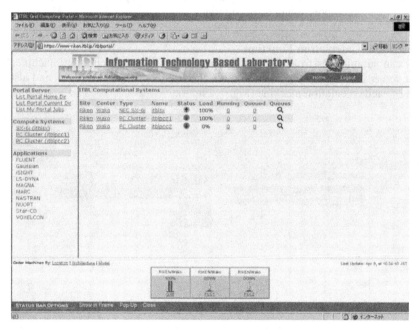

Fig. 6. Status of computation servers.

Each PC cluster uses the SCore middleware operating system with a gateway node and two or more calculation nodes. The hardware comprises, CPU: Pentium4. 2.2 GHz, main storage-capacity: RDRAM (PC-800) 1GB, disk capacity: 36.3GB, network: 100 Base-TX and 1000 Base-T. The Gigabit Ethernet is used for data communication, and 100Mbps Ethernet for. Linux is used for the Operating System. The compiler environment consists of GCC, PGI, and Intel compilers. The communication library for parallel computing can use MPI (Message Passing Interface).

The vector-type computer is an NEC SX-6i. The hardware comprises, CPU: vector processor with 8 GFLOPS theoretical peak, main memory: 8GB and disk capacity: 73GB. The SX-6i is connected to the network through a gateway server using a 1000 Base-SX network. Jobs are submitted to the PC cluster using PBS, and NQS-II for the SX-6i.

Here we describe the functions that can be used with the ITBL calculation service. Files can be uploaded to the portal server from the user's own computer. From the ITBL server, jobs can be submitted to the calculation service and the results returned to the server. Result files can then be downloaded back to the user's own computer. A functional outline is described below,

- File management Function
File handling is provided on the ITBL Portal server. Files can be uploaded and downloaded to and from the user's own computer. Moreover, we can display the contents of a file from the portal server, copy, move, delete, and compression/uncompression of a file archive.

- Job Submit Calculation Function
Users can submit jobs from files on the portal server to a calculation machine. The portal server can display the state of a job, the results of the calculation, and also the status of each calculation machine (Fig. 7, Fig. 8).

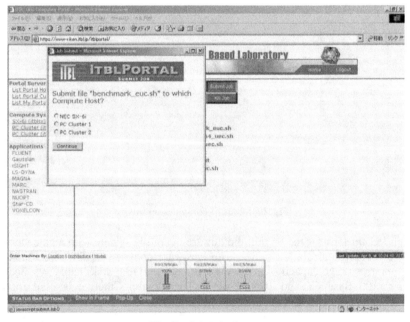

Fig. 7. Job submit screen.

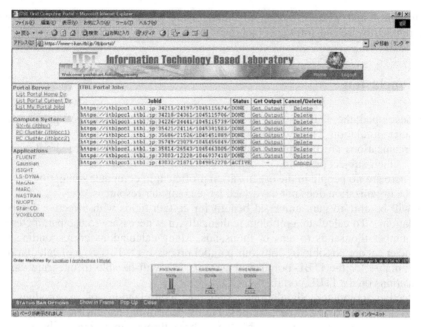

Fig. 8. Status of submitted jobs.

5 Application Showroom

The ITBL Portal is an experimental system that gives people the opportunity to validate the benefits of computer simulation in a controlled environment with applications that have already been prepared. There are two type of computer simulation user,

- Those who work from source programs

- A general user of application software

A computer simulation is possible for the former only by using the calculation resource and development language given by the calculation service.

In the case of the latter, preparation of application software in general is required. Within the same research field, a user can have access to several different kinds of application software to solve a particular problem. The user can evaluate the software in turn, but this is time consuming and does not productively use the researchers time. It is much better if there is an environment where these applications can be provided on a trial basis for evaluation. This is the vision of the "Application Showroom".

The burden to the user of the expense of time in preparing the application environment and reserving the calculation resource are taken away and maintained by the ITBL portal infrastructure.

For the application vendor, the Showroom provides an opportunity to extend the advertisement and sales promotion of the application by using a proven compute environment and without having to maintain the product at several locations and different hardware and software combinations.

6 Conclusion

We developed the ITBL portal system to remove the obstacles of performing a computer simulation. Those obstacles are as follows,

- •Organizations are unfamiliar with the availability of a computer simulation for their subject problem.
- • There are no people in the organization proficient in computer simulation.
- • The organization does not have such large computer resources.

We will be able to gain universal benefit for certain fields with effective computer simulations. To calculate a problem realistically, it is necessary to use parameter sets that number thousands to tens of thousands. Many calculations corresponding to a parameter set are possible by "obvious parallel processing". This type of calculation is very suitable for the ITBL portal grid system. We will be able to integrate various applications on the ITBL portal.

Effective computer simulations, such as optimization of a product and improvement in yield, will become possible by performing parametric studies. Using the fact that parametric studies can be done, a breakthrough in design can be made.

References

1. Grid Port: https://gridport.npaci.edu/
2. Globus: http://www.globus.org/

Evaluation of High-Speed VPN
Using CFD Benchmark

Nobuhiko Tsuruoka, Motoyoshi Kurokawa, and Ryutaro Himeno

RIKEN (The Institute of Physical and Chemical Research),
Hirosawa 2-1, Wako, Saitama, Japan, 351-0198,
(tsuruoka,motoyosi,himeno)@riken.go.jp

Abstract. We have developed high-speed VPN router based on ordinary PC. Generally, a load factor of VPN process is high for ordinary
CPU, thus a PC VPN router has been as narrow network bandwidth
device. However, the PC VPN is very low-cost against commercial VPN
routers. We propose the high-speed PC VPN router using hardware accelarator for encription. We evaluated our PC VPN router using parallel
CFD benchmark. A network bandwidth which use our PC VPN router
realized about 500 Mbps(uni-directional throughput). Since parallel CFD
Benchmark used our PC VPN router, Round Trip Time(RTT) was long,
therefore we obtained sufficient performance by applied to overlap communication with computation technique.

1 Introduction

Recently, a remote operation technology of world wide distributed computational
resources and strage systems, experimental apparatus connected by a network
has been rapidly progressed. Over the past few years, several proposes have
been made on many integrated computational resources(distributed) environment called GRID. A computer user can make unconscious use of computers
anywhere on the GRID environment[1]. The GRID environments are assumed
a unified environment for the computer user using a certification and an autho-
rization on secure network path. The GRID is an indispensable technology for
a field of High Performance Computing (HPC) research.

The Information Technology Based Laboraty (ITBL) project [2] is founded by
the Ministry of Education, Cluster, Sports, and Technology (MEXT). Project
activities concludes an agreement with a united supercomputercenter and constructs foundation technology and application programs. The foundation softwares are consisted HPC technology and netowrk technology. We have developed
the network foundation technology, especially a high performance Virtual Private Network (VPN) and a Distributed Network Intrusion Detection System
(DNIDS). The VPN is a network tunnel technology for secure private network
on the Inetnet. The DNIDS is a detection system of malicious users in a local
area network.

The network infrastructure viewpoint in the ITBL connect by secure and
high-speed VPN technology between supercomputers inside a firewall. In general, Japanese supercomputer centers are used the firewall for protection against

A. Veidenbaum et al. (Eds.): ISHPC 2003, LNCS 2858, pp. 298–306, 2003.

malicious users from Internet. As a result, users of each supercomputer are off bind of attached to organization. To use any supercomputer which exist in the ITBL, many user convenience are improved. Esentially, all scientific technical programs have suitable computer architectures. Supercomputer users may has executed unsuitable program for supercomputer attached to their orgranization.

On the other hand, interconnected supercomputers is feasible big science simulation though single supercomputer is impossible one. Under the present conditions, simulations using interconnected supercomputers was available to increase model size, although it was thoght that those simulations are no practical from long delay time (long communication time and encryption time for security) during the simulation. However, above problem is solved using a wide area broad bandwidth network like the Super Si-net and the High speeed VPN technology.

In this paper, we explain the high-speed VPN technology, and discuss network performance using VPN. We show Computational Fluid Dynamics (CFD) performance using parallel CPU benchmark and our high-speed VPN technology. Finally, we explain a further direction of our high-speed VPN router.

2   VPN

Up to the present, Internet has progressed (or increased) connecting mutually the network managed individually. As it turned out, many technological innovations and investing activity of the network technology has been carried out for Internet. Hence, connection cost of Internet is an incredible low cost.

However, private communications are always demanding (e.g., a communication between companies) ; the open architecture of the internet doesn't provide enough satisfaction for the people. When we consider the malicious use of computational resources by a third person, there is a same demand for an academic or a scientific research network which is comparatively open. In the past, a construction of a private network was easy using a dedicated line or an Asynchronous Transfer Mode(ATM) network etc. On the other hand, since the network of the dedicated line or the ATM network was terribly expensive, thus the construction of the private high-speed network was not easy.

The VPN is a generic term of the construction techniques of the private network between appointed users using the open and the low-cost network like Internet. The VPN techniques can be classified into three main groups as follows.

- establish the private network using the encryption technology on IP stack (IPsec).
- establish the private netowrk using a encryption technology on UDP/TCP session (TLS,PPTP,L2TP etc.).
- construct the private network by separating each network packet in the special network using network routers (Multi Protocol Label Switching:MPLS).

Especially, IPsec is possible to provide the network specification like ordinary network without using special network devices or protocols.

An IPsec standard has been designed by Internet Engineering Task Force (IETF)[4] defined in RFC2401.Implementation of the IPsec to IPv6 which is a next-generation IP protocol is mandatory status. Moreover, the IPsec can use also on IPv4. A goal of the IPsec is given an authentication and encryption function on TCP/IP(spread widely) stack itself. Unlike anohter encryption method for the VPN, the IPsec is possible to encrypt TCP/IP communication without to modify almost application program. Furthermore, the IPsec has the flexible framework which can respond to many encryption algorithms, without specifying specific encryption algorithm. Theoretically, the IPsec functions also with the form of Host-to-Host, Host-to-Network, and Network-to-Network throat. However, a management of the IPsec is complicated and troublesome at the moment. It has been proposed that Oppotunitistic Encryption is a solving method of the management of IPsec.

On the other hand, a weak point of the encryption for TCP session decrease network bandwidth and increase network latency. In this part, SSH tunneling is observed as an example. SSH tunneling is often used for the encryption of any protocols. However, SSH, itself is a implemented protocol on TCP. If using SSH tunneling, one TCP protocol is encrypted to other TCP protocol, TCP window-contorol is executed doubly. The problem of increase of latency and decrease bandwidth by doubley execution of the TCP window-control is pointed out by [?]. In other words, Although it is not a theoretic problem, encryption protocol of Layer 4 over that is processed on a user-land in many cases has a bad influence on network performance. As those result, It has been recognized that the VPN as TCP tunneling is suitable for low speed network and is not suitable for the application which needs high-speed data transfer.

What is important in to keep up parallel computation performance of many applications of HPC field (especially CFD application) is set to a wide bandwidth network. As has been pointed out before, it is quite likely that the delay time for encryption using IPsec dose a bad influence agaist computational performance. However, using recent high-performance personal computers(PCs), network devices and an encryption special hardware is predicted a decrease of delay time by the encryption . Accordingly, construction of the wide bandwidth VPN network using IPsec is expectable. We explain specification and performance of a high-speed PC VPN router.

3   Basic Technology of PC VPN and Distributed Computing through Routers

3.1   FreeS/WAN

We used Linux OS and FreeS/WAN[5] IPsec protocol stack in the VPN router based on PC.

We used a hardware accelarator for a encryption part of IPsec. The cryptographic part of IPsec consumes CPU power, the performance of VPN routers depends on the encryption mechanism. In this paper, we use "Cavium CN1220" as hardware cryptographic accelarator, and modified FreeS/WAN provided by Cavium. The hardware accelartor release the CPU from the cyptographic process, and the CPU devote to other process (e.g., IP packet forwarding).

We take notice of "device polling" as technology which improves performance of PC network process. A process of handling device interrpts is heavy load for operating systems. In the device polling, an operating system does active check a change of device status instead of passibly accept the interrupts. At current, FreeBSD supports device polling, Linux doesn't supports yet. We get an excellent result of VPN router experimentation with device polling. Linux intend to support device polling at next version, and thereby we expect the performance of VPN router is improved further.

3.2 Message Passing

Message passing of a wide area distributed computing using PC Cluster system has been used Ethernet in many case. In single PC Cluster case, a special network device for a interconnect (i.e. InfiniBnad or Myrinet) is used well for high bandwidth and low latency. Although above interconnect is high performance, as network device flexibility is low. For this reason, It is difficult to use for the wide area distributed computing.

On the other hand, widely used protocol like TCP or IP is not use even interconnect of the PC Cluster using Ethernet. In this case, mechanism of layer2 network (i.e. M-VIA[6], GAMMA[7] and PM/Ethernet[8]) is used. Using above mechanism of layer2 network, the bandwidth of Ethernet is the neighborhood of the theoritical peak value. However, when we consider the wide area distributed computing, communication with IP or upper layer (e.g., TCP) is indispensable, thus PC cluster interconnect protocols on Ethernet are not useable in this case.

As the message passing labrary based on TCP, MPICH[9] which is reference inplementation, LAM/MPI[10] and MP_lite[11] etc. is available. Since MPICH is strongly in cooperation with Globus, it is used well. LAM/MPI is also one of the famous implementations. In this paper, we used MP_Lite. Although MP_Lite is insufficient the function as MPI, message passing performance is very excellent. MP_Lite increase the efficiency of TCP session management and bring out the message passing performance based on TCP.

4 Result

4.1 Hardware and Software Specification

Hardware and software of PC VPN routers shows Table 1 and Table 2.

A PC cluster is composed of 8 CPUs. Its interconnect is Gigabit-ethernet. Hardware and software of PC shows Table 3 and Table 4

4.2 Throughtput of VPN

Figure 1 shows a test environment of throughput performance for PC VPN router. Throughput performance was performed to measure basic performance

Table 1. Hardware specification of PC VPN routers

CPU	Pentium Xeon 2.2 GHz
Chipset	Intel 860
Memory	RDRAM 512 MB
HDD	IDE 80GB
NIC	Intel 82546EB × 2
Hardware Accelerator	Cavium CN1220

Table 2. Software specification of PC VPN routers

OS	Linux 2.4.18
IPsec	Free/SWAN 1.96

Table 3. Hardware specification of PC Cluster

CPU	Pentium4 2.2GHz
Chipset	Intel 850
Memory	RDRAM(PC800) 2 GB
HDD	SCSI 72 GB
NIC	Intel 82546EB
RAID Disk	RAID5 560 GB
Network Switch	Catalyst 3550

Table 4. Software specification of PC Cluster

OS	Linux 2.4.10-2 (RedHat 7.1)
Compiler	PGI 4.0-2
Message Passing Library	MP Lite 2.3

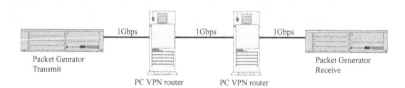

Fig. 1. Thoughput Measurement of VPN Routers

as VPN router. In this case, we obtain pure throughtput performance of PC VPN router.

The test condition shows as follows. The packet generator of transmit port transmits UDP packets and the packet generator of receiver port receives UDP packets through two PC VPN routers. The performance of packet generator can use a 1Gbps bandwidth. The amount of packets transmitted without packet drop in uni-directional communication of PC VPN routers set to throuthput performance.

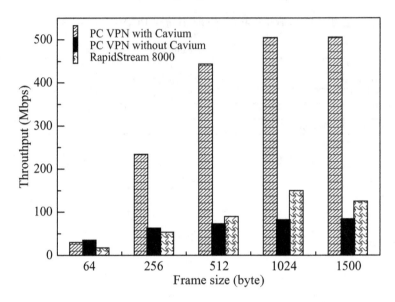

Fig. 2. Thoughput Performance of VPN Routers

The results are shown in Fig. 2. The x-axis is ethernet frame size. The y-axis is throughput performance(utni:Mega bit per second). The cavium in figure is hardware accelarator for encryption. RapidStream8000 is commercial VPN router.

It was found from the result that the bandwidth of PC VPN router with hardware accelarator was obtained about 500 Mbps, the bandwidth of PC VPN router without hardware accelarator was obtain about 70 Mbps. Performance of PC VPN router using hardware accelarator was improved about 7 times to ordinary PC VPN router and about 4 times to commercial VPN router. However, we need to consider network latency, when application performance is taken into consideration in VPN router. The results obtained were contrary to our intention about network performance. The latency of Our VPN router is about 30 ms per one VPN router.

4.3   Performance of Parallel CFD Benchmark

Figure 3 shows a test-bed of parallel CFD benchmark. We used a pair of PC Cluster through PC VPN routers, naturally, Interconnect-network of PC Clusters have 1Gbps bandwidth. In this measurement, we used 8 Processor Element (PE) (Two PC Cluster of 4 PE composition were used). PE defined one PC of the PC Cluster.

We used modified HimenoBMT[12] as Parallel CFD Benchmark. The Using MP_Lite, not available subroutine was replaced to available subroutine. In addition, the increase of communication latency was expected by use of PC VPN,

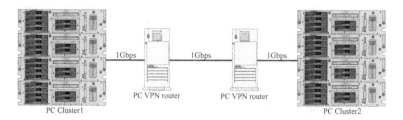

Fig. 3. Test-bed of Parallel CFD Benchmark

the veriosn which implemented the systolic communication-computation overlap method[13] was also prepared.

The subroutine which MP_Lite does not implement was replaced to avilable subroutine. In addition, the increase of communication latency was anticipanted by use of PC VPN. accordingly, we try to applied the systolic communication-computation overlap method[13]. The systolic communication-computation overlap method is a kind of method to overlap communication with computation.

The HimenoBMT is core routine of a imcompressible CFD simulation program (e.g. Marker And Cell (MAC) method). Model size of CFD simulation is $512 \times 256 \times 256$. The core routine of MAC method is the Poisson equation solver, in brief, large-scale simultaneous linear equations. Linear equation solver of the HimenoBMT has been used Jacobi method. Parallelized Jacobi method is not change convergence.

The results are shown in Fig. 4. The x-axis is case of benchmark. The y-axis is performance(utni: GFLOPS). The **ordinary** of the x-axis is ordinary benchmark, all PE were conneted one gigabit switch. The **sw-to-sw** is benchmark of direct connection gigabit switches, bypassed PC VPN router. The **vpn** is benchmark of using PC VPN router. The **vpn-overlap** is benchmark applied the systolic communication-computation overlap method using PC VPN router.

The result clearly shows that the benchmark performance was deteriorated the influence of PC VPN bandwidth and latency. The performance of **ordinary** and **sw-to-sw** is almost changeless. **vpn** degraded about 60% compared with **ordinary**. **vpn-overlap** was able to be restraied the deterioration about 20% compared with **ordinary**. When the systolic communication-computation method is applied to HimenoBMT(**vpn-overlap**), the performance has improved about 90% to **vpn**.

5   Discussion

First of all, it will be belpful to distinguish bandwidth and latency. At present, bandwidth of our PC VPN router using hardeware accelarator for encryption without high-performance hardware and software optimization was obtained equal performance of a commercial VPN router. It is clear that the bandwidth improved by using hardware accelarator for encryption. We can safely say that although PC VPN router was in the middle of development, sufficient perfor-

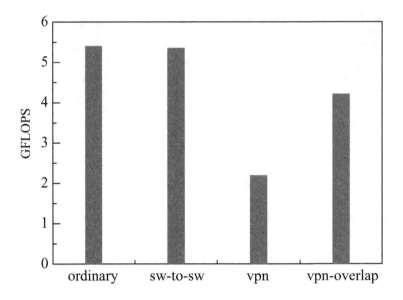

Fig. 4. Result of Parallel CFD Benchmark

mance of bandwidth was obtained. However, there is room for reconsidering network latency using PC VPN router.

In PC Cluster network condition, Round Trip Time(RTT) was about 60ms. However, Cavium CN1220 does not demonstrate the perfect performance. Although Cavium CN1220 work on 64bit/133MHz PCI-X bus, the present PC had only 64bit/66MHz PCI bus, in addition, Software (device drive and Operating system, IPsec stack) is naive yet. We have the plan of hardware upgrade for Cavium CN1220 and software optimization. It is possible that RTT becomes quite small.

The modified HimenoBMT applied to overlap communication with computation was obtained performance of 4.2 GFLOPS using 8PE. The evidence, that communication latency using our PC VPN is able to hide, can be seen in the result of **overlap**. When there is not enough bandwidth, it is thought that performance was not obtained. This result cannot be attained without network bandwidth of our PC VPN. However, the phase which surely takes the whole synchronization like reduce operations or a barrier synchronization \exists in CFD computation. In fact, when the whole synchronization performances want to improve in our test environment using the PC Cluster, a method only has shortening latency of our PC VPN.

6   Conclusion and Future Work

We developed high-speed VPN router based on ordinary PC using hardware accelarator for encryption, evaluated PC VPN router using parallel CFD benchmark. We obtained high-performance PC VPN router. Bandwidth of our PC VPN router is equal performance of a commercial VPN router. When the systolic communication-computatio method was used, HimenoBMT was measured sufficient performance in the experiment using our PC VPN router. RTT on PC VPN calls for further investigation. From these points of results, we might go on to an more detailed examination of PC VPN router, moreover, an more measurement of an actual performance in Internet.

References

1. Globus project: http://www.globus.org/
2. ITBL project: http://www.itbl.jp/
3. Pitsow Project: http://pitsaw.riken.go.jp/
4. The Internet Engineering Taks Force: http://www.ietf.org/
5. Linux FreeS/WAN: http://www.freeswan.org/
6. M-VIA (A High Performance Modular VIA for Linux): http://www.nersc.gov/research/FTG/via/
7. Chiola, G., Ciaccio, G.: Operating System Support for Fast Communications in a Network of Workstations, Technical Report DISI-TR-96-12 (1996)
8. Sumimoto, S., Tezuka, H., Hori, A., Harada, H., Takahashi, T., Ishikawa, Y.: The design and evaluation of high performance communication using a Gigabit Ethernet, International Conference on Supercomputing, 260-267 (1999)
9. MPICH: A Portable Implementation of MPI: http://www-unix.mcs.anl.gov/mpi/mpich/
10. LAM/MPM Parallel Computing: http://www.lam-mpi.org/
11. MP_Lite: http://www.scl.ameslab.gov/Projects/MP_Lite/
12. HimenoBMT: http://w3cic.riken.go.jp/HPC/HimenoBMT/
13. Kurokawa, M., Matsuzawa, T., Himeno, R., Shigetani, T.: Parallel CFD simulation using systolic communication-computation overlap, 5th International Conference and Exhibition on High-Performance Computing in the Asia-Pacific Region(HPC Asia 2001) (2001)

The Development
of the UPACS CFD Environment

Ryoji Takaki, Kazuomi Yamamoto, Takashi Yamane,
Shunji Enomoto, and Junichi Mukai

CFD Technology Center, National Aerospace Laboratory of Japan, 7-44-1
Jindaiji-higashi, Chofu, Tokyo 182-8522, Japan,
Tel.: +81-422-40-3322, Fax.: +81-422-40-3328,
{ryo,kazuomi,yamane,eno,jmukai}@nal.go.jp

Abstract. UPACS (Unified Platform for Aerospace Computational Simulation), a computational environment for CFD (Computational Fluid Dynamic) in the aerospace field, has been in development at NAL (National Aerospace Laboratory of Japan) since 1998, which includes two years of support by the ITBL (IT-Based Laboratory) project. It aims to overcome the increasing programming difficulties with recent CFD codes on parallel computers for aerospace applications. The codes mainly require applicability to complex geometries, higher accuracy, and coupling with heat conduction analysis in materials and structure analysis. UPACS successfully developed a general parallelized CFD code for aerospace applications in Japan. This paper describes in detail programming, applications, and parallel performance on the newly installed supercomputer at NAL.

1 Introduction

Several young CFD (Computational Fluid Dynamic) researchers at NAL (National Aerospace Laboratory of Japan) started the UPACS (Unified Platform for Aerospace Computational Simulation) project[1] in 1998 to create a collaborative computational environment for CFD. With the start of the ITBL (IT-Based Laboratory) project in 2001, UPACS became the core software provided for applications in the NAL ITBL environment. The development of UPACS was accelerated by the support of the ITBL project in these years. Although computational techniques used in UPACS are not new from a computer science perspective, UPACS has already shown its applicability as a general CFD software package for successfully solving actual CFD problems. It is used as a standard CFD software package at NAL to support a supersonic research airplane project[2], to collaborate in CFD code validation by industries and universities. UPACS also has a user group[1] with more than 50 users outside of NAL already registered.

---

[1] Please refer to http://www.nal.go.jp/cfd/jpn/upacs/ (in Japanese) or send an e-mail to upacs@nal.go.jp if you are interested in UPACS

A. Veidenbaum et al. (Eds.): ISHPC 2003, LNCS 2858, pp. 307–319, 2003.

We have already described UPACS briefly in other papers[1, 3, 4]. In this paper we intend to present more details about the motivation for the development of UPACS, key concepts of its flow solver and implementation. We also show the performance of parallel computations on a newly installed supercomputer at NAL, and some application examples.

2   Motivation for the Development of UPACS

CFD has begun to draw attention as a research and design tool for fluid dynamics in the aerospace field where high accuracy is required due to the rapid development of computer and numerical simulation technologies. CFD will likely be used in the following areas:

- realistic, complex configurations of aerospace vehicles,
- analysis of physically complicated flows, such as turbulent flows, combustion flows and non-equilibrium flows around re-entry vehicles,
- multi-disciplinary simulation including structure or heat conduction analysis, and
- multi-disciplinary optimization for developing higher performance vehicles.

Highly reliable CFD is particularly indispensable in multi-disciplinary optimization. Precise validation and verification should also be considered. For example, it has been suggested that aerodynamic drag prediction error should be less than 0.5%. This requirement is too difficult to meet with current CFD technology. Therefore, the previous approach that just compared CFD codes with experimental results is not enough. Therefore, we need more systematic and precise validation and verification using a standard code as a benchmark.

Many CFD codes have been developed as a result of development and validation in various segmentalized areas without collaboration. Because each program has become more complicated and required higher reliability, it has been difficult for each researcher to develop and validate CFD codes independently. Furthermore it is difficult to utilize basic algorithms commonly used in CFD codes together. Therefore significant problems appear regarding the generality, reliability and diffusion of CFD technology.

The growth in CFD development has been supported by a large increase in computer performance. Vector computers enhanced the potential of CFD and parallel computers made it possible to conduct large-scale computation and led the way to practical CFD programs. Technology advances in software such as parallel computation have made little progress compared to the advances in hardware. Parallel programming is indispensable for conducting large-scale computations because almost all high-end computers are parallel computers at present. Programmers have to parallelize their programs manually in order to get better performance out of parallel computers because auto-parallelization by compilers is not feasible yet. Therefore, knowledge of parallel processing and hardware is necessary in addition to fluid dynamics. It is greatly inefficient for each researcher to carry out difficult jobs which are not directly related to flow analysis.

These problems have occurred during the development of CFD codes with advanced demands. Thus, the UPACS project was created to overcome these difficulties and accelerate CFD research. UPACS aims to: 1) overcome difficulties in parallel computation, 2) develop a general CFD environment for aerospace applications that can be shared among CFD research scientists and engineers and provide flexibility, extensibility and portability, 3) enhance the reliability of CFD through systematic validation of numerical models and algorithms in the same environment, and 4) accelerate the advancement of CFD technology in Japan by distributing the source.

3   UPACS Solver Concepts

UPACS is a CFD environment which consists of the following:

- a CFD solver, which is the core program for carrying out CFD analysis,
- pre/post-processing programs,
- libraries for data I/O and parallel processing,
- graphic libraries, and so on.

The core CFD solver in UPACS is the most important program in UPACS and is able to analyze compressible three-dimensional flows with a multi-block structured grid on distributed shared memory multiprocessor machines. The UPACS solver is based on the following design concepts and approaches:

- Multi-block/overset structured grid method,
- Separation of CFD solvers from multi-block/overset and parallel procedures,
- Hierarchical structure and encapsulation,
- FORTRAN 90 and MPI and
- Parameter Data Base.

The details are explained in the following sections.

3.1   Multi-block/Overset Structured Grid Method

Structured grid methods are superior to unstructured methods in the accuracy and reliability of solutions for aerospace applications where a grid with a much higher aspect ratio is used to capture the physics accurately in a thin boundary layer on the aircraft's surface. Recently grids for solving Navier-Stokes equations around aircrafts actually use one-third of the grid points in the boundary layers.

It is well known that unstructured grid methods are much more adaptable to complex geometries and many researchers are attempting to improve their accuracy. Multi-block methods are a compromise between accuracy and adaptability for complicated aircraft geometries. By allowing an unstructured connection among many structured grid blocks, one can generate multi-block grids around complicated geometries with good quality in the boundary layers. Figure 1 shows an example of a multi-block grid around the NEXST-2, NAL supersonic experimental plane. The grid points between blocks are exactly matched to each other

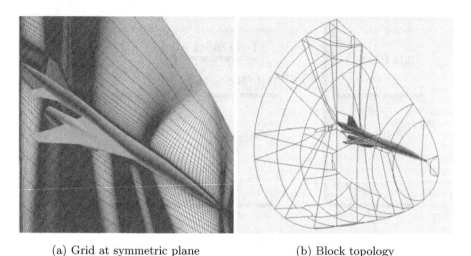

(a) Grid at symmetric plane   (b) Block topology

Fig. 1. Multi-block grid around NEXST-2

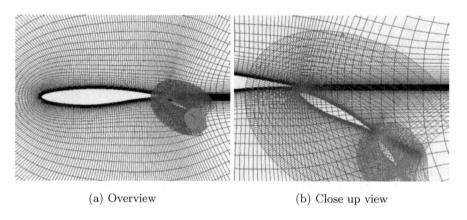

(a) Overview   (b) Close up view

Fig. 2. Overset grid around a 3-element airfoil

in the case of Fig. 1 so that the data between blocks are exchanged after the calculation in each block without any loss in the accuracy of the solutions. There
are cases when an overset grid method is advantageous at the sacrifice of accuracy. In the overset grid method, grid blocks are overlapping each other as
shown in Fig. 2. The data between blocks is exchanged after interpolation in the
overlap region. Flow solvers developed for the structured grid method can be applied in each grid block directly in the multi-block method. However, difficulties
arise in parallel computation and block connection. The size and the direction
of grid indices are different block by block because they are not determined by
the number of available CPUs. In practice, they are often determined by the
complexity of geometries. Therefore, it requires complicated data communica-

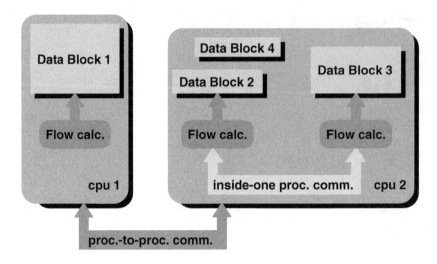

Fig. 3. Data communication for parallel and multi-block

tion in the parallel computation using the domain decomposition approach. For efficient parallel computation, the number of blocks handled by a process has to be free. This means data communication is necessary not only between processes but also within a process using memory copy as shown in Fig. 3. We generalized this complicated data communication and hide them in the multi-block layer of the UPACS solver described in the next section.

Preparation of the block connectivity information is also a problem. In many cases the connectivity is so complicated that it has to be determined automatically from grids. In UPACS, pre-process programs *createConnect3d* for point matched grids and *createOversetIndex* for overset grids, are prepared to search the connectivity of all blocks so that it outputs connection information for the solver.

3.2   Programming Issues

One of the key features of the UPACS solver is the clearly defined hierarchical structure of the program. Every procedure and all of the data is encapsulated so that the code can be shared and modified easily. The UPACS solver consists of three layers: the main loop level, multi-block level and single-block level. Parallel processes and complicated data communication are covered and completely hidden by the multi-block level. The physical models are written and can be modified in the single block level without considering the multi-block level.

To implement this concept, FORTRAN 90 was chosen as the programming language. Its module is mainly used for the encapsulation of data and procedures. There is an advantage to using object oriented programming with C++ for code development. We preferred FORTRAN 90 for the following reasons:

- the structure of CFD code has been well established and the object oriented approach is not alway necessary,
- the C++ development environment has not been well established and it is difficult for CFD researchers to write CFD code in C++ if they are interested in computational science instead of computer science.
- CFD researchers at NAL prefer FORTRAN 77 over C or C++ because FORTRAN 77 is the best choice for scientific computations on vector computers with regard to runtime performance.
- FORTRAN has more specialized functions to handle arrays, which are necessary for scientific computations, when compared to C or C++
- FORTRAN 90 has new features like modules, structures, and so on.

A parallelization method based on the domain decomposition concept using the message passing interface (MPI) is used to minimize the dependency on hardware architectures. There are several parallelization languages and libraries like High Performance Fortran(HPF), OpenMP and MPI. OpenMP is mainly used in shared memory machines and is easy to use, although there are hardware restrictions. HPF is also easy and powerful but it has not been well established in many platforms. MPI is more difficult to program with than other parallel languages such as HPF and OpenMP. However, MPI can be used in any type of parallel machines, including distributed memory and shared memory machines. Moreover, the multi-block/overset method should treat various sizes and shapes of blocks, which makes it difficult to use only HPF or OpenMP. Therefore, MPI was chosen for parallelization. Using MPI enhances the portability of UPACS, which is shown by the fact that the UPACS solver can be run on various parallel computers, from supercomputers and workstations to PC-UNIX clusters.

Figure 4 shows a simple example of the UPACS solver. Subroutines **MPI_init**, **doAllBlocks**, **doAllTransfer** and **MPI_end**, which are called in the main program, control the multi-block level process. Subroutines at the single block level are defined in a module (right hand side of the figure) and subroutines like **initialize1** are called in the multi-block level subroutines, **doAllBlocks**, as an argument.

3.3   Parameter Data Base

Another important concept necessary for code sharing and easy extension is a Parameter Data Base (PDB) to control programs. PDB is a library to handle input parameters which depend on numerical methods and target problems. Generally CFD solvers have plenty of parameters. The number of parameters usually increases with the code extension, causing difficulty for CFD codes to handle them. Actualy the UPACS solver already has almost one hundred and fifty of parameters only for the flow assuming perfect gas. PDB is prepared to overcome this difficulty. All parameters are read and stored in a data base as a style of hash, and they can be called locally anywhere in programs. New parameters can be easily added without caring I/O processes in writing new extensive subroutines. The input to PDB is usually a text file named "input.txt"

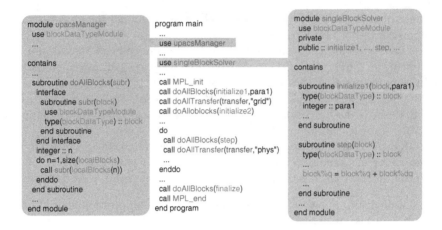

Fig. 4. Structure of the UPACS program

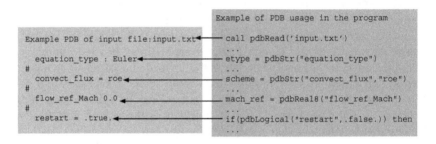

Fig. 5. Example PDB: input file and program usage

which can be edited using any text editor. The parameters in the PDB input file are written in a very simple format, which consists of a set of tag names and values separated by symbols such as "=", ":" and " "(white space). Figure 5 shows an example PDB input file and program usage. The tag name and value of a parameter are read from the input file specified in *pdbRead* and are registered in the PDB. Parameter values can be referred to by calling *pdbStr*, *pdbReal8* or *pdbLogical* with their tag names anywhere in the program. The first argument of these subroutines is the tag name of the parameter and the second argument is optional and has a default value when a parameter entry is not found in the input file. The PDB also has an intrinsic subroutine named *pdbGet*, which makes it possible to get parameter values without specifying the following parameter types: integer, character, logical, single precision and double precision.

The *PdbEditor* has been developed to help input parameter using a GUI (Graphical User Interface) because the PDB itself does not recognize interdependences and appropriate value ranges for parameters. The editor has a definition file, which contains this information, and assists in the input of suitable values for each parameter. Figure 6 shows an example of the definition file. Two

```
group:equation {
message:equation_type {equation}
choice:equation_type = NS {Euler NS RANS}

message:vis_model {viscous model}
disable:vis_model {equation_type == Euler}
choice:vis_model = full {full thin}
}
group:condition {
message:restart {restart ?}
logical:restart=.false.

message:cfl {CFL number}
real:cfl=1.0 {1 0}

message:aoa {attack angle}
real:aoa=0.0 {1 -180 180}
}
```

Fig. 6. Example of the definition file and *pdbEditor*

groups, "equation" and "condition", are defined in the definition file, which can be displayed by the *pdbEditor* as in Fig. 7 and 8.

Depending on the parameter type, the editor selects the appropriate input device from the following list: radio buttons for choice types, toggle buttons for logical types, and input fields and slide-bars for integer types and floating types, respectively. The slide-bars have a maximum and minimum value that can be specified in the definition file. In Fig. 6, "cfl" has a default value of 1 and the value range is greater than 0. "Aoa" has a default value of 0.0 and the value range is from −180 to 180. The value range is set to the boundary of the slide-bar in the editor shown in Fig. 8. The input field in the editor is switched depending on the parameter interdependency described in the definition file. Figure 6 shows how to describe the interdependency of a parameter, where the parameter "viscous_model" depends on the parameter "equation_type". The keyword *disable* suppresses the input of a value when "equation_model" is set to "Euler" The editor switches the input field based on this definition. When "Euler" is selected as the "equation_model"(Fig. 7(b)), the input field of the "viscous_model" becomes grey and will not accept a value.

4   Applications

UPACS has been used for several problems and used as a standard CFD code in some of NAL's projects. First, the parallel performance of UPACS will be evaluated. Secondly, two examples of a supersonic flow around an SST(SuperSonic Transport) and a flow around turbine blades will be presented.

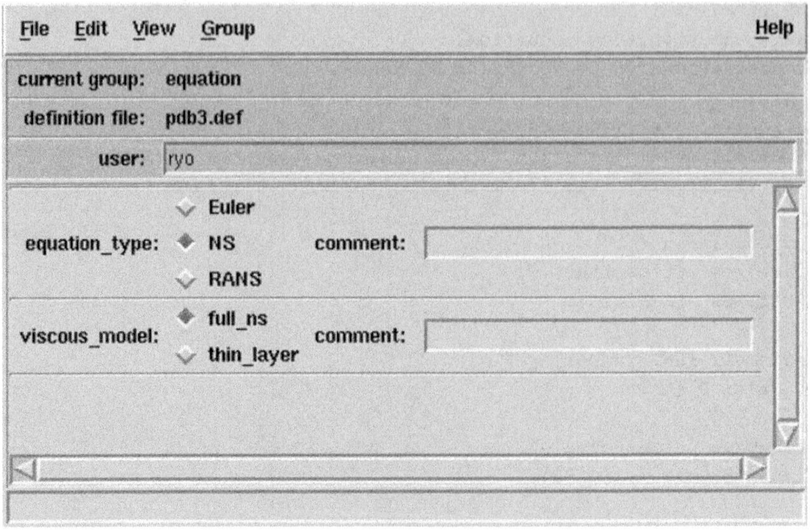

(a) If select equation_type = NS

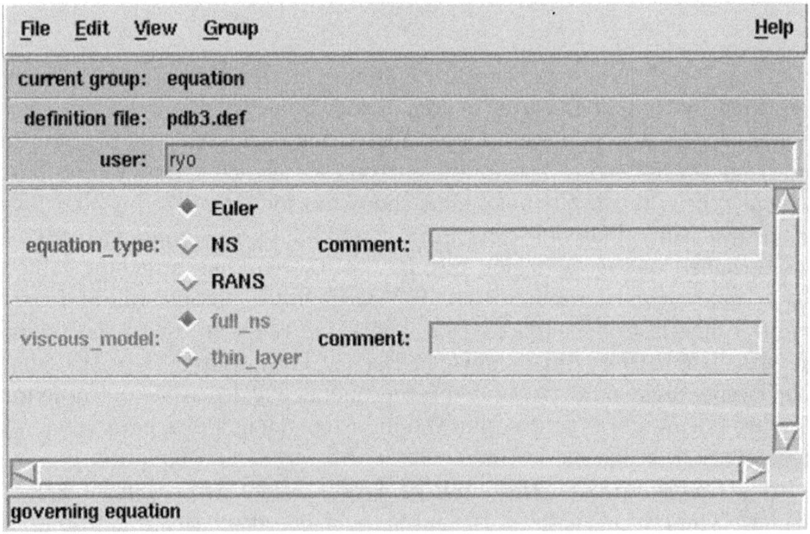

(b) If select equation_type = Euler

Fig. 7. *pdbEditor (equation group)*

4.1   Parallel Performance on CeNSS

NAL has operated a distributed memory vector parallel supercomputer called the Numerical Wind Tunnel(NWT), which consists of 166 vector processors. This epoch-making computer was then replaced by a new supercomputer called the Central Numerical Simulation System (CeNSS) in 2002[5]. CeNSS consists

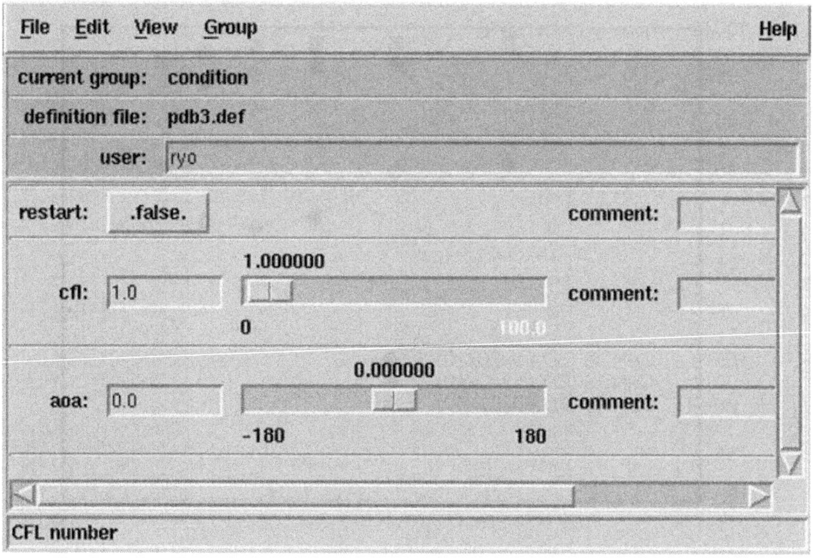

File Edit View Group   Help

current group: condition

definition file: pdb3.def

user: ryo

restart: .false.   comment:

1.000000

cfl: 1.0   comment:

0   100.0

0.000000

aoa: 0.0   comment:

-180   180

CFL number

Fig. 8. *pdbEditor* (condition group)

of 14 cabinets, where a cabinet is a unit of hardware. Each cabinet includes 128 CPUs with 256GB memory, which can be partitioned into a maximum of 4 nodes, where node is a unit determined by the operating system. All nodes are connected by a crossbar switch with a speed of 4GB/s bi-directional. The CPU is a SPARC 64V scalar chip with a 1.3GHz clock and scalar speed-up technologies. Thus the peak performance of the whole system becomes 9.3 Tflops.

Scale-up tests have been conducted to evaluate the parallel performance of the current version of UPACS on CeNSS. This test was conducted to evaluate parallel performance, where the total amount of computational work increases proportional to the number of CPUs and the test is suitable for evaluating the parallel performance of large-scale simulations. It was conducted in three dimensions, where almost all blocks except for the boundary blocks were connected three-dimensionally. The elapsed time would be constant in the ideal case. In reality, it will increase as the number of blocks increases due to the overhead of parallel computations. Figure 9 shows parallel efficiency in the scale-up test. The numbers described in the legend show the number of grid points in each block. The parallel efficiency is quite good even at the maximum number of processes (512).

The parallel efficiency is determined by several factors such as the amount of memory for each CPU, CPU performance and network performance, which is strongly related to the balance of these factors. For example, a system with relatively slow CPUs and a fast network can hide the overhead of a data transfer due to parallelization and demonstrate good parallel efficiency. Comparing the efficiency of grid of size 20^3, 40^3 and 80^3 in Fig. 9, the parallel efficiency improved with the increase in the size of the computational grid for each block. This

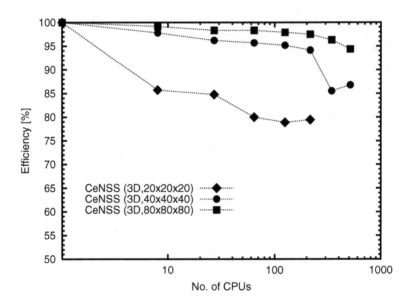

Fig. 9. Scale-up efficiency

is because when we increase the computational grid size by N times in every direction, the total number of grid points is N^3 times larger, meaning that the calculation cost for each CPU increases N^3 times. The amount of data exchanged between blocks increases by only N^2 because they are defined at the boundary face of each block. Therefore, the increase in calculation time is larger than that of the communication time when the computational grid size increases leading to higher parallel efficiency. It is also shown that the parallel efficiency exceeds 86.8% for a size of 40^3 and 94.4% for a size of 80^3 with 512 processes demonstrating a good parallel efficiency for UPACS with large-scale simulations on CeNSS. The result also shows that CeNSS has a slow CPU and a high speed network for the present UPACS implementation.

4.2   Flow around SST

NAL has been conducting a supersonic research project[2] to establish a design methodology based on CFD. The UPACS solver is expected to play an important role in the project. Figure 10 shows surface pressure distributions of the experimental vehicle, called NEXST-2, which has a very complex configuration consisting of main wing, body, tail wings, nacelles and air data sensor(ADS). The UPACS solver is able to analyze flows around such a complex configuration with a high-order-accurate scheme.

Validation and verification of CFD codes is a crucial issue in the project and several validations and verifications have been conducted for the UPACS solver. The high reliablity of the UPACS solver made it possible to point out defects in

Fig. 10. Surface pressure distributions of NEXST-2:Mach number of 1.7 and attack angle of 3 deg.

experimental data by comparing with the CFD results and to conduct mutual validation between the CFD and experiments in validation processes.

4.3   Flow around Turbine Blades

UPACS has been extended to treat multi-disciplinary problems such as conjugate simulations of flow with combustion, heat conduction or structure analysis. Figure 11 shows coupled analysis between flow and heat conduction for a 2-D turbine blade which has three circular holes with constant temperature at the inner wall[4]. The structured grid is generated not only in the flow region but also inside the turbine blades. UPACS can handle these conjugate simulations with heat conduction in materials by replacing a flow solver with a heat conduction solver in the applicable blocks.

5   Conclusion

UPACS, a computational environment for CFD in aerospace field, was presented. It has succeeded in the promotion of program and knowledge sharing between CFD researchers and engineers. The UPACS solver together with support utilities has proven its effectiveness in simulating flows around complex configurations using a multi-block/overset structured grid scheme. The parallel efficiency of the UPACS solver on CeNSS was also evaluated and showed good parallel efficiency up to 512 processes. UPACS has been successfully applied to a number of

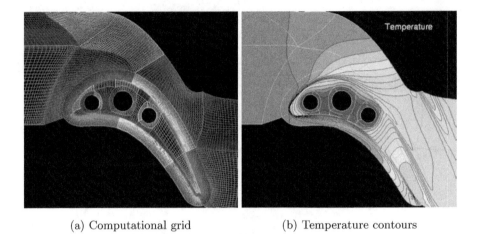

(a) Computational grid   (b) Temperature contours

Fig. 11. Coupled analysis between flow and heat conduction

aerospace simulations and demonstrated fairly good capability. Continuous development is being conducted to make sure that UPACS continues its evolution to become a better CFD environment.

References

1. Yamane, T., et. al.: Development of a Common CFD Platform – UPACS –. In: Parallel Computational Fluid Dynamics -Proceedings of the Parallel CFD 2000 Conference Trondheim, Norway (May 22-25, 2000), Elsevier Science B.V. (2001) 257–264
2. Sakata, K.: Superonic Research Program in NAL, Japan. In: 1st CFD Workshop for Supersonic Transport Design. (1998)
3. Yamazaki, H., Enomoto, S., Yamamoto, K.: A Common CFD Platform UPACS. In Valero, M., Joe, K., Kitsuregawa, M., Tanaka, H., eds.: High Performance Computing, Third International Symposium, ISHPC 2000, Tokyo, Japan, October 16-18, 2000. Proceedings. Volume 1940 of Lecture Notes in Computer Science., Springer (2000) 182–190
4. Yamane, T., Makida, M., Mukai, J., Yamamoto, K., Takaki, R., Enomoto, S.: Capability of UPACS in Various Numerical Simulations. In Matsuno, K., Ecer, A., Periaux, J., Satofuka, N., Fox, P., eds.: Parallel Computational and Fluid Dynamics – New Frontiers and Multi-Disciplinary Applications, Elsevier Science B.V. (2003) 555–562
5. Matsuo, Y., Nakamura, T., Tuchiya, M., Ishizuka, T., Fujita, N., Hirabayashi, H.O.Y., Takaki, R., et. al.: Numerical Simulator III - Building a Terascale Distributed Parallel Computing Environment for Aerospace Science and Engineering. In Matsuno, K., Ecer, A., Periaux, J., Satofuka, N., Fox, P., eds.: Parallel Computational and Fluid Dynamics – New Frontiers and Multi-Disciplinary Applications, Elsevier Science B.V. (2003) 187–194

Virtual Experiment Platform
for Materials Design

Nobutaka Nishikawa[1], Masatoshi Nihei[2], and Shuichi Iwata[3]

[1] Fuji Research Institute Corporation (FRIC),
Center for Computational Science and Engineering (CCSE),
2-3 Kandanishiki-cho, Chiyoda-ku, Tokyo, Japan,
`nisikawa@ccse.fuji-ric.co.jp`
[2] National Institute for Materials Science (NIMS),
Materials Engineering Laboratory (MEL),
1-2-1 Sengen, Tsukuba, Ibaraki, Japan,
`NIHEI.Masatoshi@nims.go.jp`
[3] University of Tokyo,
Department of Quantum Engineering and System Science,
7-3-1 Hongo, Bunkyo-ku, Tokyo, Japan,
`iwata@q.t.u-tokyo.ac.jp`

Abstract. The virtual experiment platform was newly developed, in which various components of computational materials science and engineering technologies are synthetically integrated, such as simulation, database, data analysis, visualization, and human interface. A special feature of this platform is the realization of a new concept by using the "taskflow engine" to control and to operate various components of computational technology. In this taskflow engine, the all computational modules for materials design are automatically executed based on the task scenario. Some results obtained from the calculation for materials properties are also shown.

1   Introduction

The progress of high performance computers and networking makes possible the advanced R&D in the science and technology field. Materials design is one of the recent scientific technological topics related with this research field[1][2]. The success of such materials design will depend on the progress of a computational materials science. Many components for materials design such as simulation, database, knowledge information processing, visualization and the human interface have been developed as a research target, but the integration technology into a practical system to hybridize these components has not yet been established because of the individual developments.

We developed a computational environment, called the "virtual experiment platform" based on a new integration technology. A special feature of this virtual experiment platform is the realization of a new concept by using the "taskflow" to control and operate various components of computational technologies[3][4]. The

A. Veidenbaum et al. (Eds.): ISHPC 2003, LNCS 2858, pp. 320–329, 2003.

virtual experiment platform also supports an execution of a chain of workflows for materials design by using the "taskflow engine". The virtual experiment platform is available to use through the Internet, because the all systems are implemented based on a web technology. In this paper, we will show the general specification of this virtual experiment platform and some examples of the application using this platform. This research project is progressed with coordination in the ITBL project.

2   Overview

2.1   Concept of Virtual Experiment Platform

The virtual experiment platform can realize the materials design by computing the materials properties such as constitution, composition, atomic structure, micro structure, mechanical properties and so on. To predict these materials properties, it is important to consider a hierarchical physical modeling spread widely from micro scale to macro scale. Figure 1 shows a concept of the virtual experiment technology for materials design.

The virtual experiment platform is a computer environment for realizing materials design by an integration of technologies such as simulation, database,

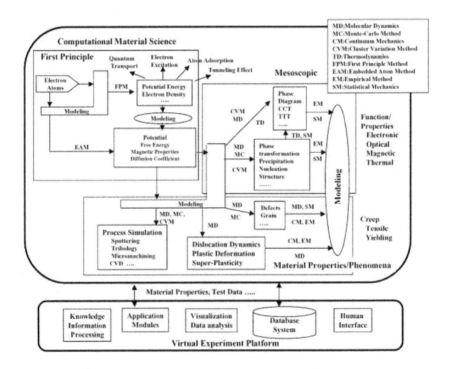

Fig. 1. Virtual experiment technology for materials design

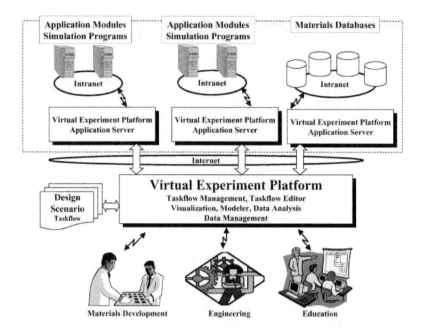

Fig. 2. The concept of virtual experiment platform for materials design

knowledge information processing, human interface, and so on. This virtual experiment platform also supports a unique taskflow that is described by using a special editor that has an Explore like view and is easily editable. User can use this editor to define task sequence and to modify the template of design scenario which is the sequence of tasks such selection of the computing module, execution of task, control of the task sequence, data transfer, visualization of results and so on. These design scenarios are stored automatically and used as a template for another user. Figure 2 shows the concept of the virtual experiment platform.

2.2 Taskflow Engine and Taskflow Editor

In this section, we describe a new concept of "taskflow" in order to integrate various technology components for materials design. Each work such execution of program and data transfer is called as a "task". A chain of workflow for materials design is defined as the "taskflow" and the execution system is a taskflow engine. In case of high temperature superalloy design, it is considered that each "task" is defined as one design work such as selection atom, prediction of constitution, prediction of microstructure, prediction mechanical property. Various analytical methods such as first-principle method, molecular dynamics, montecarlo method and thermo-dynamics and many tools such as simulation program and database can be related to each "task". This makes possible to predict various hierarchical material properties by the "taskflow engine". Because the

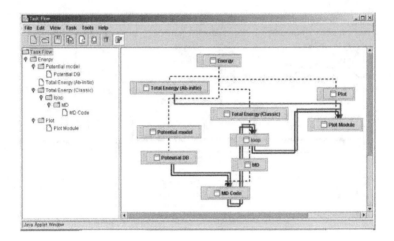

Fig. 3. Taskflow editor

"taskflow" is a description of design scenario, it is also possible to save design scenarios, design know-hows and success examples as the "taskflow".

The users can create and edit the taskflow scenario on a taskflow editor as shown in Fig. 3. By using this taskflow editor, the users can experience easily the usefulness of the computational materials science without a special help of the expert and knowledge in this research field. Figure 3 shows a taskflow for energy calculation using molecular dynamics program.

2.3   Application Modules and Tools

The application modules such a molecular dynamics, Monte Carlo, Finite Element Method are originally developed. In order to provide the parameters of simulation programs, a virtual experiment platform uses a potential parameter database that is now being developed in JST (Japan Science and Technology Corporation). In current status, a potential parameter database provides Lennard-Jones potential, Stillinger-Weber potential and EAM potential.

This platform has also some visualization tools and data analyzing tools. These tools might be available for evaluation of calculated results. To create input data for simulation, this platform has some data modeling tools such atom/molecular model and continuum model.

Figure 4 shows the example for usage of virtual experiment platform. Users will access a portal server of this platform and select a design scenario. And then they sets conditions according to the selected design scenario and execute platform. After finishing all tasks, they can use various visualization tools and data analyzing tools. In this platform, the taskflow engine will control the all executions of tasks and all data transfers described on the task scenario that users have drawn.

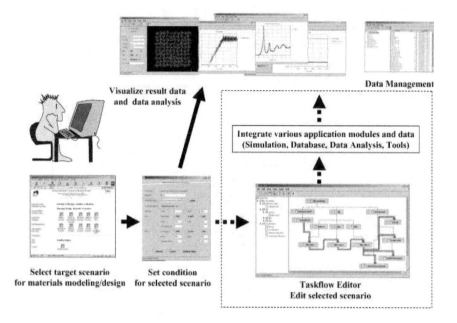

Fig. 4. Usage of virtual experiment platform

3   System Architecture and Implementation

3.1   Requirements of System

In order to realize virtual experiment on computer environment, the basic requirements to construct a virtual experiment platform are as follows.

a) Easy operation to execute combined sequence with various computational modules based on the design scenario and/or modeling scenario of client user.
b) Easy editing environment of user scenario and common use of saved scenarios.
c) Automatic data generation for the computational module.
d) Automatic data transfer and gathering among computational modules.
e) Visualization, data plotting and data analysis.
f) Flexibility to implement another new computational modules.

3.2   System Configuration

The virtual experiment platform is composed of the portal server and several application servers. The portal server provides the interface environment for user client as shown in Fig. 5. In general, the accesses to computational resources from the Internet are restricted due to the security problems, so that a portal server communicates with application servers to use the resources on each

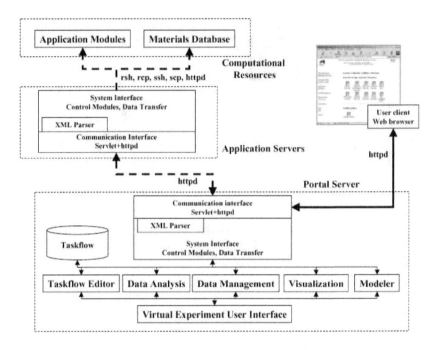

Fig. 5. System configuration of the virtual experiment platform

site. The application servers execute and control the computational module according to the demand of the portal server. The portal server is implemented of several units such as control unit based on the taskflow engine, editing unit of task scenario, data analyzing unit, data management unit, visualization unit and modeler unit. The portal server sends the demand of http protocol execution and/or data transfer to the target application server and the application server accesses and communicates the computational resources such a simulation program and database by using rsh, rcp, scp, http protocol on each site. This platform is implemented using Java class library and web technologies. Because a user interface is implemented as a Java Applet, users can use a virtual experiment platform using web browser through the Internet.

3.3   Implementation of Taskflow

A taskflow engine is implemented by using Java class library. These Java classes have three different types of class objects such Taskflow, Task and Module, respectively. The Taskflow class is a major class that manages all Tasks class objects. The Task class is a second class and has a list of sub-Task as a member. Each Task class has both previous and next process of Task objects.

The Module class is a third class and has application modules relates simulation program, database, web application, data analysis tool, visualization tool, modeler tool and data plot tool. When the Module object is activated, a related

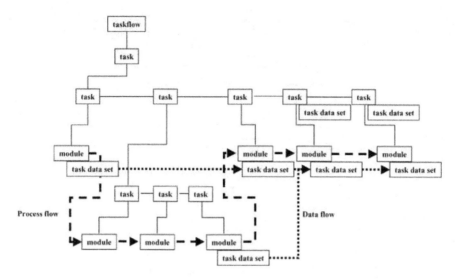

Fig. 6. Hierarchy of data structure

application is invoked on remote server. Figure 6 shows the hierarchy of data structure.

Each Task class has TaskDataSet class object as a member. TaskDataSet class has a list of TaskData class objects related to Task's data list. In case of data transfer between two Tasks and/or data gathering among Tasks, the TaskData has the information of source TaskData.

3.4   Sharing Computational Resource on the Internet

In order to share distributed computational resources on the network as mentioned in 3.2, the communication interface is implemented, which can execute applications and transfer data on a remote server by Unix-like command of Servlet. This communication interface is using http protocols for the Internet, and rsh, ssh and http protocols for the intranet as shown in Fig. 7.

3.5   Implementation of Tools

The visualization and data analyzing/managing tools were implemented by using Java and Java3D. A feature of these Java classes is the facts that are composed from two different classes of a data presentation and a user interface. A data presentation class has a physical model class and a result data class. A physical model classes has two components of atom/molecular model and continuum model in which each has geometry, structure and physical parameters. A result data classes is composed from a result of computational modules and those gathered from related data. A user interface class refers objects of data presentation

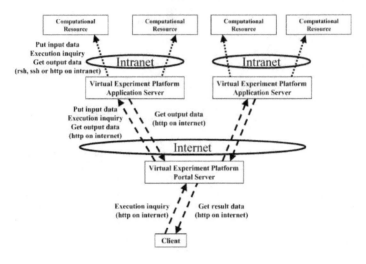

Fig. 7. Sharing computational resource on the Internet

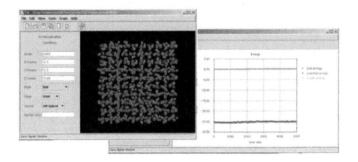

Fig. 8. Visualization tools and plotting tools

class, which allows users to operate these tools interactively through a Java applet. Figure 8 shows the examples of visualization result of molecular dynamics simulation and energy plot.

4   Experiments

4.1   Calculation of Energy and Elastic Properties

Figure 9 shows the result of calculation by using this virtual experiment platform. The energy and bulk modulus of Si single crystalline are calculated by using a molecular dynamics program. At first, users set the initial condition such as a target atom name, structure, potential model and value of lattice constant. After a platform searches the potential parameters from a database, a platform executes automatically the molecular dynamics simulation with changing values

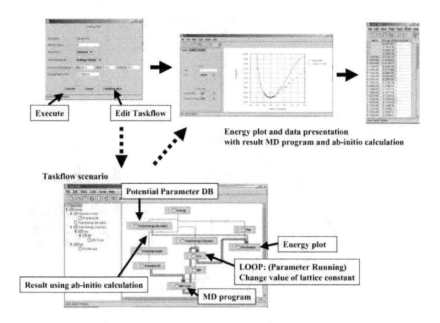

Fig. 9. Calculation of energy and elastic properties

of lattice constant as the loop task. At last, a plotting tool can show the results of calculation, if possible, together with an ab-initio calculation results in the database. By using such task scenario in this platform, the energy can be obtained easily and automatically without attention to parameter changes.

4.2 Molecular Dynamics Simulation

Figure 10 shows the example for a molecular dynamics simulation of SiO_2. The simulation procedure is firstly quenched and then after raising temperature, SiO_2 is set under controlled high temperature. The taskflow editor shown in Fig. 10 shows this process as the taskflow. A platform executes continuously these molecular dynamics simulation with data transfer between two tasks. In case of complicated problem such that various computational modules are needed, this virtual experiment platform will be effective.

5   Conclusion

The virtual experiment platform is newly developed, which is integrated many components about computational materials science and engineering technologies such as simulation, database, data analysis, visualization, and human interface. In case of complicated problems such that various computational modules are needed, this virtual experiment platform will be effective. This platform was

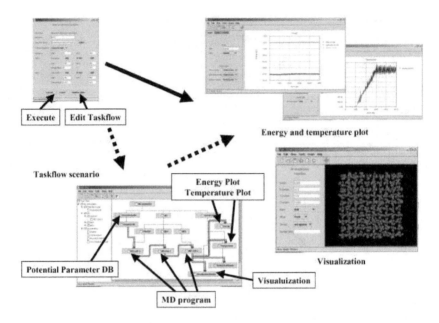

Fig. 10. Molecular dynamics simulation

applied for actual materials designs and shows the good results. The virtual experiment platform will provides the best support tools for materials researchers, materials engineers and students in a college who concern the materials design.

Acknowledgments

The authors would like to thank Dr. Masahiro Fukuda at the National Aerospace Laboratory in Japan for useful comments and encouragement.

References

1. J. Rodgers and P. Villars: Trend in Advances Materials Data: Regularities and Predictions. RMS Bulletin, Vol. 2. (1993)
2. S. Iwata: A Concept of a Virtual Production Line Produced by Integrating Database and Models of Materials. RMS Bulletin, Vol. 7. (1993)
3. N. Nishikawa, C. Nagano and H. Koike: Integration of Virtual Experiment Technology for Materials Design. Computerization and Networking of Materials Databases, ATSM STP 1311. (1997)
4. N. Nishikawa and H. Koike: Virtual Experiments System for Materials Design. European Congress on Computational Methods in Applied Sciences and Engineering (ECCOMAS2000). (2000)

Ab Initio Study of Hydrogen Hydrate Clathrates for Hydrogen Storage within the ITBL Environment

Marcel H.F. Sluiter[1], Rodion V. Belosludov[1], Amit Jain[1],
Vladimir R. Belosludov[2,3], Hitoshi Adachi[5], Yoshiyuki Kawazoe[1],
Kenji Higuchi[4], and Takayuki Otani[4]

[1] Institute for Materials Research, Tohoku University, Sendai, 980-8577 Japan,
marcel@imr.edu, http://www-lab.imr.edu/~marcel/index.html
[2] Center for Northeast Asia Studies, Tohoku University, Sendai, Japan
[3] Institute of Inorganic Chemistry, SB RAS, Novosibirsk, Russia,
[4] Center for Computational Science and Engineering,
Japan Atomic Energy Research Institute, Tokyo 110-0015, Japan,
[5] Hitachi East Japan Solutions, Ltd., Sendai, Japan

Abstract. Recently, for the first time a hydrate clathrate was discovered with hydrogen. Aside from the great technological promise that is inherent in storing hydrogen at high density at modest pressures, there is great scientific interest as this would constitute the first hydrate clathrate with multiple guest molecules per cage. The multiple cage occupancy is controversial, and reproducibility of the experiments has been questioned. Therefore, in this study we try to illucidate the remarkable stability of the hydrogen hydrate clathrate, and determine the thermodynamically most favored cage occupancy using highly accurate *ab initio* computer simulations in a parameter survey. To carry out these extraordinary demanding computations a distributed *ab initio* code has been developed using the SuperSINET with the Information Technology Based Laboratory (ITBL) software as the top-layer.

1 Introduction

1.1 Overview

In recent years, computer simulations provide increasingly attractive supplements to experimentation in research and development (R&D). In some cases, such as Scanning Tunneling Microscopy, experimental images cannot be reliably interpreted unless supported by *ab initio* electronic structure computer simulations, in other cases computer simulations can reveal trends and tendencies that are not apparent from experimental results obtained under varying conditions. For very large experimental data sets, computerized data mining may be the only feasible approach towards effective utilization. Where computer simulation supplements and enhances experimentation, it has the potential to reduce the huge investment in experimental facilities and long time periods which are required to carry out actual experiments. These arguments apply particularly to

A. Veidenbaum et al. (Eds.): ISHPC 2003, LNCS 2858, pp. 330–341, 2003.

the field of materials science. However, computer simulation of realistic materials requires high-power computing resources and a sophisticated utility environment for researchers and developers. The realization that such resources cannot be allocated to every individual materials scientist, or even to every materials science research center, is the rationale for the Information Technology Based Laboratory (ITBL) project [1] [2]. The ITBL was started in 2001 with the objective of realizing a virtual joint research environment using information technology. We have applied the ITBL system for the purpose of gaining a fundamental and comprehesive view of a new unconventional hydrogen storage material by large-scale computer simulation, which implements our specifically designed density functional electronic structure program. The ITBL project is preferred for this project over certain alternatives, such as the tools provided by the Globus collaboration [3], for several reasons. While e.g. the globus GRID toolbox is designed to run on a wide variety of platforms, it is not specifically optimized for the platforms available to us. Our parameter survey required very large scale individual runs, making it less ideal for toolkits that are optimal for large numbers of relatively low performing CPUs with limited memory. Also, the ITBL solution is practically available and deployable as a high-level top layer for job submission and output visualization whereas toolkits provide functionality through a set of 'service primitives'.

The large unitcell sizes of the proposed clathrates, the presence of first-row elements, and the delicate bonding make these calculations computationally extremely demanding. Moreover, in the current parameter survey there is a large number of nearly identical calculations, varying only in the number of hydrogen molecules in the clathrate cages and varying in the crystal lattice parameter. This makes it ideally suited for parallel supercomputers such as the PrimePower and SR8000 with large numbers of identical powerful individual nodes with large core memory. We found the PrimePower feature that memory appeared available in single addressable space very advantageous as it made it unnecessary to distribute memory for incidental extra large scale simulations.

1.2   Information Technology Based Laboratory

ITBL aims to make available to a wide group of users the resources and infrastructure for ultra large-scale computer simulations for conducting R&D. It is expected that such simulations can greatly enhance, and possibly even ultimately replace, some experimentation, and thereby improve R&D efficiency over a broad scope. Towards this goal, ITBL develops and provides systems facilitating joint research using large-scale databases and large-scale computational research facilities. ITBL enables extremely large-scale computations not possible in the past and shall provide load sharing during times of congestion by connecting and sharing supercomputers located at different research institutions and universities. In addition, ITBL aims to create an environment where different centers can use each other's unique software, databases, and hardware, as a way of enabling efficient operation and utilization of all resources, and improving their capabilities. As a part of this project, the nano-technology simulation environment has been established with ITBL.

The architecture of the ITBL system consists of several layers. The computing resources and data resources are connected by common network including from internet up to fast network such as SuperSINET that stretches the length of Japan. On this physical network, the primitive layer provides scheduling and monitoring facilities. At the intermediate layer above it, tools such as the component programming toolkit, the community toolkit, etc. exist. Using the ITBL system, we can establish a 'meta-computing environment' , such as for nano-technology simulation. Finally, the top layer encompasses the end-user interface that is provided by a Web-based technology.

1.3   Parameter Survey Computation with ITBL

We executed a large scale 'parameter survey' computation on the supercomputing network which connects several supercomputers of JAERI (Japan Atomic Energy Research Institute) (see Fig. 1). For this purpose we adapted the all-electron full-potential program TOMBO (Tohoku university Mixed Basis method) developed by the researchers at the Institute for Materials Research, Tohoku University, in cooperation with JAERI. Recently discovered [4] hydrogen hydrate clathrates which potentially can store large quantities of hydrogen in environmentally benign water were selected for study based on scientific merit, practical relevance, and computational demands. The existence of these hydrogen hydrate clathrates was not expected. Hydrogen molecules were thought to be too small

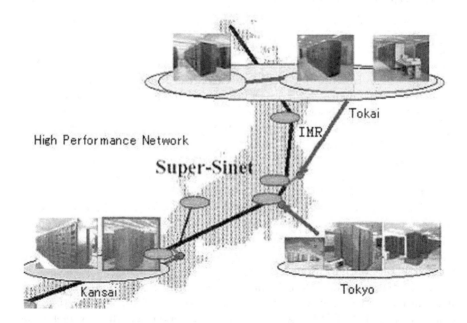

Fig. 1. Supercomputers connected by SuperSINET

to be able to stabilize an open cage-like structure consisting of hydrogen bonded water molecules [5]. Sofar, such hydrate clathrates have been stabilized always by a single guest molecule that just about ideally fit a cage. Therefore the claim that as many as four hydrogen molecules could enter and stabilize large cages and two hydrogen molecules might do the same for smaller cages met with surprise and skepticism. It was argued that such assemblies of four or two hydrogen molecules might form a 'cluster' which would act like a single molecule. It was also surmised that the intramolecular bond might be weakened to allow for bonding within such a cluster. Our objective was to test these intriguing and unconventional claims. This was done by simulating two of the most common hydrate clathrate crystal structures and fill the various types of cages with various numbers of hydrogen molecules and perform structural optimizations. The energetics of the optimized structures then reveals which cage occupancy is favored. Detailed analysis of the interactions between the hydrogen atoms in the hydrogen molecules, and of the hydrogen molecules with each other and with the water molecules in the cage structure allows us to also address the other assertions and guesses by the hydrogen hydrate clathrate discoverers [4]. The large scale computation has revealed the most stable hydrogen occupancy in the cage -like clathrate structure as well as the hydrogen molecules' location within the hydrate clathrate and the nature of the bonding in these materials. This constitutes fundamental knowledge for developing and optimizing such hydrogen storage materials.

2   ITBL-Tombo

2.1   Tombo

TOMBO is an *ab initio* program based on density functional theory. It has been developed at the Institute for Materials Research, Tohoku University, in cooperation with the Japan Atomic Energy Research Institute. TOMBO utilizes an original all-electron full-potential mixed-basis approach in which the Kohn-Sham wavefunctions are expanded in terms of atomic orbitals and planewaves. For efficiency the atomic orbitals, which derive from atomic wavefunctions, are confined to non-overlapping atomic spheres. Having localized and non-localized expansion functions for the wavefunctions enables accurate and reliable calculation of electronic properties including valence states and core states. This feature sets the mixed-basis approach apart from ordinary pseudo-potential planewave methods. Moreover, the atomic orbitals make that a very low cutoff kinetic energy for the planewave expansion suffices, typically about the same or even less than required for ultrasoft pseudopotential methods. Computation of electronic properties, structure optimization, molecular dynamics, and vibrational eigenmodes and associated spectroscopic information are available using LDA, LSDA, and GGA parametrizations of the exchange-correlation potential. We have tuned the performance of TOMBO for concerned supercomputers for the parameter survey by selecting the optimally performing mathematical libraries (eigenvalue /

eigenvectors and FFT), source code parallelization using MPI (Message Passing Interface) and by vectorizing / parallelizing the object code by specifying compiler directives.

2.2   Nano-technology Simulation Environment with ITBL

A nano-technology simulation environment has been established using TOMBO as core application. It is implemented on the ITBL system built on the network within JAERI where several supercomputers and visualization servers are connected, and this can be handled from IMR using fast SuperSINET network. The parameter survey environment consists of (a) a job entry system which enables submitting multiple jobs simultaneously, (b) a job monitoring system (shows status of each job, convergence of self-consistent calculation with its graphical views), (c) an analysis system for intermediate and final results (evolution of the total energy of each job, and visualization of other convergence criteria as well as charge density data for user-selected jobs), and providing a summarizing report. The following figure shows the architecture of this nano-technology simulation environment (Fig. 2) and the GUI on the web browser (Fig. 3, 4). The simulation environment allowed us to execute almost 100 jobs simultaneously.

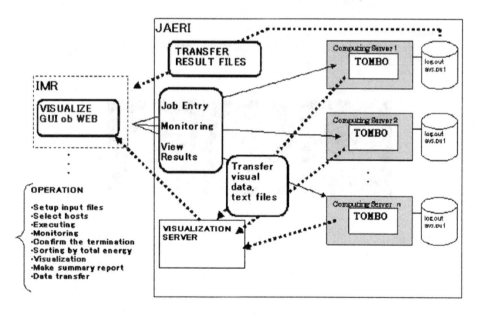

Fig. 2. The architecture of parameter survey with ITBL

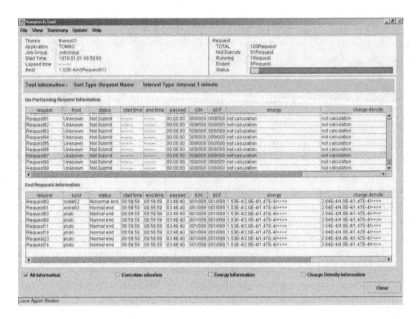

Fig. 3. Job Status view of parameter-survey

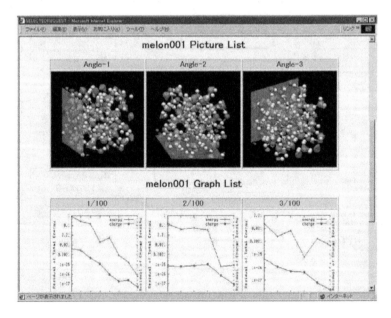

Fig. 4. Summary view (Upper:Atomic configuration and contour slice of charge density, Lower:Convergences of total energy and charge self-consistency)

3 Application to Hydrogen Hydrate Clathrates

3.1 Introduction

Hydrate clathrates were discovered about two centuries ago when chlorine gas was found to increase the melting point of ice[5]. At first hydrate clathrates were a curiosity but with the advent of piped natural gas hydrate clathrates of methane and other hydrocarbons became of great technological importance. Hydrocarbon hydrate clathrates can obstruct pipelines when gas contains water. Naturally, early research focussed on inhibiting the formation of hydrate clathrates by adding so-called inhibitors such as methanol. Later on it was realized that hydrocarbon hydrate clathrates might be a good source of natural gas and extensive deposits were discovered and are currently being exploited in Russia. Now, it is suspected that the total amount of natural gas trapped in clathrates in the deep sea and in permafrost regions is greater than the known reserves of oil[5]. Recently, the fact that hydrate clathrates can contain several hundred times their own volume in gas has drawn attention. Almost a year ago[4], for the first time, a hydrate clathrate of hydrogen gas was created. Such a clathrate hypothetically contains 500 times its volume in H_2 gas (under standard conditions 1 atm and 273 K) while stable at 145 K at a pressure of 0.1 atm. At this temperature hydrogen hydrate clathrate contains as much H_2 as pure H_2 gas at a pressure of 250 atm. This suggests that clathrates may be candidates for hydrogen storage. Aside from the important technological aspects, hydrogen hydrate clathrate is hypothesized to be the first known hydrate clathrate in which a cage in the structure is stabilized by clusters of molecules rather than by a single individual molecule. Almost all known hydrate clathrates fall structurally in the classification, cubic structure 1, cubic structure 2, and hexagonal. In all these structures the cages are much too large for an individual H_2 molecule and therefore the existence of hydrogen hydrate clathrates was not suspected. Although about a decade ago a report of a hydrogen hydrate clathrate[6] appeared, the structure discovered at that time was an ice (Ice II) with dissolved hydrogen which is not a cage-like structure and which requires much higher pressures for stability. The hydrogen hydrate clathrate discovered recently is of cubic structure 2 type. This structure has two types of cages. Both cages are much too large to be stabilized by the small individual H_2 molecules, rather, there are indications that the large (small) cage is stabilized by a cluster of 4 (2) H_2 molecules as is shown in Fig. 5. Incidentally, the multiple H_2 cage occupancy is also the reason for the large H_2 content in the clathrate. It has been surmised that the interatomic bonding between the hydrogen atoms in the H_2 molecules might be affected by enclathration and that the hydrogen atoms in a given cage form a cluster [4].

The present study seeks to answer several cardinal questions relating to the new hydrogen hydrate clathrates:

- 1) do clusters of H_2 stabilize the open cage clathrate structure?
- 2) what is the most favored occupancy of the large and small cages?

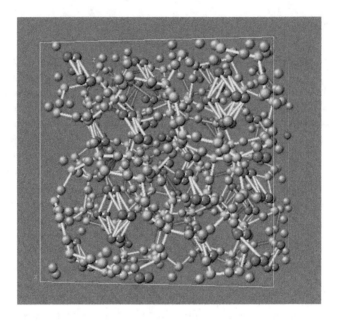

Fig. 5. optimized geometry of hydrogen clathrate hydrate (structure 2) at experimental lattice parameter of 17.047 Å. Green and blue spheres represent O and H belonging to H_2O molecules, red spheres indicate enclathrated H belonging to H_2 molecules. The bond between O atoms highlight the cage structure, and the bonds between the enclathrated H indicate the clusters formed by 4 (2) H_2 molecules in the large (small) cages.

- 3) is the chemical bond within the hydrogen molecule affected by enclathration?
- 4) do the hydrogen molecules in a given cage form a cluster?
- 5) is density functional theory capable of describing the thermodynamics fo the H_2 - hydrate clathrate system?

Below, we try to answer these questions using highly accurate state of the art electronic density functional methods and vibrational theory

3.2   Computational Details and Method

Non-spin polarized electronic structure calculations were performed with the all-electron full-potential mixed-basis approach in which the electronic wavefunctions are expanded as a linear combination of localized atom-centered orbitals and plane waves. The localized orbitals are derived from wavefunctions of non spin-polarized atoms by truncation within non-overlapping atomic spheres and subsequent ortho-normalization. The cutoff energy for the plane wave expansion of the wavefunctions was 200 eV. Studies on H_2O have shown that this rather low cutoff kinetic energy gives an accurate description. This illustrates that the mixed-basis approach is well-suited for hydrogen hydrate calculations as it does not rely on pseudopotentials. The local density approximation was used [7] and integrations over the Brillouin zone were carried out with the Γ point only. Geometries of the H_2O framework with guest H_2 molecules were optimized, see Fig. 5. The occupancy of the cages in cubic clathrate structure 1 and cubic clathrate structure 2 are varied from 0 to 5 for the large cage and from 0 to 2 for the small cage. Next the lattice parameter is varied in order to obtained the equation of state for the clathrate framework. This procedure is repeated for clathrate structures 1 and 2. A comparison of the enthalpy as a function of cage filling the optimal occupancy is deduced.

In future work we will consider also the effect of the vibrational and librational degrees of freedom on the equation of state and the relative stability as a function of cage filling and temperature. Also, a careful analysis might reveal the preference of cubic structure 2 over cubic structure 1.

3.3   Results

Starting with approximate coordinates for cubic hydrate clathrates of type 1 and 2 as determined from experiment, the atomic coordinates of these structures were optimized for various cage fillings. It should be mentioned that while the cage structures could be unique optimized, for the enclathrated H_2 molecules rotations around the centers of mass produced negligible changes in the energy confirming the quantum nature of the hydrogen rotation. The local density approximation also gave large energy separation of about 5.3 eV between occupied and unoccupied electronic states for all geometries so that the electronic and ionic degrees of freedom can be treated separately as the Born-Oppenheimer approximation implies. The relative energies are shown in Figs. 6 and 7. It is readily apparent that 4 H_2 molecules in the large cage and 2 H_2 molecules in the small cage produce the most stable cage occupancies for both cubic type 1 and type 2 structures. Figure 6 shows that for cubic structure type 1 the optimal filling is important both for the small and the large cages because cage occupancies 32 and 41 both are much less favorable than cage occupancy 42. For cubic structure type 2 the occupancy of the large cage is the most important factor because energies for cage occupancies 41 and 42 are rather close in energy, while one more or one less H_2 molecule, as in cage occupancy 52 and 32, greatly destabilizes the structure, as is evident from Fig. 7. The importance of the large cage in structure 2 is known from experiment also, where an optimal filling of the large cages is essential for the formation of this structure. In some cases the small cages can even remain empty [5].

While the optimized geometries, e.g. Fig. 5, cannot resolve the individual atomic positions of the hydrogen atoms that comprise an H_2 molecule, it is nevertheless possible to calculate the distance between centers of mass. This distance is about 2.8 Å for the most stable geometries. The H_2 distance to the nearest oxygen atom in the cage wall is about 2.3 to 2.6 Å. At these distances the interactions of H_2 molecules with other H_2 molecules and the cage wall is weak. The weak interaction is due to the large gap between occupied and unoccupied states in both the H_2 molecule and in the empty cage structure which prevents

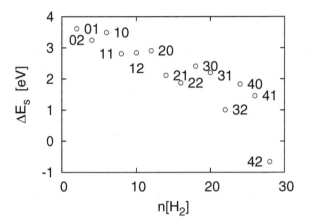

Fig. 6. Energy relative to empty cage and H_2 molecules as a function of number of H_2 molecules per cubic cell after structural optimization at experimental lattice parameter of hydrate clathrate cubic structure 1 (≈ 11.947 Å) for various hypothetical cage occupancies. The labels indicate number of H_2 molecules in large (first digit) and small (second digit) cages.

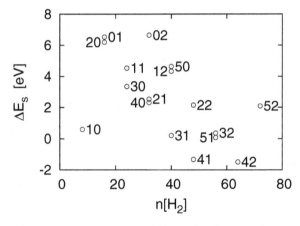

Fig. 7. Energy relative to empty cage and H_2 molecules as a function of number of H_2 molecules per cubic cell after structural optimization at experimental lattice parameter of hydrate clathrate cubic structure 2 (17.047 Å) for various hypothetical cage occupancies. The labels indicate number of H_2 molecules in large (first digit) and small (second digit) cages.

the formation of a chemical bond, therefore the molecular orbitals are little affected by placing the H_2 molecules inside the cages. Hence the only bonding mechanism is, within a local density picture, due to the overlap of the tails of the molecular wavefunctions. The exchange-correlation functional is non-linear in the charge density so that overlap creates a more negative (deeper) potential than without overlap. This is a simple local density picture for physisorption in van der Waals-like systems. Naturally, when molecules are gradually removed farther from the cage wall, the charge densities, and consequently the non-linearity of the exchange-correlation potential cease to play a role and the van de Waals interaction is the only remaining effect. At the relatively short distances of a few Å the overlap effect should still overwhelm the van de Waals interaction so that density functional theory applies. It is clear then, that 1) the interactions of the H_2 molecules with each other and with the cage wall are generally attractive and 2) that the H_2 - H_2 interactions should be much weaker than the H_2 - H_2O interaction at comparable distances because the charge density associated with the latter is much greater so that the resulting non-linearity in the exchange-correlation potential is much greater also. Calculations on isolated cages with H_2 molecules give an approximate H_2 - cage wall interaction energy of 20 to 25 meV/H_2 molecule. This compares rather nicely with the physisorption energy of 86 meV [8] for H_2 onto graphene at the center hexagon position because at the center position the overlap occurs with 6 carbon atoms. It is important to note here that the local density approximation [7] has been used and not the generalized gradient approximation which generally fails to give any bonding for this class of systems [9]. The fact that the H_2 interaction with the cage wall is much stronger than that with other H_2 molecules dispells the notion that the 4 (2) H_2 molecules in a large (small) cage form a cluster in which intramolecular bond is sacrificed for stronger intermolecular bonding [4]. Our study suggests that the H_2 molecules in a cage all bind individually to the cage wall, and that H_2 - H_2 interaction plays a role in the case of crowding only. Our calculation indicates that the interatomic distance in H_2 hardly changes when the molecule is enclathrated. Thus, as one would expect from a physisorption description of hydrogen hydrate clathrate formation, the H_2 intramolecular bonding is not affected.

3.4   Conclusions

An electronic local density functional study of cubic hydrogen hydrate clathrates was performed. It was argued that the enclathration of hydrogen molecules is based on physisorption. Therefore, (local) density functional theory can be applied to describe the interactions in the hydrogen - hydrate clathrate system. The physisorption energy was calculated to be of the order of 20 to 25 meV/H_2 molecule. The small energy involved means that the energetic barrier between absorption and desorption is small, so that little energy is lost in the storage - retrieval process of H_2 in hydrate clathrate. The optimal cage occupancy is 4 H_2 molecules in the large cage and with somewhat less preferentiallity, 2 H_2 molecules in the small cage of cubic clathrate structure 2, in agreement with the educated guess made by Mao et al. [4]. This makes for high H_2 storage capacities in H_2O of about 5 % by weight. The hydrogen molecules are individually bonded to the cage walls and do not form a cluster. The H_2 intramolecular interaction is but weakly affected by enclathration. These results confirm the structural analysis as determined by experiment, but the results also give a completely different physical picture of the bonding than what was surmised in that work.

The detailed understanding of bonding could be obtained only with extensive computer simulations. Without the high level of automation for job preparation and submission and the interactive visualization of intermediate research results, as provided by the nanotechnology environment based on the top-layer of the ITBL system, this study would have required vastly more human effort. The ability to submit runs to several supercomputers simultaneously made available computer resources previously unattainable and permits computer simulations on a new scale.

References

1. The ITBL project can be found on the world wide web at **http://www.itbl.jp/**.
2. Y.Yamaguchi and H.Takemiya, *Software Infrastructure of Virtual Research Environment ITBL*, J. Japan Soc. Computational Fluid Dynamics, **9** (2001) 83 (in Japanese); available also at **http://wwwsoc.nii.ac.jp/jscfd/j-jscfd/93/93p4.pdf**.
3. The Globus project for grid computing can be found on the world wide web at **http://www.globus.org/**.
4. W.L. Mao, H.K. Mao, A.F. Goncharov, V.V. Struzhkin, Q. Guo, J. Hu, J. Shu, R.J. Hemley, M. Somayazulu, and Y. Zhao, Science **297** (2002) 2247.
5. E.D. Sloan, *Clathrate Hydrates of Natural Gases*, (Marcel Dekker Inc., New York, 2nd ed., 1997).
6. W.L. Vos, L.W. Finger, R.J. Hemley, H.K. Mao, Phys. Rev. Lett. **71** (1993) 3150.
7. J. P. Perdew and A. Zunger, Phys. Rev. B **23** (1981) 5048.
8. J. S. Arellano, L. M. Molina, A. Rubio, and J. A. Alonso, J. Chem. Phys. **112** (2000) 8114.
9. J. S. Arellano, L. M. Molina, A. Rubio, M. J. López and J. A. Alonso, J. Chem. Phys. **117** (2002) 2281.

RI2N – Interconnection Network System for Clusters with Wide-Bandwidth and Fault-Tolerancy Based on Multiple Links

Shin'ichi Miura[1], Taisuke Boku[2], Mitsuhisa Sato[2], and Daisuke Takahashi[2]

[1] Graduate School of Science and Engineering, University of Tsukuba,
1-1-1 Tennodai, Tsukuba, Ibaraki 305-8573, Japan,
`miura@hpcs.is.tsukuba.ac.jp`
[2] Institute of Information Sciences and Electronics, University of Tsukuba,
{`taisuke, msato, daisuke`}`@is.tsukuba.ac.jp`

Abstract. This paper proposes an interconnection network system for clusters based on parallel links over commodity Ethernet to provide both wide bandwidth and high fault tolerance. In the proposed scheme, RI2N, multiple links are employed to enhance the total bandwidth in normal mode, or provide redundant connections in failure mode with automatic software control. In a current prototype implementation at the user-level in library, normal mode functionality provides a maximum bandwidth of 40 MByte/sec over four links of Fast Ethernet, and 191 MByte/sec over two links of Gigabit Ethernet.

1 Introduction

Clusters of personal computers (PCs) have been playing an increasingly important role as platforms for cost-effective high-performance computing. Although the actual efficiency of program execution is low compared to vector processing systems, the low expense and high peak performance of PCs allow the implementation of PC clusters to achieve high total numbers of processors in a system and secure TFLOPS-class peak performance.

Networks for such cluster systems have been implemented with system area network (SANs) and commodity local area networks (LANs). Myrinet [1] is one of the most popular SANs for high-performance clusters, and provides a bandwidth of more than 200 MByte/sec per link. Infiniband [2] is another promising candidate for the very near future. Despite such high performance, however, these network systems cost much more than commodity LAN such as Ethernet. Compared with the low cost of processors and memory, the cost of SAN is quite high, both in terms of network interface cards (NICs) and network switches. In particular, the price of network switches increases rapidly with the number of nodes in the system.

Commodity LAN such as Fast Ethernet or Gigabit Ethernet provides a very good cost/performance ratio. As these networks are already widely implemented for general-purpose networking, the cost of implementation is very low. Even

A. Veidenbaum et al. (Eds.): ISHPC 2003, LNCS 2858, pp. 342–351, 2003.

though the sustained performance of LANs is significantly lower than that provided by SANs, the LAN option is still very attractive when combined with low-cost processors.

In addition to cost effectiveness, the dependability or fault tolerancy is also a very important issue for large-scale clusters. The robustness of NICs and switches for commodity LAN, typically aimed at applications with low communication rate, is relatively low compared to SAN components. For example, high-speed communication between nodes on a cluster can often fail suddenly, and a SAN implementation allows the network to be recovered by a software reset of the switch without requiring a hardware fix.

We have been developing an interconnection network system for PC clusters that provides both wide bandwidth and high dependability through the use of multiple links over commodity LAN under software control. The system is called Redundant Interconnection with Inexpensive Network, or RI2N for short. The basic concept is very simple; multiple LAN links are employed to aggregate bandwidth in normal mode and provide redundancy in failure mode. Although there have been several studies on network trunking for link aggregation and hot-swappable redundant links for system failure, RI2N provides both functions in a complete system that is entirely controlled by software.

In this paper, the concept and basic design of RI2N is presented, and the performance of a prototype implementation is preliminarily evaluated.

2   Redundant Interconnection with Inexpensive Network

With the constant progress in desktop CPU performance, the performance gap between node processor power and interconnection network bandwidth in cluster computing continues to widen. Furthermore, while the cost/performance ratio for CPUs continues to improve, the same cannot be said for the interconnection network. To solve the problem of bandwidth and latency on a cluster network, SANs such as Myrinet [1], Dolphin [3], Quadrix [4] and Infiniband [2] have been introduced. However, although the performance of these network architectures is much higher than generic LAN, the very high cost of SANs has a significant impact on the total cost/performance ratio of a cluster system.

Network trunking [6] has been investigated by a number of researchers as a means of aggregating multiple LAN links over, for example, Fast Ethernet or Gigabit Ethernet to provide wider bandwidth and maintain a high cost/performance ratio for each link. Both software- and firmware-based approaches have been proposed, and some of the technologies have been commercialized. Although these schemes offer an inexpensive way of enhancing network performance, supported hardware is limited.

Another important issue for cluster networking is fault tolerance. In high-performance computing on clusters, the execution time of a job is very long, and even a small failure of the network may cause all calculation to be cancelled, wasting all the CPU time already invested in a possibly length calculation. Importantly, the reliability of an interconnection network over commodity LAN is significantly inferior to that offered by dedicated networks for traditional massively parallel processors. The lack of strong error detection and correction functions in commodity LAN limits the reliability and robustness of the network. In particular, LAN switches are highly susceptible to packet overload.

Ethernet is the most commonly used network for clusters despite the relatively poor bandwidth and high latency compared to SANs or dedicated interconnection networks. Furthermore, as both the hardware and software for LANs are designed primarily for versatile communication (e.g., TCP/IP), the load on NICs and switches when applied for high-performance cluster computing is much larger than that for which the components were designed. As a result, temporary failures such as link loss can occur frequently during computation. Although not permanent hardware failures, being fixable by simply resetting the system, such outages are a very serious obstacle to expanding the size of clusters. One effective solution is check-point/recovery [8][9] on computation, however, this scheme introduces a very large overhead for both execution time and resources such as disk and memory space.

There are actually several technologies that improve performance and fault tolerancy through the use of multiple streams. The most typical example is a redundant array of inexpensive disks, or RAID [7], which is used for high-performance fault-tolerant storage systems. RAID systems are typically implemented over multiple disks to provide both bandwidth enhancement and redundancy for failure simultaneously. The authors plan to introduce a similar concept for cluster interconnection networks, utilizing multiple links over a low-cost network (i.e., Ethernet) to provide both wider bandwidth and higher reliability without the need for SAN technology. This system is called Redundant Interconnection with Inexpensive Network, or RI2N (Fig. 1).

RI2N bundles multiple NICs on each node of a cluster, analogous to the multiple disks in RAID. As such, this technology is primarily implemented using multiple individual NICs, or multi-ports NICs (MP-NICs hereafter). MP-NICs are configured in the proposed scheme such that the bandwidth and reliability of the network can be controlled dynamically by software. The goal is to provide a user-transparent system on the network driver level.

RI2N requires each node to have an MP-NIC. For redundancy, the pathways between MP-NICs must also be separate. For example, multiple links from an

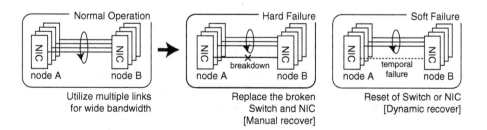

Fig. 1. System image

MP-NIC should be connected to different switches to avoid loss of network connection. This feature is not supported current standard link aggregation technology such as IEEE 802.3ad [IEEE802.3ad] which aims the same function within a single switch and NICs connected to it. Furthermore, as the purpose of RI2N is to provide a fast and safe communication environment for clusters at low cost, upon temporal failure, the system should be able to continue to execute the application at a lower performance level. A software dynamic recovery function is desirable to allow switch recovery from a soft failure, and the status of the system should be constantly monitored and reported.

3   Prototype

3.1   Overview

To design and implement the functions of RI2N, the function of wide bandwidth with multiple links is addressed first, then the function of dependable communication is added later. In the implementation of the software-controlled trunking mechanism, future extension with redundant communication for channel failure is also considered. Although the final goal is to implement driver-level RI2N functions, in this paper a user-level (library level) implementation is constructed for preliminary performance evaluation and clarification of the key issues for future practical implementation.

In the user-level implementation of wide bandwidth using MP-NICs, a library is constructed to replace the traditional TCP protocol stack (socket library). This library is described using traditional TCP, and is implemented here as a TCP/IP communication library.

3.2   Implementation Level

There are two levels of possible implementation of RI2N; user level and system level. At the user level, the system is implemented based on the normal TCP/IP socket library for ease of development and high portability. Multiple sockets corresponding to the ports on MP-NICs are used to provide aggregated bandwidth. Management of multiple sockets may be achieved by a number of methods. In this implementation, the communication performance depends heavily on the underlying single-stream TCP/IP. The basic functions in TCP/IP guaranteeing packet transfer and packet sequence may be superseded by RI2N redundancy functions. At the system level, the system is implemented at the driver level. Management of multiple links on MP-NIC is encapsulated in the device drivers, providing a very high level of user transparency. Various tuning features suitable for cluster construction may also be applicable. The implementation cost at this level is much higher than at the user level, and the target physical devices may be limited.

In this paper, a user-level implementation is attempted as basic research on MP-NIC performance and system design criteria. Based on the performance results and experiences with this system, the authors will then proceed to a system-level implementation.

A basic application programming interface (API) is employed for RI2N to ensure compatibility with the conventional TCP socket library and ease of understanding. These functions wrap regular functions to combine automatic control of MP-NICs. There are various issues associated with the basic send/receive functions as well as the establishment of connection between nodes and non-deterministic actions, etc.

The most important feature of the scheme is sending/receiving on MP-NICs. In RI2N, a user message is divided into a certain size of fragments, called *split packets*. The size of a split packet is an important parameters in the design and tuning of the system. These packets are sent from multiple ports in parallel while maintaining the packet sequence. Each split packet is given an additional header containing minimum information to maintain the sequence of packets and to detect packet failure.

Split packets are received by parallel MP-NIC ports on the receiver side. The packets are ordered in a buffer according to the header information, and after the entire message has been received and reordered, the message is passed to the application.

3.3   Dividing the Message

Two basic methods can be considered for dividing user messages into split packets for simultaneous transmission by parallel ports on MP-NICs; *block division* and *cyclic division*. In block division, a message is split into large chunks of n packets for n parallel ports, then each split packet is sent to one of the parallel ports. In cyclic division, a message is split into small chunks that are distributed to n ports in a round-robin manner.

Block division transmits larger packets of data, resulting in improved communication efficiency because of the relatively smaller size of the additional header. However, this method has disadvantages with regard to load balancing because each communication link must transmit one large packet. It is also inefficient upon packet failure because the cost of retransmission on another port is expensive due to large packet size.

On the other hand, in cyclic division, the lower efficiency of transmission of small packets may degrade the communication performance. However, dynamic load balancing can be achieved to counteract non-flat congestion over multiple network links. The retransmission cost on packet failure is also relatively small. To achieve high performance constantly, the split packet size will need to be adjusted carefully according to not only the application message size but also the system condition.

It is considered that in the present system, increased overhead can be tolerated because of the greater bandwidth of the parallel streams in RI2N. Therefore, the cyclic-division method was adopted as the data-dividing policy in consideration of load balancing and small penalty upon packet failure.

3.4   Creating Parallel Streams

Two methods are also available for transmission and reception of split packets over parallel streams and multiple ports; *multi-thread* and *select function*. To handle data transmission on multiple send/receive ports, the simplest way is to introduce multiple threads and let each handle an assigned port. Although the programming is easier, this method also introduces large overhead for thread creation, deletion and synchronization. Furthermore, although a *thread pool system* can be introduced to reduce the cost of dynamic creation and deletion of threads, it is inefficient for single-CPU nodes.

Another method for handling multiple ports without multiple threads is to utilize the *select()* system call, that is, to check the availability of ports for sending or receiving data, then perform these actions in a single thread. The concern with this method is the overhead of frequent *select()* system calls. However, this method is effective even for single-CPU nodes, and the performance of modern CPUs may overcome this apparent drawback.

Both of these approaches include performance aspects that are theoretically unclear. However, our preliminary experiment resulted in that the method with *select()* achieves better performance than multi-thread even with 2-way SMP system. Therefore, this paper presents the performance evaluation and analysis of the method which uses *select()* system call (Fig. 2).

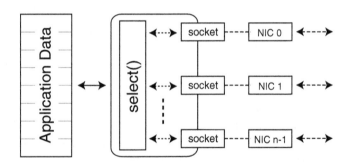

Fig. 2. Select parallel stream method

4   Performance Evaluation

4.1   General Condition

The performance of the current implementation of RI2N was examined focusing on improvement of the effective bandwidth through the use of multiple links. As described in the previous section, the effective bandwidth is strongly influenced by the chunk size of split packets under the cyclic packet division scheme. Therefore, the effective bandwidth was also examined for various chunk sizes.

For split packets, the larger chunk size theoretically improves the effective bandwidth because of the relatively small overhead on message decomposition

Table 1. Experimental environment

PC	DELL PowerEdge 1600SC (Pentium 4 Xeon 2.8 GHz 2-way SMP) × 2
NIC	Adaptec Quartet 64 on 64 bit/33 MHz PCI
	(100base-TX Ethernet × 4 ports)
	Intel PRO/1000MT on 64 bit/100 MHz PCI-X
	(1000base-T Ethernet × 2 ports)
Switch	PLANEX FMX-0248K (48 Port Fast Ethernet Switch, Layer 2)
	PLANEX FMG-24K (24 Port Gigabit Ethernet Switch, Layer 2)
OS	Linux 2.4.18

and reconstruction. Additional header information on each packet also introduces overhead. When using Ethernet, however, the actual physical packet size is limited by the maximum transmission unit (MTU), which is 1500 Bytes in general conditions. Of course, the window size and socket buffer for send/receive sockets can also be expanded, but these effects are not certain. Therefore, there will exist an optimal chunk size for split packets depending on the message size in the application.

The target network is Ethernet, and the performance of RI2N is examined first on Fast Ethernet (100base-TX), and then on Gigabit Ethernet (1000base-T) based on the results for Fast Ethernet.

Table 1 describes the experimental environment and platform. As shown here, each node has 4 ports for Fast Ethernet and 2 ports for Gigabit Ethernet, providing 50 MByte/sec and 250 MByte/sec aggregated theoretical peak performance, respectively.

In the experiment, the data passed by the application is transferred by ping-pong communication between two nodes for certain times to examine the average data transfer bandwidth according to the application data size and chunk size. In all figures, the horizontal axis represents the application data size and the vertical axis represents the transmission performance (effective bandwidth). Several chunk sizes were examined as parameters.

4.2   Performance Result

The results of experiment with four links of Fast Ethernet are shown in Fig. 3 (a). Here, 1KB of chunk size achieves good performance through all range of application data size even though it is slightly lower than that with 16KB chunk size. This result can be considered reasonable because a chunk size of 1 KByte is close to the MTU of Ethernet communication.

From these results, the overhead for invoking the *select()* system call is relatively small compared to time required to wait for completion of data transmission on Fast Ethernet with four links. This greatly simplifies implementation because it is unnecessary to change the chunk size dynamically according to the application data size.

The results over two links of Gigabit Ethernet are shown in Figure 3 (b). The performance of the system is excellent for chunk sizes with 4 and 16 KB.

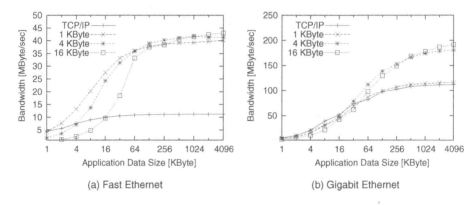

(a) Fast Ethernet

(b) Gigabit Ethernet

Fig. 3. Performance of ping-pong transfer

However, there exist multiple crossover points between the different chunk sizes, indicating that it may be necessary to select the optimal chunk size dynamically to achieve the best performance for the requested application data size. If we can neglect the small difference for small application data size, the chunk size of 4 KB seems to be the best.

The maximum performance achieved in this case is 191 MByte/sec, which is approximately 76% of the theoretical peak performance over two links. There are 3 possible factors holding back the peak performance. (1) Compared to Fast Ethernet, the data handling time including the *select()* system call is relatively high because of the high throughput of Gigabit Ethernet. (2) The bandwidth of the memory bus is not sufficient to support dual ports of Gigabit Ethernet, or the memory bandwidth could not be utilized fully. (3) The physical data transmission efficiency is not as high as expected because the MTU of Ethernet was not increased, that is, no "jumbo frame" was introduced as regular switches are not equipped with this feature.

It is also necessary to consider the effect on CPU performance due to the load of Gigabit Ethernet throughput. One solution is to introduce multi-threading in cooperation with the select method. In the select method, free sockets are processed one at a time, and as such the issue rate of actual transmission and reception is quite low. As the current trend of high-performance clusters is to use dual CPU SMP configurations for each node, it is possible to rely on dual threads to support dual streams of data t ransfer. Thus, the combination of select and multi-thread methods is worth considering. When implementing the fault tolerance feature, processing of the data stream becomes much more complicated. Incremental expansion of the current method on the Fast Ethernet implementation is planned, however, expansion appears difficult on Gigabit Ethernet. From the current experimental results, it seems that it will be necessary to reconstruct the procedure of data stream processing in RI2N to utilize the SMP feature of nodes and overlap data processing and communication.

5   Strategy for Dependability on RI2N

5.1   Detection of Failure

Two basic methods can be proposed for failure detection, depending on the implementation. For multi-thread implementation, the buffer condition for packet decomposition/reconstruction is monitored (buffer checking). For example, if the buffer contents counter has not changed for a long time, this indicates that there is no data movement on that link. For the select method, if there is no status change for a descriptor that is manipulated in the *select()* system call for a certain period of time, this is taken as indicating failure of the corresponding link (time out).

5.2   Recovering Phase

Similarly, there are two methods to ensure that all split packets have been successfully sent even in the case of link failure; use of a retransmit buffer, or a RAID5 approach. In the former, a redundant buffer is implemented on the sender side, and a copy of all transmitted packets is held in this buffer for a certain period of time. Upon failure detection, the copied packets are retransmitted over the other live link. To clear the buffer periodically, an acknowledgment from receiver side must be sent back to the sender node within a certain period (absolute time or packet counting). In the RAID5 approach, packets including redundant (parity) data are transmitted over multiple links, and the link used for parity information is changed every time. This method applies a RAID5 mechanism as used for disk redundancy. Upon link failure, it is not necessary to retransmit the lost packet because redundant data is always transmitted. This method results in a consistently higher total amount of data to be transmitted.

For the present implementation at the user lever, the RAID5 approach may represent an unacceptably large load and overhead. Additionally, TCP/IP on Ethernet has a reasonable level of reliability, and use of a retransmit buffer is considered to be a better means of providing dependability for this system. However, both of these methods will be weighed up in a future pilot study.

6   Conclusions and Future Works

In this paper, an interconnection network system called RI2N was proposed for large-scale clusters. RI2N provides wide bandwidth and dependability over multiple communication links using inexpensive commodity network media such as on Ethernet. In the future, the authors intend to implement and evaluate fault tolerancy, and examine the system's performance with real applications such as MPI and its compatibility with other systems such as cluster managements systems. Furthermore, the multi-thread implementation evaluated in this paper was based on the dynamic creation/deletion of threads, and the use of thread pooling will be examined as a more lightweight implementation. Finally, latency is a very important and will be analyzed in detail in future studies.

Acknowledgements

The authors truly thank to the members of Mega-scale Low-power Computing project. This study is partly supported by Core Research for Evolutional Science and Technology program of Japan Science and Technology Corporation (JST-CREST) and the Grant-in-Aid of Ministry of Education, Culture, Sports, Science and Technology in Japan (C-14580360).

References

1. Myricom, INC. *http://www.myri.com/myrinet/* .
2. InfiniBand Trade Association. *http://www.infinibandta.org/* .
3. Dolphin Interconnect Solutions Inc. *http://www.dolphinics.com/* .
4. Quadrix Solutions, Inc. *http://quadrix.com/* .
5. IEEE802.3ad *http://grouper.ieee.org/groups/802/3/ad/* .
6. Hiroshi Tezuka, Atsushi Hori, Yutaka Ishikawa, Mitsuhisa Sato. PM: An Operating System Coordinated High Performance Communication Library. *High-Performance Computing and Networking,* volume 1225 of Lecture Notes in Computer Science, pages 708-717. Springer-Verlag, April 1997.
7. Peter M. Chen, Edward K. Lee, Garth A. Gibson, Randy H. Katz, David A. Patterson. RAID : High-Performance, Reliable Secondary Storage. *ACM Computing Surveys,* volume 26(2), pages 145-185. 1994.
8. George Bosilca, Aurelien Bouteiller, Franck Cappello, Samir Djailali, Gilles Fedak, Cecile Germain, Thomas Herault, Pierre Lemarinier, Oleg Lodygensky, Frederic Magniette, Vincent Neri, Anton Selikhov. MPICH-V: Toward a Scalable Fault Tolerant MPI for Volatile Nodes. *ACM/IEEE International Conference on Supercomputing SC 2002.* 2002.
9. E. Elnozahy and D. Johnson and Y. Wang. A survey of rollback-recovery protocols in message-passing systems. *Technical Report CMU-CS-96-181,* Carnegie Mellon University, October 1996.

A Bypass-Sensitive Blocking-Preventing Scheduling Technique for Mesh-Connected Multicomputers

Wei-Ming Lin and Hsiu-Jy Ho

Department of Electrical Engineering, The University of Texas at San Antonio,
San Antonio, TX 78249-0665,
wlin@utsa.edu

Abstract. Among all non-blocking job scheduling techniques for mesh-connected multiprocessor systems to ensure contiguous processor allocation, Largest-Job-First (LJF) technique proves to be one of the best in achieving small latency compared to First-Come-First-Serve (FCFS) among others. The scheduling becomes "blocking" whenever a job reaches a bypass limit and it deprives the scheduling process the flexibility benefit in its non-blocking nature. Such an adverse effect is especially pronounced in LJF compared to the simpler FCFS. This paper proposes a novel scheduling technique that is capable of dynamically adjusting the priority of a job in the queue by weighing its job size with a factor based on its current bypass value. This technique ensures that the advantage of the LJF is preserved while significantly diminishing the chance for the blocking situation to occur. In our simulation results, we show that frequency in leading to "blocking" scenario is significantly reduced and thus a much improved latency ensues.

1 Introduction

Fast and efficient processor allocation and job scheduling algorithms are essential components of a multi-user multicomputer operating system. The performance of a multicomputer system depends heavily on the processor management strategy. Job scheduling is to select a job among all jobs currently in the waiting queue and to submit it to the processor allocator. Such a selection process usually is based on the jobs' processor requirement, e.g. the number of processors, the dimension of submesh for a mesh system, the expected time of processor use, etc. In a mesh system, processor allocation deals with allocating a free submesh in the whole mesh to the submesh request chosen by the job scheduler. In this paper, processor allocation is strictly referred to assigning a contiguous section of submesh to a job requesting exactly of the same size and shape.

Several non-blocking scheduling strategies have been proposed. A "non-blocking FCFS" simply repeatedly searches for the first allocatable job according to their arriving order. Largest-Job-First and Smallest-Job-First have their corresponding non-blocking version, respectively, through a search process according

A. Veidenbaum et al. (Eds.): ISHPC 2003, LNCS 2858, pp. 352–359, 2003.

to job size. *Multiqueue* [4] is another often used strategy, in which an incoming job is sent to one of several queues based on a set criteria. Chang [1] proposed a bypass-queue job scheduling strategy, a variation of the FCFS queue without the blocking problem. Jobs in a bypass-queue are checked for allocation feasibility in the order of their arrival. A job is allowed to bypass the unallocated jobs if it is the first one that can be allocated. The process continues until all the jobs in the queue have been checked or a job currently being processed departs from the system. To ensure the fairness, a threshold time is set to prevent any job from waiting beyond the threshold time. If any of the jobs remains in the queue for a time longer than the threshold, the bypassing process is disabled and jobs are served in a strictly blocking FCFS manner. In [7] D. D. Sharma et al. proposed a job scheduling algorithm which combines a priority-based scheduling policy with a submesh reservation policy. Better performance is obtained in this technique by allowing a submesh currently in use to be reserved for a waiting request. Although this method delivers significant performance improvement, its scheduling cost is substantial compared with its original method.

Note that all "non-blocking" scheduling techniques are tightly tied to the subsequent processor allocation process, i.e. the process to check for allocatability which is a time-consuming process. All these algorithms have to scan through the jobs in the queue for an allocatable one. In the worst case, all jobs in the queue need to be checked. Furthermore, quite a few scheduling algorithms establish a grading (merit) system for all feasible allocations for each job to determine the order of scheduling. In the literature, a number of contiguous submesh allocation schemes have been proposed on mesh connected multicomputers to provide a base for such a merit calculation. These include 2D Buddy strategy proposed by Li and Cheng [5], Frame Sliding(FS) scheme by Chuang and Tzeng [2], First Fit and the Best Fit strategies by Zhu [9], Adaptive Scan strategy by Ding and Bhuyan [3], boundary search approach by Sharma and Pradhan [6] and Quick Allocation strategy by Seong-Moo Yoo [8], etc. Scheduling cost for all these techniques can be too excessive to be considered practical in some cases.

Among all non-blocking techniques known for contiguous processor allocation case, Largest-Job-First (LJF) technique proves to be one of the simplest and yet outperforms others such as FCFS, SJF and even some complicated ones due to its nature in leading to smaller mesh fragmentation problem. As mentioned earlier, allowing jobs to bypass others that arrive earlier may lead to a starvation problem, and such a problem usually is more prominent in non-blocking techniques than blocking ones. A job-bypass limit that is preset to preclude starvation problem in a non-blocking platform would lead to an undesirable effect to all techniques. The scheduling process becomes a "blocking FCFS" whenever a job reaches this bypass limit and thus has to be scheduled for allocation in the next immediate turn. The more such a "mandatory blocking" situation occurs, the more degradation on its performance follows. Such an adverse effect is especially pronounced in the non-blocking LJF compared to the non-blocking FCFS due to fact that more bypassing happens in an LJF scheme. In order to relieve such an effect so as to improve overall latency, this paper proposes a novel scheduling technique based on the LJF scheme. Instead of strictly using the job size for ordering criterion, a "weighing factor" is introduced to reflect how close a job is to the bypass limit. Such a weighing factor is dynamically calculated for each job in the queue and is then used to modify the order. This technique ensures that the advantage of the LJF is preserved while significantly suppressing the chance for the blocking situation to happen. In our simulation results, we show that frequency in leading to "blocking" scenario is significantly reduced and thus a much improved latency ensues.

2   Proposed Scheduling Technique

All scheduling techniques have to be coupled with an allocation scheme to decide how a job scheduled to be allocated is allocatable and where to allocate in the mesh. All techniques, including the proposed one, in this paper are based on the same allocation method, the Boundary Search Method (BSM), proposed in [6]. The boundary search method allocates a free submesh with the largest number of busy nodes adjacent to it by checking through all candidate submeshes. The heuristic is based on the notion that if busy submeshes are adjoined to one another as much as possible, fragmentation problem in mesh would lessen. Among all the research works targeted in producing a fast allocating algorithm and at the same time to avoid internal as well as external fragmentation, the BSM consistently shows a performance better than most existing allocation approaches.

As aforementioned, among all non-blocking scheduling techniques for contiguous processor allocation, LJF has proven to be very meritorious one. The reason that an LJS technique exhibits better performance than FCFS and even other more complicated non-blocking technique is that, by having the largest job allocated first, fragmentation problem in the mesh tends to minimize and thus leads to less blocking situation. To ensure fairness, such a bypass needs to be curtailed to prevent a starvation problem from happening, a problem that may leave a small job never allocated. Usually a bypass limit (bpl) is set such that no more than bpl jobs are allowed to bypass any job. Once a job (or jobs) in the queue reaches such a limit, scheduling has to follow strictly **blocking FCFS** until all jobs' bypass values (bp) again fall below this limit. When such a "mandatory blocking" happens, all benefits from the scheduling "flexibility" in the non-blocking technique disappears, and thus longer waiting delay ensues. The more such a mandatory blocking occurs, the more adverse effect on the average waiting latency it leads to. With the bypass limit imposed, compared to FCFS, LJF has a higher tendency in leading to such a blocking scenario due to its nature in scheduling. The gain in performance from reducing fragmentation problem by LJF may easily be offset by this adverse effect from blockings. The tighter (smaller) the bypass limit is imposed, the less flexible the LJF becomes.

Our proposed technique is to be based on the LJF one, and has an added mechanism that allows the scheduler to dynamically adjust each job's priority ("job size" in the LJF case) according to its current bp value. That is, the merit criterion which is the "job size" used in a typical LJF technique is to be modified according to the job's current bp value. A general idea is to increase a job's merit according to how close its bp value is to the limit bpl, so as to diminish the chance of it reaching the limit.

Let $S(x)$ be the job size of job x, i.e. the number of processor requested by x. Also, let $bp(x)$ denote the current bypass value of x, i.e., the number of jobs that have bypassed x. Thus, the ratio $\frac{bp(x)}{bpl}$ becomes the primary weighing factor, and $0 \leq \frac{bp(x)}{bpl} \leq 1$. The following formula is used to calculate a new merit criterion, the $WS(x)$ ("Weighed job Size") for x.

$$WS(x) = S(x) \cdot [1 + M \cdot (\frac{bp(x)}{bpl})^D] \tag{1}$$

The term $[1 + M \cdot (\frac{bp(x)}{bpl})^D]$ represents the newly added weighing factor, where two following parameters are used to vary the effect of this factor in different ways:

- D: the damping factor that controls how badly the "closeness" of a bypass value to the bypass limit means. Since $0 \leq \frac{bp(x)}{bpl} \leq 1$, the higher the D is, the more tolerance for such "closeness" is imposed.
- M: the multiplying factor that changes the overall effect of the weighing compared to the original merit $S(x)$ in an additional linear multiplying fashion.

Different settings for the two parameters D and M give various degrees of "urgency" for a job that has its bypass value close to the limit. A higher M value provides a smaller job with a higher bypass value a better chance to be scheduled earlier than a larger one, while a higher D value sets a higher threshold for this to happen. An example in Fig. 1 shows how the scheduling order would change when different values are employed for these two parameters. Jobs shown in the queue are listed according to their arriving order With a higher M value, more weight is given to jobs to the front of the queue where bypass values are larger. This is clearly demonstrated by the trend from $M = 0$ to $M = 1$ (and $D = 1$) and then to $M = 2$ (and $D = 1$) that the job assigned with the first priority gradually moves to the front of the queue. On the other hand, with M being fixed, such a trend is attenuated by higher value of the damping parameter (D). This is clearly shown in the cases where the priority is gradually shifted back toward the largest job.

The proposed weighing factor is employed to prevent blocking from happening excessively by giving a smaller job with a higher bypass value a chance to be assigned a higher scheduling priority than a larger job depending how critical its bypass value is approaching the limit. Note that the scheduling order needs to be re-calculated whenever one the following situations occurs:

- the currently selected job is allocated successfully
- a new job enters into the queue

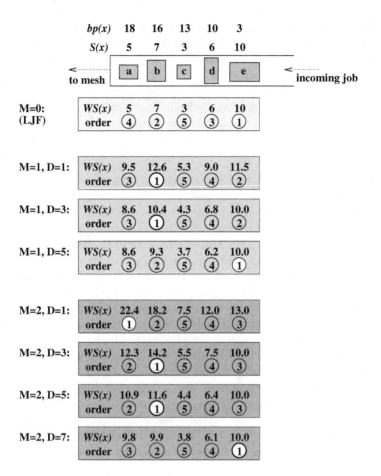

Fig. 1. Scheduling Example with Different Values for M and D Employed with $bpl = 20$

3   Simulation Results

In our simulation runs, performance of the proposed technique is compared with the non-blocking FCFS and LJF algorithms. All scheduling techniques tested are paired with the same BSM method proposed in [6] as the allocation policy, as mentioned in Section 2.1.

The mesh in our simulation is of size 200×200. Assumptions and parameters for our simulations are described as follows:

– Dimension of the submesh $m \times n$ requested by an incoming task is generated such that both m and n are uniformly distributed over $(1, 100)$.
– Service time of incoming tasks is normally distributed with a mean of μ and a standard deviation of σ.
– Inter-task arrival time is exponentially distributed with an average interarrival time of $1/\lambda$.

- 50,000 submesh requests (tasks) are simulated in each simulation run to ensure equilibrium state is reached, if possible. This is needed to make sure whether the task interarrival time $(1/\lambda)$ is too small for the task queue to maintain in an equilibrium state.
- A bypass limit (bpl) value is to denote the maximal number of tasks that are allowed to bypass any given task in the queue.
- The "Weighed Job Size" in the proposed enhanced technique is calculated based of a selection of $M = 2$ (multiplying factor) and $D = 5$ (damping factor), which combination shows an overall superior performance than others.

There are several metrics that can be used to gauge the performance of a job scheduling algorithm. The most commonly used one is the *average job waiting delay (latency)* which is the mean time for jobs waiting in the queue, and thus is used in this paper for performance comparison. Figure 2 presents the comparison results of latency among the three techniques. The proposed enhancement further improves on the LJF and continues to retain its edge over the FCFS even under a small bypass limit. When the bypass limit is set at a large value (e.g. $bpl = 30$), the proposed technique further improves on the already superior LJF technique. When the bypass limit is set at a small value (e.g. $bpl = 5$), the proposed technique prevents excessive blocking from happening (as in LJF). This ensures that, even when the task arrival rate is high, the latency is at least comparable to the superior FCFS technique. The reason for such an improvement is further illustrated using a blocking ratio as shown in Fig. 3. Let *PAMB* (Per-Allocation Mandatory Blocking) denote the ratio between the number of allocated jobs that are allocated when they have reached the bypass limit and the total number of jobs allocated. Under $bpl = 5$, although the proposed technique has a higher *PAMB* ratio than the FCFS, its intrinsic flexibility from LJF still allows it to outperform FCFS in terms of latency. Overall, the enhancement

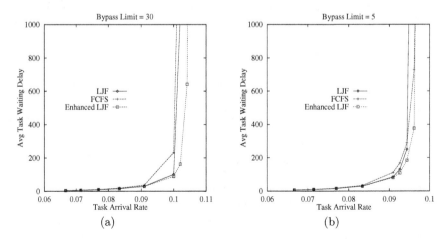

Fig. 2. Average Job Latency vs. $1/\lambda$ with $\mu = 100$, $\sigma = 25$ and $bpl = 30$ in (a) and $bpl = 5$ in (b)

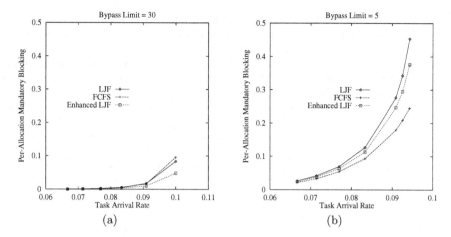

Fig. 3. Mandatory Blocking vs. $1/\lambda$ with $\mu = 100$, $\sigma = 25$ and $bpl = 30$ in (a) and $bpl = 5$ in (b)

introduced significantly reduces the number of mandatory blocking situation in LJF, which in turn leads to its sustained edge in performance.

4 Conclusion

In this paper, we propose a non-blocking job scheduling technique based on a usually outstanding technique, LJF. An extra mechanism is incorporated into the LJF to prevent excessive blocking from happening when a small bypass limit is imposed. Simulation results clearly demonstrate the efficiency of such an addition. An interesting research direction that may be extended from this research work is to have the weighing process self-adaptive to the incoming traffic, which would require a more complicated feedback adaptive control mechanism.

References

1. C.-Y. Chang and P. Mohapatra, "An integrated processor management schema for the mesh-connected multicomputer systems", *Proc. of the 1997 international conference on parallel processing*, pp. 118-121, 1997.
2. P.J. Chuang and N.F. Tzeng, "An efficient submesh allocation strategy for mesh computer systems", *Proc. Int'l Conf. Distributed Computing Systems*, pp. 256-263, Aug.1991.
3. J. Ding and L.N. Bhuyan, "An adaptive submesh allocation strategy for two-dimensional mesh connected systems", *Proc. Int'l Conf. Parallel Processing*, pp. II-192-200, Aug. 1993.
4. D. Min and M.W. Mutka, "Efficient job scheduling in a mesh multicomputer without discrimination against large jobs", *Proc. Seventh IEEE Symposium on Parallel and Distributed Processing*, pp. 52-59, 1995.

5. K. Li and K.H. Cheng, "A two-dimensional buddy system for dynamic resource allocation in a partitionable mesh connected system", *J. Parallel and Distributed Computing*, Vol. 12, pp. 79-83, May 1991.
6. D.D. Sharma and D.K. Pradhan, "A fast and efficient strategy for submesh allocation in mesh-connected parallel computers", *Proc.Symp. Parallel and Distributed Processing*, pp. 682-689, Dec. 1993.
7. D.D. Sharma and D.K. Pradhan, "Job scheduling in mesh multicomputers", *IEEE transactions on parallel and distributed systems*, Vol. 9, No. 1, Jan. 1998.
8. S.M. Yoo, etc, "An efficient task allocation scheme for 2D mesh architectures", *IEEE Transactions on Parallel and Distributed Systems*, Vol. 8, No. 9, Sept. 1997.
9. Y. Zhu, "Efficient processor allocation strategies for mesh-connected parallel computers", *Journal of Parallel and Distributed Computing*, Vol. 16, pp. 328-337, Dec. 1992.

Broadcast in a MANET
Based on the Beneficial Area*

Gui Xie and Hong Shen

Graduate School of Information Science,
Japan Advanced Institute of Science and Technology, JAIST,
Tatsunokuchi, Ishikawa 923-1292, Japan,
{g-xie,shen}@jaist.ac.jp

Abstract. Broadcasting is a common operation in mobile ad hoc networks (MANET). A straightforward approach to perform broadcast in such a mobile environment is by flooding, which will result in the so-called broadcast storm problem with serious redundant rebroadcasts, contentions and collisions. Many methods have been proposed to solve this problem, whose theoretical foundation is to prohibit the rebroadcast when the non-redundant transmission area, namely beneficial area, is too small. It's an open problem to compute the beneficial area for a receiving host when the number of the sending hosts reaches 3 or more. In this paper, we identify this problem by showing how rapidly the complexity of computing the beneficial area rises with respect to the number of sending hosts when a broadcast task is being performed. We propose an efficient approximation method to compute the beneficial area by recursively filling a grid matrix, which can approach the actual one at an arbitrary accuracy. Moreover, we statistically estimate the exponential relationship between the beneficial area and the number of sending hosts, from which we can assert with high possibility that the beneficial area would decrease to zero when the number of the sending hosts reaches six or more.

1   Introduction

MANET is a mobile ad hoc network, which consists of a set of mobile hosts that may communicate with one another and roam around at their will. The network is decentralized, where no administration devices or base stations are supported. In such an ad hoc network, each node operates not only as a host but also as a router. The applications of MANET's range from military use in battlefields, personnel coordinate tools in emergency disaster relief, to interactive conferences temporarily formed using handheld devices.

Broadcasting to all nodes is a common operation in MANET, which has extensive applications, such as route query [1–3] error routes erasure [4], multicast in fast moving ad hoc networks [5] and so on [6]. In a MANET, when a mobile

---

* This work is supported by Japan Society for the Promotion of Science (JSPS) Grant-in-Aid for Scientific Research (B) under Grant No.14380139.

A. Veidenbaum et al. (Eds.): ISHPC 2003, LNCS 2858, pp. 360–367, 2003.

host sends a packet, all of its neighbors will receive the packet. So when a source node in a MANET broadcasts a package, a flood tree will be formed by the source node and the other forward nodes for broadcasting that package. Blind flooding is a straightforward approach for broadcasting in a MANET. In blind flooding, a host in a wireless ad hoc network, on receiving a broadcast package for the first time, has the obligation to rebroadcast the package.

Many different algorithms for broadcasting in MANETs have been developed, which can be classified into three categories [7]: probability-based methods, neighborhood-knowledge-based methods and beneficial-area-based methods. In probabilistic scheme, nodes rebroadcast with a predetermined probability. When the probability is 100%, this scheme is identical to blind flooding. In neighborhood- knowledge-based methods, the node utilizes the information of its neighbors (one-hop or two-hop) to reduce the redundancy of broadcasting by flooding. For example, Flooding with self-pruning [8] requires that each node uses the knowledge of its 1-hop neighbors to rebroadcast the messages, and the dominating pruning (DP) algorithm [9] uses 2-hop neighbors information to do the job. In beneficial-area-based scheme, a node can evaluate beneficial area using all the received redundant transmissions. A receiving host can rebroadcast the received packet in a non-empty beneficial area without introducing any redundant transmission. So the beneficial area is a good quantity to determine whether the receiving host should continue to rebroadcast the received packet. If the beneficial area were too small, the rebroadcasting process on the host would be dropped. Several methods based on this idea have been proposed [7], such as distance-based scheme, counter-based scheme, location based scheme, and cluster-based scheme.

Finding a mathematical formula to compute the beneficial area for a receiving host directly is an open problem. As the number of sending hosts increases over 3, the complexity of computing the beneficial area would become so intolerable that we could assert that it's impossible to get a formula to compute it. So we have to find an efficient approximation method to compute it. In [7], a convex polygon test scheme was proposed to approximately compute the beneficial area, which allows the host to rebroadcast only if it is not located in the convex polygon formed by the locations of previously sending hosts. This method computes the beneficial area so coarsely that the polygon test could permit a host to rebroadcast even when the actual beneficial area approaches zero.

The rest of this paper is organized as follows. Section 2 defines the beneficial area and gives the main principle of the beneficial area based scheme for broadcasting in MANETs. In Section 3, we compute the beneficial area directly by the locations of sending hosts. Also presented in this section are an efficient approximation method to compute the beneficial area by grid matrix filling and a statistical exponential relationship between the beneficial area and the number of sending hosts. We conclude the paper in the last section.

2 Preliminaries

For a receiving host in a MANET, when it receives the same rebroadcast packet for k ($k \geq 0$) times from k sending hosts, the beneficial area for that receiving host, denoted by B_k, is defined to be the area within the range of the receiving host's transmission that isn't covered by the total range of the k senders' transmissions. As depicted in Fig. 1, the shaded area is the beneficial area for the receiving host h_r when it has received a rebroadcast packet for k times from the k sending hosts h_i ($i = 1, 2, ..., k$). A host can benefit by rebroadcasting the received packet in a non-empty beneficial area as such rebroadcasting will not introduce redundant transmission. Let the transmission range of each sender be a disk with radius 1 and S_i be the area of the disk space of the ith sending host h_i located at coordinate (x_i, y_i), we have

$$B_k = S_r - S_r \cap \left(\bigcup_{i=1}^{k} S_i \right) \tag{1}$$

where S_r is the area of the receiving host h_r's disk space and S_i denotes the area of host h_i's disk space.

Obviously, $B_k \in [0, \pi]$ and the beneficial area is dependant on the number and locations of the senders. It is more likely for the beneficial area to decrease when there are more sending hosts which send the rebroadcast packets to the receiving host. The distribution of the senders with respect to the receiver's position also affects the value of the beneficial area. When the receiver hasn't received any rebroadcast packets ($k = 0$), the value of the beneficial area reaches the maximum π, which is the area of the total disk space of the receiving host's transmission. When the receiver has received a rebroadcast packet for k times from k senders ($k \geq 1$), the worst case of the beneficial area will occur after the entire range of the receiver's transmission is covered by the senders' transmission range. The value of the beneficial area in this case would be zero. The best case of the beneficial area for $k \geq 1$ will occur when all the senders are located at the same position of the edge of the receiver's transmission range, which is equivalent to the situation of only one sending host because of the senders' superposition. This kind of case will be discussed in detail in the next section.

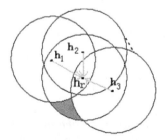

Fig. 1. The beneficial area that can be covered from a rebroadcast when host h_r rebroadcasts the received packets from k sending hosts

There is no redundant transmission inside beneficial area. So the beneficial area is an appropriate termination criterion for rebroadcasting on a receiving host. For conceptual simplicity, we usually normalize it as follows

$$\tilde{B}_k = \frac{B_k}{\pi} \tag{2}$$

where $\tilde{B}_k \in [0, 1]$.

We can compute the normalized beneficial area when the receiver has received the same rebroadcast packet k times from k sending hosts. Then we compare the value of the normalized beneficial area with a predetermined threshold $T(0 \leq T \leq 1)$ to determine whether the receiving host should rebroadcast or not. This beneficial-area-based broadcast scheme is presented in detail in [7]. Obviously, computing BR is a key process in the above scheme, which will be discussed in detail in the paper. Another important quantity in the above scheme is the threshold of the beneficial area. This scheme in [7] chooses a static predetermined threshold, but we can also use a dynamic threshold, which is adaptive to the status of the receiving host and the environment around it.

3 Beneficial Area Computation

Computing the beneficial area is to compute the intersections among multiple disks of the receiving and sending hosts, which is too difficult to have a closed mathematical formula when the number of disks is greater than three. We first derive the closed formula to compute the beneficial area when the number of disks are only two and then present an approximation method to compute it.

3.1 One Sending Host

Consider the simplest scenario with two disks when $k = 1$ as depicted in Fig. 3, where the sending host h_1 broadcasts a packet to the receiving host h_r and host h_r decides to rebroadcast the packet.

Without loss of generality, let the location of host h_r be the origin, and the line across the centers of disks h_r and h_1 be the abscissa axis. Suppose the coordinate of host h_1 is $(x_1, 0)$ and both of the disks have the same radius unit one. In such a setting, the shaded region illustrated in Fig. 2 is the beneficial area for the receiving host h_r.

Thus, the area of this shaded region can be computed as follows:

$$B_1 = \pi - 4 \int_{x_1/2}^{1} \sqrt{1 - x^2} \, dx \tag{3}$$

where the first term is the total area of one disk space and the second term is the intersection area between the two disks. Coordinate x_1 is in the range $[0, 1]$ because the sending host h_1 is located inside the disk of the host h_r. Obviously, when $x_1 = 1$, B_1 reaches the maximum value of $(\frac{\pi}{3} + \frac{\sqrt{3}}{2})$, which is about 61%

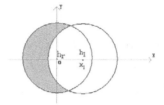

Fig. 2. Beneficial area between two hosts

of the total circular area of host h_r's disk space. The value of B_1 decreases as host h_1 approaches host h_r. If the two hosts are located at the same position, the beneficial area becomes empty.

3.2   More than One Sending Hosts

We increase the number of disks to three ($k = 2$) by adding another sending host into Fig. 2. Computing the beneficial area for a receiving host by a mathematical formula is so complex, even though there are only three hosts ($k = 2$). It is hence very difficulty to find a closed formula to compute it. So we have to find an approximation method to express the beneficial area when there area more than 3 hosts.

Suppose the receiving host h_r receives the same rebroadcast packet from k ($k \geq 1$) sending hosts $h_1, h_2, ..., h_k$, whose reordered version with respect to their arrival time is $h_1^*, h_2^*, ..., h_k^*$, where host $h_i^* \in \{h_i | 1 \leq i \leq k\}$ and the packet from h_i^* arrives at host h_r before that from h_{i-1}^*. Let B_i^* be the beneficial area for the receiving host h_r when it has received the ith rebroadcast packet from the sending host h_i^* ($i = 1, 2, ..., k$). The beneficial area B_k equals B_k^* after host h_r receives all the rebroadcast packets from the k senders. Let B_0^* be the initial beneficial area before host h_r starts to receive the rebroadcast packets. Obviously, B_i^* ($i = 0, 1, ..., k$) is a non-increasing sequence, which means $B_{i-1}^* \geq B_i^*$. Let T denote the operator that compute B_i^* when we got the previous beneficial area B_{i-1}^*. As we know the value of B_0^* is π if we let the radius of disks be unit one, we can compute the beneficial area recursively as follows

$$B_0^* = \pi; B_i^* = T(B_{i-1}^*) \tag{4}$$

where $i = 1, 2, ..., k$.

Here, we give an approximation method to do the job of T. As shown in the Fig. 3, we fill the disk space of the receiving host h_r by grids. We can partition the horizontal and vertical diameters into $2N$ segments with the same width $\Delta = \frac{1}{N}$, respectively. We get $4N^2$ grids, which covers the range of the host h_r's transmission. Then, we create a 2-D matrix $M[-N : N, -N : N]$, whose element $M[i, j]$ corresponds to the cell $[i, j]$. Apparently, the center of the cell $[i, j]$ locates at $(\frac{(2i-1)}{2}\Delta, \frac{(2j-1)}{2}\Delta)$ in the given system of coordinates depicted in Fig. 3. If the center of the cell $[i, j]$ is located inside the disk h_r's space, we set

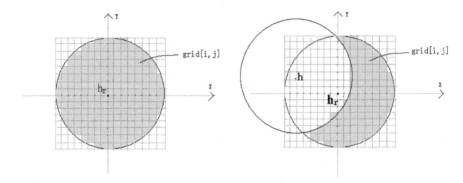

Fig. 3. Approximation the disk   **Fig. 4.** Approximation the beneficial area

its corresponding $M[i, j]$ to be one, otherwise to be zero, after which we get an initial matrix whose all non-zero elements can be thought as an approximation to the initial beneficial area B_0^* for the receiving host h_r.

Just as Eq. (4) does, by using the grid matrix, we can recursively compute the beneficial area. As depicted in Fig. 4, after the host h_r received a broadcast packet from a host h, we set all the values of the $M[i, j]$ to be zero, whose correspondent grids $[i, j]$ are covered by the disk space of host h. Repeat the process recursively until host h_r receives all the broadcast packets from the senders. Then count the non-zero elements in the matrix M. Suppose the number is m. The beneficial area for host h_r after receiving the rebroadcast packets from k sending hosts is approximately computed by

$$B_k = B_k^* = m\Delta^2 = \frac{m}{N^2} \tag{5}$$

Certainly, we can approach the actual beneficial area as closely as possible by increasing N to infinity. In a wireless environment, we have to make the tradeoff between accuracy and computation complexity with respect to N. If we increase N, the approximate value of the beneficial area approaches more closely to the actual one, but the need for computation resources such as time and storage would also increase.

Next, we estimate the relationship between the beneficial area and the number of sending hosts. Let the k senders uniformly distribute in the disk space of the receiver h_r. We then use a random number generator to generate the uniformly distributed locations of the k senders and compute the beneficial areas by the above approximation method. Let the radius be one unit. Through extensive experimentation, we find that the beneficial area has an exponential relationship with the number of hosts as shown in the following expression:

$$\frac{B(N)}{\pi} = \alpha_1 e^{\beta_1 N} + \alpha_2 e^{\beta_2 N} \tag{6}$$

where $N = k + 1$.

By nonlinear curve fitting, we can estimate the parameters of the above expression approximately to get a relationship equation as follows

$$B(N) \approx \pi(1.2027e^{-0.8744N} + 1.1078e^{-0.7403N}) \tag{7}$$

The next figure illustrates the relationship we've got. In the figure, the bold solid curve corresponds to the Eq. (7) and the other dashed curves are the results of three separate experiments using the presented approximation method. You can see, the beneficial area decreases so rapidly that it is approximately empty after the number of the sending hosts reaches six or more.

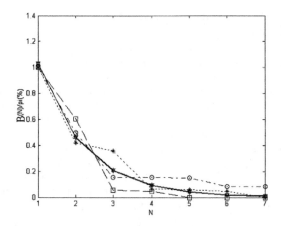

Fig. 5. The beneficial area's exponential decreasing trend

4   Conclusion

This paper addressed an important problem of computing the beneficial area for rebroadcast to prevent from the broadcast storm. We established mathematical expressions for the beneficial area when the number of hosts in a MANET is not greater than two. We showed that the computational complexity rises with the number of the hosts, and presented a grid matrix filling scheme to compute the beneficial area approximately, when there are more than three hosts in the MANET. Our approximation scheme can approach the actual value of the beneficial area at an arbitrary accuracy. Through analysis and simulations, we found an exponential relationship between the beneficial area and the number of hosts. Our simulation results assert with high possibility that the beneficial area would approach zero when the number of hosts reaches seven or more. When we want to use the beneficial area to determine whether the host would rebroadcast the received signal, we can compute it with the proposed new approximation method or compute it with the exponential relationship directly. Moreover, in the beneficial-area-based scheme, we can use not only a static predetermined threshold to prevent from the broadcast storm, but also a dynamic one, which is adaptive to the status of the receiver and the environment around it. Future research could be on how to compute the dynamic threshold of the beneficial area and incorporate these beneficial-area-based schemes with other MANET protocols, such as reliable broadcast, multicast, and routing protocols. Analyzing the relationship between the beneficial area and the number of the hosts in other distributions other than the uniform one used in our paper, for example Gassian, Laplacian, and etc., is also an important research topic in the future.

References

[1] D.B. Johnson and D.A. Maltz, *Mobile Computing*, chapter "Dynamic Source Routing in Ad-Hoc Wireless Networks," pp. 153-181, Kluwer Academic, 1996
[2] M.R. Perlman and Z.J. Haas, "Determining the Optimal Configuration of the Zone Routing Protocol," *IEEE J. Selected Areas in Comm.*, vol. 17, no. 8, pp. 1394-1414, Feb. 1999
[3] C. Perkins and E.M. Royer, "Ad-Hoc On-Demand Distance Vector Routing," *Proc. Second IEEE Workshop Mobile Computing Systems and Applications (WMCSA)*, pp. 90-100, Feb. 1999.
[4] V.D. Park and M.S. Corson, "Temporary-Ordered Routing Algorithm (TORA) Version 1: Functional Specification," *Internet Draft*, 1997.
[5] C. Ho, K. Obraczka, G. Tsudik, and K. Viswanath, "Flooding for Reliable Multicast in Multihop Ad Hoc Networks," *Proc. ACM Int'l Workshop Discrete Algorithms and Methods for Mobile Computing'99*, pp. 64-71, Aug. 1999.
[6] Wei Lou and Jie Wu, "On Reducing Broadcast Redundancy in Ad Hoc Wireless Networks," *IEEE Transaction On Mobile Computing*, vol. 1, no.2, April-June 2002.
[7] Sze-Yao Ni, Yu-Chee Tseng, Yuh-Shyan Chen, and Jang-Ping Sheu, "The Broadcast Storm Problem in A Mobile Ad Hoc Network," *In Proceedings of The Fifth Annual ACM/IEEE International Conference on Mobile Computing and Networking*, Aug. 1999.
[8] Brad Willianms and Tracy Camp, "Comparison of Broadcasting Techniques for Mobile Ad Hoc Networks," *In Proceedings of The Third ACM International Symposium on Mobile Ad Hoc Networking and Computing (MOBIHOC 2002)*.
[9] H. Lim and C. Kim, "Multicast tree construction and flooding in wireless ad hoc networks," *In Proceedings of the ACM International Workshop on Modeling, Analysis and Simulation of Wireless and Mobile Systems (MSWIM)*, 2000.

An Optimal Method for Coordinated En-route Web Object Caching*

Keqiu Li and Hong Shen

Graduate School of Information Science,
Japan Advanced Institute of Science and Technology,
1-1 Tatsunokuchi, Ishikawa, 923-1292, Japan,
keqiu,shen@jaist.ac.jp

Abstract. Web caching is an important technology for saving network bandwidth, alleviating server load, and reducing the response time experienced by users. In this paper, we consider the object caching problem of determining the optimal number of copies of an object to be placed among en-route caches on the access path from the user to the content server and their locations such that the overall net cost saving is maximized. We propose an object caching model for the case that the network topology is a tree. The existing method can be viewed as a special case of ours since it considers only linear topology. We formulate our problem as an optimization problem and the optimal placement is obtained by applying our dynamic programming-based algorithm. We also present some analysis of our algorithm, which shows that our solution is globally optimal. Finally, we describe some numerical experiments to show how optimal placement can be determined by our algorithm.

Key words: Web caching, optimization problem, dynamic programming, algorithm.

1   Introduction

Web caching is an important technology for saving network bandwidth, alleviating server load, and reducing the response time experienced by users. An overview on web caching can be found in [1, 9]. Generally, there are three architectures for web caching: hierarchical caching [3, 4], distributed caching [4, 8], and en-route caching [2, 7]. In this paper, we focus on en-route caching in which caches are placed in the intermediate nodes on the access path from the user to the content server, and are referred to as en-route caches. When a client request passes through an en-route cache, the object will be sent to the client if there is a copy of that object stored in the cache. Otherwise, the request will be forwarded upstream until it can be satisfied. Since all requests are routed on the path from the user to the content server, en-route caching can reduce the additional

* This work is supported by Japan Society for the Promotion of Science (JSPS) Grant-in-Aid for Scientific Research under its General Research Scheme (**B**) **Grant No. 14380139.**

A. Veidenbaum et al. (Eds.): ISHPC 2003, LNCS 2858, pp. 368–375, 2003.

network bandwidth consumption and network delay necessary for cache misses. The performance of en-route caching mainly depends on two factors: where the en-route caches are placed and how the objects are cached.

Cache cooperation is an important approach for improving web performance. Cooperation among the caches can be performed in two dimensions: horizontal and vertical. For horizontal cooperation, caches are placed around the content server and have similar distances to the content server[5]. For vertical cooperation, caches are geographically located throughout the network and have different distances to the content server [10]. The cooperation between parent and child proxies for hierarchical caching was discussed, but there was no analytical modelling [10]. Our model for object caching is applicable to both of the cases.

Determining the appropriate number of copies of an object and placing them in suitable en-route caches are challenging tasks. The interaction effect between object placement and replacement further complicates the problem. Most of the existing work was concentrated on either object placement or replacement at individual caches only and are not optimal [7]. Little work has been done on web object caching[6]. In this paper, we consider the coordinated en-route object caching problem of determining the optimal number of copies of an object to be placed among the en-route caches on the access path from the user to the content server and their locations such that the overall net cost saving is maximized. We propose an object caching model for the case that the network topology is a tree. In [7], an optimization problem is presented for linear topology, and a dynamic programming-based algorithm is proposed to solve the optimization problem; thus, the problem in [7] can be viewed as a special case of ours. We formulate our problem as an optimization problem and the optimal placement is obtained by applying our dynamic programming-based algorithm. We also present some analysis of our algorithm, which shows that our solution is globally optimal.

The rest of the paper is organized as follows. Section 2 describes an object caching model and formulates our problem. Section 3 presents our dynamic programming-based algorithm and some analysis of our algorithm. Section 4 summarizes our work.

2 Problem Formulation

In this section we propose the object caching model and formulate our problem. The network we use in this paper is modelled by a tree $T_w = (V, E)$, where $V = \{v_i, i = 1, 2, \cdots, n\}$ is the set of routers, each of which is associated with an en-route cache, $E = \{e_i, i = 1, 2, \cdots, m\}$ is the set of network links, and $w \in V$ is the root of the tree. Without loss of generality, we assume that there is only one content server at the root and clients' requests can be eventually satisfied by this server. A client's request goes along the path from the client to the server and is satisfied by the first node on the path to the server that has a copy of the requested object. The response is transmitted to the client along the same path. For simplicity, we assume that the links are bi-directional. An example of such a tree topology is shown in Figure 1. Our model can be easily

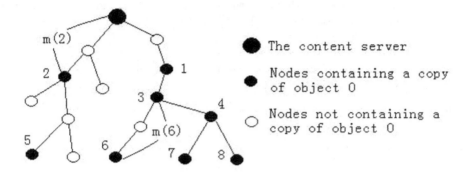

Fig. 1. En-Route Web Caching

extended to the case in which en-routes are associated with a subset of nodes only if we include the nodes in the graph with en-route caches. For object O, we associate every edge $(u, v) \in E$ with a nonnegative cost $c(u, v)$, which can be network latency, bandwidth consumption, processing cost at the cache, or some combination of these measures. If a request goes through multiple edges, the total cost is considered as the summation of the cost on each edge.

The object caching problem has a trivial solution if the cache sizes are infinite. Objects can be stored in every cache to minimize the total access cost. However, due to the limited cache space, one or more objects may need to be removed from the cache when a new object is inserted. Removing an object from a node makes all the nodes previously accessed it now have to go to another node holding the object at a higher level, hence, increases its access cost, which is referred to as cost loss in this paper. Likewise, storing a copy of an object at a node decreases its access cost, which is referred to as cost saving. In this paper, we consider both cost saving and cost loss.

Before presenting our optimization problem, we give some notations and definitions. Let $P \subseteq V$ be a subset of nodes, at each of which a copy of object O is placed. For every node $v \in V$, $D(v)$ denotes the set of all nodes that are the descendants of node v, and $C(v)$ denotes the set of all nodes that are the children of node v. For any two nodes $u, v \in V$, $E[u \to v]$ denotes the set of all edges on the path between u and v and $V[u \to v]$ denotes the set of all nodes on the path including u and v. Let $f(v)$ denote the access frequency of object O defined by the number of requests to access for object O (including the requests initiated at node v itself), obviously $f(v) \geq \sum_{w \in C(v)} f(w)$. Let $m(v)$ be the miss penalty of object O with respect to node v, which is given by $m(v) = \sum_{(u_1, u_2) \in E[v \to v']} c(u_1, u_2)$, where v' is the nearest higher level node of v that stores a copy of object O (see Figure 1). Therefore, the cost saving for node $v \in P$ denoted by $s(v)$ is defined as $s(v) = \left(f(v) - f'(v) \right) m(v)$, where $f'(v)$ is the total access frequency of object O that can still be satisfied by the original caches that belong to downstream of node v if the copy of object O at node v is removed. For example, in Figure 1, $f'(1) = f(3)$, $f'(2) = f(4)$, $f'(3) = f(4) + f(5)$, $f'(8) = 0$. Let $l(v)$ be the cost loss for storing a copy of object O at node v. Thus, the net cost saving for a single node v denoted by $g(v)$ is defined as $g(v) = s(v) - l(v) = \left(f(v) - f'(v) \right) m(v) - l(v)$. Based on Equation (**??**), we can present the object caching problem as an optimization problem as follows:

$$\max_{P} G(T, P) = \max_{P} \left\{ \sum_{v \in P} \left[\left(f(v) - f'(v) \right) m(v) - l(v) \right] \right\} \qquad (1)$$

where $G(T, P)$ is the overall net cost saving for placing copies of an object at all the nodes in P. Therefore, our objective is to find the optimal placement P^* that maximizes $G(T, P)$.

3   Dynamic Programming-Based Algorithm

In this section, we first present our dynamic programming-based algorithm by which we can obtain the optimal solution to our problem, and then give some analysis of our algorithm.

3.1   Algorithm Realization

Suppose that $A_w \subseteq D(w)$ is a subset of nodes, by Equation (1), our problem is represented by the following equation:

$$\max_{A_w} G(T_w, A_w) = \max_{A_w} \left\{ \sum_{v \in A_w} \left[\left(f(v) - f'(v) \right) m(v) - l(v) \right] \right\} \qquad (2)$$

So our objective is to find such a subset of nodes, A_w, that maximizes $G(T_w, A_w)$. Suppose that A_w^* is the optimal solution to equation (2), which means we should store a copy of an object on each node in A_w^*, and $G(T_w, A_w^*)$ represents the maximum overall net cost saving for our object caching problem. In this paper, we define A_w^* as the optimal placement with respect to tree T_w

Let $T_{w,x}$ be a subtree of T_w, whose root is w, at which a copy of an object is stored and whose node net is $V[w \to x] \cup D(x)$, where $x \in D(w)$. Similarly, we denote the relevant optimal solution by $A_{w,x}^*$, which is the optimal subset of nodes, at each of which a copy of an object should be placed, and $G(T_{w,x}, A_{w,x}^*)$ means the maximal overall net cost saving for the object caching problem with respect to tree $T_{w,x}$. In this paper, we define $A_{w,x}^*$ as the optimal placement with respect to tree $T_{w,x}$

Before presenting our dynamic programming-based algorithm, we give the following theorems.

Lemma 1. *For T_w, if $C(w) = \{w_1, w_2, \cdots, w_m\}$, then we have $G(T_w, \cup_{i=1}^m A_{w,w_i}) = \sum_{i=1}^m G(T_{w,w_i}, A_{w,w_i})$, where $A_{w,w_i} \subseteq D(w_i) \cup \{w_i\}$, $i = 1, 2, \cdots, m$.*

Proof. Since $A_{w,w_i} \subseteq D(w_i) \cup \{w_i\}$, we have $\cup_{i=1}^m A_{w,w_i} \subseteq \cup_{i=1}^m (D(w_i) \cup \{w_i\}) = D(w)$. Since $A_{w,w_i} \cap A_{w,w_j} = \phi$ for $i \neq j$, by the definition of $G(T_w, A_w)$, we have

$$G(T_w, \cup_{i=1}^m A_{w,w_i}) = \sum_{v \in \cup_{i=1}^m A_{w,w_i}} \left[\left(f(v) - f'(v) \right) m(v) - l(v) \right]$$

$$= \sum_{i=1}^m \sum_{v \in A_{w,w_i}} \left[\left(f(v) - f'(v) \right) m(v) - l(v) \right]$$

$$= \sum_{i=1}^m G(T_{w,w_i}, A_{w,w_i}).$$

Theorem 1. *For a tree T_w, if $C(w) = \{w_1, w_2, \cdots, w_m\}$, then we have $A_w^* = \cup_{i=1}^m A_{w,w_i}^*$.*

Proof. For $A_{w,w_i}^* \subseteq D(w_i) \cup \{w_i\}$, we have $\cup_{i=1}^m A_{w,w_i}^* \subseteq \cup_{i=1}^m (D(w_i) \cup \{w_i\}) = D(w)$. Since A_w^* is a solution that maximizes $G(T_w, A_w)$, we have $G(T_w, A_w^*) \geq G(T_w, \cup_{i=1}^m A_{w,w_i}^*)$. Let $A'_{w,w_i} = A_w^* \cap (D(w_i) \cup \{w_i\})$, then we have $A_w^* = \cup_{i=1}^m A'_{w,w_i}$. Obviously, $A'_{w,w_i} \subseteq D(w_i) \cup \{w_i\}$, so we have $G(T_{w,w_i}, A'_{w,w_i}) \leq G(T_{w,w_i}, A_{w,w_i}^*)$, since $A_{w,w_i}^* \subseteq D(w_i) \cup \{w_i\}$ is the optimal placement with respect to tree T_{w,w_i}. By Lemma 1, we have

$$G(T_w, A_w^*) = G(T_w, \cup_{i=1}^m A'_{w,w_i}) = \sum_{i=1}^m G(T_{w,w_i}, A'_{w,w_i})$$

$$\leq \sum_{i=1}^m G(T_{w,w_i}, A_{w,w_i}^*) = G(T_w, \cup_{i=1}^m A_{w,w_i}^*)$$

Therefore, we have $G(T_w, A_w^*) = G(T_w, \cup_{i=1}^m A_{w,w_i}^*)$, so we have $A_w^* = \cup_{i=1}^m A_{w,w_i}^*$.

Theorem 2. *For a tree $T_{w,x}$, if $C(x) = \{x_1, x_2, \cdots, x_k\}$, then we have $A_{w,x}^* = A_x^* \cup \{x\}$ or $A_{w,x}^* = \cup_{i=1}^m A_{w,x_i}^*$.*

Proof. It is easy to get $A_x^* \cup \{x\} \subseteq (D(x) \cup \{x\})$ and $\cup_{i=1}^m A_{w,x_i}^* \subseteq (D(x) \cup \{x\})$. For node x, there are two possibilities. One case is that a copy of an object is stored at node x, i.e. $x \in A_{w,x}^*$; and the other case is that no copy of an object is placed there, i.e. $x \notin A_{w,x}^*$. (1) First, we prove $A_{w,x}^* = A_x^* \cup \{x\}$ for $x \in A_{w,x}^*$. Let $A'_{x,x_i} = A_{w,x}^* \cap (D(x_i) \cup \{x_i\})$, then we have $\cup_{i=1}^k A'_{x,x_i} = A_{w,x}^* - \{x\}$. So $A_{w,x}^* = \cup_{i=1}^k A'_{x,x_i} \cup \{x\}$. Thus, we have

$$G(T_{w,x}, A^*_{w,x}) = G(T_{w,x}, \cup_{i=1}^{k} A'_{x,x_i} \cup \{x\})$$

$$= \sum_{v \in (\cup_{i=1}^{k} A'_{x,x_i} \cup \{x\})} \left[\left(f(v) - f'(v) \right) m(v) - l(v) \right]$$

$$= \sum_{v \in \cup_{i=1}^{k} A'_{x,x_i}} \left[\left(f(v) - f'(v) \right) m(v) - l(v) \right] + \left[\left(f(x) - f'(x) \right) m(x) - l(x) \right]$$

$$= \sum_{i=1}^{k} \sum_{v \in A'_{x,x_i}} \left[\left(f(v) - f'(v) \right) m(v) - l(v) \right] + \left[\left(f(x) - f'(x) \right) m(x) - l(x) \right]$$

$$= \sum_{i=1}^{k} G(T_{x,x_i}, A'_{x,x_i}) + \left[\left(f(x) - f'(x) \right) m(x) - l(x) \right]$$

$$= G(T_x, \cup_{i=1}^{k} A'_{w,x_i}) + \left[\left(f(x) - f'(x) \right) m(x) - l(x) \right]$$

$$\leq G(T_x, A^*_x) + \left[\left(f(x) - f'(x) \right) m(x) - l(x) \right]$$

On the other hand, we have

$$G(T_{w,x}, A^*_{w,x}) \geq G(T_{w,x}, A^*_x \cup \{x\})$$

$$= \sum_{v \in A^*_x} \left[\left(f(v) - f'(v) \right) m(v) - l(v) \right] + \left[\left(f(x) - f'(x) \right) m(x) - l(x) \right]$$

$$= G(T_x, A^*_x) + \left[\left(f(x) - f'(x) \right) m(x) - l(x) \right]$$

Therefore, we have $G(T_{w,x}, A^*_{w,x}) = G(T_{w,x}, A^*_x \cup \{x\})$, so $A^*_{w,x} = A^*_x \cup \{x\}$ for $x \in A^*_{w,x}$. (2) Now, we prove $A^*_{w,x} = \cup_{i=1}^{m} A^*_{w,x_i}$ for $x \notin A^*_{w,x}$. Let $A'_{w,x_i} = A^*_{w,x} \cap (D(x_i) \cup \{x_i\})$, similarly, we have $A^*_{w,x} = \cup_{i=1}^{k} A'_{w,x_i}$, therefore, we have

$$G(T_{w,x}, A^*_{w,x}) = G(T_{w,x}, \cup_{i=1}^{k} A'_{w,x_i})$$

$$= \sum_{v \in \cup_{i=1}^{k} A'_{w,x_i}} \left[\left(f(v) - f'(v) \right) m(v) - l(v) \right]$$

$$= \sum_{i=1}^{k} \sum_{v \in A'_{w,x_i}} \left[\left(f(v) - f'(v) \right) m(v) - l(v) \right]$$

$$= \sum_{i=1}^{k} G(T_{w,x_i}, A'_{w,x_i}) \leq \sum_{i=1}^{k} G(T_{w,x_i}, A^*_{w,x_i}) = G(T_{w,x}, \cup_{i=1}^{k} A^*_{w,x_i})$$

Since $G(T_{w,x}, A^*_{w,x}) \geq G(T_{w,x}, \cup_{i=1}^{k} A^*_{w,x_i})$, we have $G(T_{w,x}, A^*_{w,x}) = G(T_{w,x}, \cup_{i=1}^{k} A^*_{w,x_i})$. So $A^*_{w,x} = \cup_{i=1}^{k} A^*_{w,x_i}$ for $x \notin A^*_{w,x}$. Hence, the theorem is proven.

Corollary 1. *For tree $T_{w,x}$, if $C(x) = \{x_1, x_2, \cdots, x_k\}$ and $G(T_{w,x}, \cup_{i=1}^{k} A^*_{w,x_i}) \geq G(T_{w,x}, A^*_x \cup \{x\})$, then we have $A^*_{w,x} = \cup_{i=1}^{k} A^*_{w,x_i}$; otherwise, we have $A^*_{w,x} = A^*_x \cup \{x\}$.*

Now, we can give our dynamic programming-based algorithm as follows.

Algorithm: Optimal Object Caching Algorithm for Tree Networks
Step 1. Initialization: $A_w^* = \phi$ and $G(T_w, A_w^*) = 0$;
Step 2. end condition: if $D(w) = \phi$ then return;
Step 3. recursive procedure
for $v \in C(w)$ do
    if $D(v) = \phi$ then
        if $f(v)c(w,v) - l(v) > 0$ then
            $A_{w,v}^* = \{v\}$
        else
            $A_{w,v}^* = \phi$
    else
        for $x \in C(v)$ do
            if $\sum\limits_{x \in C(v)} G(T_{w,x}, A_{w,x}^*) \geq G(T_v, A_v^*) + (f(v) - f'(v))c(w,v) - l(v)$ then
               $A_{w,v}^* = \cup_{x \in C(v)} A_{w,x}^*$
            else
               $A_{w,v}^* = A_v^* \cup \{v\}$ (According to Theorem 2 and Corollary 1)
$A_w^* = \cup_{v \in C(w)} A_{w,v}^*$ (According to Theorem 1)

3.2 Algorithm Analysis

From our algorithm, we can know that every cache should maintain some information on the objects, including object size, access frequency, update frequency, and miss penalty of the object with the associated node. Fortunately, it is not necessary to store the information of an object at all the nodes. The following theorem describes an important property of our algorithm.

Theorem 3. *Suppose A_w^* is the optimal placement with respect to tree T_w, then we have $f(v)c(v,w) - l(v) \geq 0$, $\forall\, v \in A_w^*$.*

Proof. Suppose there exists $x \in A_w^*$ that satisfies $f(x)c(x,w) - l(x) < 0$, then we have

$$
\begin{aligned}
G(T_w, A_w^*) &= \sum_{v \in A_w^*} \left[\left(f(v) - f'(v)\right)m(v) - l(v)\right] \\
&= \sum_{v \in (A_w^* - \{x\})} \left[\left(f(v) - f'(v)\right)m(v) - l(v)\right] + \left[\left(f(x) - f'(x)\right)c(x,x') - l(x)\right] \\
&< \sum_{v \in (A_w^* - \{x\})} \left[\left(f(v) - f'(v)\right)m(v) - l(v)\right] + \left[f(x)c(x,w) - l(x)\right] \\
&< \sum_{v \in (A_w^* - \{x\})} \left[\left(f(v) - f'(v)\right)m(v) - l(v)\right] = G(T_w, A_w^* - \{x\})
\end{aligned}
$$

which contradicts the fact that A_w^* is the optimal placement with respect to tree T_w. Hence, the theorem is proven.

From Theorem 3, we can easily see that we should consider placing an copy of an object only among the caches where the object caching is locally beneficial.

Here, locally beneficial means that the net cost saving is greater than zero if we put only one copy of an object among the caches. It is easy to see that the time complexity for our algorithm is less than $O(n^2)$.

4   Concluding Remarks

In this paper, we addressed the problem of object caching to determine the optimal number of copies of an object to be placed among en-route caches on the path from the user to the content server and the locations where the copies should be placed such that the overall net cost saving is maximized. We proposed a mathematical model for this problem for the case that the network topology is a tree. In our model, according to the information stored at each node, we can optimally decide where the requested object should be cached and what can be removed to make room for it, if necessary. We formulated our problem as an optimization problem and developed a dynamic programming-based algorithm to solve it. The analysis of our algorithm showed that the solution produced by our algorithm is globally optimal.

Acknowledgements

The authors appreciate the help of Robert. B. DiGiovanni, PhD, a Visiting Professor of Technical Communication, Japan Advanced Institute of Science and Technology, in editing the manuscript.

References

1. B. D. Davison: A Web Caching Primer. IEEE Internet Computing. **5** (2001) 38–45
2. P. Krishnan, D. Raz, and Y. Shavitt: A Web Caching Primer. IEEE/ACM Transaction on Networking. **8** (2000) 568–582
3. M. Rabinovich and O. Spatscheck: Web Caching and Replication. Addison-Wesley. (2002)
4. P. Rodriguez, C. Spanner, and E. W. Biersack: Analysis of Web Caching Architectures: Hierarchical and Distributed Caching. IEEE/ACM Transaction on Networking. **9** (2001) 404–418
5. K. W. Ross: Hash Routing for Collections of Shared Web Caches. IEEE Network. **11** (1997) 37–44
6. P. Scheuermann, J. Shim, and R. Vingralek: A Case for Delay-Conscious Caching of Web Documents. Computer Network and ISDN Systems. **29** (1997) 997–1005
7. X. Tang and S. T. Chanson: Coordinated En-Route Web Caching. IEEE Transactions on Computers. **51** (2002) 595–607
8. X. Tewari, M. Dahlin, H. M. Vin, and J. S. Kay: Design Considerations for Distributed Caching on the Internet. Proc. 19th IEEE Int'l Conf. Distributed Computing Systems (ICDCS). (1999) 273–284
9. J. Wang: A Survey of Web Caching Schemes for the Internet. ACM SIGCOMM Computer Comm. Rev. **29** (1999) 36–46
10. P. S. Yu and E. A. MacNair: Performance Study of a Collaborative Method for Hierarchical Caching in Proxy Servers. Computer Networks and ISDN Systems. **30** (1998) 215–224

An Improved Algorithm
of Multicast Topology Inference
from End-to-End Measurements

Hui Tian and Hong Shen

Graduate School of Information Science,
Japan Advanced Institute of Science and Technology,
Tatsunokuchi, Ishikawa, 923-1292, Japan,
{hui-t,shen}@jaist.ac.jp

Abstract. Multicast topology inference from end-to-end measurements
has been widely used recently. Algorithms of inference on loss distribution show good performance in inference accuracy and time complexity.
However, to our knowledge, the existing results produce logical topology
structures that are only in the complete binary tree form, which differ in
most cases significantly from the actual network topology. To solve this
problem, we propose an algorithm that makes use of an additional measure of hop count. The improved algorithm of incorporating hop count
in binary tree topology inference is helpful to reduce time complexity
and improve inference accuracy. Through comparison and analysis, it is
obtained that the time complexity of our algorithm in the worst case
is $O(l^2)$ that is much better than $O(l^3)$ required by the previous algorithm. The expected time complexity of the algorithm is estimated at
$O(l \cdot \log_2 l)$, while that of the previous algorithm is $O(l^3)$.

Key words: Multicast, topology inference, end-to-end measurement,
hop count.

1 Introduction

As the use of IP multicast sessions becomes widespread, the potential benefits
derived from topological information on multicast distribution trees becomes
increasingly critical. In order to infer the performance characteristics of the internal links [10] without cooperation of network elements in the path such as
packet loss rates, [5, 7], packet delay distributions [13], and packet delay variance
[11], advance knowledge of the multicast tree topology is necessary. Moreover,
knowledge of the multicast topology can be very useful for multicast applications
[13].

Much work has been devoted to multicast topology discovery recently. Since
topology inference algorithms based on end-to-end measurements are particularly advantageous in that they require no support from internal nodes, many
algorithms have been developed in this way. [14], [13] and [6] work out an algorithm based on the sole loss observations at receivers for inferring the topology of

A. Veidenbaum et al. (Eds.): ISHPC 2003, LNCS 2858, pp. 376–384, 2003.

a multicast tree and prove its correctness. Algorithms based on delay measurements instead were proposed for identifying multicast topologies. In [10] several algorithms were specified based on delay performance measures such as link utilization, delay average and delay variance. An adaptive multicast topology inference algorithm proposed in [11] uses joint packet loss and link delay measurements. Other methods for identification of network topology from end-to-end measurements can be found in [3], [4] and [12].

Apparently, it's very difficult, if not impossible, to infer the actual physical topology precisely. Any inferred topology tree, referred to as logical tree, may differ at some degree from the actual physical tree. Inferred logical topologies by the existing algorithms based on end-to-end measurements can differ significantly from the actual multicast network. Our purpose of multicast topology inference in this paper is to obtain an inferred topology tree that is closer to the physical tree and also more efficient than the existing results.

The main idea of our new algorithm is to incorporate an additional parameter of hop count in the process of inference to guide inference direction and reduce inference time. The value of hop count of a packet can be obtained directly from the TTL field when received without additional cost. Topology inference by the previous algorithms fails to identify long paths with no branch. Our improved algorithm can identify these paths and obtain a topology that is much closer to actual topology structure. The improved algorithm, namely HBLT-Binary Loss Tree Classification Algorithm, is based on Binary Loss Tree Classification Algorithm (BLT). HBLT shows less time complexity and more accurate performance than BLT.

The paper is organized as follows. Multicast network modeling is introduced in Section 2. The improved algorithm and its analysis are given in section 3. Section 4 concludes the paper.

2   Muticast Network Modeling

Let $T = (V, L)$ denote a logical multicast tree with node set V and link set L. The root node 0 is the source of probes, and $R \subset V$ denotes the set of leaf nodes representing the receivers. Each non-leaf node k has a set of children node $d(k) = \{d_i(k) \,|\, 1 \leq i \leq n_k\}$, and non-root node k has a parent $p(k)$. The link $(p(k), k) \in L$ can be simply denoted by link k. U are said to be siblings if they have the same parent. The loss of probe packets on the basic multicast tree is modeled by a set of mutually independent Bernoulli processes. Losses are therefore independent for different links and different packets. Assume that each probe packet is successfully transmitted across link k with probability α_k. The process of each probe packet down the tree is described by an independent copy of a stochastic process $X = (X_k)_{k \in V}$. $X_0 = 1, X_k = 1$ if the probe packet reaches node $k \in V$ and 0 otherwise.

A logical multicast tree comprises the branch points of the physical tree, and logical links between them. The logical topology induced by a physical topology T is formed from T after all internal nodes with only one child have been

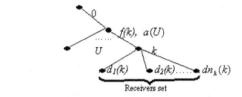

Fig. 1. Tree Model

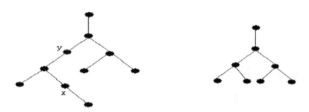

Fig. 2. The left tree is the physical tree, the right one is the inferred logical tree

collapsed into their parent recursively (as defined in [13]). Thus a physical tree can be converted to a logical tree. An example of binary physical and logical trees is shown in Fig. 2, where in the logical tree node x and y are removed and the links terminating at them are integrated to one link. Therefore, a logical multicast tree has the property that every node has at least two descendants, apart from the root node (which have only one) and the leaf-nodes (which have none). On the other hand, nodes in the physical multicast tree may have only one descendant, such as node x and y.

If we infer the topology of the physical tree in Fig. 2 by use of previous algorithms, the logical tree shown in Fig. 2 is obtained. It's obvious that the logical tree inferred by those algorithms is much different from the actual physical tree.

3   The Improved Algorithm – HBLT

Because the binary tree-based grouping algorithms such as BLT provide good combination of accuracy and computational simplicity [10, 13], our improved algorithm is based on them. The detailed BLT Algorithm can be found in [13]. The key function $B(U)$ used in BLT is the loss distribution at a set of nodes U that is minimized when U is a set of siblings. Firstly, BLT finds the siblings from all the receivers. Then substitute a composite node that represents their parent for the siblings. It continues on finding other siblings from the rest receivers set and the parent node by comparing $B(U)$ within every two nodes. Iterating this procedure reconstructs the tree. In practice, it is proved that $B(U)$ can be estimated with $\hat{B}(U)$ [13].

$$\hat{B}(U) = \frac{\sum_{i=1}^{n} \hat{X}_{u_1}^{(i)} \sum_{i=1}^{n} \hat{X}_{u_2}^{(i)}}{n \sum_{i=1}^{n} \hat{X}_{u_1}^{(i)} \hat{X}_{u_2}^{(i)}} \tag{1}$$

Here $U = \{u_1, u_2\}$; n is the number of probe packet; $(X_k^{(i)})_{k \in R}^{i=1,\dots,n}$ denotes the measured outcomes observed at the receiver arising from each of n probes.

The topology tree inferred by BLT is much different from the actual physical topology. Those nodes with single child are all removed, so are the links terminating at such nodes. Hop count measurements accounted into BLT overcome such drawback. The improved algorithm HBLT shows better time complexity than BLT in the worst case and expected case.

3.1   Description of HBLT

Hop count at a node is defined to be the total number of devices a given piece of probe packet passes through from the source to that node. In a multicast tree, hop count can be computed from the TTL value read directly from the received packet at the receivers. It shows the number of the routers every probe packet encountered from the root to the destination. For internal nodes, the value of hop count is easily computed by degression. Thus it only requires little overheads to obtain the value of hop count in the procedure of inference.

Inference starts from the nodes with the maximum hop count and proceeds in a bottom up fashion. For the internal nodes, those with different values of hop count can be all inferred, including those having only one child. And the links terminating at them can also be identified. It works as follows:

1. Input: The set of receivers R, number of probes n, receiver traces $(X_k^{(i)})_{k \in R}^{i=1,\dots,n}$;
2. $R' := R$, $V' := R$; $L' := \Phi$, $m = \max(k.hop)$, $W_i = \Phi$, $i = 1, \dots, m$;
   $//$ W_i is a set of all nodes with hop number being i, m is initialized as the maximum value of hop count in all nodes of the multicast tree. $//$
3. for $k \in R$, do
4. $\hat{B}(k) := n^{-1} \sum_{i=1}^{n} X_k^{(i)}$;
5.   for $i = 1, \dots, m$ do
6.   if $(k.hop = i)$ then $W_i = W_i \cup \{k\}$
7.   end for;
8.   for $i = 1, \dots, n$ do $\hat{X}_k^{(i)} = X_k^{(i)}$; end do;
9. end for;
10. while $m > 1$ do
11.   while $W_m \neq \Phi$ do
12.   search u_{j,w_m} to minimize $\hat{B}(u_{i,w_m}, u_{j,w_m})$
13.   if $(\hat{B}(U) >$ an coherence value) then $U = u_{i,w_m}$;
14.   replace u_{i,w_m}, u_{j,w_m} with U $//$ U is the parent node of u_1, u_2 $//$
15.   for $i = 1, \dots, n$ do $\hat{X}_U^{(i)} = \bigcup_{u \in U} \hat{X}_u^{(i)}$; end for;
16. $U.hop = m - 1$;
17. $V' = V' \cup \{U\}$; $W_m = W_m \backslash U$; $W_{m-1} = W_{m-1} \cup \{U\}$;

18.   for each $u \in U$ do $L' := L' \cup \{(U, u)\}$; $\hat{\alpha}_u := \hat{B}(u)/\hat{B}(U)$; end for;
19.   end while;
20. $m = m - 1$;
21. end while;
22. $V' = V' \cup \{0\}$; $L' = L' \cup \{0, U\}$;
23. $\hat{\alpha}_U = \hat{B}(U)$; $\hat{\alpha}_0 = 1$;
24. Output: Loss tree $((V', L'), \hat{\alpha})$.

Each probe packet has a TTL value field that records the hop count from the root to each receiver. Because two siblings must have the same values of hop count, grouping is performed within the sets with the same value of hop count. Find siblings in the set of nodes with the maximum value of hop count m firstly. Then let their parent's hop count be m-1, loss distribution be the "or" operation of the siblings. The siblings are removed from the node set W_m, at the same time the parent node is added to the node set V' and W_{m-1}, and the link from the parent to the node is added to link set L'. Iterate this procedure to find the siblings with the hop count m-1. The topology inference is finished until the set with hop count is zero.

3.2   Algorithm Comparison and Analysis

For simplicity, suppose the following conditions:

1) The total number of the binary tree's nodes is n.
2) The total number of leaf nodes is l.
3) Only trees whose internal nodes all have two children are considered. Clearly, $n = 2 * l - 1$.
4) The total level of the binary tree is h, then $\log_2 l + 1 \leq h \leq l$.
5) The number of nodes at level i is a_i, $1 \leq i \leq h$. Then the equations $2 \leq a_i \leq 2a_{i-1}$, $a_1 + a_2 + \ldots + a_i + \ldots + a_h = n$, $2 \leq i \leq h$ hold.

For BLT algorithm, time complexity is determined by the total number of leaf nodes. It is denoted by $T_{\text{BLT}}(l)$.

$$T_{\text{BLT}}(l) = \sum_{k=2}^{l} k * (k - 1)/2 = \frac{1}{6}l^3 - \frac{1}{6}l \qquad (2)$$

For HBLT algorithm, time complexity is computed by following equation.

$$T_{\text{HBLT}}(l) = T_{\text{HBLT}}(a_1, a_2, \ldots a_h) = \sum_{k=2}^{h} a_k^2 \qquad (3)$$

3.2.1   Time Complexity Comparison in The Worst Case
Because HBLT will be in the worst case if the tree is a complete binary tree. In such case, for h, l, a_i, $h = \log_2 l + 1$ and $a_i = 2^{i-1}$, $1 \leq i \leq h$ hold.

Then, the result is obtained from equation (3).

$$T_{\text{HBLT}}(l) = \frac{4}{3}l^2 - \frac{4}{3} \tag{4}$$

From equations (2) and (4), we get that the worst case of HBLT improves much. Time complexity of HBLT is $O(l^2)$, while that of the previous algorithm BLT is $O(l^3)$.

3.2.2 Expected Time Complexity Comparison

Though from the worst case's analysis, HBLT has improved a lot. Actually HBLT shows very high speed in procedure of operating, which can be proved in calculation of average time complexity.

With given total nodes number n of a binary tree, it is well known that Catalan number gives a very large total number of all different binary trees. The number of structurally different binary tree can also be calculated. While time complexity analysis to our algorithm only concerns heights of the trees and different number of nodes at each level, regardless of trees' concrete structure. Even if many trees have different structures, the time costs of algorithm for them are similar if their heights and combination of nodes at all levels are the same.

Lemma. The time complexity of HBLT depends only on the values of $a_1, a_2, \ldots,$ a_h, not on their order of appearance in the tree.

Proof. $T_{\text{HBLT}}(a_1, a_2, \ldots a_h) = a_1^2 + a_2^2 + \cdots + a_i^2 + \cdots + a_h^2$, therefore the value of T_{HBLT} is unchanged if any elements are swapped.

Assume the number of nodes at each level be the integer power of 2, $a_i = 2^k$. Let $l = 2^r$. Under this condition, there are still too many different binary trees. We cannot take every case into consideration. So sampling from all trees is an efficient way to calculate average time cost. And the lemma above is very helpful in calculation.

The trees are included by this rule. Only leftist trees are considered. We begin with the best case tree for HBLT, the height of the tree is the total number of leaves, that is, $h = l$. Then fold the highest and thinnest tree to reduce its height, and make it be a fatter leftist tree as twice as before in bottom up way. Let i denote folding times. Folding is stopped until a complete binary tree is formed.

Fig. 3 gives folding procedure when leaves' number is 8. After twice folding, the tree changes from the best case to the worst case. Folding is performed with level number reduced by 1 every time. There are many possibilities for constructing a binary tree every time height decreases 1. But as presented in Lemma, time costs are the same for every different binary tree. From Fig. 3 (b), it is obvious no matter whose children node m and n are, the tree has the same time cost. So it's unnecessary to compute every kind of such binary trees. Till the tree forming the structure as Fig. 4 (d), one folding is finished, i becomes 1 at this time. Iterating reducing height and folding, a complete binary tree will be constructed. For every case in the procedure, time costs are calculated. Then

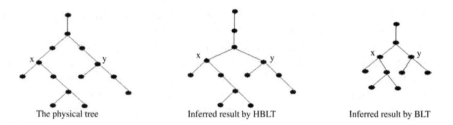

Fig. 3. Result Comparison

the total time cost divided by total number of trees will give the expected time complexity.

The total number of trees sampled in this way is denoted by TN, and total time cost by TC.

$$TN = \sum_{i=0}^{r-2}(2^{r-i-1} - 1) = 2^r - r - 1 = l - \log_2 l - 1 \tag{5}$$

$$TC = \sum_{i=0}^{r-2} \sum_{h=2^{r-(i+1)}+i+1}^{2^{r-i}+i-1} \left[(2^{r-i} + i - h) \cdot 2^{2(1+(i+1))} \right.$$
$$\left. + \sum_{k=1}^{h-(2^{r-i}+i-h)-(i+1)} 2^{2(i+1)} + \sum_{k=2}^{i} 2^{2k} \right] \tag{6}$$

$$= 3l^2 \log_2 l - \frac{13}{2}l^2 + 6l + \frac{4}{3}r + \frac{4}{9} \tag{7}$$

$$\text{Average Cost} = TC/TN = O(l \cdot \log_2 l) \tag{8}$$

Thus, the expected time complexity analysis of HBLT is $O(l \cdot \log_2 l)$. Since the time complexity of BLT is only relevant to the number of leaf nodes, all these sampling trees have the same number of leaves, the expected time complexity of BLT $O(l^3)$, which is far worse than HBLT.

Such a folding model misses some binary trees, but binary trees included in calculation represent all cases influencing much to HBLT from the best case tree to the worst case tree. Therefore, the result calculated by such sampling can be used as and estimation of actual expected time complexity to compare with the previous algorithm.

3.3   Analysis on Accuracy

As shown in [5, 7], the Bernoulli model assumption is violated for the presence of spatial and temporal correlation between packet losses. Especially in the case that the number of probes n isn't large enough, topology inference will give rise to some error. Some pseudo-siblings with the minimized $\hat{B}(U)$ inferred in BLT aren't true siblings in reality. But different values of hop count in HBLT are helpful to distinguish such pair of pseudo-siblings. So with this criterion in inferring, the algorithm shows better performance in accuracy.

BLT obtains a logical topology structure that has been proved to be much different from the actual physical tree because all nodes with only one child and related links fail to be identified. While HBLT can infer such nodes and links accurately in general because these single-child nodes have different values of hop count. For example, as to the physical tree in Fig. 2, if inferred topology by HBLT a topology structure similar to the actual topology tree is identified without any node or link being removed. However, BLT only gets the logical structure as shown in Fig. 2. It is evident that inferred logical tree by use of BLT changes a lot from the actual physical tree for those nodes with only one child being deleted and the links terminating at them being collapsed to one link. Therefore, HBLT excels BLT in inferring the actual physical topology in that the inferred result is far closer to the true topology.

However, inference by use of HBLT may fail to infer some internal links in some kind of physical tree. Fig.4 gives an example.

If two internal nodes (such as x and y in Fig. 4) have no siblings whose common parent node is very near to them, and internal links connecting them are symmetric, the inference for this part of the tree may be performed inaccurately. But the whole topology structure is sure to be much closer to the actual physical topology than topology result by BLT. This is evident in the comparison between the inferred results in Fig. 4.

4   Summary

Improved algorithms for inference of binary loss tree and non-binary loss tree have been proposed in this paper. In the improved algorithms HBLT and HGLT, an additional parameter hop count is used to achieve better inference accuracy and time complexity. For HBLT, it has been obtained that the time complexity of HBLT in the worst case is $O(l^2)$, which is much less than $O(l^3)$ required by BLT. And through analysis, it is obtained that the expected time complexity estimation is $O(l \cdot \log_2 l)$, which is far better than that of BLT $O(l^3)$. Our method of considering hop count into topology inference may also be extended to non-binary trees inference and other parameters based algorithms. Similar improvements in performance are expected to be obtained in future work.

References

[1] A. Adams, T. Bu, R. Caceres, B. G. Duffield, T. Fricdman, J. Horowitz, F. Lo Presti, S. B. Moon, V. Paxson, D. Towsley, "The Use of End-to-End Multicast Measurements for Characterizing Internal Network Behavior", IEEE Communicaiotns Magazine, May 2000.
[2] Azer Bestavros, John Byers, Khaled Harfoush, "Inference and Labeling of Metric-Induced Network Topologies", IEEE Infocom' 2002, New York, June, 2002.

[3] Rui Castro, Mark Coates and Robert Nowak, "Maximum Likelihood Identification from End-to-end measurements", in DIMACS Workshop on Internet Measurement, Mapping and Modeling, DIMACS Center, Rutgers University, Piscataway, New Jersey, February 2002.

[4] Mark Coates, Rui Castro, Robert Nowak, Manik Gadhiok, Ryan King, Yolanda Tsang, " Maximum likelihood network topology identification from edge-based unicast measurements", SIGMETRICS 2002

[5] R. Caceres, N. G. Duffield, J. Horowitz, D. Towsley, " Multicast-based Inference of Network-Internal Loss Characteristics", IEEE Trans. On Information Theory, VOL. 45, NO. 7, Nov. 1999.

[6] R. Caceres, N. G. Duffield, J. Horowitz, F. Lo Presti, D. Towsley, "Loss based Inference of Multicast Network Topology", in Proc. 1999 IEEE Conference on Decision and Control, Phoenix, AZ, December 1999.

[7] R. Careres, N. G. Duffield, H. Horowitz, D. Towsley and T. Bu, " Multicast-Based Inference of Network Internal Loss Characteristics: Accuracy of Packet Loss Estimation", Proc. Of Infocom'99. New York, NY, Mar. 1999.

[8] N.G. Duffield, J. Horowitz, F. Lo Presti, "Adaptive Multicast Topology Inference", in Proc. IEEE Infocom 2001, Anchorage, Alaska, April 22-26, 2001.

[9] N. G. Duffield, J. Horowitz, F. Lo Presti, D. Towsley, "Multicast Topology Inference from Measured End-to-End Loss", IEEE Trans. Information Theory, 2002.

[10] N.G. Duffield, J. Jorowitz, F.Lo Presti, D. Towsley, "Multicast Topology Inference from End-to-End Measurements", in ITC Seminar on IP Traffic Measurement, Modeling and Management, Monterey, CA, Sept. 2000.

[11] N. G. Duffield, F. Fo Presti, " Multicast Inference of Packet Delay Variance at Interior Network Links", Pro. IEEE Infocom 2000, Tel Aviv, March 2000.

[12] Jangwon Lee, Gustavo de Veciana, "Resource and Topology Discovery for IP Multicast unsing a Fan-out Decrement Mechanism", IEEE Infocom'2001.

[13] F. Lo Presti, N. G. Duffield, J. Horowitz and D. Towsley, " Multicast-Based Inference of Network-Internal Delay Distributions", IEEE/ACM Transactions on Networking, Vol. 10, No. 6, Dec. 2002.

[14] S. Ratanasamy & S. McCanne, "Inference of Multicast Routing Tree Topologies and Bottleneck Bandwidths using End-to-end Measurements", Proc. IEEE Inforcom'99, New York, NY(1999)

Chordal Topologies
for Interconnection Networks*

Ramón Beivide[1], Carmen Martínez[1], Cruz Izu[2], Jaime Gutierrez[1],
José-Ángel Gregorio[1], and José Miguel-Alonso[3]

[1] Universidad de Cantabria, Avenida de los Castros s/n, 39005 Santander, Spain,
mon@atc.unican.es
[2] The University of Adelaide, 5005 Adelaide, South Australia
[3] Univ. del País Vasco UPV/EHU. P. M. Lardizabal, 1, 20018 San Sebastián, Spain

Abstract. The class of dense circulant graphs of degree four with optimal distance-related properties is analyzed in this paper. An algebraic
study of this class is done. Two geometric characterizations are given,
one in the plane and other in the space. Both characterizations facilitate
the analysis of their topological properties and corroborate their suitabi-
lity for implementing interconnection networks for distributed and parallel computers. Also a distance-hereditary non-disjoint decomposition of
these graphs into rings is computed. Besides its practical consequences,
this decomposition allows us the presentation of these optimal circulant
graphs as a particular evolution of the traditional ring topology.

1 Introduction

Ring topologies have frequently been used to implement local and campus area
networks as well as other interconnection subsystems for diverse digital devices.
In order to improve their relatively poor performance and survivability, extensions of rings such as Chordal Rings of degree three and four have been considered in the literature [1],[5]. The Torus is another low-degree regular network,
based on a collection of rings, often used to interconnect highly coupled parallel
systems. Some of the latest parallel computers using this topology are the IBM
BlueGene/L supercomputer [2] and different multiprocessor servers based on the
Alpha 21364 microprocessor [13].

Interconnection networks can be modelled by graphs with vertices representing their processing elements and edges representing the communication links
between them. The performance and robustness of any network are mostly determined by the topological characteristics of its underlying graph.

Circulant graphs have deserved significant attention during the last decades,
the traditional ring and the complete graph topologies belong to this class of
graphs. Circulants of different degrees constituted the basis of some classical

---

* This work has been partially supported by the Spanish CICYT project TIC2001-
0591-C02-01 and by the Spanish Ministry of Education under grants FP-2001-
2482,PR2002-0043 and BFM2001-1294.

A. Veidenbaum et al. (Eds.): ISHPC 2003, LNCS 2858, pp. 385–392, 2003.

distributed and parallel systems [15], [7]. The design of certain data alignment networks for complex memory systems have also relied on circulant graphs [16].

A class of circulant graphs of degree four with minimal topological distances, denoted as *Midimew* networks, was presented in [4] as a basis for building optimal interconnection networks for parallel computers. One *Midimew* network exists for any given number of nodes, N, which is completely defined by means of a single parameter, as described later. These graphs are *optimal* because they have the minimum average distance among all circulant graphs of degree four; consequently, the diameter of a *Midimew* graph is also minimum. In addition, these graphs are regular, vertex-symmetric and maximally connected, which means that they possess a very high degree of fault tolerance. Optimal VLSI layouts for these networks have also been explored [9] and some instances of *Midimew* graphs have recently been used as a basis for designing networks for massively parallel computers [17]. It is known that some instances of *Midimew* graphs are isomorphic to Chordal Rings of degree four, thus sharing their topological distance properties; however, there exist infinite values of N for which *Midimew* networks have smaller diameter and smaller average distance [5].

In this paper, we focus exclusively on dense circulant graphs of degree four. For every integer k, there exists a dense optimal circulant graph composed of $2k^2 + 2k + 1$ nodes, being k the diameter of the graph. In section 2 we prove that any optimal dense circulant graph of degree four with diameter k is isomorphic to a Chordal Ring whose chord has length $2k + 1$. In particular, the corresponding *Midimew* graph is also isomorphic to this Chordal Ring. Section 3 is devoted to show two geometric characterizations of these optimal graphs that facilitate the analysis of their properties, especially those related to topological distances. As a by-product of the above-mentioned geometric characterizations, we present in the last section a distance-hereditary decomposition of dense optimal Chordal Rings into a set of traditional rings. All the rings belonging to this set have the same number of nodes and their diameter corresponds to the diameter of the Chordal Ring in which they are embedded. Moreover they preserve the minimal routing of the original circulant graph.

2   Dense Optimal Chordal Rings of Degree Four

A *circulant graph* with N vertices and jumps $\{j_1, j_2, \ldots j_m\}$ is an undirected graph in which each vertex n, $0 \leq n \leq N$, is adjacent to all the vertices $n \pm j_i$, with $1 \leq i \leq m$. We denote this graph as $C_N(j_1, j_2, \ldots j_m)$. The family of circulant graphs includes the complete graph and the cyclic graph (ring) among its members.

We say that a circulant is *dense* if it has the maximum possible number of nodes for a given diameter k. Thus, if k is a positive integer, a dense ring or $C_N(1)$ of degree two has $N = 2k + 1$ nodes. There is another non-dense ring with $2k$ nodes. When adding another jump to the list, every integer k defines a family of $4k$ optimal (with minimum average distance and therefore minimum diameter) $C_N(j_1, j_2)$ graphs or *Midimew* networks of diameter k, with $j_1 = b-1$ and $j_2 = b$, where $b = \sqrt{\frac{N}{2}}$, [4]. The dense member of this family contains $2k^2 + 2k + 1$ nodes. The other $4k - 1$ values of N in the family correspond to non-dense graphs.

Two graphs G and H are *isomorphic* $(G \cong H)$ if there exists a bijective mapping between their node sets which preserves adjacency. Is easy to prove the following:

Lemma 1. *Let k be a positive integer, $N = 2k^2 + 2k + 1$. If $C_N(c, d)$ is an optimal circulant graph, then $C_N(c, d) \cong C_N(k, k + 1)$.*

Now, we consider the Chordal Ring $C_N(1, 2k + 1)$. This Chordal Ring is isomorphic to the dense *Midimew* graph and hence, it is also optimal. To prove this claim we are going to use the following well-known theorem [1]:

Theorem 1. *Let N be a natural number. We have $C_N(j_1, j_2) \cong C_N(i_1, i_2)$ if and only if there exists an integer u such that $gcd(u, N) = 1$ and $u\{\pm j_1, \pm j_2\} = \{\pm i_1, \pm i_2\}$ mod N.*

Hence, the element $u = (k + 1)^{-1} = -2k$ provides the adequate isomorphism between the two graphs.

For the rest of the paper we are going to consider $C_N(1, 2k+1)$ Chordal Rings as the representative members of the class of dense optimal circulant graphs of degree four.

3   Geometrical Characterization of Optimal Chordal Rings

In this section we will present two geometrical characterizations of $C_N(1, 2k+1)$ optimal graphs, in two and three dimensions, that lead to two physical implementations of interconnection networks based on these graphs.

Let us begin with the bi-dimensional geometric characterization. The vertex-symmetry of graphs allows their analysis from any vertex; node 0 will be used in the rest of the paper. In these graphs, there are $4d$ different nodes at distance d from node 0, with $1 \leq d \leq k$, and consequently, the total number of nodes of the graph is $N = k^2 + (k + 1)^2$. It is known that certain families of graphs can be fully represented by plane tessellations when their nodes are associated with regular polygons [7]. If each node in a $C_N(1, 2k + 1)$ graph is associated with a unitary square with its four neighbours attached to its sides as in [15], the graph can be characterized by a serrated or discrete square tile of area N, like the one in Fig. 1. This tile tessellates the plane, and the network wrap-around links are defined according to the periodical pattern dictated by such a tessellation.

A popular geometric proof of the Pythagorean theorem, which is clearly reflected in the equation $N = k^2 + (k + 1)^2$, shows the square of the hypotenuse of a right-angled triangle of legs k and $k + 1$ as composed of four copies of such a right-angled triangle plus a central square with side $k + 1 - k = 1$. The tile in Fig. 1 can be seen as a geometrical proof of a discrete version of the Pythagorean Theorem.

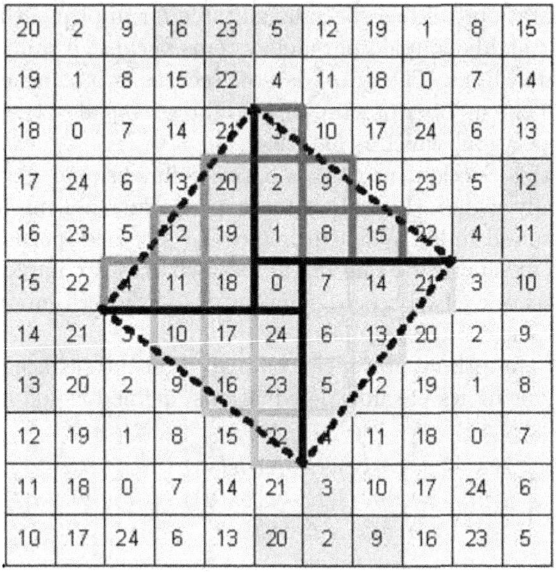

20 2 9 16 23 5 12 19 1 8 15
19 1 8 15 22 4 11 18 0 7 14
18 0 7 14 21 3 10 17 24 6 13
17 24 6 13 20 2 9 16 23 5 12
16 23 5 12 19 1 8 15 22 4 11
15 22 4 11 18 0 7 14 21 3 10
14 21 3 10 17 24 6 13 20 2 9
13 20 2 9 16 23 5 12 19 1 8
12 19 1 8 15 22 4 11 18 0 7
11 18 0 7 14 21 3 10 17 24 6
10 17 24 6 13 20 2 9 16 23 5

Fig. 1. Pythagorean Interpretation of a $C_{25}(1,7)$ Graph.

In this discrete version, the right-angled triangles with legs k and $k+1$ have been replaced by discrete right-angled triangles with both legs of length k, in which the hypotenuse adopts a serrated shape. The area of one of these right-angled triangles coincides with the area of its discrete version. Then, every dense optimal $C_N(1, 2k+1)$ graph is characterized by four copies of a discrete right-angled triangle with both legs of length k, plus a central unitary square. Such geometric interpretation, in terms of right-angled triangles, simplifies the study of the distance-related properties of these graphs.

Related also to another geometrical proof of the Pythagorean Theorem, a new layout for $C_N(1, 2k+1)$ graphs can be obtained as shown in Fig. 2.

7 14 21 3 10 17 24
6 13 20 2 9 16 23
5 12 19 1 8 15 22
4 11 18 0

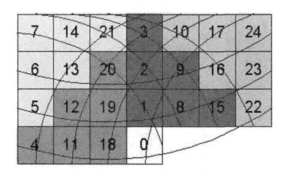

Fig. 2. Mesh-connected $C_{25}(1,7)$ Graph with Minimum Number of Wrap-around Links.

This representation can be especially suitable for implementing interconnection networks for highly coupled parallel systems because it minimizes the number of wrap-around links. The graph is now seen as two attached squares with sides k and $k+1$. The number of wrap-around links is equal to the semi-perimeter of the resulting polygon, which is $3k + 2$.

The rest of this Section introduces a three-dimensional graph transformation that naturally avoids the unbalance between internal mesh-like links and external wrap-around links. When technological advances permit it, our three-dimensional layout should be one of the best choices for implementing an interconnection network based on circulant graphs with both minimal topological and physical distances among nodes.

The nodes of an optimal graph can be placed on the surface of a Torus with radii R and r by using its parametric equations, defined as follows:

$$\begin{cases} x(u,v) = cos(v)(R - rcos(u)) \\ y(u,v) = sin(v)(R - rcos(u)) \quad u,v \in [0, 2\pi] \\ \quad z(u,v) = rsin(u) \end{cases}$$

If we define, as in [9], $u_j = \frac{2\pi}{N}j$ and $v_j = \frac{2\pi(2k+1)}{N}j$, then every node j of the graph, with $j \in \{0, 1, \ldots, N-1\}$, corresponds to a point T_j of the Torus surface by using the mapping $T_j = \Psi(j) = (x(u_j, v_j), y(u_j, v_j), z(u_j, v_j))$. The points $T_j = (t_{x(j)}, t_{y(j)}, t_{z(j)})$, being t_x, t_y, t_z their Cartesian coordinates, are located on the helix $T(s; 1, 2k + 1) = (x(u_s, v_s), y(u_s, v_s), z(u_s, v_s))$, where $s \in [0, N-1]$. Similarly, the same points $T_j = (t_{x(j)}, t_{y(j)}, t_{z(j)})$, are located on the curve $T(s; 2k + 1, -1)$, $s \in [0, N-1]$. The intersection points of $T(s; 1, 2k + 1)$ and $T(s; 2k + 1, -1)$ are the points T_j, in which we locate every node j of the graph. Actually, both curves define two disjoint Hamiltonian paths, which are always embedded in this class of graphs. When $R = 2r$, as shown in Fig. 3 for $C_N(1, 2k + 1)$ with $k = 3$, the length of all the graph edges is the same, being this length the closest value to the Euclidean distance between two neighbour nodes [9].

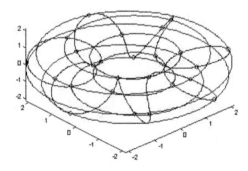

Fig. 3. 3D Torus Embedding of the $C_{25}(1, 7)$ Graph.

This embedding of the $C_N(1, 2k + 1)$ graph on a Torus surface can lead to a 3D network implementation with all the physical links having the same minimum length and never crossing among them. In addition, the underlying graph has the minimal topological distances among all the circulants of degree four. Some designs for futuristic applications as the ones introduced in [10], suggest the use in the few next years of cylindrical structures and free-space optical communications for implementing extreme interconnection networks.

4   Optimal Dense Chordal Rings as a Set of Rings

It can be shown that dense optimal $C_N(1, 2k + 1)$ graphs can be decomposed into a non-disjoint collection of $2N$ rings that preserves the distance-related properties of the original graph. These rings, or circulants of degree two, are also dense and their diameter is the diameter of the circulant in which they are embedded.

In order to obtain this set of rings and avoiding repeated cases in the study, only are considered rings obtained by taking positive steps through jumps $\{1, 2k + 1\}$ or by taking positive steps through jumps $\{1, -(2k + 1)\}$. Therefore, for the rest of the paper, a $(2k + 1)$-ring embedded on a $C_N(1, 2k + 1)$ is a cycle such that, if $\{n_1, n_2, \ldots, n_{2k+1}\}$ is its set of nodes, then one of the following assertions must be true:

- $n_{i+1} = n_i + 1 \bmod N$ or $n_{i+1} = n_i + (2k+1) \bmod N$, $\forall i \in \{1, 2, \ldots, 2k+1\}$.
- $n_{i+1} = n_i + 1 \bmod N$ or $n_{i+1} = n_i - (2k+1) \bmod N$, $\forall i \in \{1, 2, \ldots, 2k+1\}$.

To simplify the notation we represent a $(2k + 1)$-ring as a vector whose coordinates are in the set $\{0, 1\}$ or in $\{0, -1\}$, depending on the type of $(2k+1)$-ring. If we are considering a $(2k + 1)$-ring in a $C_N(1, 2k + 1)$, 0 represents a (positive) step in jump 1, 1 represents a step in $2k+1$ and -1 a step in $-(2k+1)$. For example, the 7-tuple $(1, 0, 0, 1, 0, 0, 1)$ denotes the 7-ring $\{0, 7, 8, 9, 16, 17, 18\}$ in $C_{25}(1, 7)$.

Using this notation above, we consider the following sets of $(2k + 1)$-rings:

- $A_1^k = \{\overline{\lambda_1}, \overline{\lambda_2}, \ldots, \overline{\lambda_k}\}$ where $\overline{\lambda_i} = (\overbrace{0, \ldots, 0}^{i}, \overbrace{1, \ldots, 1}^{k}, \overbrace{0, \ldots, 0}^{k+1-i})$, for $i = 1, 2, \ldots, k$

- $A_2^k = \{\overline{\mu_1}, \overline{\mu_2}, \ldots, \overline{\mu_k}\}$ where $\overline{\mu_i} = (\overbrace{-1, \ldots, -1}^{i}, \overbrace{0, \ldots, 0}^{k}, \overbrace{-1, \ldots, -1}^{k+1-i})$, for $i = 1, 2, \ldots, k$.

Then, for every node n of $C_N(1, 2k + 1)$ there exists an element in $A_1^k \cup A_2^k$ containing n. This set of $2k$ rings gives us a complete picture of the graph from node 0. An example of this rings from node 0 in $C_{25}(1, 7)$ can be seen in Fig. 4.

Now, we consider the union of subsets of $(2k + 1)$-rings for all the nodes in the graph. It can be shown that this union has $2N$ different elements (each ring from node 0 is repeated exactly k times). Hence, the graph has been decomposed

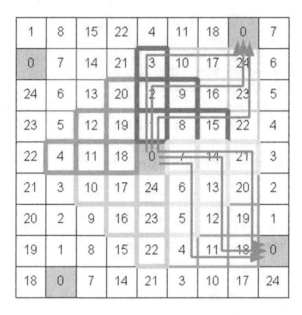

Fig. 4. Characteristic Set of Rings for Node 0 on a $C_{25}(1,7)$ Graph.

into a set of $2N$ $(2k+1)$-rings, or dense rings, providing a complete picture of the graph.

In a recent paper of the authors [12], it was also proved that, for any integer k, there is a non-dense $C_{2k^2}(1, 2k-1)$ graph, which can be seen as a collection of $2N = 4k^2$ $2k$-rings, or non-dense rings. The value of the integer k corresponds to the diameter of both graphs and totally defines both topologies.

Finally, we can conclude that some specific members of the family of optimal circulant graphs of degree four are a particular evolution of the traditional ring topology or circulants of degree two.

References

1. A. Adám. "Research problem 2-10". J. Combinatorial Theory, 393, pp. 1109-1124, 1991.
2. NR. Adiga et al. "An Overview of the BlueGene/L Supercomputer", Supercomputing 2002, November 2002.
3. B.W. Arden and H. Lee. "Analysis of Chordal Ring Networks". IEEE Transactions on Computers, Vol. C-30, No. 4, pp. 291-295, 1981.
4. R. Beivide, E. Herrada, J.L. Balczar and A. Arruabarrena. "Optimal Distance Networks of Low Degree for Parallel Computers". IEEE Transactions on Computers, Vol. C-40, No. 10, pp. 1109-1124, 1991.
5. J.-C. Bermond, F. Comellas and D.F. Hsu. "Distributed Loop Computer Networks: A Survey". Journal of Parallel and Distributed Computing, Vol. 24, pp. 2-10, 1995.
6. B. Bose, B. Broeg, Y. Known and Y. Ashir. "Lee Distance and Topological Properties of k-ary n-cubes". IEEE Transactions on

7. W.J. Computers, Vol. 44, No. 8, pp. 1021-1030, 1995. Bouknight, S.A. Denenberg, D.E. McIntyre, J.M. Randall, A.H. Sameh and D.L. Slotnick. "The Illiac IV System". Proc. IEEE, Vol. 60, No. 4, pp. 369-388. 1972.

8. M.A. Fiol, J.L. Yebra, I. Alegre and M. Valero. "A Discrete Optimization Problem in Local Networks and Data Alignment". IEEE Transactions on Computers, Vol. 36, No. 6, pp. 702-713, 1987.

9. K. Huber. "Codes over Tori". IEEE Trans. on Information Theory, Vol. 43, No. 2, pp. 740-744. March 1997.

10. L.E. LaForge, K.F. Korver and M.S. Fadali. "What Designers of Bus and Network Architectures Should Know about Hypercubes", IEEE Transactions on Computers, Vol. 52, No. 4, pp. 525-544, April 2003.

11. C. M. Lau and G. Chen. "Optimal Layouts of Midimew Networks". IEEE Transactions on Parallel and Distributed Systems, Vol. 7, No. 9, pp. 954-961, 1996.

12. C. Martínez, R. Beivide, C. Izu and J. Gutierrez. "Distance-Hereditary Embeddings of Circulant Graphs". Proc. of IEEE Int. Conf. on Information Technology: Coding and Computing (ITCC-2003). pp. 320-324. Las Vegas, 2003.

13. S. Mukherjee, P. Bannon, S. Lang, A. Spink and D. Webb. "The Alpha 21364 Network Architecture". IEEE Micro, Vol. 22, No. 1, pp. 26-35, 2002.

14. V. Puente, C. Izu, J.A. Gregorio, R. Beivide, J.M. Prellezo and F. Vallejo. "Rearranging Links to Improve the Performance of Parallel Computers: The Case of Midimew Networks". Proc. of International Conference on Supercomputing. ICS'2000. pp. 44-53. 2000.

15. R. S. Wilkov. "Analysis and Design of Reliable Computer Networks". IEEE Trans. On Communications, Vol. 20, 660-678.1972.

16. C.K. Wong and D. Coppersmith. "A Combinatorial Problem Related to Multimodule Memory Organizations". Journal of the ACM, Vol. 21, No. 3, pp. 392-402. 1974.

17. Y. Yang, A. Funashashi, A. Jouraku, H Nishi, H Amano and T. Sueyoshi, "Recursive Diagonal Torus: An Interconnection Network for Massively Parallel Computers". IEEE Transactions on Parallel and Distributed Systems, Vol. 12, No. 7, pp. 701-715, 2001.

Distributed Location of Shared Resources and Its Application to the Load Sharing Problem in Heterogeneous Distributed Systems

Satoshi Fujita and Shigeaki Tagashira

Department of Information Engineering,
Graduate School of Engineering,
Hiroshima University

Abstract. In this paper, we propose a distributed algorithm for solving a resource location problem in distributed systems. The proposed algorithm is fully distributed in the sense that it assumes no centralized control, and has a remarkable property such that it can always find a target node satisfying a certain property, if any. The result of simulations implies that: (1) the performance of the underlying load sharing scheme can be significantly improved by increasing the preciseness of a node location, and (2) in the proposed scheme, the average number of inquiries per location is bounded by a very small value (e.g., only two inquiries are enough even when the underlying system consists of 100 nodes).

1 Introduction

In this paper, we propose a new load sharing algorithm for distributed systems that is fully distributed in the sense that it assumes no centralized control by a collection of dedicated servers. In addition, the proposed algorithm has a remarkable property such that it can *always* locate a target node for task migration, if any. It is worth noting that no previous distributed load sharing schemes fulfill such an important property except for the naive server-based schemes, although the possibility of missing a "good" target node is expected to be a main obstacle in constructing (nearly) optimal load sharing schemes.

In the proposed algorithm, a locating of a target node is conducted in the following manner: (1) each node keeps a "fragment" of information concerned with the current status of the system; and (2) a node locates a target node by collecting an appropriate subset of those fragments. The algorithm is an application of the distributed algorithm for maintaining a *dynamic set* proposed by the authors in [5], and it was proved that the number of fragments to be collected is at most $O(\log n)$ per location in an amortized sense, where n is the number of nodes in the given network. The objective of the current paper is to examine the effectiveness of the node location algorithm as a part of load sharing mechanism. By the result of simulations, we found that: (1) the performance of load sharing algorithm can really be improved by increasing the preciseness of a node location, and (2) in the proposed load sharing algorithm, the average

A. Veidenbaum et al. (Eds.): ISHPC 2003, LNCS 2858, pp. 393–401, 2003.

number of inquiries per location is bounded by a very small value (e.g., two or three even for $n = 100$) for the whole range of the system load.

The remainder of this paper is organized as follows. Section 2 introduces necessary notations and definitions. Section 3 overviews related works, including two typical distributed node location algorithms. Section 4 describes an outline of our proposed algorithm. An experimental comparison of the proposed algorithm to previous ones is given in Section 5. Finally, Section 6 concludes the paper with future problems.

2   Preliminaries

Let $V = \{0, 1, \cdots, n-1\}$ be a set of nodes representing a set of host computers. In the following, we use terms "host" and "node" interchangeably. Nodes in V can directly communicate with each other by using an underlying packet routing mechanism. The time for transmitting a datum of size K from node i to node j is given by $\alpha_{i,j} + \beta_{i,j} \times K$, where $\alpha_{i,j}$ corresponds to the communication latency, and $1/\beta_{i,j}$ corresponds to the communication bandwidth. Note that for short messages, the transmission time is approximated by $\alpha_{i,j}$ and for long messages of size K, it is approximated by $\beta_{i,j} \times K$. A task is dynamically created on a host, and can be executed on any host in V, where tasks waiting for their turn on a host form a FIFO queue (i.e., a ready queue) on it. A task in a ready queue can start its execution when all predecessors in the same queue complete their execution; i.e., we assume a *single-tasking* environment. Let $\tau(t)$ denote the execution time of task t on host 0. In the following, we assume that the execution time of any task t on host $v \in V - \{0\}$ is given by $C_v \times \tau(t)$ for some constant C_v (> 0); i.e., we assume a related model [2] for heterogeneous task execution. The **response time** of a task is defined to be the elapsed time from when the task is created at a node to when the execution of the task completes. Note that when the task is executed on a remote host, the response time includes migration time besides waiting and execution times.

The **load** of node v, denoted by $load(v)$, is defined as the number of tasks in the ready queue of node v. The status of a node is classified into the following three states, light, medium, or heavy, by using two thresholds T_L and T_U ($T_L \leq T_U$), as follows: the state of a node v is light if $load(v) < T_L$; medium if $T_L \leq load(v) < T_U$; and heavy if $load(v) \geq T_U$. Those two thresholds could be given to each node independently in heterogeneous systems; e.g., several computers with higher performance should be given higher thresholds than others, in order to take into account the difference of the computational power. In what follows, we call a node in the light state simply a "light node" (we use a similar representation for nodes in the medium or heavy state). Note that the above definitions of load and state merely consider the number of tasks in the ready queue, and do not explicitly reflect the execution time of each task. In the experimental study given in Section 5, however, two thresholds will be associated with execution times, in such a way that T_L roughly corresponds to the response time of 1 seconds, and T_U roughly corresponds to the response time of 4 seconds, for each host computer.

The response time of a task can be reduced by using a task migration mechanism, if the task is held by a heavy node and the cost for migrating the task to a light node is small enough. Since the main objective of this paper is to propose an efficient node location algorithm that could be used in *universal* load sharing situations, in the following, we merely assume the existence of an appropriate task migration mechanism, and will omit the detail of the implementation; the difference of implementations can be reflected to our model as the "ratio" of the communication time to the execution time.

3   Related Work

Load sharing algorithms can be classified into the following three types, by their initiation policy for the location phase [9]; i.e., the **sender-initiated policy**, in which overloaded nodes initiate a location of an underloaded node, the **receiver-initiated policy**, in which underloaded nodes initiate a location of an overloaded node, and the **symmetrically-initiated policy**, in which both overloaded and underloaded nodes initiate their location process.

In the following, we focus on the following two load sharing algorithms as the reference [1, 4]. Let L_p (≥ 1) be a given parameter called probe limit. The first algorithm, referred to as RANDOM, is based on a randomization. The algorithm proceeds as follows:

Algorithm RANDOM (Sender-Initiated)
If the state of a node v becomes heavy, it repeats the following operation at most L_p times or until a task migration occurs: *Node v selects a node u from V randomly, and sends an inquiry to node u to examine if the state of u is light or not; If it is light, node v migrates a task to u.*

The second algorithm is based on the notion of the working set, in which each node v is associated with three working sets Sender_v, $\mathsf{Receiver}_v$, and OK_v[1].

Algorithm WORKING_SET (Sender-Initiated)
If a node v becomes heavy, it sends inquiries to candidate nodes in the following order, until a task migration occurs or the number of inquiries exceeds a probe limit: *from the head of* $\mathsf{Receiver}_v$, *from the tail of* OK_v, *and from the tail of* Sender_v; Receiver u of the message replies the state of u to v; if it is heavy (resp. medium), node v moves u to the head of Sender_v (resp. OK_v) in its local lists; otherwise, node v migrates a task to u. If a node v becomes light, it broadcasts the fact to all nodes w in V. Upon receiving the message from v, node w moves v to the head of $\mathsf{Receiver}_w$ in its local lists.

---

[1] Sender_v is intended to contain candidates for heavy nodes, $\mathsf{Receiver}_v$ is intended to contain candidates for light nodes, and OK_v is intended to contain candidates for medium nodes. $\mathsf{Receiver}_v$ is initialized to contain all nodes in V except for v, and Sender_v and OK_v are initialized to be empty.

4   Proposed Algorithm

4.1   Overview

Recall that V denotes the set of hosts in the given (heterogeneous) network. We first consider a graph G with node set V and edge set E, where two nodes i, j in V are connected by an edge in E if and only if $\beta_{i,j} \leq TH_1$ and $\alpha_{i,j} \leq TH_2$ hold for some predetermined thresholds TH_1 and TH_2. By definition, any two nodes connected by an edge in G are connected by a communication channel with a short latency $(\leq TH_2)$ and a broad bandwidth $(\geq 1/TH_1)$ in the given network. We then calculate a *clique cover* of G; i.e., obtain a set of subsets of nodes $\mathcal{V} = \{V_1, V_2, \ldots, V_k\}$ such that V_i is a complete subgraph of G and $V_1 \cup V_2 \cup \cdots \cup V_k = V$. Note that $V_i \cap V_j = \emptyset$ may not hold, in general. In the proposed method, the following algorithm for achieving a load sharing is applied to *all* subsets in $\mathcal{V}$ simultaneously; i.e., a node in $V_i \cap V_j$ participates to the load sharing for subset V_i, and at the same time, participates to the load sharing for subset V_j. In the following, without loss of generality, we assume that $G = (V, E)$ is covered by a single clique; i.e., G is a complete graph.

Let $U \in V$ be a subset of nodes satisfying some property $\mathcal{P}$. In the proposed algorithm, we let U be the set of light nodes under the sender-initiated policy, and that of heavy nodes under the receiver-initiated policy. In other words, the problem of locating a target node in the given network is restated as the problem of finding an element in a subset U, where U is expected to be maintained appropriately as the state of nodes changes dynamically; e.g., under the sender-initiated policy, a node should be "inserted" to U when the state of the node becomes light from medium, and should be "deleted" from U when its state becomes medium from light.

4.2   Load Sharing Algorithm

In [5], we have proposed a distributed algorithm for maintaining such a dynamic set in a fully distributed manner; more concretely, our algorithm supports the following three operations concerned with the maintenance of the dynamic set: i.e., (1) Insert to insert the caller to the dynamic set; (2) Delete to delete the caller from the dynamic set; and (3) Find to find an element in the dynamic set. By using those three operations, we can describe three load sharing algorithms, as follows:

Algorithm DYNAMIC_SET (Sender-Initiated)

- Initialize local variables of each node in such a way that $U = V$ holds; i.e., $\mathsf{state}_v := \mathsf{idle}$ and $\mathsf{next}_v := v + 1 \pmod{n}$ for each $v \in V$, and the token is initially given to node 0 $(\in V)$.
- If the state of a node becomes light, the node calls Insert to insert itself to the dynamic set.
- If the state of a node changes from light to medium, the node calls Delete to remove itself from the dynamic set.

– If the state of a node v becomes heavy, the node calls Find to locate a light node; if there exists a light node u in U, then node v migrates a task to u and terminates; otherwise (i.e., if $U = \emptyset$), it terminates without invoking task migration.

Receiver-initiated, and symmetrically-initiated algorithms can be described in a similar manner.

5   Simulation

5.1   Analytic Model

The simulation model used in the experiment is described as follows. Each node in V is modeled by a queuing center [7], where an "enqueue" corresponds to the arrival of a task to the node, and a "dequeue" corresponds to the execution of a task. The rate for creating a task on a node is assumed to follow a Poisson distribution with parameter λ [6], and the service time of each task is assumed to follow the distribution given in [6, 8]; i.e., (1) tasks that are shorter than or equal to 1 second follow an exponential distribution with parameter μ; and (2) tasks that are longer than 1 second follow the distribution such that the probability that a task will run for more than t seconds is proportional to t^k, with k in the range $-1.25 < k < -1.05$. In the experiments, we fixed parameter μ to 2.66 for fitting the percentage of such short tasks to 94 % [6], and k to -1 as is recommended in [8] as a default case.

In the following, parameter λ is referred to as the **system load** (the reader should notice that it is different from the "load" of a node). The system load takes the same value for all nodes in the system. The time required for an inquiry and a reply is fixed to 0.03 seconds, since the time required for an inquiry is dominated by the time for collecting the status of the receiver node, and by the result of our preliminary experiments, it generally takes 0.03 seconds on typical Unix workstations. The time required for broadcasting a short message that is used in WORKING_SET is fixed to 0.03 seconds, as well. The time required for migrating a task is fixed to 0.3 seconds as is assumed in [6], although it is actually dependent on the task size. In the following, we assume that the time for executing the code of load sharing algorithms is negligible compared with the communication time, and unless otherwise stated, the number of nodes is fixed to 100 and the probe limit is fixed to 5.

Two thresholds T_L and T_U are determined as $T_L = 2$ and $T_U = 3$ by the result of preliminary experiments (we omit the detail of the experiments in this extended abstract). In the following, besides those algorithms described in Sections 3 and 4, we consider the following two extreme cases in order to clarify the goodness (or badness) of the algorithms; The first case corresponds to a situation with no task migrations. This extreme case will be referred to as M/M/1, since it can be modeled by n independent M/M/1 queues. The second case corresponds to a situation in which a task created on a node u is enqueued into the ready queue of a node v with the lowest load, where an enqueue involves a task migration time if $u \neq v$. In what follows, we call it SHORTEST[2].

5.2   Fundamental Comparison of Location Algorithms

In the remainder of this section, we evaluate the effectiveness of the proposed algorithm DYNAMIC_SET in terms of the response time of each task and the number of inquiries issued per location phase. A comparison of algorithms under the sender-initiated policy is shown in Figure 1, where Figure 1 (a) illustrates the average response time, and Figure 1 (b) illustrates the average number of inquiries issued per location phase. In both figures, the horizontal axis represents the system load λ.

From Figure 1 (a), we can observe that DYNAMIC_SET outperforms the other schemes for $0.5 < \lambda \leq 0.7$, although it is slightly worse than the others for $0.05 \leq \lambda \leq 0.5$; e.g., when $\lambda = 0.65$, the average response time of DYNAMIC_SET is only 25 % of WORKING_SET and 55 % of RANDOM. This phenomenon is due to the high preciseness of a node location in DYNAMIC_SET; i.e., in the scheme, unlike other schemes, a heavy node can always locate a light node, as long as such a node exists in the system. The effect of such a precise location is expected to be maximized when both heavy and light nodes exist in the system and the number of heavy nodes exceeds the number of light nodes. For larger λ, however, the curves in the figure get complicated, which is probably due to the "blocking" of successors by a long task, that is inevitable under the single-tasking environment. In fact, in the simulation, more than 0.01 % of the tasks have an execution time of more than 500 seconds, and such a blocking behavior is frequently observed (there is a task whose execution is blocked for more than 500 seconds). The average number of inquiries issued per location phase is given in Figure 1 (b). From the figure, we can observe that although DYNAMIC_SET exhibits almost

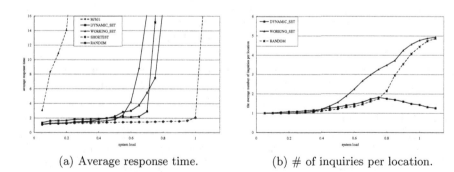

(a) Average response time.   (b) # of inquiries per location.

Fig. 1. Comparison of sender-initiated algorithms ($n = 100$).

---

[2]   Since the load of a node is defined by the number of tasks rather than the actual waiting time, the selection of a node with a lowest load does not always cause an earliest start time of migrated tasks; i.e., SHORTEST is not optimal in this sense.

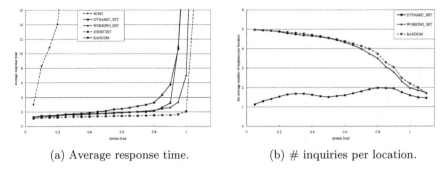

(a) Average response time. (b) # inquiries per location.

Fig. 2. Comparison of receiver-initiated algorithms ($n = 100$).

the same behavior with RANDOM for $0.05 \leq \lambda \leq 0.75$, the number of inquiries gradually decreases for larger λ, in contrast to the other two schemes.

A comparison of algorithms under the receiver-initiated policy is shown in Figure 2. From Figure 2 (a), we can observe that DYNAMIC_SET exhibits a worse performance than the other two schemes for the whole range of λ. A reason of this phenomenon could be explained as follows: (1) DYNAMIC_SET involves a (slightly) more overhead than the other two schemes, (2) for small λ, the expected number of tasks in a ready queue is also small, which can bound the response time of each task to be small enough, and (3) for large λ, the preciseness of a node location does not help to reduce the response time, since under such a situation, *any* location algorithm is expected to locate a heavy node very easily. The second point we can observe from the figure is that the system load at which an avalanche of slow-down occurs shifts to the right-hand side by changing the initiation policy from sender-initiated to receiver-initiated. A reason of this phenomenon could be explained by the difference of the number of location phases that cause a delay of all tasks in the queue; i.e., for large λ, almost all nodes initiate a node location for every event (of "enqueue" and "dequeue") under the sender-initiated policy, while under the receiver-initiated policy, such an initiation rarely happens. Although WORKING_SET exhibits the best performance among others, it is because of the frequent broadcasts issued by heavy nodes, and such a superiority will be reduced if we use a refined model for inter-processor communication (we are assuming that each broadcast takes 0.03 seconds regardless of the number of receivers). Figure 2 (b) shows the average number of inquiries under the receiver-initiated policy. We can observe that the number of inquiries is bounded by two under DYNAMIC_SET, although under the other two schemes, it approaches to the probe limit by decreasing λ to zero.

A comparison of algorithms under the symmetrically-initiated policy is shown in Figure 3. From the figure, we can observe that although three schemes exhibit almost the same response time for $0.05 \leq \lambda \leq 0.75$, DYNAMIC_SET apparently outperforms the other schemes for larger λ; e.g., even when $\lambda = 0.9$, the average response time is less than 5 seconds, although that of the other schemes exceeds 50 seconds (it corresponds to a speedup of more than 10 times). As for the

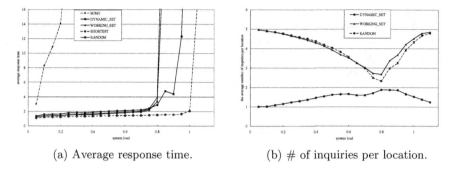

(a) Average response time.   (b) # of inquiries per location.

Fig. 3. Comparison of symmetrically-initiated algorithms ($n = 100$).

number of inquiries, three schemes exhibit a "combined" feature of the sender-initiated and receiver-initiated policies (see Figure 3 (b)).

6   Conclusion

In this paper, we proposed a distributed algorithm for solving a resource location problem, that generally arises in solving the dynamic load sharing problem in distributed systems. The result of simulations implies that in the proposed algorithm, the number of inquiries per location is bounded by a very small value for the whole range of the system load; e.g., it is bounded by 2 even when the number of nodes is 100. As a future problem, we should conduct simulations under heterogeneous environments. An implementation on an actual distributed environment is also an important direction for future research.

Acknowledgements

This research was partially supported by the Japan Society for the Promotion of Science Grant-in-Aid for Scientific Research (C), 13680417.

References

1. R. D. Blumofe and C. E. Leiserson. Scheduling multithread computations by work stealing. *JACM*, 46(5):720–748, 1999.
2. A. Borodin and R. El-Yaniv. *Online Computation and Competitive Analysis*. Cambridge University Press, 1998.
3. C. Catlett and L. Smarr. Metacomputing. *CACM*, 35:44–52, 1992.
4. D. L. Eager, E. D. Lazowska, and J. Zahorjan. Adaptive load sharing in homogeneous distributed systems. *IEEE Trans. Software Eng.*, 662–675, May 1986.
5. S. Fujita and M. Yamashita. Maintaining a dynamic set of processors in a distributed system. *Proc. 10th WDAG LNCS*, 1151:220–233, 1996.
6. M. Harchol-Balter and A. B. Downey. Exploiting lifetime distributions for dynamic load balancing. In *Prof. ACM SIGMETRICS '96 Conference on Measurement & Modeling of Comput. Syst.,*, pages 13–24. ACM, 1996.

7. L. Kleinrock. *Queuing Systems, Vol. 1.* Theory New York: Wiley, 1976.
8. W. E. Leland and T. J. Ott. Load-balancing heuristics and process behavior. In *Prof. ACM SIGMETRICS '86 Conference on Measurement & Modeling of Comput. Syst.,*, pages 54–69. ACM, 1986.
9. N. G. Shivaratri, P. Krueger, and M. Singhal. Load distributing for locally distributed systems. *IEEE Computer*, 25(12):33–44, December 1992.

Design and Implementation
of a Parallel Programming Environment
Based on Distributed Shared Arrays

Wataru Kaneko, Akira Nomoto, Yasuo Watanabe,
Shugo Nakamura, and Kentaro Shimizu

Department of Biotechnology, The University of Tokyo,
1-1-1 Yayoi, Bunkyo-ku, Tokyo 113-8657, Japan,
{wataruk,nomoto,wa,shugo,shimizu}@bi.a.u-tokyo.ac.jp,
http://www.bi.a.u-tokyo.ac.jp/

Abstract. We have developed a parallel programming environment called the "Distributed Shared Array" (DSA), which provides a shared global array abstract across different machines connected by a network. The DSA provides array-specific operations and fine-grained data consistency based on arrays. It allows a user to explicitly manage array area allocation, replication, and migration. This paper also describes the use of our model for gene cluster analysis, multiple alignment, and molecular dynamics simulation. Global arrays are used in these applications for storing the distance matrix, alignment matrix and atom coordinates, respectively. Large array areas, which cannot be stored in the memory of individual machines, are made available by the DSA. Scalable performance of DSA was obtained and compared to that of conventional parallel programs written in MPI.

1 Introduction

Multiple computer systems that consist of many machines of various configurations such as massively parallel processors, cluster computers and personal computers are used widely as hardware platform of scientific computing. Such hardware platform contains both shared-memory and distributed-memory architectures and various topologies of interconnection networks. It is important to design a uniform parallel programming environment which realizes efficient and transparent data sharing in such heterogeneous multiple computer systems.

There are two primary programming models for parallel programming environments: message-passing model and shared-memory model. The message-passing model is widely used and is implemented by message-passing libraries such as MPI [1] and PVM [2]. It enables programmers to explicitly distribute data and balance the computation load, but such programming is difficult in some applications. The shared-memory model simplifies programming by hiding communication details, but the conventional programming environments based on the shared-memory model, namely, software distributed shared-memory

A. Veidenbaum et al. (Eds.): ISHPC 2003, LNCS 2858, pp. 402–411, 2003.

(DSM) system, often provide little control over communication and computation costs. We are now developing a parallel programming environment, called "Distributed Shared Array" (DSA), which realizes both simple programming and efficient data sharing. The DSA provides a shared global array abstract across different machines connected by a network.

The DSA differs from the conventional DSM system, in that its operations are based on arrays. Unlike the page-based DSM [10], fine-grained data consistency is maintained for each array element or each region of an array. To achieve the efficient data sharing, a programmer manages data consistency explicitly so that he or she can select application-specific consistency models. Data replication and migration are also supported to promote optimization for desired applications by data reuse and locality of reference. Their programming interface allows users to specify the regions of an array to be replicated or migrated. Such partial replication and migration are the features of our system which have not been supported by the conventional object-based DSM systems [7][8][9].

Many scientific applications require a lot of memory space for computation. Because one computer can only have a limited amount of memory, it is often necessary to use the memory distributed among the computers connected by a network. DSA enables to access such distributed memory areas through the uniform interface of distributed shared array. In DSA, disk accesses are performed only when the local memory and the remote memories of neighbor nodes are fully used.

2   System Design

2.1   Programming Model of DSA

The DSA provides global arrays that are shared by multiple machines in a network (Figure 1). It is a location-transparent entity whose physical storage areas may be distributed among any machines in a network; users can access global arrays in the same manner irrespective of their physical location. The DSA is based on a programming model in which concurrent processes are synchronized explicitly and communicate by executing operations on the shared array as shown in Section 2.2. Since in the DSA a programmer specifies explicit calls to functions accessing global array, it is easy to construct arrays that are too large for the memory of each machine and achieve such explicit management by the programmer. Linear algebra functions such as matrix-matrix multiplication and inverse matrix calculation are implemented on global arrays.

In DSA, when a user accesses an array, he or she defines a region, a part of the array to be fetched to the local node. The system creates a replica (local copy) of that region if the region has not been allocated in the node. Thus, an array is partially replicated in DSA. The DSA provides acquire and release operations for data consistency and they are peformed on the region. Migration is also performed on region basis.

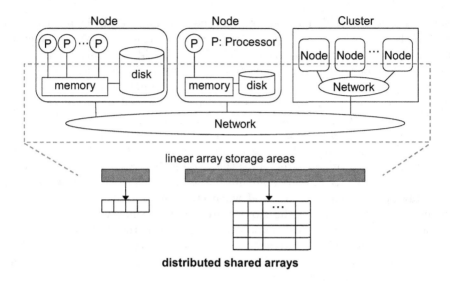

distributed shared arrays

Fig. 1. Overview of the DSA system

2.2   Programming Interface

The DSA system provides the interface of a global array as the C++ class library. Table 1 shows the basic interface. **Array** class represents a global array; its dimension and element type are parameterized. The element type can be a user-defined type, but it must be scalar type like **int, float**, etc. The constructor and destructor are used to create and delete an array, respectively. The size of an array is specified as arguments in the constructor and the DSA system allocates its storage on demand distributedly.

Read and write operations against array elements are performed on a local replica of an array. The **getReplica** is used to obtain a replica when a user

Table 1. Array operations

Array::Array()	Constructor to create a global array.
Array::~Array()	Destructor to delete a global array.
Array::getReplica(region)	Return the ReplicaHandle of specified region.
Array::operator()(indices)	Access to the specified element.
ReplicaHandle::acquire(read|write)	Acquire a lock of the corresponding region.
ReplicaHandle::release()	Release a lock of the corresponding region.
ReplicaHandle::~ReplicaHandle()	Release if being acquired, and allow the system to discard this replica.
ReplicaHandle::setUpdatePriority(prio)	Set the update priority.
ReplicaHandle::setStoragePriority(prio)	Set the storage priority.
AsyncHandle::wait()	Wait until the corresponding async operation completes.

defines the region. This method returns a `ReplicaHandle` object to identify a replica. For consistency of replicas, `acquire` and `release` must be called at the start and end of access, respectively. They provide the single-writer-and-multi-readers lock mechanism against corresponding region; other regions can be accessed concurrently. To overlap computation and communication easily, these methods are executed asynchronously and return `AsyncHandle` object that is used to wait for completion of operation.

In DSA, each replica has an update priority which defines the order of replica updates. It is specified by the `setUpdatePriority`. A replica with the highest-priority is updated immediately after the write lock is released; A replica with the lowest-priority is not updated until it is acquired. The storage areas of an array are dynamically allocated when creating its replica in one or more nodes. The deallocation of the storage areas is performed when they are not needed. The order of the deallocation is determined by the storage priority. For a replica that will be used in a short time, a higher strorage priority should be given to the replica. This priority can be specified by the programmer by using `setStoragePriority` method.

The system ensures that there exists at least one up-to-date replica for any region to assure consistency.

2.3   System Structure

Figure 2 shows the system structure of the DSA. Our implementation adopts a centralized control to manage shared resources; it consists of a single master node and multiple worker nodes. The master node has the following modules: The node manager monitors dynamic usage of resources in each node such as CPU usage and free memory space, and provides this information to control policies of replica management. The replica catalog maintains locations and attributes of all replicas of the array. In the DSA system, each array is given a

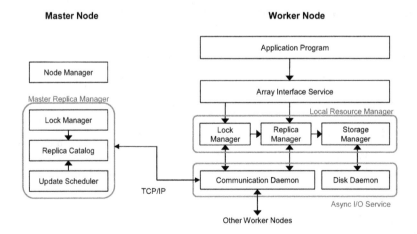

Fig. 2. System architecture of DSA

unique global identifier which is used to locate the physical replica. The lock manager maintains the locked regions of arrays and performs mutual exclusion of lock-requests. The update scheduler schedules the updates of replicas and issues the requests to worker nodes for exchanging updated data with each other; the replica consistency is maintained by this module.

Each worker node has the following modules: The array interface service provides an interface which an application program uses to access the array elements in the local replicas. The lock manager and the replica manager manage the locks and the replicas requested by local threads, respectively. The storage manager manages the storage areas for replicas. The async I/O service provides asynchronous access to disk and network, and schedules the I/O requests from blocked threads with high priority.

We implemented the DSA in the C++ language with the POSIX threads. The array library is thread-safe and can be used in combination with the MPI library. TCP/IP is used as the basis for control information exchange and data transfer.The DSA is now running on a cluster of Pentium III / 700 MHz dual processor nodes connected by an Ethernet 100base-T network. The system runs on the Linux 2.2.12 operating system.

3   Applications

3.1   Gene Cluster Analysis

Gene cluster analysis is used to classify genes into groups with similar functions by comparing gene expression patterns (sequences of gene expression values for individual tissues) to clarify the functional relationships among genes. We used an algorithm called UPGMA, or Unweighted Pair Group Method with Arithmetic mean [3], which iteratively joins the two nearest clusters (or groups of pattern sequences) creating a new cluster, until only one cluster remains. The distance between clusters is defined as the average of distances between constituent pattern sequences. UPGMA is characterized by use of a distance matrix that holds distances between two distinct genes. Obviously a distance matrix with size N^2 is needed to cluster N genes.

In our cluster analysis, we used more than 18,000 genes and a 1.2 GB distance matrix. Because such a distance matrix is too large to fit into the memory of each node of our cluster machine, we used the DSA system to store the distance matrix. Our cluster machine comprises six nodes, each of which has 256 MB memory. We partitioned the distance matrix into nodes; each node had to store only about 206 MB. The matrix was then put into the memory. In comprehensive genome analysis, more than 30,000 genes must be analyzed. The DSA system will be useful by cluster machines with sufficiently large memory.

Figure 3 shows a log-log plot of computation time in gene cluster analysis using the above cluster machine when the number of genes (N) is 7,000 and 18,000. The solid line shows the ideal (perfectly linear) speed increase. We mean by the ideal speed increase that the computation time with P processors is $1/P$ of that with a single processor. Computation of cluster analysis of 18,000 genes

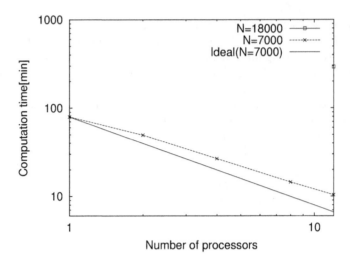

Fig. 3. Performance of gene cluster analysis

can not be computed with less than 12 processors as a result of lack of memory. With 12 processors, the speed increase was 7.7 times. Scalable performance was obtained for the number of processors shown in this figure.

3.2   Multiple Sequence Alignment

Multiple alignment is used to find the relationship between sequences of nucleotides or amino acids. It is widely used for biological analysis such as structural and functional analysis of proteins, phylogenetic analysis, and so on [3].

Dynamic programming (DP) is known as an optimal algorithm of multiple alignment. However, the computational cost of multiple alignment based on DP is $O(2^N L^N)$, where N is the number of sequences to be aligned and L is an average length of sequences. Multiple alignment based on DP also requires a large memory area as a N-dimensional score matrix whose size is $O(L^N)$. For example, DP calculation of four sequences of amino acids which have 130 residues, on an average, requires about 1 GB of memory.

We used DSA for calculating multiple alignment based on DP. For parallel computation, we developed a block-based wave front algorithm in which the alignment matrix is divided into blocks and the computation of each block is executed in parallel after all preceding blocks are executed. Figure 4 shows a log-log plot of the computation time in multiple alignment for four sequences of amino acids, all of which have 80 residues. Multiple alignment of protein sequences of 130 residues can only be computed with 12 processors. In this case, about 8.8 times speed increase was achieved.

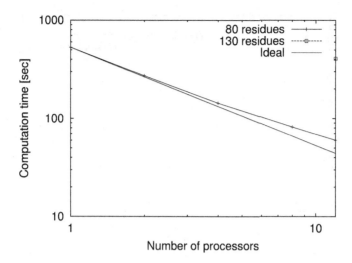

Fig. 4. Performance of multiple alignment

3.3   Parallel Molecular Dynamics Simulation

In molecular dynamics (MD) simulation, a continuous process is disaggregated into discrete timesteps, each of which is a combination of two processes: force calculation (calculating forces from the evaluated conformational energies) and atom update (calculating new coordinates of the atoms). There are two commonly used parallel MD simulation algorithms: replicated data (RD) and space decomposition (SD) [4]. In RD, each node has complete copies of atom positions (coordinates); it updates these positions and calculates forces on each node. In SD, the simulation domain is usually broken into subdomains (cells); then each node computes forces only on the atoms in its subdomain. Each node requires only that information from the nodes assigned to neighboring subdomains. Our simulation used the SD algorithm; each node had information only on the atom positions in the subdomain. By this approach, we can simulate a large system that has too many atoms to store in the memory of each node.

We ran an MD simulation for a system consisting of one BPTI (bovine pancreatic trypsin inhibitor) protein surrounded by water molecules and comprising a total of 16,735 atoms. The simulation domain was divided into 125 ($= 5 \times 5 \times 5$) subdomains. Figure 5 shows results of the molecular dynamics simulation. When the number of processors was 12 (6 nodes), the performance of our SD algorithm based on the DSA was about twice as good as the performance of the RD algorithm. As the number of nodes increased, our DSA-based SD simulation became more efficient than RD simulation. The RD simulation requires global synchronization when broadcasting the force to all nodes with `MPI_Allreduce()`, which does not allow for computation and communication to be overlapped (performed concurrently). By our measurement of details of the execution time for 12 pro-

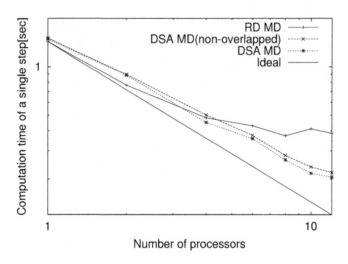

Fig. 5. Performance of molecular dynamics simulation

cessors, communication time constitutes more than 70% of the execution time in RD simulation.

Such overlapped processing can be implemented by use of asynchronous operations described in Section 2.2. Figure 5 also shows results of overlapped and non-overlapped processing. In overlapped processing, computation starts immediately after related signals have arrived without waiting for all signals. In non-overlapped processing, computation starts after all signals have arrived. As we can see, overlapped processing improves computation efficiency by about 6.4% with 12 processors. Conventional SD simulation can suffer deficiency as a result of the load imbalance caused by non-uniform atomic densities, but our SD simulation reduces such deficiency by dividing the unit of computation, namely by dividing the force calculation process into two subprocesses based on the cell pair decomposition technique [4].

4   Related Work

There exist several works on systems that support network-wide, shared global arrays whose physical entities are distributed and shared. Global Array/Disk Resident Array (GA/DRA) [5][6] is most related to this paper. The GA exposes programmers to Non-Uniform Memory Access (NUMA) characteristics of parallel computers, providing an array-based programming model both at the memory (GA) and the secondary storage levels (DRA). In the GA, the programmer explicitly specifies data transfer between a global array space and local storage. Our system, DSA, has a completely different design; memory hierarchy and data location are transparent to programmers so that the system is responsible for the adaptation to various hardware-configurations and dynamic resource changes. In the DSA, the programmers can access global arrays in the same manner irrespective of whether the physical location of array areas is remote or local.

To the best of our knowledge, Xu et al. [11] first introduced the term of "distributed shared array," and presented a prototype system as a proof-of-concept. Their implementation is based on Java; the adaptation to local resources is up to the virtual machine. Our system is implemented as a runtime system that runs directly on top of the operating system, so it can directly monitor accesses to arrays and use of local resources. Our system supports multi-dimensional arrays and asynchronous operations while Xu's system only supports single-dimensional arrays and synchronous operations. Region-based consistency is the feature of our system, and invalidation or update consistency protocol can be selected while Xu's system only supports invalidation protocol. Unlike Xu's distributed shared array, our system provides the facilities of user-controlled replication and migration, which are useful for efficient data sharing in large-scale multiple computer systems.

5   Conclusions

We described a parallel programming environment that provides programmers with a shared global array abstract across different machines connected by a network. Our system is designed to support large-scale scientific applications with high computation cost that require so much memory area that the memory of one machine is insufficient. The system implements efficient data sharing on a distibuted system of machines with various configurations. For such an environment, data replication and migration are important to improve efficiency. In addition, their policies can be determined using resource information about each node.

We found that programs can be described easily using an array-oriented communication infrastructure and programming interface. Our system can be used effectively for gene cluster analysis and molecular dynamics simulation. Existing application programs can also expected to be run on our system with small modifications: programs based on the shared memory model can be used on our system by modifying localized portions of array accesses. We are now applying our system to calculation of second derivatives (Hessian) of conformational energy, which can be used for Newton-Raphson energy minimization, normal mode analysis, and Monte Carlo simulation. Our systems are inferred to be effective for these applications.

This work was supported by Grant-in-Aid for Scientific Research on Priority Areas (C) "Genome Information Science" and "Informatics Studies for the Foundation of IT Evolution" from the Ministry of Education, Culture, Sports, Science and Technology of Japan.

References

1. Message Passing Interface Forum, *MPI: A Message-Passing Interface Standard* 1.1 edition, 1995.
2. Geist, A., et al.: *PVM: A Users' Guide and Tutorial for Networked Parallel Computing* MIT Press, 1994
3. Durbin, E., Eddy S., Krogh A., Mitchison G.: *Biological sequential analysis*, Cambridge University Press, 1998.
4. Kal, L., et al.: NAMD2: Greater scalability for parallel molecular dynamics, *Journal of Computational Physics*, 151, 283-312, 1999.
5. Nieplocha, J., Harrison R.J., Littlefield, R.J.: Global Arrays: A nonuniform memory access programming model for high-performance computers, *The Journal of Supercomputing*, 10, 197-220, 1996.
6. Nieplocha, J., Foster, I.: Disk Resident Arrays: An array-oriented I/O library for out-of-core computations, *In Proceedings of IEEE Conference on Frontiers of Massively Parallel Computing*, 196-204, 1996.
7. Carter, J.B., Bennett J.K., Zwaenepoel, W.: Implementation and performance of MUNIN, *In Proceedings of 13th ACM Symposium on Operating Systems Principles*, 152-164, 1991.
8. Bershad, B.N., Zekauskas, M.J., Sawdon, W.A.: Midway: Shared Memory Parallel Programming with Entry Consistency for Distributed Memory Multiprocessors, Technical Report CMU-CS-91-170, Carnegie Mellon University, 1991.
9. Castro, M., Guedes, P., Sequeira, M., Costa M.: Efficient and flexible object sharing, *In Proceedings of 1996 International Conference on Parallel Processing*, 128-137, 1996.
10. Keleher, P., Cox, A.L., Dwarkadas, S., Zwaenepoel, W.: TreadMarks: Shared memory computing on networks of workstations, *IEEE Computer*, 29, 2, 18-28, 1996.
11. Xu, C., Wims, B., Jamal, R.: Distributed shared array: an integration of message passing and multithreading on SMP clusters, *In Proceedings of the 11th IASTED International Conference on Parallel and Distributed Computing and Systems*, 305-310, 1999.

Design and Implementation
of Parallel Modified PrefixSpan Method

Toshihide Sutou, Keiichi Tamura, Yasuma Mori, and Hajime Kitakami

Graduate School of Information Sciences, Hiroshima City University,
3-4-1 Ozukahigashi, Asaminami-ku, Hiroshima-shi, Hiroshima 731-3194, Japan,
{tosihide,ktamura,mori,kitakami}@db.its.hiroshima-cu.ac.jp

Abstract. The parallelization of a Modified PrefixSpan method is proposed in this paper. The Modified PrefixSpan method is used to extract the frequent pattern from a sequence database. This system developed by authors requires the use of multiple computers connected in local area network. This system, which has a dynamic load balancing mechanism, is achieved through communication among multiple computers using a socket and an MPI library. It also includes multi-threads to achieve communication between a master process and multiple slave processes. The master process controls both the global job pool, to manage the set of subtrees generated in the initial processing and multiple slave processes. The results obtained here indicated that 8 computers were approximately 6 times faster than 1 computer in trial implementation experiments.

1 Introduction

A motif is a featured pattern in amino acid sequences. It is assumed to be related to a function of the protein that has been preserved in the evolutionary process of an organism. The amino acid sequence is composed of 20 kinds of alphabets. The featured pattern, which includes wild cards, is discovered from frequent patterns extracted in the sequences. However, the existing pattern-extraction algorithms, which include multiple-alignment, statistical method, and PrefixSpan method [1], present some problems. These algorithms are neither functional nor fast for the discovery of motifs from large-scale amino-acid sequences.

To solve the functional problem, a Modified PrefixSpan method [2] was proposed, which introduces a mechanism to limit the number of wild cards to the PrefixSpan method. The Modified PrefixSpan method was successful in the deletion of extra patterns and in the reduction of computational time for about 800 amino acid sequences, each of which ranges in length from 25 to 4,200 characters.

To achieve faster pattern extraction from large-scale sequence databases, the parallelization of the Modified PrefixSpan method by the use of multiple computers connected in local area network was used and is explained in this paper. Parallelization consists of dividing frequent pattern-extraction processing by the Modified PrefixSpan method into multiple k-length (explain in Section 2) frequent pattern-extraction processing to achieve dynamic load balancing.

A. Veidenbaum et al. (Eds.): ISHPC 2003, LNCS 2858, pp. 412–422, 2003.
© Springer-Verlag Berlin Heidelberg 2003

One computer, which manages parallel processing, divides the entire extraction into multiple k-length frequent-pattern extractions. The k-length frequent-pattern extraction is distributed to the computers, which perform parallel processing. As a feature of the Modified PrefixSpan method, the workload of each k-length frequent pattern-extraction processing is different. After a k-length frequent pattern is extracted by a computer, the step is repeated. Other computers continue processing, one after another, while a designated computer extracts a k-length frequent pattern. As a result, the load balancing is dynamic.

The Modified PrefixSpan was implemented based on a k-length frequent-pattern extraction on an actual PC cluster. The method with a dynamic load-balancing mechanism is achieved by communication among multiple computers that make use of a socket and an MPI library. Moreover, this method includes the use of multi-threads to achieve communication. Our result indicated that 8 computers were approximately 6 times faster than 1 computer in the trial implementation experiments.A verification experiment confirmed the effective parallel effect.

The remainder of this paper is organized as follows: Section 2 contains related work about motif discovery. Section 3 introduces the Modified PrefixSpan method. Section 4 describes the method of parallelization for the Modified PrefixSpan method and reports the design and implementation of the parallel Modified PrefixSpan method. Section 5 contains the experimental results, and Section 6 is the conclusion.

2   Related Work about Motif Discovery

In this section, related work for motif discovery methods in the field of molecular biology is described.

Let L be the average sequence length for the set of sequences, $\{S_1, S_2,..,S_N\}$, defined over some alphabet Σ. Multiple alignment requires a large amount of time to find only one optimal solution, since the time complexity for the multiple alignment is $O(L^N)$. In spite of the significant time it takes, multiple alignment does not reveal other motifs that include the set of sequences.

In order to solve the problem, Timothy et al. developed a new system named MEME [3], which has a two-step approach. First, multiple alignment is applied to sub-sequences that are decided by parameters given by the user, such as motif length, W, and the position of the motif. Second, the expectation maximization algorithm, which is a statistical method, is repeatedly applied to the sub-sequences processed in the first step. After the expectation maximization algorithm is executed N times, frequent patterns are extracted from the sub-sequences. The time complexity of MEME is less than the multiple alignment $O((NM)^2 W)$. However, the frequent patterns extracted by MEME do not include any wild cards. Therefore, it is difficult to find the relationship between them and motifs including such motif databases as PROSITE [4] and Pfam [5].

The Pratt system [6] proposed by Jonassen et al. can extract frequent patterns P, including wild cards from the set of segments B_w in which each segment named a W-segment has a sub-sequence of length W and is one of all the sub-sequences constructed from the sequences. The length W given by the user signifies the maximum length of frequent patterns extracted by the system. To be able to detect patterns near the ends of sequences, W-segments are also constructed at the ends by padding the sequences with W-1 dummy symbols that are not in Σ. The Pratt system uses an algorithm that recursively generates (k+1)-length frequent patterns from k-length frequent patterns in B_w, where 1 $\leq$ k. If the maximum length that a useful pattern can have is not predictable for users, there is a problem, namely, that the Pratt system overlooks patterns that are longer than the length W.

The TEIRESIAS algorithm [7] [8] [9] to solve the problem has a two-phase approach. First, the scanning phase finds the set of elementary frequent patterns that include exactly L alphabet characters and have pattern length U, where $L \leq U$. Second, the convolution phase reconstructs the maximal frequent pattern from the all-elementary frequent patterns gathered in the first phase. The TEIRESIAS algorithm finds elementary frequent patterns that include the maximum number $(L\text{-}U)$ of wild-card symbols in total. However, TEIRESIAS does not afford the user the capability to directly specify the maximum size V of a wild-card region between alphabet characters included in any elementary frequent pattern. Moreover, it does not follow any principle when the user tries to enter the number L of alphabet characters. The main objective of TEIRE-SIAS is to find the maximal frequent patterns. The main objective is not to find whole frequent patterns. Under the circumstances, TEIRESIAS is not easy for users to use.

The Modified PrefixSpan method [2] proposed here is more useful than the PrefixSpan method [1] and allows the user to specify the maximum number V of symbols that are included in the wild-card region appearing between alphabet characters. Moreover, it is also capable of finding whole frequent patterns. The method does not require specifying the number L of alphabet characters for users, since it does not need to find the elementary frequent patterns. Moreover, it is not needed for the construction of the W-segments that restrict the maximum length of the frequent patterns. In this way, the method is more convenient than the previously stated methods of TEIRESIAS and Pratt.

3   Modified PrefixSpna Method

A Modified PrefixSpan method needs a minimum support ratio and a maximum number of wild cards. The support ratio indicates the percentage of a pattern, where the support count of the pattern is defined as the number of elements in a set of sequences including the pattern. The existing PrefixSpan method cannot generate frequent patterns including some wild cards. For instance, "A***B" and "A****B" are considered to be a pattern named "AB" in the PrefixSpan method, where * indicates one wild-card symbol. However, the Modified PrefixS-pan method distinguishes pattern "A***B" from "A****B." These two patterns are represented as "A3B" and "A4B," respectively, where each value indicates the number of wild cards.

The PrefixSpan method generates (k+1)-length frequent patterns from each k-length frequent pattern in a set of sequences. The last character of each (k+1)-length frequent pattern is found from one of the characters that exist among the next position of a k-length frequent pattern and the last position in the sequencers. On the other hand, the Modified PrefixSpan method finds the last character from one of the characters that exist from the next position of a k-length pattern to the maximum length V of the wild-card symbols. The Modified PrefixSpan method is narrower in the range of frequent-pattern extractions than the PrefixSpan method. Therefore, the Modified PrefixSpan method can anticipate a speed-up with more accuracy than the PrefixSpan method because of the long sequence data.

First of all, the Modified PrefixSpan method extracts the 1-length frequent patterns which fill the support ratio from sequences. Next, this method extracts the 2-length frequent patterns from one 1-length frequent pattern. In fact, this method extracts the (k+1)-length frequent pattern from k-length frequent pattern, where 1 k. For instance, Figure 1 shows frequent patterns that are extracted in the two sequences, where the minimum support rate is 100 percent and the number of wild cards is 3.

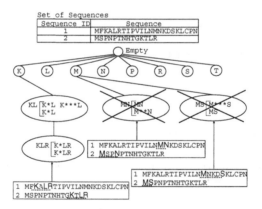

Fig. 1. Extraction of Freqent Pattern of Modified PrefixSpan Method

The Modified PrefixSpan method is shown in the tree structure. In Figure 1, this method extracts 1-length frequent patterns, "K, L, M, N, P, R, S, T," which fill the support ratio. Next, this method extracts 2-length frequent patterns from 1-length frequent patterns, which are extracted. When the 1-length frequent pattern is "K," the 2-length frequent patterns are "K*L" in Figure 1. Because of the same number of wild cards in the two sequences, the 2-length frequent pattern, "K1L," is extracted. Next, this method extracts 3-length frequent patterns from the 2-length frequent pattern, "K*L." The extracted 3-length frequent pattern is "K*LR." In this frequent pattern, the wild card between "L" and "R" is 0. In a word, this is the same number of wild cards. Therefore, the k-length frequent patterns, which are extracted from the 1-length frequent pattern "K," are 2-length "K1L" and 3-length "K1LR." Next, when the 1-length frequent pattern is "M," the 2-length frequent patterns are "MN" and "M**N" in Table 1 and Figure 1. Because of the different number of wild cards in the two sequences, the 2-length frequent pattern, "MN," is not extracted. The 2-length frequent pattern "MS" is also not extracted because of the different number of wild cards. Thus, the Modified PrefixSpan method divides the branch sequentially and extracts the frequent pattern.

Table 1. Experimental Environment of the Cluster Machine

8 Personal Computers	CPU 450MHz Intel PentiumIII Processor
	(L1 cache 16k, L2 cache 512k)
	Memory 128MB
	DISK 6G
Network	100 M-bit Ethernet Switching HUB
OS	Redhat Linux 8.0
Compiler	Intel C/C++ complier 6.0
MPI	mpich-1.2.4

4   Parallel Modified PrefixSpan Method

4.1   Parallel Frequent Pattern Extraction

In the parallelization of the Modified PrefixSpan method, first of all, one computer extracts from the 1-length to the k-length frequent pattern using a breadth-first search, where k indicates the value of depth given by the user. Hereafter, the searched-for arbitrary depth is called the threshold. Each process receives some of the k-length frequent patterns and keeps processing in order to generate longer frequent patterns than the length k. The Modified PrefixSpan method processes the search for the tree structure. By extracting multiple k-length frequent patterns, k-length frequent-patterns extraction processing is processed independently of other extraction processing.

Figure 2 shows the flow of processing when the threshold is 2 and the number of processes is four. Hereafter, multiple k-length frequent patterns, which are nodes of each subtree, are called jobs.

Amino acids correspond to the letters "A, C, D, E, F, G, H, I, K, L, M, N, P, Q, R, S, T, V, W, Y" of the alphabet. In other words, 20 letters of the alphabet are used, excluding "B, J, O, U, X, Z."

In Figure 2, because the threshold is a maximum of two, the combination of alphabets of the amino acid shows 400 jobs, which are 20 letters times 20 letters. When the jobs are extracted, for instance, with four computers, each computer will do 100 jobs on the average.

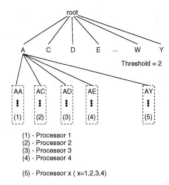

Fig. 2. Parallel Tree Search for Threshold 2

4.2   Design and Implementation

The process, which generates the job and manages other processes, is called the master process. This process uses Multithreading to manage the job dynamically. This process generates threads of number of other processes. The other processes except the master process are called the slave processes. The slave process extracts the frequent pattern by the Modified PrefixSpan method. Each slave process communicates with each thread of master process. The management of multiple computers uses the MPI library. The sending and receiving of the master process data of each slave-process uses the socket. The selection of a job from the master process and communication between the master and slave processes are implemented in parallel.

Figure 3 shows the processing flow of the parallel Modified PrefixSpan method.

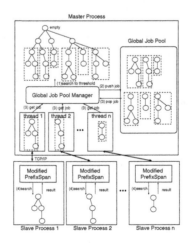

Fig. 3. Processing Flow of the Parallel Modified PrefixSpan

The processing steps in Figure 3 are completed in the following order:

(1) The master process accomplishes the frequent-pattern extraction processing at the threshold.
(2) Multiple jobs generated from (1) are inserted in the Global Job pool, which stores jobs.
(3) The thread count are the same as the number of slave processes. Each threads takes jobs that are in the Global Job pool, and sends to slave process by using the TCP/IP.
(4) Each slave process extracts the frequent pattern from the data that has been sent by the Modified PrefixSpan method.
(5) When the end of the calculation of the slave process is expected and the end is confirmed, the steps are repeated from (2).

5 Evaluation

We implemented the detailed design shown in Section 4 on an actual PC cluster. Next, we conducted a verification experiment.

First of all, the cluster machine consists of 8 personal computers. Table 1 shows the experimental environment of the cluster machine.

The sequence data used for the experiment consists of a data set that includes a motif named Kringle and a data set that includes a motif named Zinc Finger. Table 2 shows the detail of these data sets. These data sets are offered by PROSITE.

Table 2. Detail of Sequence

	Data Records	Total Length(byte)	Average Length(byte)	Maximum Length(byte)	Minimum Length(byte)
Kringle	70	23385	334	3176	53
Zinc Finger	467	245595	525	4036	34

We used small-scale sequence databases, because this experiment is a verification. This verification experiment shows the execution time and the performance ratio in Figure 4 when the number of computers is increased from two to eight. The performance ratio means improvement ratio in a performance.

Each parameter is as follows: (1) Kringle data set, threshold 2, minimum support ratio 40, and wild-cards number 7, and (2) Zinc Finger data set, threshold 2, minimum support ratio 40, and wild-cards number 5, for this experiment.

To achieve efficient parallel processing, a processor of No. 1 slave process runs master process on the same processor. In this case, when the number of computers is N, there are N+1 processes in our experimental environments. As a result, all processors extract frequent patterns from jobs.

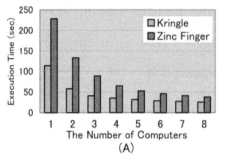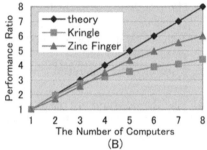

Fig. 4. (A) Execution time. (B) Performance Ratio

Figure 4(B) shows the number of computers in the x-axis and the performance ratio in the y-axis. The theoretical value of Figure 4(B) is to be shortened to $1/N$, where N is the number of computers in paralel.

From the result in Figure 4(B), the performance ratio is about 2 times when 2 computers are used, 3.5 times when 4 computers are used, and 6 times when the maximum of 8 computers is used. Efficient parallel processing shortens the execution time to $1/N$ if parallelization is maintained with N computers. When executing with 8 computers, the execution time will theoretically become $1/8$. In this experiment, the execution time is about $1/6$ at maximum. It is assumed that the cause of the difference is the overhead of communication. To explain the difference, Figure 5 shows the computation time and the number of patterns which is extracted of each slave process when 8 computers are used. And, Figure 6 shows the communication time and total received data from master process of each slave process when 8 computers are used.

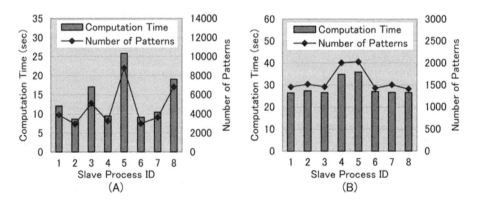

Fig. 5. Computation Time and Number of Patterns of 8 Computers (A) Kringle. (B) Zinc Finger

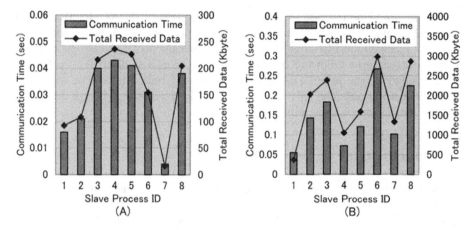

Fig. 6. Communication Time and Total Received Data of 8 Computers (A) Kringle. (B) Zinc Finger

The execution time of each process is computation time plus communication time. Computation time is proportional to the number of patterns which is extracted in Figure 5. And, communication time is proportional to the total receiving data in Figure 6. The slave processes shown in Figures 6 vary in their communication times. The overhead of the communication is assumed to be the cause of the difference between execution and theory. However, this cause is not the overhead of the communication. The data sent is, at most, thousands of kilobytes. Therefore, we understand that execution time is almost computation time.

The slave processes shown in Figures 5 vary in their computation times. From Figure 5 and 7, computation time is not proportional to communication time. In a word, it is unrelated between the number of patterns which is extracted and the total received data from master process (for example, No.4 and No.6 slave processes in Zinc Finger data set). This resulted in an incomplete load balancing in the system. In Figure 5(A), the earliest process with respect to the execution time is about three times faster than the latest process on the condition that 8 computers process the Kringle data set. On the other hand, the Zinc Finger data set is smaller than that of the Kringle data set at the execution time of each slave process; therefore, the performance ratio was better than that in the Kringle data set. When the parameters, which are the support ratio, wild cards, and threshold, were changed, there was at greater difference at the execution time of each slave process. As a result, on occasion, an extremely slow slave process would occur in the experiment.

The reason that there was a greater difference at the execution time of each slave process can be attributed to the jobs. The number of frequent patterns extracted from each job is different. A previous study has shown the possibility of dividing a job when the job size is known [10]. This report shows a high performance ratio. However, the Modified PrefixSpan method cannot determine accurately how long it takes to extract the generation of the job, for the reason why there is no relation between the number of patterns which is extracted and the total received data from master process. Consequently, because the number of frequent patterns following the job is not uniform, it is difficult to calculate the workload of the job. To achieve the best load balancing automatically, a method of automatically calculating the workloads of the k-length frequent-pattern extractions will be implemented.

6   Conclusions

In order to achieve faster pattern extractions from a large-scale sequence database, a parallel Modified PrefixSpan method has been proposed. This method is based on k-length frequent extraction. The PC cluster appeared to be effective, according to the results of the verification experiment. The amino acid sequence used by the verification experiment was a small-scale sequence. It will be necessary to verify the results by using a variety of amino acid sequences in the future. The problem identified by the verification experiment is that workloads are not to be distributed when there is an extremely heavy job. Future plans to complete the efficient parallel Modified PrefixSpan method and to evaluate its performance are underway.

Acknowledgements

We thank Associate Professor Susumu Kuroki of Hiroshima City University for his valuable suggestions. This work was supported in part by a Hiroshima City University Grant for Special Academic Research (General Studies, No.3106) and a Grant-in-Aid for Scientific Research (C) (2) from the Japanese Society for the Promotion of Science.

References

1. Jian Pei, Jiawei Han, Behzad Mortazavi-Asl, Helen Pinto: PrefixSpan: Mining Sequential Patterns Efficiently by Prefix-Projected Pattern Growth, Proc. of International Conference on Data Engineering (ICDE 2001), IEEE Computer Society Press, p215-224, 2001.
2. Hajime Kitakami, Tomoki Kanbara, Yasuma Mori, Susumu Kuroki and Yukiko Yamazaki: Modified PrefixSpan method for Motif Discovery in Sequence Databases, M. Ishizuka and A. Sattar (Eds.): PRICAI 2002, Vol. 2417, Springer-Verlag, pp. 482-491, 2002.
3. Timothy L. Bailey and Charles Elkan, Fitting a Mixture Model by Expectation Maximization to Discover Motifs in Biopolymers, Proceedings of the Second International Conference on Intelligent Systems for Molecular Biology, pp. 28-36, AAAI Press, Menlo Park, California, 1994.
4. Amos Bairoch, Philipp Bucher, and Kay Hofman: The PROSITE Database: Its Status in 1995, Nucleic Acids Research, Vol.24, pp.189-196, 1996.

5. Erik L.L. Sonnhamer, Sean R. Eddy, and Richard Durbin: Pfam: A Comprehensive Database of Proteins, Vol.28, pp.405-420, 1997.
6. Inge Jonassen, John F. Collins, and Desmond G. Higgins: Finding Flexible Patterns in Unaligned Protein Sequences, Protein Science, pp.1587-1595, Cambridge University Press, 1995.
7. Isidore Rigoutsos and Aris Floratos: Combinatorial Pattern Discovery in Biological Sequences: The TEIRESIAS Algorithm, BIOINFORMATICS, Vol.14 No.1, pp.55-67, 1998.
8. Isidore Rigoutsos and Aris Floratos: Motif Discovery without Alignment or Enumeration, Proceedings of Second Annual ACM International Conference on Computational Molecular Biology (RECOMB 98), pp.221-227, March 1998.
9. Aris Floratos and Isidore Rigoutsos: On the Time Complexity of the TERIESIAS Algorithm, IBM Research Report, RC 21161(94582), April 1998.
10. Takuya Araki, Hitoshi Murai, Tsunehiko Kamachi, and Yoshiki Seo: Implementation and Evaluation of Dynamic Load Balancing Mechanism for a Data Parallel Language, Information Processing Society of Japan: Vol. 43 No. SIG 6(HPS5) Transactions on High Performance Computing System, pp.66-75, Sep. 2002.

Parallel LU-decomposition on Pentium Streaming SIMD Extensions

Akihito Takahashi, Mostafa Soliman, and Stanislav Sedukhin

The University of Aizu, Aizu-Wakamatsu City, Fukushima 965-8580, Japan,
{s1080142,d8031102,sedukhin}@u-aizu.ac.jp

Abstract. Solving systems of linear equations is central in scientific computation. In this paper, we focus on using Intel's Pentium Streaming SIMD Extensions (SSE) for parallel implementation of LU-decomposition algorithm. Two implementations (non-SSE and SSE) of LU-decomposition are compared. Moreover, two different variants of the algorithm for the SSE version are also compared. Our results demonstrate an average performance of 2.25 times faster than the non-SSE version. This speedup is higher than 1.74 times the speedup of Intel's SSE implementation. The source of the speedup is highly reusing of loaded data by efficiently organizing SSE instructions.

Keywords: Gaussian elimination, streaming SIMD, parallel processing, performance evaluation, data reusing

1   Introduction

Solving systems of linear equations $Ax = b$ is central in scientific computation [1], where A is a coefficient matrix, b is a vector which specifies the right-hand side of the system of equations, and x is a vector of unknown values. A common procedure for solving these systems is to factor the matrix A into the product of a lower-triangular matrix and an upper-triangular matrix (LU-decomposition) such that $LUx = b$, the solution is then found by forward/backward substitution technique [2]. To accelerate the execution time of LU-decomposition, peak computer performance should be used; moreover, memory operations per floating-point operation (FLOP) should be reduced by reusing of loaded data.

Intel's Pentium SSE [3–5] allows one floating-point instruction to be performed on four pairs of single-precision (32-bit) floating-point numbers simultaneously. The four floating-point operations are performed on data stored in registers called XMM. The XMM registers are 128-bit wide, which can hold four 32-bit floating-point data. There are eight XMM registers, which can be directly addressed by using the register names: XMM0, XMM1, ..., XMM7.

In this paper, we will discuss implementation of LU-decomposition algorithm using SSE, as well as the advantages and disadvantages of SSE. The rest of the paper is organized as follows. Section 2 provides some theoretical background for LU-decomposition algorithm and a discussion of the first implementation

A. Veidenbaum et al. (Eds.): ISHPC 2003, LNCS 2858, pp. 423–430, 2003.

(non-SSE version). In Section 3, comparisons between two implementations of different formulations of the algorithm are discussed. Experimental results are given in Section 4. Finally, Section 5 concludes the paper.

2   LU-decomposition

Gaussian elimination transforms an original matrix A into the product of two triangular matrices L and U. The matrices L and U have the same $n \times n$ dimension as A, where L is an unit lower-triangular matrix and U is an upper-triangular matrix. The algorithm for Gaussian elimination with partial pivoting can be described in MATLAB language as follows:

```
for i = 1 : n
    piv(i) = i
end
for k = 1 : n − 1
    for i = k + 1 : n
        if abs(A(piv(i), k)) > abs(A(piv(k), k))
            tmp = piv(k)
            piv(k) = piv(i)
            piv(i) = tmp
        end
    end
    if A(piv(k), k)) == 0
        return
    end
    v = 1/A(piv(k), k)
    for i = k + 1 : n
        A(piv(i), k) = A(piv(i), k) * v
    end
    for i = k + 1 : n
        for j = k + 1 : n
            A(piv(i), j) = A(piv(i), j) − A(piv(i), k) * A(piv(k), j)
        end
    end
end
```

The algorithm above is so-called kij-variant of Gaussian elimination, which is one of the several possible formulations [2]. In the kij-variant, an input matrix A is accessed on a row-by-row basis in the inner loop, which is suitable for C language. Our SSE implementation is based on this kij-variant. On the other hand, Intel's implementation [6] is based on kji-variant given by exchanging the indexes i and j:

```
for i = 1 : n
    piv(i) = i
end
for k = 1 : n − 1
    for i = k + 1 : n
        if abs(A(piv(i), k)) > abs(A(piv(k), k))
            tmp = piv(k)
            piv(k) = piv(i)
            piv(i) = tmp
        end
    end
    if A(piv(k), k)) == 0
        return
    end
    v = 1/A(piv(k), k)
    for i = k + 1 : n
        A(piv(i), k) = A(piv(i), k) ∗ v
    end
    for j = k + 1 : n
        for i = k + 1 : n
            A(piv(i), j) = A(piv(i), j) − A(piv(i), k) ∗ A(piv(k), j)
        end
    end
end
```

Both of these variants involve $2n^3/3 + O(n^2)$ FLOPs and $4n^3/3 + O(n^2)$ memory references, i.e., two memory references per FLOP.

3   Parallel LU-decomposition on SSE

Rank-1 update [7] of the matrix A in LU-decomposition

$$A(k + 1 : n, k + 1 : n) = A(k + 1 : n, k + 1 : n) − A(k + 1 : n, k) ∗ A(k, k + 1 : n)$$

can be described at vector level as follows:

```
/* the kji-variant */
for j = k + 1 : n
    A(k + 1 : n, j) = A(k + 1 : n, j)−
    A(k + 1 : n, k) ∗ A(k, j)
end
```

```
/* the kij-variant */
for i = k + 1 : n
    A(i, k + 1 : n) = A(i, k + 1 : n)−
    A(i, k) ∗ A(k, k + 1 : n)
end
```

SSE provides floating-point instructions which can be performed on four single-precision data in parallel. Since an XMM register can hold a 4-long row vector, each variant can be presented as follows:

```
/* the kji-variant */
for j = k + 1 : 4 : n
    A(k + 1 : 1 : n, j : j + 3) =
    A(k + 1 : 1 : n, j : j + 3)−
    A(k + 1 : 1 : n, k) * A(k, j : j + 3)
end
```

```
/* the kij-variant */
for i = k + 1 : 1 : n
    A(i, k + 1 : 4 : n) =
    A(i, k + 1 : 4 : n)−
    A(i, k) * A(k, k + 1 : 4 : n)
end
```

Both packing a 4-long vector (e.g. $A(k, k + 1 : 4 : n)$) into an XMM register and unpacking the result of the computation into a proper location of the memory are performed by *movups* instruction [4]. Scalar data (e.g. $A(k+1 : 1 : n, k)$) is stored in an XMM register by two instructions: *movss* and *shufps*. Multiplication and subtraction are operated by *mulps* and *subps*, respectively. Data copy between XMM registers is achieved by *movaps* instruction.

Although the packing/unpacking operation can be performed more quickly by *movaps* instruction, *movups* is used instead of *movaps* for following reasons: firstly, *movaps* requires that its argument be aligned on a 16-byte (128-bit) boundary; secondly, since an input matrix A is processed along the diagonal as $i = k + 1$ and $j = k + 1$, the vector data is not always aligned on 16-byte boundaries; and thirdly, *movups* does not require 16-byte alignment. However, a disadvantage of the packing/unpacking operation by *movups* is an increased overhead compared with *movaps*. As a result, the performance improved by SIMD computations tends to decrease for larger matrices due to the increased packing/unpacking overhead [8]. Therefore, reuse of data stored in XMM registers is important.

The following algorithms show computations of four rows and eight columns in each iteration in order to reserve data reuse. Note that since the kji-variant proceeds column-wise as from $(k+1)$-th row to n-th row sequentially, we describe simply the kji-variant with $A(k + 1 : 1 : n, j : j + 7)$ instead of $A(k + 1, j : j + 7)$, $A(k + 2, j : j + 7)$, $A(k + 3, j : j + 7)$, ..., and so on.

```
/* the kji-variant */
for j = k + 1 : 8 : n
    A(k + 1 : 1 : n, j : j + 7) =
    A(k + 1 : 1 : n, j : j + 7)−
    A(k + 1 : 1 : n, k) * A(k, j : j + 7)
end
```

```
/* the kij-variant */
for i = k + 1 : 4 : n
    A(i, k + 1 : 8 : n) =
    A(i, k + 1 : 8 : n)−
    A(i, k) * A(k, k + 1 : 8 : n)
    A(i + 1, k + 1 : 8 : n) =
    A(i + 1, k + 1 : 8 : n)−
    A(i + 1, k) * A(k, k + 1 : 8 : n)

    A(i + 2, k + 1 : 8 : n) =
    A(i + 2, k + 1 : 8 : n)−
    A(i + 2, k) * A(k, k + 1 : 8 : n)

    A(i + 3, k + 1 : 8 : n) =
    A(i + 3, k + 1 : 8 : n)−
    A(i + 3, k) * A(k, k + 1 : 8 : n)
end
```

Table 1. The number of SSE instructions

Instruction	Latency	kji-variant	kij-variant	Ratio(kji/kij)
movss(load)	1	$n(n-1)(2n-1)/48$	$n(n-1)/2$	$(2n-1)/24$
movaps(reg. to reg.)	1	$n(n-1)(2n-1)/48$	$n(n-1)(2n-1)/32$	$2/3$
shufps	2	$n(n-1)(2n-1)/48$	$n(n-1)/2$	$(2n-1)/24$
movups(load)	4	$n(n-1)(n+1)/12$	$5n(n-1)(2n-1)/96$	$8(n+1)/5(2n-1)$
subps	4	$n(n-1)(2n-1)/24$	$n(n-1)(2n-1)/24$	1
mulps	5	$n(n-1)(2n-1)/24$	$n(n-1)(2n-1)/24$	1
movups(store)	5	$n(n-1)(2n-1)/24$	$n(n-1)(2n-1)/24$	1

Underlined parts indicate data to be reused. Single-underline indicates constant scalar/vector data, and double-underline shows data commonly used. As it can be seen from the algorithms above, the kij-variant provides more data reuse.

The kij-variant can be implemented with eight XMM registers: XMM0-XMM3 for holding scalars, XMM4 and XMM5 for 8-long vectors, and XMM6 and XMM7 for computation results. On the other hand, for kji-variant, six XMM registers are used: XMM0 and XMM7 for holding 8-long vectors, XMM1 and XMM3 for both scalars and computation results, and XMM2 and XMM4 for computation results. Therefore, for the kij-variant, all XMM registers are utilized.

In order to estimate the difference in performance for both variants, we counted the number of different SSE instructions as shown in Table 1. According to Table 1, the number of arithmetic (*subps*, *mulps*) and store operations (*movups*) is the same. However, the number of load operations (*movss*) for the kij-variant is smaller than that for the kji-variant. Therefore, less execution time for the kij-variant is expected than that for the kji-variant. Fig. 1 shows the kji/kij-ratio of the total number of SSE instructions, which are saturated at about 1.1. Roughly, we can expect that the kij-variant will provide performance improvement of 10% compared with that of the kji-variant. Moreover, from the number of load/store operations as shown in Table 1, we examined the num-

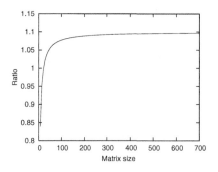

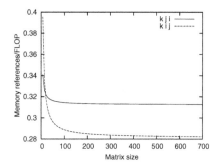

Fig. 1. kji/kij-ratio of the total number of SSE instructions

Fig. 2. Memory references/FLOP

ber of memory references per FLOP. Fig. 2 illustrates the memory references per FLOP for both variants. A number of memory references per FLOP which is smaller than two indicates a degree of data reuse because LU-decomposition without reusing data involves two memory references per FLOP (see Section 2). Referring to Fig. 2, the number of memory references of the kij-variant is improved 10% over that of the kji-variant. Therefore, a performance improvement total of 20% by the kij-variant is expected.

4   Experimental Results

Our experiments were carried out on a 500MHz Pentium III processor running Windows 2000 operating system. The processor cycles were measured by the time-stamp counter with *rdtsc* instruction [9]. From the number of cycles, the running time can be calculated as follows:

$$\text{Running time} = \frac{\text{Processor cycle}}{\text{Frequency in MHz}} \ (microseconds)$$

The input matrices used in our experiments were random, nonsingular, dense, and square matrices, varying from $n = 10$ to 700. The experimental results were obtained by the following algorithms:

- LU-decomposition algorithm, without SSE (see Section 2).
- The kji-variant of the algorithm for SSE described in Intel Application Note AP-931 [6].
- The proposed kij-variant of the algorithm for SSE.

Fig. 3 shows the running times for the kji-variant and the kij-variant of parallel LU-decomposition. As we can see, the running time for the kij-variant is less than that for the kji-variant. As shown in Fig. 4, the peak speedup ratios for the kji-variant and the kij-variant are 2.82 and 2.93 respectively when the matrix size is 60×60. However, the speedup tends to decrease for larger matrix sizes. In fact, when the matrix size is 700, the speedup for the kji-variant and

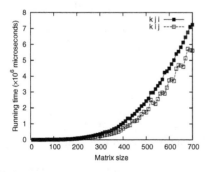
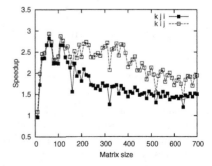

Fig. 3. Running times in microseconds   **Fig. 4.** Speedup over sequential execution

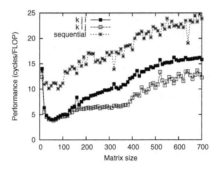

Fig. 5. Performance in cycles/FLOP

the kij-variant decrease to 1.51 and 1.95, respectively. It is likely that the packing/unpacking overhead affects the performance [8]. On the average, the speedup ratios are 2.25 for the kij-variant, and 1.74 for the kji-variant. The source of the speedup is efficient reuse of loaded data as well as maximum use of XMM registers, which results in decreasing the number of cycles per FLOP as shown Fig. 5. In the case where the size of matrix is 60, the number of cycles per FLOP for the kji-variant and the kij-variant are 3.98 and 3.83 as a peak performance, respectively. Referring to Fig. 5, when the size of matrix is 700, the numbers of cycles per FLOP for the kji-variant and the kij-variant are about 15.8 and 12.3, respectively; therefore, the saturations of the number of cycles per FLOP for the kji-variant and the kij-variant seem to be around these values.

5   Conclusions

We have shown a difference in performance for the kji-variant and the kij-variant of parallel LU-decomposition. Our results demonstrate an average performance of 2.25 times faster than the non-SSE version. Moreover, this speedup is higher than 1.74 times the speedup of Intel's implementation. The source of the speedup is highly reusing of loaded data by efficiently using SSE instructions. One of the important considerations in utilizing SSE is the overhead associated with packing/unpacking operations. Compared with the kji-variant, the kij-variant provides more efficient reuse of data as well as use of XMM registers.

References

1. Dongarra, J.J., Duff, I.S., Sorensen, D.C., Van der Vorst, H.A.: Solving Linear Systems on Vector and Shared Memory Computers. SIAM Publications (1991)
2. Golub, G.H., Van Loan, C.F.: Matrix Computations. 3rd edn. The Johns Hopkins University Press (1996)
3. Raman, S.K., Pentkovski, V., Keshava, J.: Implementing Streaming SIMD Extensions on the Pentium III Processor. IEEE Micro, Vol. 20, No. 4, (2000) 47–57

4. IA-32 Intel Architecture Software Developer's Manual - Volume 2 Instruction Set Reference. Intel Corporation (2003)
5. Intel Architecture Optimization Reference Manual. Intel Corporation (1999)
6. Streaming SIMD Extensions - LU Decomposition. Intel Application Note AP-931, Intel Corporation (1999)
7. Dongarra, J.J., Du Croz, J., Hammarling, S., Hanson, R.J.: An Extended Set of Fortran Basic Linear Algebra Subprograms. ACM Transactions on Mathematical Software, Vol. 14, No. 1, (1988) 1–17
8. Fung, Y.F., Ercan, M.F., Ho, T.K., Cheung, W.L.: A Parallel Solution to Linear Systems. Microprocessors and Microsystems, Vol. 26, No. 1, (2002) 39–44
9. Using the RDTSC Instruction for Performance Monitoring. Intel Application Note, Intel Corporation (2000)

Parallel Matrix Multiplication and LU Factorization on Ethernet-Based Clusters

Fernando G. Tinetti, Mónica Denham, and Armando De Giusti

Laboratorio de Investigación y Desarrolo en Informática (LIDI),
Facultad de Informática,
UNLP,
50 y 115, 1900 La Plata, Argentina,
Tel.: +54 221 4227707, Fax: +54 221 4273235,
{fernando,mdenham,degiusti}@lidi.info.unlp.edu.ar

Abstract. This work presents a simple but effective approach for two *representative* linear algebra operations to be solved in parallel on Ethernet-based clusters: matrix multiplication and LU matrix factorization. The main objectives of this approach are: simplicity and performance optimization. The approach is completed at a lower level by including a broadcast routine based directly on top of UDP to take advantage of the Ethernet physical broadcast facility. The performance of the proposed algorithms implemented on Ethernet-based clusters is compared with the performance obtained with the ScaLAPACK library, which is taken as having highly optimized algorithms for distributed memory parallel computers in general and clusters in particular.

Keywords: Cluster Computing, Parallel Linear Algebra Computing, Communication Performance, Parallel Algorithms, High Performance Computing.

1   Introduction

The field of parallel computing has evolved in several senses, from computing hardware to algorithms and libraries specialized in particular areas. The processing basic hardware used massively in desktop computers has also increased its capacity in orders of magnitude and, at the same time, it has reduced its costs to end users. On the other hand, costs associated to a specific (parallel) computer cannot always be afforded, and remained *almost constant* since many years ago. From this point of view, clusters have the two main facilities needed for parallel computing: 1) Processing power, which is provided by each computer (CPU-memory), 2) Interconnection network among CPUs, which is provided by LAN (Local Area Network) hardware, usually Ethernet [7] at the physical level.

Traditionally, the area of problems arising from linear algebra has taken advantage of the performance offered by available (parallel) computing architectures. LAPACK (Linear Algebra PACKage) [1] and BLAS (Basic Linear Algebra

A. Veidenbaum et al. (Eds.): ISHPC 2003, LNCS 2858, pp. 431–439, 2003.

Subroutines) [3] definitions represent the main results of this effort. Parallel algorithms in the area of linear algebra have been continuously proposed mainly because of their great number of (almost direct) applications and large number of potential users. Two immediate examples are: 1) Matrix multiplication has been used as a low-level benchmark for sequential as well as parallel computers [6] or, at least, as a BLAS benchmark. In some way or another, results are still being reported in relation to matrix multiplication performance in parallel and sequential computers, and 2) LU factorization is used to solve dense systems of linear equations, which are found in applications such as [5] airplane wing designs, radar cross-section studies, supercomputer benchmarking, etc.

In the field of parallel (distributed) linear algebra, the ScaLAPACK (Scalable LAPACK) library has been defined [2]. ScaLAPACK is no more (and no less) than LAPACK implementing high quality (or optimized) algorithms for distributed memory parallel computers. Matrix multiplication and LU factorization will be analyzed in detail for Ethernet-based clusters, and they will also be taken as representatives of the whole LAPACK-ScaLAPACK libraries.

The main project of which this work is part of focuses on a methodology to parallelize linear algebra operations and methods on Ethernet-based clusters, taking into account the specific constraints explained above. The methodology involves optimizing algorithms in general, and communication patterns in particular, in order to obtain optimized parallel performance in clusters interconnected by Ethernet.

2   Algorithms for Matrix Multiplication and LU Factorization

Being parallel matrix multiplication and LU factorization implemented in very well known, accepted, and optimized libraries such as ScaLAPACK, is it necessary to define *another* parallel matrix multiplication and LU factorization? Parallel algorithms selected for implementation and inclusion in ScaLAPACK are assumed to have optimized performance in parallel computers with distributed memory and good scalability. However, as the performance is usually strongly dependent on hardware, the algorithms selected are also influenced by the *current* distributed memory parallel computers at the time ScaLAPACK was defined (the '90s). Those computers follow the guidelines of the *traditional* distributed memory parallel (super)computers, i.e.: 1) High-performance CPUs, the best for numerical computing, and sometimes designed *ad hoc* for this task, 2) Memory directly attached to each processor, with message passing necessary for interprocess-interprocessor communication, and 3) Low latency (startup)-High throughput (bandwidth) processor interconnection network. But clusters are not (at least *a priori*) parallel computers.

As regards clusters interconnected by Ethernet networks, the interconnection network implies strong performance penalties when processes running on different computers of the cluster are communicated. From this point of view, parallel algorithms have not been designed *for* the parallel architecture of the clusters, they have been rather used as designed for distributed memory parallel computers. Furthermore, there has been and there is a strong emphasis on leveraging the interconnection network, so as to approach the interconnection network of the traditional parallel (super)computers.

Taking into account the characteristics of clusters with Ethernet network interconnection, some guidelines have been devised in order to design parallel algorithms for linear algebra operations and methods:

- SPMD (Single Program, Multiple Data) parallel execution model, which is not new in the field of linear algebra numerical computing. Also, the SPMD model is very useful for simplifying the programming and debugging task.
- The Ethernet hardware defined by the IEEE standard 802.3 [7] *almost directly* leads to: 1)Unidimensional data distribution, given that every interconnected computer is connected at the same "logical bus" defined by the standard, and 2) Broadcast based intercommunication among processes, given that the Ethernet logical bus makes broadcast one of the most *natural* messages to be used in parallel applications. Also, every cabling hardware is IEEE 802.3 compliant and, thus, broadcast has to be physically implemented.

2.1   Matrix Multiplication

The basic parallel matrix multiplication algorithm ($C = A \times B$) proposed is derived from the algorithm already published in [10] for heterogeneous clusters, and can be considered as a *direct* parallel matrix multiplication algorithm [11]. It is defined in terms of data distribution plus the program that every computer has to carry out for the parallel task. The (one-dimensional) matrix distribution is defined so that: 1) Matrices A and C are divided into as many row blocks as computers in the cluster, and every computer holds one row block (row block partitioning [8]); computer P_i has row blocks $A^{(i)}$ and $C^{(i)}$, 2) Matrix B is divided into as many column blocks as computers in the cluster, and every computer holds one column block (column block partitioning [8]); computer P_i has column block $B^{(i)}$, 3) Each computer P_i has only a fraction of the data needed to calculate its portion of the resulting matrix $C^{(i)}$, i.e. $C^{(i_i)} = A^{(i)} \times B^{(i)}$.

Fig. 1 shows the pseudocode of the program on each computer, which has only local computing plus broadcast communications for processor P_i, where 1) C^(i_j) stands for $C^{(i_j)}$, A^(i) for $A^{(i)}$, and B^(j) for $B^{(j)}$, and 2) The operations send_broadcast_o and recv_broadcast_o are used to send and receive broadcast data in "background", i.e. overlapped with local processing in the computer, where available.

2.2   LU Factorization

Currently used parallel LU factorization algorithm is derived from the combination of [5]: a) the traditional right-looking LU factorization algorithm, b)

```
if (i == 0)
    send_broadcast B^(i)
for (j = 0; j < P; j++)
{
    if (j != i)
    {
        recv_broadcast_b B^(j)
        if ((j+1) == i)
            send_broadcast_b B^(i)
    }
    C^(i_j) = A^(i) * B^(j)
}
```

Fig. 1. Matrix Multiplication with Overlapped Communications.

data processing made by blocks (block processing), and c) two dimensional matrix data partitioning. Following the guidelines stated previously, the selected matrix partitioning is chosen following the so-called row block matrix distribution [8]. To avoid the unacceptable performance penalty of unbalanced workload maintaining the one-dimensional data distribution among computers, the matrix is divided in many more blocks than computers and assigned cyclically to processors. In this way: 1) One dimensional data partitioning is made, there is no defined neighborhood for each processor according to its position in a processor mesh, and 2) When the number of blocks is large enough ($bs \ll p$), every processor will have data to work with in most of the iterations.

2.3   Broadcasting Data Using UDP

The algorithms explained are *strongly* dependent on the broadcast performance. In particular, the broadcast performance of PVM [4] is not acceptable, since it is implemented as multiple point-to-point messages from the sender to each receiver when processes are running on different computers. In general, MPI [9] implementations do not necessarily have to optimize broadcast performance and/or optimize the availability of the Ethernet logical bus since the MPI standard itself does not impose any restriction and/or characteristic on any of the routines it defines.

When the previously mentioned broadcast-based guideline is used for parallelization, the design and implementation of an optimized version of the broadcast communication between processes of parallel programs is highly encouraged. Ethernet networks are based on a logical bus which can be reached from any attached workstation [7], where the method to access the bus is known as multiple random access. Basically, in every data communication, the sender floods the bus, and only the destination/s will take the information from the channel. The destination/s might be a group of computers and, in this case, the sender uses a multicast/broadcast address, sending data only once. All the referenced computers will take the data from the channel at the same time.

3   Experimental Work

The proposed parallel algorithms (for matrix multiplication and LU factorization) plus the new broadcast routine optimized for Ethernet interconnected clusters are used in two homogeneous clusters of eight PCs each: the PIII cluster and
the Duron Cluster. The characteristics of PCs in each cluster are summarized in
Table 1. Mflop/s was taken by running the best matrix multiplication sequential algorithm on single precision floating point matrices. The interconnection
network on each cluster is a fully switched 100 Mb/s Ethernet. Algorithms performance is compared with that obtained by the ScaLAPACK library, which is
installed on both clusters on top of MPICH. A *third* cluster is considered, which
will be called 16-PC: both previous clusters interconnected and used as a cluster
with sixteen *homogeneous* PCs of 580 Mflop/s with 64 MB each.

Table 1. Clusters Characteristics.

Cluster	CPU	Clock	Mem	Mflop/s
PIII	Pentium III	700 MHz	64 MB	580
Duron	AMD Duron	850 MHz	256 MB	1200

3.1   Matrix Multiplication Performance

The algorithm proposed for parallel matrix multiplication on Ethernet-based
clusters (Section 2.1 above) was implemented and compared with the one provided by the ScaLAPACK library. Fig. 2 shows the best five ScaLAPACK performance values in terms of speedup on the PIII cluster and the speedup obtained for the parallel matrix multiplication proposed in this. Two matrices
of 5000x5000 single precision floating point numbers are multiplied. Given that
ScaLAPACK performance depends on some parameters values, a range of values
was used for testing ScaLAPACK matrix multiplication performance: a) Square
blocks: 32x32, 64x64, 128x128, 256x256, and 1000x1000 elements, and b) Processors meshes: 1x8, 8x1, 2x4, and 4x2 processors on the clusters with 8 PCs,

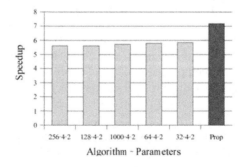

Fig. 2. Matrix Multiplication on PIII Cluster.

and 2x8, 8x2, and 4x4 processors on the cluster with 16 PCs. Each bar of Fig. 2 is labeled according to the algorithm and parameters used (when required). ScaLAPACK matrix multiplication is identified by the values for the three parameters on which it depends upon: processors per row, processors per column, and block size. Performance of the parallel matrix multiplication proposed is labeled as "Prop".

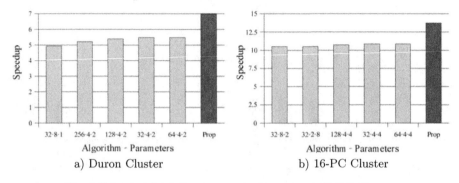

a) Duron Cluster   b) 16-PC Cluster

Fig. 3. Matrix Multiplication on the Duron and 16-PC Cluster.

Fig. 3 a) shows the best five ScaLAPACK performance values in terms of speedup on the Duron cluster and the speedup obtained for the parallel matrix multiplication proposed in this work, and matrices of 10000x10000 elements. Also, Fig. 3 b) shows the best five ScaLAPACK performance values in terms of speedup on the 16-PC cluster as well as the speedup obtained for the parallel matrix multiplication proposed in this work. Matrices are of 8000x8000 elements. Table 2 summarizes performance results on the three clusters.

Table 2. Matrix Multiplication Performance Summary.

Cluster	Sc-Best	Prop-Best	% Better Perf.	Opt. Perf.
PIII	5.9	7.2	23%	8
Duron	5.5	7.0	27%	8
16-PC	10.9	13.7	26%	16

3.2   LU Matrix Factorization Performance

Similar tests for parallel LU factorization were carried out on the same clusters. In this case, the proposed parallel algorithm is also defined in terms of a (row) blocking size according to Section 2.2 above. The blocking sizes for ScaLAPACK as well as for the proposed algorithm were: 32, 64, 128, 256, and 1000. The proposed algorithm was tested with other intermediate blocking sizes (as blocks of 100 rows) in order to verify its dependence on row block size.

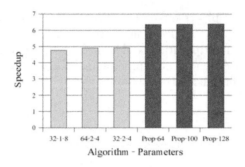

Fig. 4. LU Factorization on PIII Cluster.

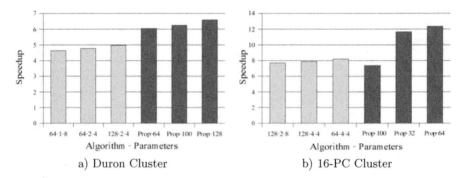

a) Duron Cluster   b) 16-PC Cluster

Fig. 5. LU Matrix Factorization on the Duron and 16-PC Cluster.

Table 3. LU Matrix Factorization Performance Summary.

Cluster	Sc-Best	Prop-Best	% Better Perf.	Opt. Perf.
PIII	5.0	6.4	30%	8
Duron	5.0	6.6	32%	8
16-PC	8.2	12.3	50%	16

Fig. 4 shows the best three performance (speedup) values for each of the algorithms used on the PIII cluster. The proposed parallel LU matrix factorization algorithm is labeled "Prop-*rbs*" where *rbs* is the row blocking size used in the experiment. On each experiment, a matrix of 9000x9000 single precision elements is factorized into L and U. Fig. 5 a) shows the performance (speedup) values on the Duron cluster. On each experiment, a matrix of 20000x20000 single precision elements is factorized into L and U. Also, Fig. 5 b) shows the performance (speedup) values on the PC-16 cluster where a matrix of 13000x13000 single precision elements is factorized into L and U on each experiment. Table 3 summarizes performance results on the three clusters.

4 Conclusions and Further Work

The main conclusion derived from performance results shown in the previous section is that: there is enough room for optimization of linear algebra algorithms on Ethernet-based clusters. Optimizations are focused to optimize resource usage on Ethernet-based clusters. Proposals on this work can be summarized as: 1) A few guidelines to parallelize linear algebra applications, with two main objectives: simple parallel algorithms, and optimized performance on Ethernet-based clusters. 2) Two parallel algorithms have been successfully designed and implemented: matrix multiplication and LU matrix factorization. 3) It has been shown by experimentation that it is possible to obtain better performance than that obtained by the algorithms implemented in the highly optimized ScaLAPACK library. 4) It is possible to have optimized performance on Ethernet-based clusters without imposing fully switched interconnections. In this way, the whole cost of the interconnection network is greatly reduced for clusters with large number of computers, where the switching cost increases more than linearly when the number of computers grows.

Even when broadcast messages based on UDP are expected to have very good scalability, it is important to experiment on clusters with many computers to verify this assumption. In this sense, experiments should be carried out on cluster with tens and hundreds of computers interconnected by Ethernet. Also, it should be highly beneficial to experiment on the recently proposed Gb/s Ethernet in order to analyze scalability and performance gains. From the point of view of linear algebra applications, it is necessary to continue with optimization of other methods and/or operations as those included in LAPACK-ScaLAPACK.

References

1. Anderson E. *et al.*, "LAPACK: A Portable Linear Algebra Library for High-Performance Computers", Proceedings of Supercomputing '90, pages 1-10, IEEE Press, 1990.
2. Blackford L. *et al.* ScaLAPACK Users' Guide, SIAM, Philadelphia, 1997.
3. Dongarra J., J. Du Croz, S. Hammarling, R. Hanson, "An extended Set of Fortran Basic Linear Subroutines", ACM Trans. Math. Soft., 14 (1), pp. 1-17, 1988.
4. Dongarra J., A. Geist, R. Manchek, V. Sunderam, "Integrated pvm framework supports heterogeneous network computing", Computers in Physics, (7)2, pp. 166-175, April 1993.
5. Dongarra J., D. Walker, "Libraries for Linear Algebra", in Sabot G. W. (Ed.), High Performance Computing: Problem Solving with Parallel and Vector Architectures, Addison-Wesley Publishing Company, Inc., pp. 93-134, 1995.
6. Hockney R., M. Berry (eds.), "Public International Benchmarks for Parallel Computers", Scientific Programming 3(2), pp. 101-146, 1994.
7. Institute of Electrical and Electronics Engineers, Local Area Network - CSMA/CD Access Method and Physical Layer Specifications ANSI/IEEE 802.3 - IEEE Computer Society, 1985.
8. Kumar V., A. Grama, A. Gupta, G. Karypis, Introduction to Parallel Computing. Design and Analysis of Algorithms, The Benjamin/Cummings Publishing Company, Inc., 1994.

9. Message Passing Interface Forum, "MPI: A Message Passing Interface standard", International Journal of Supercomputer Applications, Volume 8 (3/4), 1994.
10. Tinetti F., A. Quijano, A. De Giusti, E. Luque, "Heterogeneous Networks of Workstations and the Parallel Matrix Multiplication", Y. Cotronis and J. Dongarra (Eds.): EuroPVM/MPI 2001, LNCS 2131, pp. 296-303, Springer-Verlag, 2001.
11. Wilkinson B., Allen M., Parallel Programming: Techniques and Applications Using Networked Workstations and Parallel Computers, Prentice-Hall, Inc., 1999.

Online Remote Trace Analysis of Parallel Applications on High-Performance Clusters

Holger Brunst[1,2], Allen D. Malony[1], Sameer S. Shende[1], and Robert Bell[1]

[1] Department for Computer and Information Science,
University of Oregon,
Eugene, USA,
{brunst,malony,sameer,bertie}@cs.uoregon.edu
[2] Center for High Performance Computing,
Dresden University of Technology,
Germany,
brunst@zhr.tu-dresden.de

Abstract. The paper presents the design and development of an online remote trace measurement and analysis system. The work combines the strengths of the TAU performance system with that of the VNG distributed parallel trace analyzer. Issues associated with online tracing are discussed and the problems encountered in system implementation are analyzed in detail. Our approach should port well to parallel platforms. Future work includes testing the performance of the system on large-scale machines.

Keywords: Parallel Computing, Performance Analysis, Performance Steering, Tracing, Clusters

1 Introduction

The use of clusters for high-performance computing has grown in relevance in recent years [5], both in terms of the number of platforms available and the domains of application. This is mainly due to two factors: 1) the cost attractiveness of clusters developed from commercial off-the-shelf (COTS) components and 2) the ability to achieve good performance in many parallel computing problems. While performance is often considered from a hardware perspective, where the number of processors, their clock rate, the network bandwidth, etc. determines performance potential, maximizing the price/performance ratio of cluster systems more often requires detailed analysis of parallel software performance. Indeed, experiences shows that the design and tuning of parallel software for performance efficiency is at least as important as the hardware components in delivering high "bang for the buck."

However, despite the fact that the brain is humankind's most remarkable example of parallel processing, designing parallel software that best optimizes available computing resources remains an intellectual challenge. The iterative process of verification, analysis, and tuning of parallel codes is, in most cases,

A. Veidenbaum et al. (Eds.): ISHPC 2003, LNCS 2858, pp. 440–449, 2003.

mandatory. Certainly, the goal of parallel performance research has been to provide application developers with the necessary tools to do these jobs well. For example, the ability of the visual cortex to interpret complex information through pattern recognition has found high value in the use of trace analysis and visualization for detailed performance verification. Unfortunately, there is a major disadvantage of a trace-based approach – the amount of trace data generated even for small application runs on a few processors can become quite large quite fast. In general, we find that building tools to understand parallel program performance is non-trivial and itself involves engineering tradeoffs. Indeed, the dominant tension that arises in parallel performance research is that involving choices of the need for performance detail and the insight it offers versus the cost of obtaining performance data and the intrusion it may incur.

Thus, performance tool researchers are constantly pushing the boundaries of how measurement, analysis, and visualization techniques are developed to solve performance problems. In this paper, we consider such a case – how to overcome the inherent problems of data size in parallel tracing while providing online analysis utility for programs with long execution times and large numbers of processes. We propose an integrated online performance analysis framework for clusters that addresses the following major goals:

- Allow online insight into a running application.
- Keep performance data on the parallel platform.
- Provide fast graphical remote access for application developers.
- Discard uninteresting data interactively.

Our approach brings together two leading performance measurement and analysis tools: the TAU performance system [4], developed at University of Oregon, and the VNG prototype for parallel trace analysis, developed at Dresden University of Technology in Germany.

The sections below present our system architecture design and discuss in detail the implementation we produced for our in-house 32-node Linux cluster. We first describe the TAU performance system and discuss issues that affect its ability to produce traces for online analysis. The section following introduces VNG and describes its functional design. The integrated system combining TAU and VNG is then presented. The paper concludes with a discussion of the portability and scalability of our online trace analysis framework as part of a parallel performance toolkit.

2 Trace Instrumentation and Measurement (TAU)

An online trace analysis system for parallel applications depends on two important components: a parallel performance measurement system and a parallel trace analysis tool. The TAU performance system is a toolkit for performance instrumentation, measurement and analysis of parallel programs. It targets a general computational model consisting of shared-memory compute nodes where multiple contexts reside, each providing a virtual address space shared by multiple threads of execution. TAU supports a flexible instrumentation model that applies at different stages of program compilation and execution [3]. The instrumentation targets multiple code points, provides for mapping of low-level execution events to higher-level performance abstractions, and works with multithreaded and message passing parallel computation models. Instrumentation code makes calls to the TAU measurement API. The measurement library implements performance profiling and tracing support for performance events occurring at function, method, basic block, and statement levels during program execution. Performance experiments can be composed from different measurement modules (e.g., hardware performance monitors) and measurements can be collected with respect to user-defined performance groups. The TAU data analysis and presentation utilities offer text-based and graphical tools to visualize the performance data as well as bridges to third-party software, such as Vampir [2] for sophisticated trace analysis and visualization.

2.1   TAU Trace Measurement

TAU's measurement library treats events from different instrumentation layers in a uniform manner. TAU supports message communication events, timer entry and exit events, and user-defined events. Timer entry and exit events can apply to a group of statements as well as routines. User-defined events are specific to the application and can measure entry/exit actions as well as atomic actions. TAU can generate profiles by aggregating performance statistics such as exclusive and inclusive times spent in routines, number of calls, etc. These profiles are written to files when the application terminates. With the same instrumentation, TAU can generate event traces, which are time-ordered event logs. These logs contain fixed-size trace records. Each trace record is a tuple comprising of an event identifier, a timestamp, location information (node, thread) and event specific parameters. The trace records are periodically flushed to disk when private in-memory buffers overflow, or when the application executes a call to explicitly flush the buffers. TAU also writes event description files that map event identifiers to event properties (such as event names, group names, tags, and event parameter descriptions).

2.2   Dynamic Event Registration

For runtime trace reading and analysis, it is important to understand what takes place when TAU records performance events in traces. The first time an event takes place in a process, it registers its properties with the TAU measurement library. Each event has an identifier associated with it. These identifiers are generated dynamically at runtime as the application executes, allowing TAU to track only those events that take actually occur. This is in contrast to static schemes that must predefine all possible events that could possibly occur. The main issue here is how the event identifiers are determined. In a static scheme, event IDs are drawn from a pre-determined global space of IDs, which restricts the scope of performance measurement scenarios. In our more general and dynamic scheme, the event identifiers are generated on-the-fly, local to a context. Depending on the order in which events first occur, the IDs may be different for the same event (i.e., events with the same name) across contexts. When event streams are later merged, these local event identifiers are mapped to a global identifier based on the event name. The TAU trace merging operation is discussed in section 4.3.

3   Remote Trace Data Analysis (VNG)

The distributed parallel performance analysis architecture applied in this paper has been recently designed by researchers from the Dresden University of Technology in Dresden, Germany. Based on the experience gained from the development of the performance analysis tool Vampir [1], the new architecture uses a distributed approach consisting of a parallel analysis server running on a segment of a parallel production environment, and a visualization client running on a remote graphics workstation. Both components interact with each other over a socket based network connection. In the discussion that follows, the parallel analysis server together with the visualization client will be referred to as *VNG*. The major goals of the distributed parallel approach are:

1. Keep performance data close to the location where they were created.
2. Analyze event data in parallel to achieve increased scalability.
3. Provide fast and easy remote performance analysis on end-user platforms.

VNG consists of two major components: an analysis server (**vngd**) and visualization client (**vng**). Each is supposed to run on a different platform. Figure 1 shows a high-level view of the VNG architecture. Boxes represent modules of the components whereas arrows indicate the interfaces between the different modules. The thickness of the arrows gives a rough measure of the data volume to be transferred over an interface, whereas the length of an arrow represents the expected latency for that particular link.

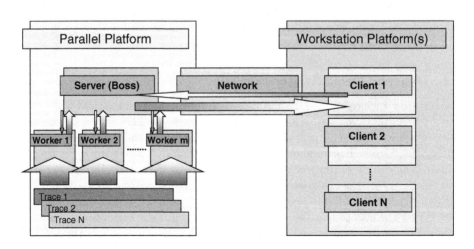

Fig. 1. VNG Architecture

On the left hand side of Figure 1 we can see the analysis server, which is intended to execute on a dedicated segment of a parallel machine. The reason for this is two-fold. First, it allows the analysis server to have closer access to the trace data generated by an application being traced. Second, it allows the server to execute in parallel. Indeed, the server is a heterogeneous parallel program, implemented using MPI and pthreads, which consists of worker and boss processes. The workers are responsible for trace data storage and analysis. Each of them holds a part of the overall trace data to be analyzed. The bosses are responsible for the communication to the remote clients. They decide how to distribute analysis requests among the workers. Once the analysis requests are completed, the bosses also merge the results into a single packaged response that is to be sent to the client.

The right hand side of Figure 1 depicts the visualization client(s) running on a local desktop graphics workstation. The idea is that the client is not supposed to do any time consuming calculations. It is a straightforward sequential GUI implementation with a look-and-feel very similar to performance analysis tools like Vampir and Jumpshot [6]. For visualization purposes, it communicates with the analysis server according to the user's preferences and inputs. Multiple clients can connect to the analysis server at the same time, allowing simultaneous distributed viewing of trace results.

4   Enabling Online Trace Analysis

Up to now, TAU and VNG could only provide post-mortem trace analysis. The inherent scalability limitations of this approach due to trace size motivated us to evaluate the extensions and modifications needed to enable online access to trace data. The following discusses the technical details of our proposed online tracing solution.

4.1   Overview

Figure 2 depicts the overall architecture of our online trace analysis framework. On the far left we see a running application instrumented by TAU. The inserted probes call the TAU measurement library which is responsible for the trace event generation. Periodically, depending on external or internal conditions (see section 4.2 for details), each application context will dump its event trace data. In our present implementation, this is done to disk via NFS on a dedicated, high-speed network separate from the MPI message passing infrastructure. An independent trace writing network is a nice feature of our cluster that reduces tracing intrusion on the application.

Concurrently, an independent TAU process runs on a dedicated node and merges the parallel traces streams. Its responsibility is to produce a single globally-consistent trace with synchronized timestamps. This stream is then passed to the VNG analysis server which is intended to run on a small subset of dedicated cluster nodes. From there, pre-calculated performance profiles

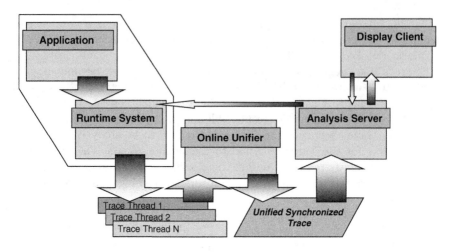

Fig. 2. Online Analysis Architecture

and event timelines are sent to the visualization client running on the user's local computer. This multi-layer approach allows to event data to be processed online and in parallel, close to its point of origin on the cluster. Furthermore, the interactive access to the data, coupled with runtime instrumentation and measurement control, allows the detailed analysis of long production runs.

4.2   Triggers for Trace Dumping

The interactive character of online trace analysis requires a strategy for triggering the TAU runtime system to dump trace data. We considered the four different options:

1. **Buffer size driven:** The tracing library itself makes the decision when to store trace data. One approach is to dump data whenever the internal memory buffers are full. It is easy to implement and does not require any external interaction with the measurement library. Unfortunately, it may produce large amounts of data when not combined with any filtering techniques. Also, the different contexts (i.e., processes) will produce the trace data at different times, depending on when their buffers fill up.
2. **Application driven:** The tracing library could provide the application with an API to explicitly dump the current trace buffers. (Such an API is already available in TAU.) The advantage of this approach is that the amount and frequency of trace dumping is entirely under the control of the application. We expect this approach to be desired in many scenarios.
3. **Timer driven:** To make triggering more periodic, a timer could be used to generate an interrupt in each context that causes the trace buffers to be dumped, regardless of the current amount of trace data they hold. In theory, this is simple to implement. Unfortunately, there is no general, widely portable solution to interrupt handling on parallel platforms.

4. **User driven:** Here the user decides interactively when to trigger the dump process. Assuming the user is sitting in front of a remote visualization client (e.g., **vng**), the trigger information needs to be transported to the cluster and to the running application (i.e., the trace measurement system). Again, this requires some sort of inter-process signaling. From the options we discussed so far, we regard this approach to be the most challenging, but also the most desirable.

For the work presented here, we implemented the buffer size and application driven triggering mechanisms. These are generally termed "push" models, since they use internal decision strategies to push trace data out to the merging and analysis processes. In contrast, the "pull" models based on timer or user driven approaches require some form of inter-process signalling support. Our plan was to first use the simpler push approaches to validate the full online tracing system before implementing additional pull mechanisms.

4.3   Background Merging and Preparation

Previously, TAU wrote the event description files to disk when the application terminated. While this scheme was sufficient for post-mortem merging and conversion of event traces, it could not be directly applied for online analysis of event traces. This was due to the absence of event names that are needed for local to global event identifier conversion. To overcome this limitation, we have re-designed our trace merging tool, TAUmerge, so it executes concurrently with the executing application generating the trace files. From each process's trace file, TAUmerge reads event records and examines their globally synchronized timestamps to determine which event is to be recorded next in the ordered output trace file. When it encounters a local event identifier that it has not seen before, it reads the event definition file associated with the given process and updates its internal tables to map that local event identifier to a global event identifier using its event name as a key. The trace generation library ensures that event tables are written to disk before writing trace records that contain one or more new events. A new event is defined as an event whose properties are not recorded in the event description file written previously by the application. This scheme, of writing event definitions prior to trace records, is also used by the TAUmerge tool while writing a merged stream of events and event definitions. It ensures that the trace analysis tools down the line that read the merged traces also read the global event definitions and refresh their internal tables when they encounter an event for which event definitions are not known.

4.4   Trace Reader Library

To make the trace data available for runtime analysis, we implemented the TAU trace reader library. It can parse binary merged or unmerged traces (and their respective event definition files) and provides this information to an analysis tool using a trace analysis API. This API employs a callback mechanism where the tool registers callback handlers for different events. The library parses the trace and event description files and notifies the tool of events that it is interested in, by invoking the appropriate handlers with event specific parameters. We currently support callbacks for finding the following:

- Clock period used in the trace
- Message send or receive events
- Mapping event identifiers to their state or event properties
- Defining a group identifier and associated group name
- Entering and leaving a state

Each of these callback routines have event specific parameters. For instance, a send event handler has source and destination process identifiers, the message length, and its tag as its parameters. Besides reading a group of records from the trace file, our API supports file management routines for opening, closing a trace file, for navigating the trace file by moving the location of the current file pointer to an absolute or relative event position. It supports both positive and negative event offsets. This allows the analysis tool to read, for instance, the last 10000 events from the tail of the event stream. The trace reader library is used by VNG to analyze the merged binary event stream generated by an application instrumented with TAU.

4.5   Remote Steering and Visualization

Online trace analysis does not necessarily require user control if one looks at it as a uni-directional information stream that can be continuously delivered to an application developer. However, the high rate at which information is likely to be generated in a native tracing approach suggests the need for advanced user control. Ideally, the user should be able to retrieve new trace data whenever he thinks it is necessary. At the VNG display client, this could be accomplished by adding a "retrieve" option that allows the user to specify how many recent events are to be loaded and analyzed by the analysis server. Such a request could translate all the way to trace generation. Even more sophisticated online control of the runtime system by the application developer would include:

- Enabling and disabling of certain processes
- Filtering of unwanted events
- Switching between a profile only mode and a detailed trace mode

While our infrastructure was designed to provide the necessary communication methods for those control requests, our present implementation allows for the trace analyzer to only work with whatever trace data is presently available from the TAU trace reader interface. This is not a limitation for our current purposes of validating the system implementation.

4.6   System Partitioning

Figure 3 depicts how our cluster system might be partitioned for online trace analysis. Together with the master node, we dedicated four additional processors

(29-32) for the purpose of trace data processing. Parallel applications are run on the rest of the machine (processors 01 to 28). The cluster nodes are interconnected by a redundant gigabit Ethernet network each with its own network switch. The two networks allow us to keep message passing communication independent from NFS traffic, although our approach does not require this explicitly.

During an application run, every worker process is equipped with an instance of the TAU runtime library. This library is responsible for collecting trace data and writing it to a temporary file, one per worker process. The TAU merger process on processor 32 constantly collects and merges the independent streams from the temporary files to a unified representation of the trace data. At this point, the analysis process of VNG takes over. Whenever a remote user client connects to the server on the master node, the VNG worker processes come into action on the processors 29 to 31 in order to update their data or perform new analysis requests. During the entire process, trace data stays put in the cluster file system. This has tremendous benefits in GUI responsiveness at the client. Furthermore, the VNG analyzer communicates with the remote visualization client in a highly optimized fashion, guaranteeing fast response time even when run over a long distance connection with low bandwidth and high latencies (see Results).

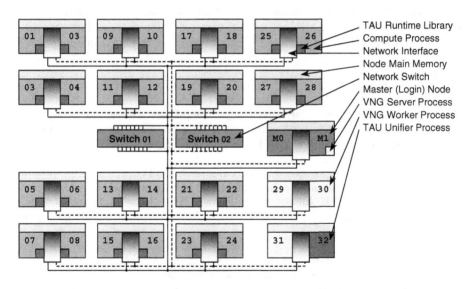

Fig. 3. System Partitioning

5 Conclusion

There is growing interest in monitoring parallel applications and interacting with the running program as the computation proceeds. Some efforts are focussed on computational steering and support the creation of sensors to observe execution dynamics and actuators to change program behavior. Consistent with these directions, our attention is towards online performance evaluation using tracing as a measurement approach. The purpose is to offer the user the same rich functionality as off-line trace analysis, but without the penalties of large trace data management.

However, the development of a general-purpose online trace analysis system is difficult, especially if it is to be portable and scalable. The work presented here is a first step toward this goal. Combining the strengths of the TAU and VNG tools, we demonstrated a full-path, working system that allows interactive trace generation, merging, analysis, and visualization. In its present form, the work is quite portable across parallel platforms, as it is based on already portable existing tools and the file system and inter-process communication interfaces used are standard.

Our next step in this research is to conduct scalability performance tests. We expect the file system-based trace merging approach will suffer at some point. To address this problem at higher levels of parallelism, we are considering the use of the recent work on MRNet, a multi-cast reduction network infrastructure for tools, to implement a tree-based parallel version of the TAU merger.

References

1. Brunst, H., Nagel, W. E., Hoppe, H.-C.: Group-Based Peformance Analysis for Multithreaded SMP Cluster Applications. In: Sakellariou, R., Keane, J., Gurd, J., Freeman, L. (eds.): Euro-Par 2001 Parallel Processing, No. 2150 in LNCS, Springer, (2001) 148–153
2. Brunst, H., Winkler, M., Nagel, W. E., Hoppe H.-C.: Performance Optimization for Large Scale Computing: The Scalable VAMPIR Approach. In: Alexandrov, V. N., Dongarra, J. J., Juliano, B. A., Renner, R. S., Kenneth Tan, C. J. (eds.): Computational Science – ICCS 2001, Part II, No. 2074 in LNCS, Springer, (2001) 751–760
3. Lindlan, K.A., Cuny, J., Malony, A.D., Shende, S., Mohr, B., Rivenburgh, R., Rasmussen, C.: Tool Framework for Static and Dynamic Analysis of Object-Oriented Software with Templates. Proceedings SC'2000, (2000)
4. Malony, A., Shende, S.: Performance Technology for Complex Parallel and Distributed Systems. In: Kotsis, G., Kacsuk, P. (eds.): Distributed and Parallel Systems From Instruction Parallelism to Cluster Computing. Proc. 3rd Workshop on Distributed and Parallel Systems, DAPSYS 2000, Kluwer (2000) 37–46
5. Meuer, H.W., Strohmaier, E., Dongarra J.J., Simon H.D.: Top500 Supercomputing Sites, 18th Edition, (2002) http://www.top500.org
6. Zaki, O., Lusk, E., Gropp, W., Swider, D.: Toward Scalable Performance Visualization with Jumpshot. The International Journal of High Performance Computing Applications 13(3) (1999) 277–288

Performance Study
of a Whole Genome Comparison Tool
on a Hyper-Threading Multiprocessor[*]

Juan del Cuvillo[1], Xinmin Tian[2], Guang R. Gao[1], and Milind Girkar[2]

[1] Department of Electrical and Computer Engineering, University of Delaware,
Newark, DE 19716, USA
[2] Intel Compiler Laboratory, SSG/EPG, Intel Corporation,
3600 Juliette Lane, Santa Clara, CA 95052, USA

Abstract. We developed a multithreaded parallel implementation of a
sequence alignment algorithm that is able to align whole genomes with
reliable output and reasonable cost. This paper presents a performance
evaluation of the whole genome comparison tool called ATGC — Another
Tool for Genome Comparison, on a Hyper-Threading multiprocessor. We
use our application to determine the system scalability for this particular type of sequence comparison algorithm and the improvement due
to Hyper-Threading technology. The experimental results show that despite of placing a great demand on the memory system, the multithreaded
code generated by Intel compiler yields to a 3.3 absolute speedup on a
quad-processor machine, with parallelization guided by OpenMP pragmas. Additionally, a relatively high 1st level cache miss rate of 7-8%
and a lack of memory bandwidth prevent logical processors with hyper-
threading technology enabled from achieving further improvement.

1 Introduction

Multithreading with architecture and microarchitecture support is becoming increasingly commonplace: examples include the Intel Pentinum 4 Hyper-Threading Technology and the IBM Power 4. While using this multithreaded hardware
to improve the throughput of multiple workloads is straightforward, using it to
improve the performance of a single workload requires parallelization. The ideal
solution would be to transform serial programs into parallel programs automatically, but unfortunately this is notoriously difficult. However, the OpenMP
programming model has emerged as the de facto standard of expressing parallelism since it substantially simplifies the complex task of writing multithreaded
programs on shared memory systems. The Intel C++/Fortran compiler supports OpenMP directive- and pragma-guided parallelization, which significantly
increases the domain of applications amenable to effective parallelization. This

---

[*] This work was partially supported by NSF and DOE: NSF through the NGS program, grant 0103723; DOE grant DE-FC02-01ER25503.

A. Veidenbaum et al. (Eds.): ISHPC 2003, LNCS 2858, pp. 450–457, 2003.

paper focuses on the parallel implementation and performance study of a whole genome comparison tool using OpenMP on the Intel architecture.

Over the last decades, as the amount of biological sequence data available in databases worldwide grows at an exponential rate, researchers continue the never-ending quest for faster sequence comparison algorithms. This procedure is the core element of bioinformatic key applications such as database search and multiple sequence comparison, which can provide hints to predict the structure, function and evolutionary history of a new sequence. However, it has not been until recently that whole genomes have been completely sequenced, opening the door to the challenging task of whole genome comparison. Such an approach to comparative genomics has a great biological significance. Nonetheless, it can not be accomplished unless computer programs for pair-wise sequence comparison deal efficiently with both execution time and memory requirements for this large-scale comparison. With these two constraints in mind, a new wave of algorithms have been proposed in the last few years [1]. Some have showed to achieve good execution time at the expense of accuracy and they require a large amount of memory [3]. Others, based on heuristics such as "seed-extension" and hashing techniques offer a better trade off between execution time and memory consumption, and are the most commonly used nowadays [7]. More recently, the so called Normalized Local Alignment algorithm has been presented [1]. This iterative method uses the Smith-Waterman algorithm, which provides the accuracy other methods might not have, and a varying scoring system to determine local alignments with a maximum degree of similarity. It also solves the shadow and mosaic effects but increases the algorithm complexity. However, we have showed that by means of parallelization, the execution time for the Smith-Waterman algorithm decreases almost linearly, and hence the original algorithm itself as well as others based on the dynamic programming technique such as the NLA, become affordable for small and medium size genomes [2, 5, 6].

We developed an OpenMP implementation of the affine gap penalties version of the Smith-Waterman algorithm [10]. We also plugged our implementation into the NLA algorithm framework, which consists of an additional few lines of code, to verify the affordability of this method. The parallelization is achieved by means of OpenMP pragmas, more specifically, an *omp parallel pragma* construct that defines a parallel region, which is a region of the program that is to be executed by multiple threads in parallel. The algorithm's recurrence equation results in a relatively regular application, with data dependencies that easily fit into a consumer-producer model. Communication among threads under such a model is performed by shared rotating buffers. At any given time, a buffer is owned by a single thread. The contents of a buffer are consumed at the beginning of the computation stage, and once a new result is produced by the current owner, the buffer is released and its ownership automatically granted to the thread beneath it. As a consequence, neither locks nor critical sections are needed to prevent data-race conditions, a limiting factor in many parallel applications. Our OpenMP implementation runs on a hyper-threading multiprocessor system in

---

[1] A comprehensive study of related work can be found in the author's Master thesis.

which shared memory enables the mentioned features, i.e. easy programmability and minimum interthread communication overhead.

The rest of the paper is organized as follows. Section 2 describes our multithreaded parallel application. The results from a performance study based on this implementation are presented in section 3, and our conclusions in section 4.

2 Parallel Computation of a Sequence Alignment

A parallel version of the sequence comparison algorithm using dynamic programming must handle the data dependences presented by this method, yet it should perform as many operations as possible independently. The authors have showed that the similarity matrix calculation can be efficiently parallelized using fine-grain multithreading [2, 5, 6]. The implementation described in this paper is based upon that but it exploits parallelism by using OpenMP pragmas.

2.1 Hyper-Threading Technology

Intel's hyper-threading technology [4] is the first architecture implementation of the Simultaneous Multi-Threading (SMT) proposed by Tullsen [9]. By making a single physical processor appears as two logical processors, the operating system and user application can schedule processes or threads to be run in parallel instead of sequentially. Such an implementation of hyper-threading technology represents less than a 5% increase in die size and maximum power requirements. Nevertheless, it allows instructions from two independent streams to persist and execute simultaneously, yielding more efficient use of processor resources, improving performance.

As an example, Figure 1(a) shows a traditional multiprocessor system with two physical processors that are not hyper-threading technology-capable. Figure 1(b) shows a similar multiprocessor system with two physical processors and hyper-threading support. With two copies of the architectural state on each physical processor, the second system appears to have four logical processors.

2.2 An OpenMP Parallel Implementation

Our multithreaded implementation divides the scoring matrix into strips and each of these, in turn, into rectangular blocks. Generally speaking, it assigns

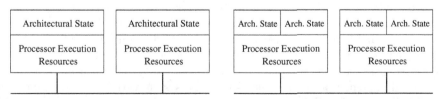

(a) Hyper-threading non-capable processors (b) Hyper-threading-enabled processors

Fig. 1. Traditional and HT-capable multiprocessor systems

the computation of each strip to a thread, having to 2 independent threads per physical processor, i.e. one thread per logical processor.

A thread iterates over blocks in a strip by means of a for loop. At each iteration, the computation of a block is followed by a communication phase in which the thread sends the scores and other information from the last row of its block to the thread beneath. In this way, alignments that cross processors' boundaries can be detected. Furthermore, since memory is shared by all processors this task can be accomplished efficiently without copying data.

Given the sizes of the input sequences, the scoring matrix can not be kept in memory throughout the computation as it is usually done when comparing short sequences. Instead, a pool of buffers is allocated at the beginning of the program execution and these buffers are used repeatedly by all threads to perform the computation. Each buffer holds only the scores of a single row of a block, requiring each thread two of these to compute a block.

Before the computation starts, a single buffer is assigned to each thread but the first, which gets all that is left in the pool. The first thread is the only with two buffers available and, based on data dependencies, the only that can start. When it finishes computing the first block, it makes the buffer containing the scores for the block's last row available to the thread below. To achieve this, it updates a variable that accounts for the number of buffers assigned to the second thread. Actually, this counter is updated in such a way it tells the thread that owns it which iteration should be performed with the data it contains. In other words, a thread just waits [2] until the producer signals it by updating this counter that says which iteration the consumer should be starting. Additionally, since this counter is only written by one thread, the producer, and read by the thread below, the consumer, no locking mechanism needs to be implemented.

At the end of the communication phase, the first thread starts the next iteration taking another buffer from the pool and reusing the one left from the previous iteration. Meanwhile the second thread, which owns two buffers now, starts computing its first block. When threads get a buffer to work with, it means that the data dependencies are satisfied. They then can start computing the corresponding block. Buffers released by the last thread are assigned back to the first thread so they can be reused.

Since each block does not represent exactly the same amount of work, as execution proceeds some threads might be idle waiting for the thread above to release a buffer. Having a pool of buffers allows a thread to work ahead, assuming a buffer is available, when an iteration is completed earlier. As an example of this behavior, the first thread can initially work ahead since buffers are available from the pool. It does not have to wait for a buffer to be used by all threads and become available again before starting the second iteration. That would have serialized the computation and made our implementation quite inefficient.

A snapshot of the computation of the similarity matrix using our multi-threaded implementation on a quad-processor system with hyper-threading support is illustrated in Figure 2. A thread is assigned to each horizontal strip and

---

[2] To avoid wasting cycles on this spin wait, the PAUSE instruction is used.

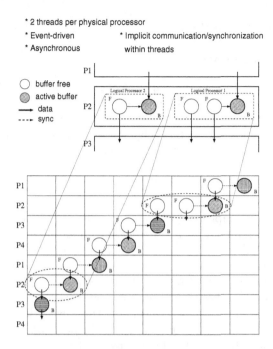

Fig. 2. Computation of the similarity matrix on a SMP system

the actual computation is done on the buffers labeled B(usy). Data or buffers assigned to the thread beneath are labeled F(ree). The figure shows the computation of the main anti-diagonal of the matrix. The arrows indicate data or buffer availability signals. For example, processor 2 sends data (downward arrows) to processor 3 and receives data from processor 1 [3]. Within a thread, that is, between blocks on the same strip, a synchronization signal is implicit (horizontal arrows) as results from one iteration are available to the next without explicit communication.

The number of possible alignments grows exponentially with the length of the sequences compared. Therefore, we can not simply report all the alignments. Instead we are interested in selecting only alignments with high scores. On each node, as each strip of the scoring matrix is calculated, scores above a given threshold are compared with the previous highest scores stored in an table. The number of entries in the table corresponds to the maximum number of alignments that a node can report. Among other information, the table stores the cells' position where an alignment starts and ends. This feature allows us to produce a plot of the alignments found. A point worth noticing is that high score alignments are selected as the similarity matrix is calculated, row by row, since the whole matrix is not stored in memory.

---

[3] Actually, threads 1 and 2 have worked faster than thread 3, which has now an additional F(ree) buffer ready to start working on without delay.

3   Performance Analysis

3.1   The System Configuration

The experiments described in this paper were carried out on a 4-way Intel Xeon
processor at 1.4GHz with Hyper-Threading support, 2GB memory, a 32KB L1
cache (16KB instruction cache and 16KB two-way write-back data cache), a
256KB L2 cache, and a 512KB L3 cache. All programs were compiled with the
Intel C++/Fortran OpenMP compiler version 7.0 Beta [8].

3.2   Results

Our OpenMP parallel implementation was used to align human and mice mitochondrial genomes, human and *Drosophila* mitochondrial genomes, and human
herpesvirus 1 and human herpesvirus 2 genomes. We run each comparison several times to consider the effects of the number of threads, block size, and having
hyper-threading enabled or disabled in the program execution time.

Figure 3 reports the absolute speedup achieved for the genome comparisons
mentioned above. Each comparison runs under three execution modes: SP, single
processor with a single thread, QP HT off, 4 threads running on 4 processors
with hyper-threading disabled, QP HT on, hyper-threading support enabled and
8 threads running on 8 logical processors. For each execution mode several block

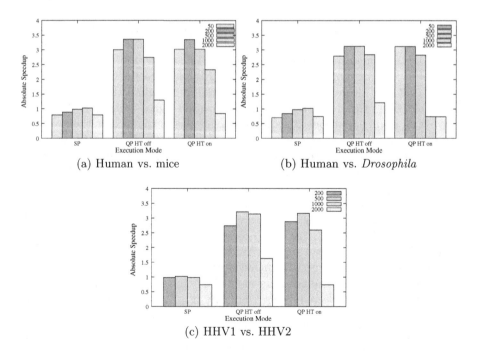

(a) Human vs. mice   (b) Human vs. *Drosophila*

(c) HHV1 vs. HHV2

Fig. 3. Absolute speedup for mitochondrial genome comparisons

Table 1. Workload characteristics for the human vs. mouse and human vs. Drosophila genome comparisons with block width 1,000

	SP	HT off	HT on	SP	HT off	HT on
Execution Time (seconds)	30	9	9	28	9	9
Instructions retired (millions)	24,547	42,606	44,819	24,429	42,967	42,347
μops retired (millions)	23,175	52,376	54,418	20,551	49,942	53,134
Trace cache delivery rate	74.40%	102.45%	87.96%	68.08%	98.11%	100.00%
Load accesses (millions)	9,990	17,035	17,873	9,816	16,658	16,662
L1 cache load misses (millions)	712	1,365	1,236	758	1,349	1,222
L2 cache load misses (millions)	32	290	27	30	338	29
L1 cache miss rate	7.13%	8.01%	7.62%	7.72%	8.09%	7.33%
L2 cache miss rate	0.32%	1.70%	0.15%	0.30%	2.03%	1.74%
L3 cache miss rate	0.02%	0.02%	0.04%	0.02%	0.02%	0.05%
Branches retired (millions)	3,111	5,255	5,528	3,318	5,823	5,473
Mispredicted branches (millions)	112	220	217	138	335	231
Branch misprediction rate	3.60%	4.19%	3.90%	4.16%	5.75%	4.22%

sizes are tested as well. The first observation we can make from both three figures is that a 3.3 absolute speedup is achieved for the three test cases. Second, performance does not improve when hyper-threading support is enabled. Runs with 8 threads on 8 logical processors report basically the same execution time as runs with 4 threads on 4 physical processors without hyper-threading support.

Table 1 summarizes the workload characteristics for the human and mouse and human and *Drosophila* genome comparisons, respectively. The first point worth noticing is the similarity between both tables. The *Drosophila* mitochondrial genome is longer than mice mitochondrial genome. Therefore, the amount of computation required to fill a larger similarity matrix is larger as well. However, the amount of work required by the second comparison to keep track of a fewer number of alignments seems to compensate the total amount of work required by each workload. Another issue is the relatively high 1st level cache miss rate, always between 7% and 8%. Although we try to exploit as much locality as possible by selecting an appropriate block width, this parameter can not be set to an arbitrarily small value. When we do so performance decreases because of the small computation-communication ratio. Execution resources might be shared efficiently between logical processors running independent threads. However, if the memory bus can not keep up with the application demands (remember the high cache miss rate) instructions will not be able to proceed normally through the pipeline since store and load buffers are waiting for memory accesses initiated by previous instructions to complete. For this application, many of these instructions represent a memory operation. However, memory operations have to be queued because the memory bandwidth is limited (the same as what four threads see in the QP HT off execution mode). Once the load and store buffers are full, instructions can not be issued faster, regardless of hyper-threading, since resources are unavailable.

4   Conclusions

We have developed a multithreaded parallel implementation of a dynamic programming based algorithm for whole genome comparison that runs on an SMP system and meets the requirements for small and medium size genomes.

The experimental results show that despite being a memory bound application, which places a great demand on the memory system, the multithreaded code generated by Intel compiler yields to a 3.3 absolute speedup on a quad-processor machine, with parallelization guided by OpenMP pragmas. Additionally, a relatively high 1st level cache miss rate of 7-8% and a lack of memory bandwidth prevent logical processors with hyper-threading technology enabled from achieving further improvement.

As future work we intend to investigate why the hyper-threading does not bring additional performance gain. In particular, we will focus on the overhead caused by the loop with the PAUSE instruction, which accounts for 5% of the execution time, and four load-store operations that account for more than 90% of the L2 cache misses. Should we tune our application such that hyper-threading effectively reduces these two factors, we can expect a significant performance improvement.

References

1. A. N. Arslan et al. A new approach to sequence comparison: Normalized sequence alignment. *Bioinformatics*, 17(4):327–337, 2001.
2. J. del Cuvillo. Whole genome comparison using a multithreaded parallel implementation. Master's thesis, U. of Delaware, Newark, Del., Jul. 2001.
3. A. L. Delcher et al. Alignment of whole genomes. *Nucleic Acids Res.*, 27(11):2369–2376, 1999.
4. D. T. Mar et al. Hyper-threading technology architecture and microarchitecture. *Intel Tech. J.*, 6(1):4–15, Feb. 2002.
5. W. S. Martins et al. Whole genome alignment using a multithreaded parallel implementation. In *Proc. of the 13th Symp. on Computer Architecture and High Performance Computing*, Pirenópolis, Brazil, Sep.10–12, 2001.
6. W. S. Martins et al. A multithreaded parallel implementation of a dynamic programming algorithm for sequence comparison. In *Proc. of the Pacific Symp. on Biocomputing*, pages 311–322, Mauna Lani, Haw., Jan. 3–7, 2001.
7. S. Schwartz et al. PipMaker — A web server for aligning two genomic DNA sequences. *Genome Res.*, 10(4):577–586, April 2000.
8. X. Tian et al. Intel OpenMP C++/Fortran compiler for hyper-threading technology: Implementation and performance. *Intel Tech. J.*, 6(1):36–46, Feb. 2002.
9. D. M. Tullsen et al. Simultaneous multithreading: Maximizing on-chip parallelism. In *Proc. of the 22nd Ann. Intl. Symp. on Computer Architecture*, pages 392–403, Santa Margherita Ligure, Italy, Jun. 1995.
10. M. S. Waterman. *Introduction to Computational Biology: Maps, Sequences, and Genomes*. Chapman and Hall, 1995.

The GSN Library and FORTRAN Level I/O Benchmarks on the NS-III HPC System

Naoyuki Fujita and Hirofumi Ookawa

National Aerospace Laboratory, 7-44-1 Jindaiji-higashi Chofu, Tokyo 182-8522 Japan,
Phone:+81-422-40-3318, FAX:+81-422-40-3327
{fujita,ookawa}@nal.go.jp

Abstract. The data generated by a High Performance Computing (HPC) server is very large. For example, a combustion chemical reaction flow simulation generates hundreds gigabytes data. In order to analyze this enormous amount of data, a huge HPC server and huge pre/post systems are both necessary. In this paper, we have benchmarked the I/O performance of two applications on *NS-III*. One application environment is the GSN library visualization system and the other is FORTRAN I/O library on *NS-III*. *NS-III* is the name of a supercomputer system located at National Aerospace Laboratory. That system has a nine-teraflops compute system, a huge visualization system, and a high-speed mass storage system. These servers are connected with GSN, and/or crossbar-network. In addition to benchmark results, we also demonstrate the user-programming interface of the GSN library that realizes a file sharing.

Keyword. I/O Benchmark, Scheduled Transfer Facility, Memory Copy, GSN, File Sharing

1 Introduction

In 2002, the National Aerospace Laboratory of Japan (NAL) introduced the *Numerical Simulator III System (NS-III)*. *NS-III* has three subsystems. The first subsystem is a compute system named the *Central Numerical Simulation System (CeNSS)*, the second is a visualization system named the *Central Visualization System (CeViS)*, and the third is a mass storage system named the *Central Mass Storage System (CeMSS)*. In the field of computational fluid dynamics, huge scale numerical simulations – e.g.: simulations on combustion flow with chemical reactions – have recently become possible. However, these simulations require high-speed file I/O systems that can operate on files at a rate of about one gigabyte per second, and these simulations also require high-performance pre/post systems (e.g. visualization system). We designed *NS-III* to meet these requirements. The first design point is that I/O system should have no bottleneck. We already verified this point in previous works.[1],[2] The second design point is that the I/O system should have an interface that has sufficient throughput to handle HPC data. In this paper, we verify this second characteristic and show the result for two I/O benchmarks. The first benchmark was executed from the

A. Veidenbaum et al. (Eds.): ISHPC 2003, LNCS 2858, pp. 458–467, 2003.

GSN library; the second one was executed from FORTRAN I/O library. From these benchmarks, we recognized that the GSN library interface accessed files at a rate of about 593(read)/582(write) megabytes per second and FORTRAN I/O library accessed files with about 1822(read)/1732(write) megabytes per second throughput. In addition, we show the user-programming interface of the GSN library.

2   Numerical Simulator III (NS-III)

At the National Aerospace Laboratory of Japan, we introduced *Numerical Simulator I* System with a VP400 compute system that performed Navier-Storks equation based numerical simulation in the '80s. In the '90, we built *Numerical Simulator II* system with a *Numerical Wind Tunnel* compute server and performed a parametric study calculation on complete aircraft aerodynamics simulation. Then in 2002, we introduce the *Numerical Simulator III* System. The *NS-III* has a nine-teraflops compute system, a 3D visualization system, and a high-speed mass storage system. We are going to perform multidisciplinary numerical simulations, such as unsteady flow analysis, and so forth, on the *NS-III*. Figure 1 shows an overview of *NS-III*. *CeViS* is a visualization system for the *NS-III*. It has six graphic pipelines. *CeMSS* is a mass-storage system on *NS-III*. According to our estimate of requirements for *CeMSS* [1], *CeMSS* has to have a throughput of about one gigabyte per second. However, a single storage device has several megabytes per second throughput. So *CeMSS* should be a parallel I/O system.

There are some large scale storage systems in the world [3],[4],[5]. The major differences between *CeMSS* and other large scale storage systems are the followings: *CeMSS* is a local file system of compute system, instead most others are autonomous file and/or storage system, and *CeMSS* uses Gigabyte System Network (GSN) for physical interface, instead most others use Fiber Channel.

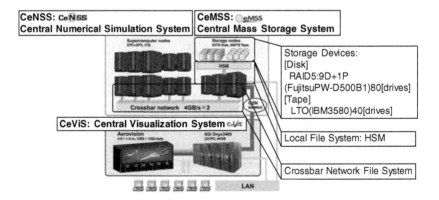

Fig. 1. Numerical Simulator III and it's Mass Storage System Overview

3   The GSN Library Programming Interface

We describe the programming interface of the GSN library before presenting the benchmarks. GSN [6] is the highest bandwidth and lowest latency interconnect standard. It provides full duplex dual 6400 megabit (800 megabyte) per second of error-free, flow controlled data. The proposed ANSI standard provides for interoperability with Ethernet, Fiber Channel, ATM, HIPPI-800 and other standards [7].

In short, it is a specialized interface, but it provides a C language system call-like interface. Therefore, when programmers use this interface, the only change necessary is to replace standard system calls to the GSN one in the source code. The programmer should also define two runtime environment variables: STF_REMOTE_PATH and STF_REMOTE_HOST. Figure 2 shows a sample program. The GSN library has two types of programming interface groups, which will be explained later.

```
#include<stdio.h>
#include<stflib.h>

main() {
        int *data, fp, ret;

        /* Get Transfer Buffer */
        data=stf_malloc(sizeof(int)*256);

        /* open file */
        fp=stf_open("/stf2/remote/test/data",O_RDONL);

        if(fp<-1) {
                /* Handle Error */
                switch(fp) {
                case -1:
                        perror("local file open failed");
                        break;
                case -2:
                        perror("remote file open failed");
                        break;
                case -3:
                        perror("STF Daemon not found");
                        break;
                }
                exit(-1);
        }
        /* read file */
        ret = stf_read(fp, data, sizeof(int)*256);

        if(ret<-1) {
                /* Handle Error */
                switch(ret) {
                case -1:
                        perror("readfailed");
                        break;
                case -2:
                        perror("memory isn't stflib's mem.");
                        break;
                }
                stf_close(fp);
                exit(-1);
        }

        /* Close file */
        ret = stf_close(fp);

        if(ret) {
                /* Handle Error */
                perror("closeerror");
                exit(-1);
        }

        /* Release Transfer Buffer */
        if(stf_free(data)) {
                printf("stf_freeerror\n");
                exit(-1);
        }
}
```

Fig. 2. Sample program of the GSN library interface

3.1   Function Outline

This interface is made up of some components shown in Fig. 3.

STF library: An interface that supports high-speed local and remote file transfer for *CeViS* user program. STF stands for Scheduled Transfer Facility.

STF Daemon: A daemon that delivers remote files to the STF library. A remote file is accessed with the HSM local file system. HSM stands for Hierarchical Storage Management.

Local file: A file system directly connected to *CeViS*

Remote file: A file system connected to a remote system (i.e., *CeMSS*).

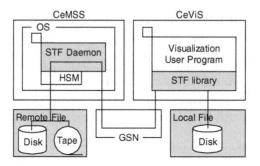

Fig. 3. The GSN library interface components

3.2 Features

In this section, we itemize the GSN library features. We sometimes use "STF library" instead of "GSN library".

STF Library Features: 1) Works correctly even when called in thread
2) Delivered as not a static library but a dynamic library
3) Supports GSN striping

STF Daemon Features: 1) Uses client (*CeViS*) account information
2) Error code is converted to client (*CeViS*) system code
3) Has restricted file system area access control

3.3 STF Library Overview

The STF library has two interface groups: the File interface and the Memory interface. Figure 4 shows these interfaces. In order to use these interfaces, two environment variables, STF_REMOTE_PATH and STF_REMOTE_HOST, must be defined. Figure 5 shows example of environment variable usage.

The File interface is analogous to a standard C language system call interface. This interface handles file operations (i.e., open and close file operations); however, the Memory interface is unfamiliar. When we use file operation functions, the operating system and the standard file operation library initiate some

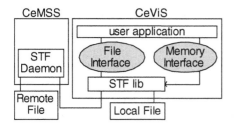

Fig. 4. Two views of the GSN library interface

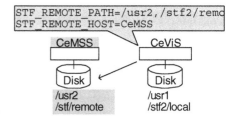

Fig. 5. Sample of environment variable usage

memory copy operations. This memory copy operation is one of the factors that decrease the file transfer rate. The Memory interface solves this memory copy problem. Before using the File interface, we can allocate memory space for the File interface operations. User applications and the STF library use this allocated memory space. Because of this memory allocation, the operating system and STF library do not require a memory copy operation between user applications and the STF library.

In the next two sections, we demonstrate some C language functions from the GSN library.

3.3.1   File Operation in the File Interface

1) stf_open

[Function] Opens an assigned file. If the upper path of a file path matches STF_REMOTE_PATH, operates on a remote file, otherwise operates on a local file.

[Usage]   `int stf_open(const char *path, int oflag, /* mode_t mode */ ...);`

2) stf_close

[Function] Closes the file

[Usage]   `int stf_close(int fildes);`

3) stf_read

[Function] Reads a specified file into a buffer

[Usage]   `ssize_t stf_read(int fildes, void *buf, size_t nbyte);`

4) stf_write

[Function] Writes buffer contents to specified file

[Usage]   `ssize_t stf_write(int fildes, const void *buf, size_t nbyte);`

5) stf_lseek

[Function] Locates a pointer in a specified file

[Usage]   `off_t stf_lseek(int fildes, off_t offset, int whence);`

6) stf_creat

[Function] Creates a new file and opens the file

[Usage]   `int stf_creat(const char *path, mode_t mode);`

7) stf_fstat

[Function] Gathers status information from a specified file

[Usage]   `int stf_fstat(int fildes, struct stat *buf);`

3.3.2   Memory Operations in the Memory Interface

1) stf_malloc

[Function] Allocates memory, which is used for data transfer in the STF library.

[Usage]   `void *stf_malloc(size_t size);`

2) stf_free

[Function] Releases memory, which is used for data transfer in the STF library.

[Usage]   `int stf_free(void *ptr);`

Table 1. Differences between standard library and the GSN library

	The GSN library	Standard lib	Purpose
File like I/F	stf_open	open(2)	Opens a file
	stf_close	close(2)	Closes a file
	stf_read	read(2)	Reads data
	stf_write	write(2)	Writes data
	stf_lseek	lseek(2)	Locates a pointer
	stf_create	create(2)	Creates a new file
Memory like I/F	stf_malloc	malloc(3)	Allocates memory
	stf_free	free(3)	Releases memory
	stf_calloc	calloc(3)	Allocates memory with zero clear
	stf_realloc	realloc(3)	Re-allocates memory
	stf_valloc	valloc(3)	Allocates memory fit to page boundary
	stf_memalign	memalign(3)	Allocates memory with designating boundary
	stf_cvmalloc_error	cvmalloc_error(3)	Obtains error info.

There are some other functions that operate on the memory space: stf_calloc, stf_realloc, stf_valloc, stf_memalign, stf_cvmalloc_error. These memory operations are self-explanatory, so we do not describe them here [8].

3.4 Differences between Standard Library and the GSN Library

Table 1 is a comparison table between standard library and the GSN library.

4 Benchmarks

In this chapter, we describe the results from two benchmarks. The first one is the GSN library benchmark and the second one is the FORTRAN I/O library benchmark. In Fig. 6, present benchmark's measuring points for benchmarks: 1 and 2 and former measuring point: A and B are shown. Measuring point is a place where the benchmark code is executed.

4.1 Prior Benchmark Results

Before discussing the new results of our benchmarking, we present some prior benchmarking results in Fig. 7. Result (a) was measured at "Measuring Point B" and result (b) was measured at "Measuring Point A" in Fig. 6. This is a parallel I/O storage device characteristic; (a) is the disk result and (b) is the tape result. Using result (a), we determine that this mass storage system – *CeMSS* – has a capacity to handle I/O about 1.7 – 1.9 gigabyte/sec. We set this as a condition for the current benchmarks; that is to say, we used 80 stripe disks to measure the STF library and FORTRAN I/O library benchmarking. Result (b) is provided to demonstrate *CeMSS*'s tape I/O capacity. *CeMSS* is a Hierarchical Storage Management (HSM) system. So if the disk cache is full, files on the disk are moved to the tape. Result (b) shows the disk-to-tape transfer rate.

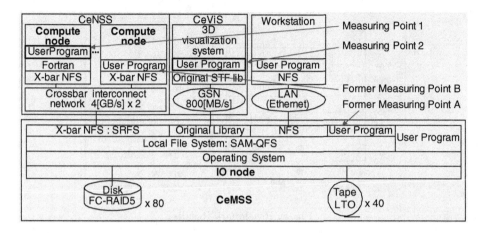

Fig. 6. I/O benchmarking point

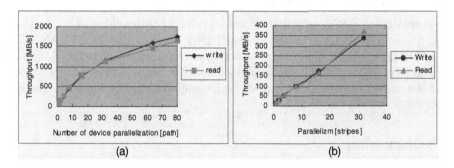

Fig. 7. Storage Device Parallel I/O Characteristics

Figure 8 includes the prior benchmark results. The "*SAM-QFS* (Write/ Read)" and "*SRFS* (Write/Read)" lines are prior benchmark results. *SAM-QFS* is a *NS-III* local file system. It manages the disk and tape hierarchical storage system. *SRFS* is a crossbar network file system between *CeNSS* and *CeMSS*. *SAM-QFS* was benchmarked at "Measuring Point A" and *SRFS* was benchmarked at "Measuring Point B" in Fig. 6.

4.2   The GSN Library Benchmark

In Fig. 8, the "STF Write" and "STF Read" lines are the GSN library benchmark results. They are the bottom two lines in Fig. 8. This benchmark is not sensitive to the user I/O size, which is the size of each transaction, compared to other benchmarks as they are almost completely horizontal lines.

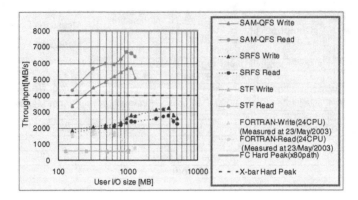

Fig. 8. Benchmark Results

4.3 FORTRAN I/O Library Benchmark

The "FORTRAN-Write(24CPU)" and "FORTRAN-Read(24CPU)" lines on Fig. 8 are FORTRAN I/O library benchmark results. The FORTRAN-Write and FORTRAN-Read lines exhibit a maximum throughput of 1822MB/s and a minimum of 763MB/s. The peak performance for write actions occurs at user I/O size of 640 megabytes, and peak performance of read action is at user I/O size of 320 megabytes.

5 Discussion

At first, file share is possible using the GSN library between *CeNSS* and *CeViS* with minimal changes to the source program. It is also important to note that this file sharing function is realized between different file system: Solaris and IRIX. As can be seen in Fig. 8, the GSN library I/O throughput shows a nearly flat rate of about 550 megabytes per second. The local file system on *CeViS* has a RAID5 Fiber Channel disk, whose peak throughput is 100 megabytes per second. So the GSN library I/O throughput is five or more times faster than this local file system. So the GSN library I/O interface is useful on not only for file sharing but also for throughput. This throughput advantage is carried out by the Memory interface, which decreases memory copy.

FORTRAN-write throughput, which is measured at "Measuring Point 1", is between 795 to 1732 megabytes per second. As we can see in Fig. 8's "SRFS Write" line, the throughput measured at "Measuring Point B" in Fig. 6 is between 1855 to 3250 megabytes per second. The difference between "Measuring Point 1" and "Measuring Point B" is whether it is a FORTRAN environment or not. A FORTRAN environment has a FORTRAN buffer. When the user performs a write in FORTRAN, the FORTRAN environment copies user data from user data buffer to the FORTRAN buffer. This copy operation and the FORTRAN buffer size affect I/O throughput. Figure 9 shows another FOR-TRAN view benchmark result. This figure shows throughput trend with one

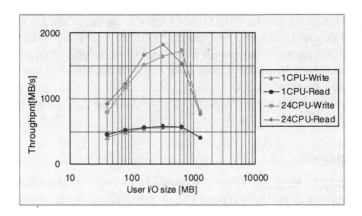

Fig. 9. Memory Copy Effect

CPU and 24 CPU FORTRAN process. In this case, the number of CPUs affects the memory copy throughput between the user buffer and the FORTRAN memory buffer. Figure 9 demonstrates that the single CPU memory copy data transfer rate capacity is about 500 megabytes per second, but when we use 24 CPUs, the capacity will increase up to about 1800 megabytes per second with an I/O size of 256 megabytes.

FORTRAN read throughput, which is measured at "Measuring Point 1", is between 763 to 1822 megabytes per second. As we can see in Fig. 8's "SRFS Read" line, the throughput measured at "Measuring Point B" in Fig. 6 is between 1724 to 2774 megabytes per second. The FORTRAN environment causes this difference, which is similar to the difference in the FORTRAN write operation.

While this file system has about 6700 megabytes per second read operation, memory-to-memory copy operation capacity is about 500 megabytes per second when a single CPU performs the operation. So in order to utilize this file system's high-speed I/O function, a FORTRAN user should do a numerical simulation with multiple CPUs.

6   Conclusion

We described the GSN library user program interface, and benchmarked the GSN library environment and FORTRAN I/O library environment. The GSN library's file I/O throughput was about 550 megabytes per second. We recognized that the GSN library I/O interface is 5+ times faster than local file system because the local file system of *CeViS* has about 100 megabytes per second peak throughput. The GSN library's file system provides a heterogeneous operating system file share function. This high-speed I/O is realized by using a GSN connection and the Memory interface, which speeds up memory copy operation. A high-speed heterogeneous file sharing system can be achieved at low cost.

The FORTRAN I/O library file write/read throughput is about 1800 megabytes per second. When we use the FORTRAN compiler, we determined that one should pay attention to the FORTRAN buffer in order to obtain fast file I/O operation. Because of the effect of the FORTRAN buffer, file I/O throughput reaches the ceiling with the memory-to-memory copy operation. As multiple CPUs are employed to perform numerical simulations, the memory bandwidth bottleneck can be avoided.

For future work, we are planning to measure multiple I/O accesses and crossbar striping file I/O throughput.

Acknowledgment

We are grateful to Katsumi Yazawa, Kazuhiro Kanaya, Toshihiko Kai, and Katsutoshi Yamauchi of Fujitsu people for their work measuring the performance of *NS-III*.

References

1. N. Fujita, Y. Matsuo: High-Speed Mass Storage System of Numerical Simulator III and its Basic I/O Performance Benchmark. Parallel Computational Fluid Dynamics, Elsevier Science B.V. (2003) 157-164
2. N. Fujita, H. Ookawa: Storage Devices, Local File System and Crossbar Network File System Characteristics, and 1 Terabyte File IO Benchmark on the "Numerical Simulator III". Proc. of 20th IEEE Symposium on Mass Storage Systems and Technologies/11th NASA Goddard Conference on Mass Storage Systems and Technologies (2003) 72-76
3. T. Jones, B. Sarnowska, F. Lovato, D. Magee, J. Koth: Architecture, Implementation, and Deployment of a High Performance, High Capacity Resilient Mass Storage Server(RMSS). Proc. of 18th IEEE Symposium on Mass Storage Systems and Technologies/9th NASA Goddard Conference on Mass Storage Systems and Technologies (2001) 41-65
4. N. T. B. Stone, J. R. Scott, et al.: Mass Storage on the Terascale Computing System. Proc. of 18th IEEE Symposium on Mass Storage Systems and Technologies/9th NASA Goddard Conference on Mass Storage Systems and Technologies (2001) 67-77
5. P. Andrews, T. Sherwin, V. Hazlewood: High-Speed Data Transfer via HPSS using Striped Gigabit Ethernet Communications. Proc. of 18th IEEE Symposium on Mass Storage Systems and Technologies/9th NASA Goddard Conference on Mass Storage Systems and Technologies (2001) 127-133
6. HiPPI-6400/Gigabyte System Network. http://public.lanl.gov/radiant/research/architecture/hippi.html
7. High Performance Networking Forum: http://www.hnf.org/
8. If you are interested in these functions, please refer "CeNSS User Reference Guide" on https://censs.nal.go.jp/

Large Scale Structures of Turbulent Shear Flow via DNS

Shin-ichi Satake[1], Tomoaki Kunugi[2], Kazuyuki Takase[3],
Yasuo Ose[4], and Norihito Naito[1]

[1] Department of Applied Electronics, Tokyo University of Science,
2641 Yamazaki, Noda, Chiba, 278-8510, Japan,
satake@te.noda.tus.ac.jp
[2] Department of Nuclear Engineering, Kyoto University,
Yoshida, Sakyo-ku, Kyoto 606-8501, Japan,
kunugi@nucleng.kyoto-u.ac.jp
[3] Japan Atomic Energy Research Institute,
Tokai Naka, Ibaragi, 319-1195, Japan,
takase@popsvr.tokai.jaeri.go.jp
[4] Yamato System Engineer,
1-17-1 Hitachi, Ibaraki, 317-0063, Japan,
ose@atom.tokai.jaeri.go.jp

Abstract. A direct numerical simulation (DNS) of turbulent channel flow has been carried out to understand the effects of Reynolds number. In this study, the Reynolds number for channel flow based on a friction velocity and channel half width was set to be constant; $Re_\tau = 1100$. The number of computational grids used in this study was $1024 \times 1024 \times 768$ in the $x-$, $y-$ and $z-$directions, respectively. The turbulent quantities such as the mean flow, turbulent stresses and the turbulent statistics were obtained via present DNS. Large scale turbulence structures visualized by paralleled AVS/Express appear in whole region. The structures are merged by small scales structures

1   Introduction

A fully developed channel and pipe flow are one of turbulent wall-bounded flow, which have been investigated in many researchers. In the first stage, although the researches were only experiments, in recent years, direct numerical simulation (DNS) is performed by development of a computer and detailed database is offered. The pioneering DNS of turbulent channel flow (Kim et al., 1987)[1] and the turbulent pipe flow (Eggeles et al., 1994)[4] were performed from Re_τ=180. In recently, the channel turbulent flow for Re_τ=650(Iwamoto et al., 1999)[8], Re_τ=800(Miyamoto et al., 2002) [2] were performed. Furthermore, the high Reynolds number (Re_τ=1050) equivalent to the experiment value of Laufer (1954)[3] was carried out by the authors group (Satake et al., 2000)[7]. The aim of this study is to investigate the Reynolds number dependence in turbulent channel flow. Specifically, in order to consider the Reynolds number effect of

A. Veidenbaum et al. (Eds.): ISHPC 2003, LNCS 2858, pp. 468–475, 2003.

turbulent channel and pipe flow about the case beyond Re_τ=1000, the turbulent channel flow of Re_τ=1100 is performed, and authors' existing pipe data of Re_τ=1050 (Satake et al., 2000) is compared.

2   Numerical Procedure

We have developed that the DNS code with cylindrical coordinates can numerically solve the continuity and momentum equations using the radial momentum flux formulation (Satake and Kunugi,[5],[6]) . In present study, however the spatial descritization in the homogeneous directions is changed from a second-order finite volume scheme to spectral method. The coordinate system is also changed from cylindrical to Cartesian grid system. In this code, the spectral method is used to compute to the special derivative in the stream and spanwise direction. The grid point used to compute the nonlinear terms had 1.5 times finer resolution in the directions to remove aliasing errors. For y-direction, a second-order finite volume method on a staggered mesh system is only applied to the spatial derivatives on the stretched gird. The incompressible Navier-Stokes and continuity equations described in Cartesian coordinate are integrated in time by using the fractional-step method. The second-order Crank-Nicholson scheme is applied to the viscous terms treated implicitly and a modified third-order Runge-Kutta scheme is used for other terms explicitly. The Helmholtz equation for viscous terms and pressure Poisson equations are only solved by using tridiagonal matrix technique in Fourier space. In Fig. 1, the domain decomposition and the transpose algorithms were applied. This code can be treated up to 805300000-grid system on 32 processors of the VPP5000 machine. In this code, the spectral method is used to compute to the special derivative in the stream and spanwise direction. The continuity and incompressible Navier-Stokes equations are

$$\frac{\partial u_i}{\partial x_i} = 0, \tag{1}$$

$$\frac{\partial u_i}{\partial t} + \frac{\partial u_i u_j}{\partial x_j} = -\frac{\partial p}{\partial x_i} + \nu \frac{\partial^2 u_i}{\partial x_i^2} \tag{2}$$

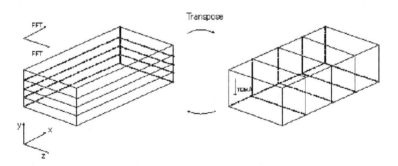

Fig. 1. Computational domain

In Fourier space, the spacial derivative operator for continue equation are computed:

$$\widehat{\frac{\partial u_i}{\partial x_i}} = ik_x\widehat{u} + \frac{d\widehat{v}}{dy} + ik_z\widehat{w}, \tag{3}$$

where the k_x, k_z are wave number in the stream and spanwise directions, respectively. The derivative in wall normal direction is computed by the second order finite difference scheme at staggered grid arrangement. The stream and span wise components located at the pressure point. The wall normal component only shifted to the half mesh size. The second-order finite difference method in y-direction on a staggered mesh system is applied to the stretched grid by hyperbolic tangent function. For the second derivative, the same manner can be applied and the term is expressed as:

$$\widehat{\frac{\partial^2 u}{\partial x^2}} = -k_x^2\widehat{u} \tag{4}$$

In Fourier space, the nonlinear terms in the streamwise direction are

$$\widehat{\frac{\partial u_i u_j}{\partial x_j}} = ik_x\widehat{uu} + \frac{d\widehat{uv}}{dy} + ik_z\widehat{uw} \tag{5}$$

However, $\widehat{uu}, \widehat{uv}, \widehat{uw}$ must be calculated in physical space. Because the grid point used to compute the nonlinear terms had 1.5 times finer resolution in the directions to remove aliasing errors.

These equations in time are integrated by using the fractional-step method (Dukowicz and Dvinsky, 1992)[9]. The second-order Crank-Nicholson scheme is applied to the viscous terms treated implicitly and a modified third-order Runge-Kutta scheme (Spalart et al., 1991)[10] is used for other terms explicitly. The Helmholtz equation for viscous terms and pressure Poisson equations are only solved by using tridiagonal matrix technique in Fourier space.

3   Computational Conditions

The number of grid points, the Reynolds number and grid resolutions summarized in Table 1. To perform a DNS with high Reynolds number in this study, a grids size of 1024 × 1024 × 768 is adopted for 224GB main memory as 32 PEs on a vector-parallel computer Fujitsu VPP 5000 at JAERI. The Reynolds number based on the friction velocity, viscosity and the channel half width (δ) is assumed to be 1100. The periodic boundary conditions are applied to the streamwise (x) and the spanwise (z) directions. As for the wall normal direction (y), non-uniform mesh spacing specified by a hyperbolic tangent function is employed. The number of grid points is 1024 × 1024 × 768 in the $x-$, $y-$ and $z-$directions, respectively. The all velocity components imposed the non-slip condition at the wall.

Table 1. Computational condition

Case	Grid numbers	Δx^+	Δz^+	Δy^+_{wall}	Δy^+_c
$Re_\tau = 1100$	$1024 \times 1024 \times 768$	16.8	8.59	0.177	4.5

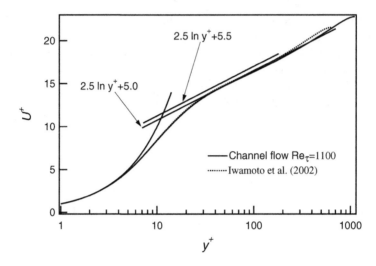

Fig. 2. Mean velocity profile

4   Results and Discussions

Mean velocity profiles are shown in Fig. 2. Iwamoto et al. (2002)[8]found that the effect of Reynolds number for channel flow is at Re_τ=110,150,300,400,650. The logarithmic profile at Re_τ=650 appears clearly. The profile is coincident with the $y^+ = 2.5 \ln y^+ + 5.0$. The present result also attends the equation. In the logarithmic region, the result of pipe flow located at between $y^+ = 2.5 \ln y^+ + 5.0$ and $y^+ = 2.5 \ln y^+ + 5.5$.

Figure 3 shows the total and Reynolds shear stress. The total shear stress is entirely straight line and symmetric profile. The situation derived at fully developed turbulent state. The velocity fields become completely convergence. Figure 4 shows the distributions of velocity fluctuations. It is interesting that the streamwise component at the peak location is close to the wall. For the distributions of wall-normal velocity fluctuations, the values at peak location for channel at Re_τ=1100 is large. The similar tendency also appears in the spanwise component. Thus, the turbulence behavior in the DNS with High Reynolds number tends to be more isotropic than that low Reynolds number, the streamwise component near wall region saturated, and the behavior of the normal and spanwise velocity fluctuating components near wall region is more enhance near wall region. Figure 5 shows the second invariant of velocity gradient tensor (Q^+ <0.008).The value corresponds to the vortical motion. The structures appears at the near wall region. Figure 6 shows the contour of the low speed streaky

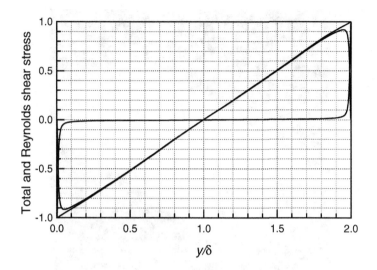

Fig. 3. Total and Reynolds shear stress

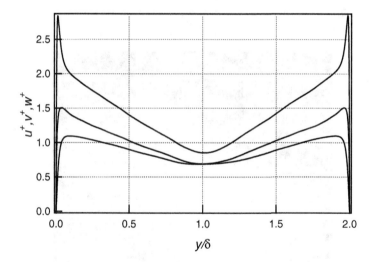

Fig. 4. Velocity fluctuations

structures $u^+ < -3.0$ at the several view point. It is normalized by and u_τ. The volume visualized obtained as full volume ($L_x^+ =17278, L_y^+ =2200, L_z^+=6911$) . The many small streaky structures exist in large streaky structures. The width of the large streaky structures are larger than 1000, located at away from the wall. Figure 7 shows the contour of the low speed streaky structures from top view. A few merged large streaks elongated to the channel center away from the wall. A characteristic size of the large streaky structures to the streamwise

Fig. 5. The second invariant of velocity gradient tensor ($Q^+ <$ 0.008)

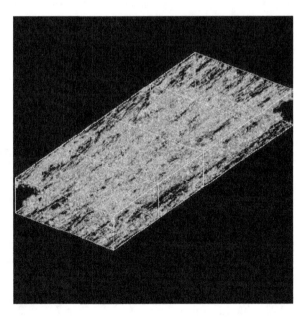

Fig. 6. The contour of the low speed streaky structures ($u^+ <$ -3.0)

Fig. 7. The contour of the low speed streaky structures ($u^+ <$-3.0)

direction is even larger than the half of the channel width. Almost large structures located in $y^+ > 200$, correspond to the wake region in the mean velocity profile. In the region, the scales of the fluid motion are different from the near wall region.

5   Conclusion

The DNS for the channel flow with High Reynolds number was carried out. The result is compared with previous DNS data for channel flow. Although further computations will be necessary to confirm the current turbulent statistics, the effect of Reynolds number can be predicted via this DNS. The streamwise component near wall region is saturated. The other components are more enhanced. The DNSs for channel have been continued. More detailed results with higher Reynolds number by Earth Simulator, the discussion will be reported at presentation.

References

1. Kim, J., Moin, P., and Moser, R.D.,1987, " Turbulence statistics in fully developed channel flow at low Reynolds number, J. Fluid Mech., Vol. 177, pp. 133.
2. Miyamoto, T., Tanahashi, M. and Miyauchi, T. 2002," Scaling law of fine scale eddies in high Reynolds number turbulent channel flow,"Proc. of 16 th Computational Fluid Dynamics Conf., in Tokyo, E12-2.

3. Laufer, J. ,The structure of turbulence in fully developed pipe flow, *NACA report 1174*,1954.
4. Eggels, J.G.M., Unger, F., Weiss, M.H., Westerweel, J. Adrian, R.J., Friedrich, R., and Nieuwstadt, F.T.M., Fully developed turbulent pipe flow: comparison between direct simulation and experiment, *J. Fluid Mech., Vol. 268*, pp. 175–209,1994.
5. Satake, S. and Kunugi, T. *Direct numerical simulation of turbulent pipe flow, Bulletin JSME Vol.64(in Japanense)*, pp. 65–70,1998.
6. Sakate, S. and Kunugi, T. *Direct numerical simulation of an impinging jet into parallel disks, Int. J. Numerical Methods for Heat and Fluid Flow*, 8, 768–780, 1998.
7. Satake, S., Kunugi, T. and Himeno, R., 2000, " High Reynolds Number Computation for Turbulent Heat Transfer in a pipe flow," Lecture Notes in Computer Science 1940, High Performance Computing, M. Valero et al. (Eds.), Springer-Verlag Berlin Heidelberg, pp. 514-523.
8. Iwamoto, K., Suzuki, Y., and Kasagi, N., 2002,"Database of fully developed channel flow,"THTLAB Internal Report, N0. ILR-0201,see http://www.thtlab.t.u-tokyo.ac.jp.
9. Dukowicz, J. K. and Dvinsky, A. S. *Approximate factorisation as a high order splitting for the implicit incompressible flow equations, J. Comp.,Phys., Vol.102, No.2,*, pp. 336–347, 1992.
10. Spalart, P.R.., Moser, R. D. and Rogers, M.,M. Spectral methods for the Navier-Stokes equations with one infinite and two periodic directions, *J. Comp. Phys., 96*, pp. 297–324, 1991.

Molecular Dynamics Simulation of Prion Protein by Large Scale Cluster Computing

Masakazu Sekijima[1], Chie Motono[1], Satoshi Yamasaki[2],
Kiyotoshi Kaneko[3], and Yutaka Akiyama[1]

[1] Computational Biology Research Center (CBRC),
National Institute of Advanced Industrial Science and Technology,
2-41-6 Aomi, Koto-ku, Tokyo 135-0064, Japan
[2] Department of Biotechnology,
The University of Tokyo,
1-1-1 Yayoi, Bunkyo-ku, Tokyo 113-8657, Japan
[3] Department of Cortical Function Disorders,
National Institute of Neuroscience,
National Center of Neurology and Psychiatry
and Japan Science and Technology Corporation (CREST),
4-1-1 Ogawa-Higashi, Kodaira, Tokyo 187-8502, Japan

Abstract. Molecular dynamics (MD) simulation is important theme in the parallel computing. MD simulations are widely used for simulating the motion of molecules in order to gain a deeper understanding of the chemical reactions, fluid flow, phase transitions, and other physical phenomena due to molecular interactions. In this study, we performed molecular dynamics simulations on monomeric and dimeric HuPrP at 300K and 500K for 10 ns to investigate the differences in the properties of the monomer and the dimer from the perspective of dynamic and structural behaviors. Simulations were also undertaken with Asp178Asn and acidic pH known as a disease-associated factor.

1 Introduction

Transmissible spongiform encephalopathies (TSEs) are neurodegenerative diseases attributable to the structural transformation of cellular prion (PrP^C) to its anomalous isoform (PrP^{Sc}). In humans, these diseases include kuru, Creutzfeldt-Jacob disease (CJD), fatal familial insomnia (FFI), and Gerstmann-Straussler-Scheinker syndrome (GSS), in sheep, scrapie, and in cattle, bovine spongiform encephalopathy (BSE). The most important aspect of prion diseases is the conformational transition of PrP^C to PrP^{Sc}, both of which are isoforms with identical amino acid sequence. However, comparison of their secondary structures shows that PrP^C is 42% helical with a very low (∼3%) β-sheet content, PrP^{Sc}, on the other hand, consists of 30% α-helices and 43% β-sheets. While the precise physiological role of PrP^C, and the chemical difference between PrP^C and PrP remain unknown, it appears that their differences are conformational.

The three-dimensional structures of monomeric PrP^C from various sources have been determined by NMR spectroscopy[16] and found to be very similar

A. Veidenbaum et al. (Eds.): ISHPC 2003, LNCS 2858, pp. 476–485, 2003.

Fig. 1. Schematic ribbon diagram of HuPrPc. (a) monomer, (b) dimer.

among many species. The N-terminal region (residues 23-124) is flexible, and the C-terminal region (residues 125-228) that contains the globular domains is well s tructured. All of these structures contain intramolecular disulfide bridges, three α-helices, and a short double-stranded β-sheet (Fig. 1(a)). Recent X-ray crystallographic studies determined the dimeric form of human PrP^C [7]. The dimer is the result of three-dimensional swapping of the C-terminal helix 3 and rearrangement of the disulfide bonds (Fig. 1(b)). Several mutations in the primary structure of PrP^C are known to segregate in variety of TSEs[15]. In this study, we selected the Asp178Asn (D178N) mutation known to be associated with FFI (Met129/Asn178). Recombinant forms of human and murine PrP^C manifest a pH-dependent conformational change in the pH range from 4.4 to 6, a loss of helix, and a gain of strands. Lower pH accelerated conversion in a cell-free conversion assay. Thus, acidic pH may play a role in facilitating the conformational change that ultimately results in the formation of PrP^{Sc}.

More recent conformational conversion models focus on intra- and intermolecular disulfide bonds[18, 19]. Some experiments have suggested that intramolecular disulfide bonds in PrP^C are required for its conversion to PrP^{Sc}[12]. To weaken these disulfide bonds a hypothetical molecular chaperone may be necessary [6].

Dimerization is usually required for proteins to evolve oligomeric proteins[11]. With respect to PrP, Meyer et al. [10] reported a monomer-dimer equilibrium under native conditions in a fraction of PrP^C from bovine brain. Others have suggested that 3D domain swapping-dependent oligomerization is an important step in the conformational change of PrP^C to PrP^{Sc} [7, 18]. However, the function and dynamics of the dimeric form of PrP^C remain to be elucidated. Molecular dynamics (MD) simulations are widely used to simulate the motion of molecules in order to gain a deeper understanding of the chemical reactions, fluid flow, phase transitions, and other physical phenomena due to molecular interactions. Rapidly increasing computational power has made MD simulation a powerful tool for studying the structure and dynamics of biologically important molecules. Day et al. [2] have shown that by increasing the temperature, protein unfolding can be accelerated without changing the pathway of unfolding and that this method is suitable for elucidating the details of protein unfolding at minimal computational expense. With these methods, one can obtain proper trajectories that reflect the conformational and dynamic characteristics of molecules at each time point during simulation. Most reported MD simulations of PrP^C have been reported [13, 20], involved short simulation times of less than 2 ns, or were performed using the AMBER ff94 force field, and all of the previously reported simulation targets were the monomer. We now report the first MD simulation of the dimeric PrP^C conformation. The aim of our study was to assess differences in the functions and dynamics of the PrP monomer and dimer. We performed eight 10-ns MD simulations of PrP^C dimer and monomer using the AMBER ff96 potential under different experimental conditions, a temperature of 300K and 500K, D178N mutant, and acidic pH.

2   Materials and Methods

All simulations were performed with the AMBER 7 program package [1], using the ff96 force field. The starting structures were HuPrP^C, entry 1QM2 (residues 125-228) as a monomer model and 1I4M (Chain A: residues 119-226; Chain B: residues 227-334) as a dimer model in the Brookhaven Protein Data Bank (PDB). There are disulfide bonds between Cys179-Cys214 in the monomer and between Cys179-Cys311 and Cys214-Cys287 in the dimer. The systems were surrounded with a 20 Ålayer of TIP3P water molecules and neutralized by sodium ions, using the LeaP module of AMBER 7. The number of solvent water molecules and counter ions in each system are shown in Table 1. The protocols have been described [17]. All simulations were performed on the MAssively parallel computer for Genome Informatics (Magi) Cluster running SCore 4.1 [4] at CBRC. Secondary structures were analyzed using DSSP [5], and images of simulated proteins were generated using MOLMOL [8].

Table 1. Simulation Conditions

	No. of ions	No. of water molecules
monomer (WT)	Na$^+$ 3	6814
dimer (WT)	Na$^+$ 8	9374
monomer (D178N)	Na$^+$ 2	6807
dimer (D178N)	Na$^+$ 6	9373
monomer (acidic pH)	Cl$^-$ 15	7168
dimer (acidic pH)	Cl$^-$ 28	9714

3   Results and Discussion

3.1   Simulation Stability

Fig. 2 shows the C$_\alpha$ root mean square deviations (RMSDs) from the initial structures of globular domains of HuPrP. In this paper, we define residues 129-223 [including strand 1 (S1), helix 1 (H1), strand 2 (S2), helix 2 (H2), and helix 3 (H3)] of the monomer and dimer (chain A) as globular domains. As we

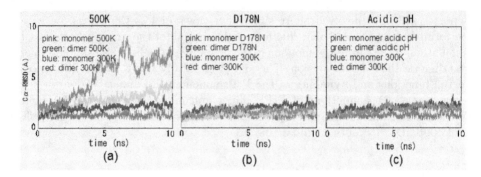

Fig. 2. Root mean square deviations (RMSDs) of C_α from the initial structures.

encountered few differences in the C_α RMSDs of each dimeric subunit, averaged data were used to present our results. In Figs. 2(a)-(c), simulation data at 300K are shown as control data. In the simulation at 300K, the C_α RMSD of both the monomer and the dimer remained relatively low for a duration of 10 ns although the monomer deviated from the initial structure more than the dimer. The average RMSD values of the monomer and dimer in the last 5 ns were 2.18 Åand 1.27 Å, respectively. In Fig. 2(a), at 500K, the C_α RMSDs of the monomer increased and reached 9.01 Åat 6.6 ns. In contrast, the C_α RMSDs of the dimer increased gently; the peak deviation was 4.7 Åat 9.2 ns. The average RMSD values of the monomer and dimer in the last 5 ns were 6.63 Åand 3.23 Å, respectively, indicating that the monomer increased faster than the dimer. This tendency was a characteristic common to simulations at 300K and 500K. In Figs. 2(b) and (c), under conditions of D178N and acidic pH, the C_α RMSDs of both the monomer and the dimer showed the same tendency as they did at 300K. At D178N, the peak values of C_α RMSDs were 2.76 Åand 1.32 Å, respectively. The average C_α RMSDs of the monomer and dimer in the last 5 ns were 1.72 Åand 1.44 Å, respectively, indicating that little conformation change occurred in the protein tertiary structure. At acidic pH, the monomer and dimer peak C_α RMSDs were 2.86 Åat 4.45 ns and 2.81 Åat 9.94 ns, respectively. The average RMSD values of the monomer and dimer in the last 5 ns were 2.01 Åand 2.05 Å.

3.2 Secondary Structure Evolution

Figs. 3 and 4 show the secondary structure evolution during simulation as determined by the DSSP program. Figs. 5 and 6 are ribbon illustrations of snap shots of the trajectories. Fig. 3(a) and 5(a) depict simulation results of monomer HuPrP at 300K. Although residues 152-156 of the H1 region formed a 310 helix or H-bonds over a 0.0 to 7.0 ns period, after 8.0 ns they formed an α-helix. Other secondary structure elements (S1, S2, H2, and H3) were retained throughout the simulation, however, several elongated S1 and S2 were observed until 4.0 ns (see the snapshot at 3.0 ns in Fig. 5(c)). As shown in Figs. 3(b) and 5(b), at about 2.0 ns at 500K, the monomer began to unfold in the β-sheet and at

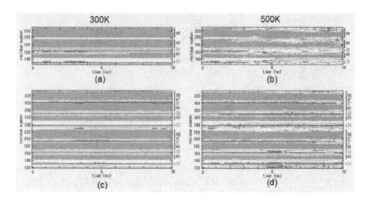

Fig. 3. Secondary structure as a function of simulation time determined with DSSP.

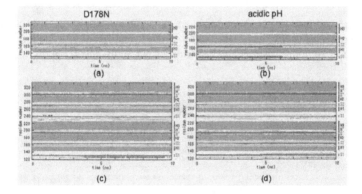

Fig. 4. Secondary structure as a function of simulation time determined with DSSP.

the C-terminus of H2 and H3. It appears that the degradation of the helices corresponds with the increase in C_α RMSD observed from 2.0 ns to 4.0 ns (Fig. 2(a)). We noted subsequent changes in the secondary structure at 4.0 - 6.0 ns: (1) the transient formation of non-native β-sheets at residues 129-130 and 222-223 and residues 132-133 and 159-160 and their unfolding, (2) the unfolding of the C-terminus of H2, and (3) the unfolding and refolding of H1 (Fig.5 (d)). These changes produced a rapid increase in the C_α RMSD of the monomer to 7.2 Åat 4.6 ns (Fig. 2(a)). Although the simulation at 500K was denaturation simulation, we can consider the results as conformational search at 500K. Fig. 5(d) shows the denaturation state of H1 at 4.55 ns, the elongated S1 and S2, and the additional β-sheet at 4.65 ns. Glockshuber et al. [3] and Korth et al.[9] demonstrated that the structure of H1 is different between PrP^C and PrP^{Sc} and suggested this region might form β-sheet. There were notable changes in secondary structure elements from 6.0 ns to 7.0 ns. There were some instances of fluctuation in the loop and the C_α RMSD reached 9.01 Åat 6.6 ns. At 300K and 500K, comparison with the monomer revealed that the dimer contained two

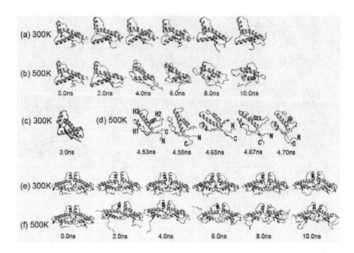

Fig. 5. Temporal history of the monomer at (a) 300K and (b) 500K. (c) One snapshot of the monomer at 300K. (d) Details of temporal history of the monomer at 500K around 4.6 ns. Temporal history of the dimer at (e) 300K and (f) 500K.

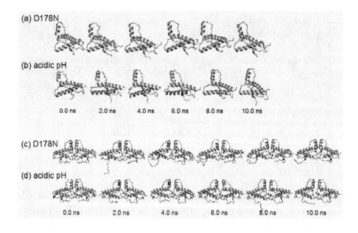

Fig. 6. Temporal history of (a) the monomer at D178N, (b) the monomer at acidic pH, (c) the dimer at D178N, and (d) the dimer at acidic pH.

additional structural elements, helices H' (residues 194-197 and 302-305) and a β-sheet S' (residues 191-193 and 299-301), that formed subunit interfaces (Fig. 3(c) and (d)). At 300K, all elements including S' and H' were retained throughout the simulation, although there was slight disruption at some points (Fig. 3(c)). The C-terminus of H1 tended to form a 310 helix. In H2, H3, and H', there were several H-bonds. At 500K, the C-terminus of helices crumbled like that of monomer (Fig. 3(d)). Our results imply that in both the monomer and the dimer there is a tendency for H1, H2, and H3 to unfold and that they share conformational vulnerability in these regions. Although in both the monomer and dimer we noted similar tendency for the denaturation of several regions, the dimeric form retained a remnant of the initial structure (Fig. 5(e)). S1, S2, and S' were retained throughout the simulation. In fact, as shown in Fig. 2 (a), the C_α RMSDs of the dimer increased more slowly than those of the monomer. Inter-subunit interactions of H', S', and H1 and its molecular size (weight) may contribute to solidity of the dimer.

At D178N, residues 167-169 and 275-277 (residues 167-169 in Chain B) in the dimer were mainly H-bonds, however, the helices were similar to those seen at 300K (Figs. 4(a) and (c)). In both the monomer and dimer we noted several elongations of β-sheets, these were more pronounced in the monomer (Figs. 4(a), 4(c), 6(a), and 6(c)).At acidic pH, several H-bonds were formed at the C-terminus of H1 in the dimer,however, the helices were similar to those seen at 300K (Figs. 4(b) and (d)). In both the monomer and dimer we noted several elongations of β-sheets, these were more pronounced in the monomer (Figs. 4(b), 4(d), 6(b), and 6(d)).

3.3   Positional RMSD from the Average Structure

As we were unable to detect major differences in the results obtained at 300K, D178N, and acidic pH, we present our results obtained with simulations at 300K and 500K. Fig. 7 shows C_α RMSDs from the mean structure as a function of residue number and is suitable for describing flexibility differences among the residues. As there were few differences between the RMSD profiles of Chains A and B in the dimer, only Chain A is depicted. In order, the C_α RMSDs increased for the dimer at 300K, the monomer at 300K, the dimer at 500K, and the monomer at 500K. Fluctuations at positions 1 (Arg 136), 2 (Phe141), 3 (Tyr157), and 4 (Asp168) were far larger than were fluctuations of other residues. The residues exhibiting the large fluctuations correspond with the loop regions and at 300K, only loop regions manifested fluctuations. Interestingly, a region adjacent

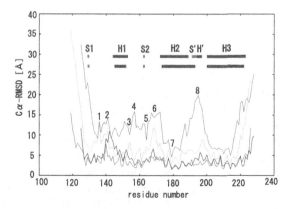

Fig. 7. C_α root mean square deviations (RMSDs) from the average structure as a function of residue number.

to position 4 (residues 169-171) is a putative binding site for protein X [6] and NMR showed it to be flexible. Resonances of the loop are not observed in HuPrP due to conformational exchange. At position 6 (Gly195), the C_α RMSDs in the monomer increased to 19.9 Åat 500K. At the same temperature, the same residue of the dimer increased to only 5.96 Å. This is consistent with Fig. 3(d) which shows that S' and H' stabilized this region. The smallest fluctuations were observed at positions 5 (Cys179) and 7 (Cys214) in H2 and H3 where residues Cys179 - Cys214 of the monomer, and Cys179 - Cys322 and Cys214 - Cys287 of the dimer are connected by disulfide bridges and contribute to the stabilization of neighboring regions. It appears that H2 and H3 form a relatively stable core of the protein and MD simulations of the prion from Syrian hamster indicated that the remainder of the protein has a degree of conformational plasticity[13]. Studies that mapped antibodies to various epitopes on PrP also support the hypothesis that a core of the molecule containing H2 and H3 remains intact after the conversion of PrP^C to PrP^{Sc}.

4   Conclusion

Ours is the first reported exploration of the dynamics of dimeric PrP^C, residues 119-226, using MD simulation to assess whether the dimer is essential for the conformational transition of PrP^C to PrP^{Sc}. Our results showed that denaturation of helices and elongation of the β-sheet were common to both the monomer and dimer. However, additional secondary structure elements formed in the dimer might result in the greater retention of dimeric than monomeric tertiary structure. Our results suggest that if dimerization plays an important role in the transition from PrP^C to PrP^{Sc}, some factors are required to enhance it. At present, we cannot rule out the possibility that dimerization of HuPrP is a necessary step in the transition from PrP^C to PrP^{Sc}. Efforts are underway in our laboratory to perform simulations of PrP 27-30, residues 90-231, to gain a better understanding of the underlying process(es) of conformational change from PrP^C to PrP^{Sc}.

5   Summary

PrP exists in not only a monomeric but also a dimeric form [7, 10]. Recent models suggest that dimerization plays an important role in the conformational change of PrP^C to PrP^{Sc} [18]. Although earlier MD simulations have yielded information on the monomeric form of PrP^C, data on its dimeric form remain scarce. To elucidate the conformational change of PrP^C to PrP^{Sc}, the dynamics of the dimeric form of PrP^C must be known. Therefore, we performed totally monomeric 40 ns simulations and dimeric 40 ns simulation in various conditions. Our conclusion that the monomer started denaturing earlier than the dimer is based on results we obtained in our study of C_α RMSDs from the initial structures (Fig. 2), of secondary structure evolution during simulation (Fig. 3), and of structures representative of conformational changes (Fig. 5). Our results also showed that α-helices in both the monomer and dimer denatured in a similar manner (Fig. 3). As the rate of protein denaturation is molecular weight-dependent, the greater retention of dimeric than monomeric tertiary and secondary structures is expected. However, we observed that in the dimer, the helices were denatured more readily while the tertiary structure was retained more than in the monomer. This suggests that dimer interface, H' helices (residues 194-197 and 302-305), and an S' β-sheet (residues 191-193 and 299-301), play an important role in the inhibition of tertiary structure of denaturalization. Although the discussion above was based on the simulation at 500K which was performed only one time for each prion model, those results potentially have biological importance. In our simulations, S1 and S2 in the dimer and especially the monomer, tended to elongate to the C-terminus side and the N-terminus side under most of the experimental conditions, respectively (Figs. 4, 5, 6). This suggests that the monomeric form of PrP^C is more likely to gain β-sheets. Our results suggest that if dimerization plays an important role in the transition from PrP^C to PrP^{Sc}, some factors are required to enhance it. Kaneko et al. [6] posited the existence of a molecular chaperone, protein X, and Tompa et al. [18] proposed a disulfide reshuffling model that is based on contacts between PrP^C- and PrP^{Sc} dimers and disulfide rearrangement(s). Our simulations were performed mainly on the well-ordered part of HuPrP^C (residues 125-228 in the monomer and 119-226 and 227-334 in the dimer, termed the "C-terminal region"). In addition, N-terminal residues 90-124, truncated in the present model, are required for α-helix to β-sheet transition and for prion disease infectivity [12, 14]. Current simulation models will continue to yield insights into the structure, functions, and dynamics of PrP and work is continuing in our laboratory to elucidate the dynamics, structural change(s), and other factors that involve the monomeric and dimeric forms of PrP.

References

1. Case, D.A., D.A. Pearlman, J.W. Caldwell, T.E. Cheatham III, J. Wang, W.S. Ross, C.L. Simmerling, T.A. Darden, K.M. Merz, R.V. Stanton, A.L. Cheng, J.J. Vincent, M. Crowley, V. Tsui, H. Gohlke, R.J. Radmer, Y. Duan, J. Pitera, I. Massova, G.L. Seibel, U.C. Singh, P.K. Weiner, and P.A. Kollman. 2002. AMBER 7. University of California, San Francisco.
2. Day R., B. Bennion, S. Ham, and V. Daggett. 2002. Increasing temperature accelerates protein unfolding without changing the pathway of unfolding. J. Mol. Biol. 322:189-203.
3. Glockshuber R., S. Hornemann, R. Riek, G. Wider, M. Billeter, K. Wuthrich. 1997. Three-dimensional NMR structure of a self-folding domain of the prion protein PrP(121-231). Trends in Biochemical Sciences. 22:241-2
4. Hori, A., H. Tezuka, Y. Ishikawa, N. Soda, H. Konaka, and M. Maeda.1996. Implementation of gang-scheduling on workstation cluster. In D. G. Feitelson and L. Rudolph, editors, IPPS'96 Workshop on Job Scheduling Strategies for Parallel Processing. volume 1162 of Lecture Notes Computer Science. 76-83. Springer-Verlag.

5. Kabsch, W., and C. Sander. 1983. Dictionary of protein secondary structure: pattern recognition of hydrogen-bonded and geometrical features. Biopolymers. 22:2577-2637.

6. Kaneko, K., L. Zulianello, M. Scott, C.M. Cooper, A.C. Wallace, T.L. James, F.E. Cohen, and S.B. Prusiner. 1997. Evidence for protein X binding to a discontinuous epitope on the cellular prion protein during scrapie prion propagation. Proc. Natl. Acad. Sci. USA. 94:10069-10074.

7. Knaus, K.J., M. Morillas, W. Swietnicki, M. Malone, W.K. Surewicz, V.C. Yee. 2001. Crystal structure of the human prion protein reveals a mechanism for oligomerization. Nat. Struct. Biol. 8:770-774.

8. Koradi, R., M. Billeter, and K. Wuthrich. 1996. MOLMOL: a program for display and analysis of macromolecular structures. J. Mol. Graphics. 14:51-55.

9. Korth C., B. Stierli, P. Streit, M. Moser, O. Schaller, R. Fischer, W. Schulz-Schaeffer, H. Kretzschmar, A. Raeber, U. Braun, F. Ehrensperger, S. Hornemann, R. Glockshuber, R. Riek, M. Billeter, K. Wuthrich, B. Oesch. 1997. Prion(PrP^{Sc}) -specific epitope defined by a monoclonal antibody. Nature. 390:74-77

10. Meyer, R.K., A. Lustig, B. Oesch, R. Fatzer, A. Zurbriggen, and M. Vandevelde. 2000. A monomer-dimer equilibrium of a cellular prion protein (PrP^C) not observed with recombinant PrP, J. Biol. Chem. 275:38081-38087.

11. Monod, J., J. Wyman, and J.-P. Changeux. 1965. On the nature of allosteric transitions: a plausible model. J. Mol. Biol. 12:88-118.

12. Muramoto, T., M. Scott, F.E. Cohen, and S.B. Prusiner. 1996. Recombinant scrapie-like prion protein of 106 amino acids is soluble. Proc. Natl. Acad. Sci. USA. 93:15457-15462.

13. Parchment, O., and J. Essex. 2000. Molecular dynamics of mouse and Syrian hamster PrP: implication for activity. Proteins. 38:327-340.

14. Prusiner, S.B. 1982. Novel proteinaceous infectious particles cause scrapie. Science. 216:136-144.

15. Prusiner, S.B. 1996. Trends Biochem Sci. 1996 Molecular biology and pathogenesis of prion diseases. 21:482-7.

16. Riek, R., S. Hornemann, G. Wider, M. Billeter, R. Glockshuber, K. Wuthrich. 1996. NMR structure of the mouse prion protein domain PrP(121-321). Nature. 382:180-182.

17. Sekijima, M., C. Motono, S. Yamasaki, K. Kaneko, Y. Akiyama. 2003. Molecular Dynamics Simulation of Dimeric and Monomeric Forms of Human Prion Protein: Insight into Dynamics and Properties. Biophysical J. 85:1-10.

18. Tompa, P., G.E. Tusnady, P. Friedrich, and I. Simon. 2002. The role of dimerization in prion replication. Biophysical J. 82:1711-1718.

19. Welker, E., W.J. Wedemeyer, and H.A. Scheraga. 2001. A role for intermolecular disulfide bonds in prion diseases? Proc. Natl. Acad. Sci. USA. 98:4334-4336.

20. Zuegg, J., and J.E. Greedy. 1999. Molecular dynamics simulations of human prion protein: importance of correct treatment of electrostatic interactions. Biochemistry. 38:13862-13876.

OpenMP/MPI Hybrid vs. Flat MPI on the Earth Simulator: Parallel Iterative Solvers for Finite Element Method

Kengo Nakajima

Research Organization for Information Science and Technology (RIST),
2-2-54 Naka-Meguro, Meguro-ku, Tokyo 153-0061, Japan,
nakajima@tokyo.rist.or.jp

Abstract. An efficient parallel iterative method for finite element method has been developed for symmetric multiprocessor (SMP) cluster architectures with vector processors such as the Earth Simulator. The method is based on a three-level hybrid parallel programming model, including message passing for inter-SMP node communication, loop directives by OpenMP for intra-SMP node parallelization and vectorization for each processing element (PE). Simple 3D linear elastic problems with more than 2.2×10^9 DOF have been solved using 3×3 block ICCG(0) method with additive Schwarz domain decomposition and PDJDS/CM-RCM reordering on 176 nodes of the Earth Simulator, achieving performance of 3.80 TFLOPS.

1 Introduction

1.1 SMP Cluster Architecture and Hybrid Parallel Programming Model

Recent technological advances have allowed increasing numbers of processors to have access to a single memory space in a cost-effective manner. As a result, symmetric multiprocessor (SMP) cluster architectures have become very popular as teraflop-scale parallel computers, such as the Accelerated Strategic Computing Initiative (ASCI, currently "Advanced Simulation and Computing (ASC)") [17] machines and the Earth Simulator [18].

In order to achieve minimal parallelization overhead, a multi-level hybrid programming model [2, 3, 6, 12] is often employed for SMP cluster architectures. The aim of this method is to combine coarse-grain and fine-grain parallelism. Coarse-grain parallelism is achieved through domain decomposition by message passing among SMP nodes using a scheme such as Message Passing Interface (MPI) [20], and fine-grain parallelism is obtained by loop-level parallelism inside each SMP node by compiler-based thread parallelization such as OpenMP [21].

Another often used programming model is the single-level flat MPI model [2, 3, 6, 12], in which separate single-threaded MPI processes are executed on each processing element (PE). The advantage of a hybrid programming model

A. Veidenbaum et al. (Eds.): ISHPC 2003, LNCS 2858, pp. 486–499, 2003.

over flat MPI is that there is no message-passing overhead in each SMP node. This is achieved by allowing each thread to access data provided by other threads directly by accessing the shared memory instead of using message passing. However, a hybrid approach usually requires more complex programming.

Although a significant amount of research on this issue has been conducted in recent years [2, 3, 6, 12], it remains unclear whether the performance gains of this hybrid approach compensate for the increased programming complexity. Many examples show that flat MPI is rather better [2, 3, 6, 12], although the efficiency depends on hardware performance (CPU speed, communication bandwidth, memory bandwidth), features of applications, and problem size [13].

1.2   Overview

In 1997, the Science and Technology Agency of Japan (now, the Ministry of Education, Culture, Sports, Science and Technology, Japan) began a five-year project to develop a new supercomputer, the Earth Simulator [18]. The goal of the project was to develop both hardware and software for earth science simulations. The Earth Simulator has SMP cluster architecture and consists of 640 SMP nodes, where each SMP node consists of 8 vector processors. The peak performance of each PE is 8 GFLOPS, and the overall performance of the system is 40 TFLOPS. Each SMP node of the Earth Simulator has 16 GB of memory, corresponding to 10 TB for the entire system. The present study was conducted as part of research toward developing a parallel finite-element platform for solid earth simulation, named GeoFEM [19].

In this paper, parallel preconditioned conjugate-gradient (CG) iterative solvers with incomplete Cholesky (IC) factorization have been developed for finite-element method using a three-level hybrid parallel programming model on the Earth Simulator. Individual PE of the Earth Simulator is a vector processor, therefore third-level of parallelism for vector processing should be considered in addition to the two levels, OpenMP and MPI. Following three levels of parallelism are considered:

- Inter-SMP node MPI for communication
- Intra-SMP node OpenMP for parallelization
- Individual PE compiler directives for vectorization

In flat MPI approach, communication among PEs through MPI and vectorization for individual PE have been considered for the Earth Simulator. In the hybrid parallel programming model, the entire domain is partitioned into distributed local data sets [7, 9, 10], and each partition is assigned to one SMP node. On the contrast, each partition corresponds to each PE in the flat MPI.

In order to achieve efficient parallel/vector computation for applications with unstructured grids, the following three issues are critical:

- Local operation and no global dependency
- Continuous memory access
- Sufficiently long innermost loops for vectorization

A special reordering technique proposed by Washio et. al. [16] has been integrated with parallel iterative solvers with localized preconditioning developed in the GeoFEM project [9, 19] in order to attain local operation, no global dependency, continuous memory access and sufficiently innermost long loops.

In the following part of this paper, we give an overview of parallel iterative solvers in GeoFEM, special reordering techniques for parallel and vector computation on SMP nodes, and present the results for an application to 3D solid mechanics on the Earth Simulator. Developed hybrid parallel programming model has been compared with the flat MPI programming model. Finally, recent results of solid earth simulations are shown.

2   Parallel Iterative Solvers in GeoFEM

2.1   Distributed Data Structures

A proper definition of the layout of the distributed data structures is an important factor determining the efficiency of parallel computations with unstructured meshes. The local data structures in GeoFEM are node-based with overlapping elements, and as such are appropriate for the preconditioned iterative solvers used in GeoFEM (Fig. 1) [9, 19].

2.2   Localized Preconditioning

The incomplete lower-upper (ILU)/Cholesky (IC) factorization method is one of the most popular preconditioning techniques for accelerating the convergence of Krylov iterative methods. Factorization by backward/forward substitution is repeated in each iteration. This factorization requires global data dependency, and is not suitable for parallel processing in which the locality is of utmost importance. The localized ILU used in GeoFEM is a pseudo ILU preconditioning method that is suitable for parallel processors. The ILU operation is performed

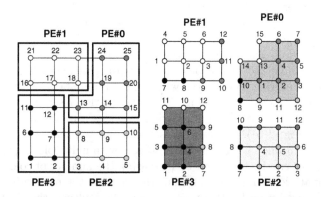

Fig. 1. Node-based partitioning into fout PEs [9, 19].

locally for a coefficient matrix assembled on each processor by zeroing out components located outside the processor domain. This localized ILU provides data locality on each processor and good parallelization because no inter-processor communications occur during ILU operation. This idea is originally from the incomplete block Jacobi preconditioning method [1, 15]. In order to stabilize localized ILU preconditioning, additive Schwarz domain decomposition (ASDD) for overlapped regions [15] has been introduced.

3   Reordering Methods for Parallel/Vector Performance on SMP Nodes

3.1   Cyclic Multicolor – Reverse Cuthil McKee Reordering

In order to achieve efficient parallel/vector computation for applications with unstructured grids, the following three issues are critical, (1) Local operations and no global dependency, (2) Continuous memory access and (3) Sufficiently long innermost loops for vector computation [10]. For unstructured grids, in which data and memory access patterns are very irregular, the reordering technique is very effective for achieving highly parallel and vector performance, especially for factorization operations in ILU/IC preconditioning.

The popular reordering methods are Reverse Cuthill-McKee (RCM) and multicoloring [14]. RCM method is a typical level set reordering method. In Cuthill-McKee reordering, the elements of a level set are traversed from the nodes of the lowest degree to those of the highest degree according to dependency relationships, where the degree refers to the number of nodes connected to each node. In RCM, permutation arrays obtained in Cuthill-McKee reordering are reversed. RCM results in much less fill-in for Gaussian elimination and is suitable for iterative methods with IC or ILU preconditioning. Multicoloring (MC) is much simpler than RCM. MC is based on an idea where no two adjacent nodes have the same color.

In both methods, elements located on the same color (or level set) are independent. Therefore, parallel operation is possible for the elements in the same color (or level set) and the number of elements in the same color (or level set) should be as large as possible in order to obtain high granularity for parallel computation or sufficiently large length of innermost loops for vectorization.

RCM (Fig. 2(a)) reordering provides fast convergence of IC/ILU-preconditioned Krylov iterative solvers, yet with irregular numbers of the elements in each level set. For example in Fig. 2(a), the 1st level set is of size 1, while the 8th level set is of size 8. Multicoloring provides a uniform element number in each color (Fig. 2(b)). However, it is widely known that the convergence of IC/ILU-preconditioned Krylov iterative solvers with multicoloring is rather slow [4]. Convergence can be improved by increasing the number of colors because of fewer incompatible local graphs [4] , but this reduces the number of elements in each color. The solution for this trade-off is cyclic multicoloring (CM) on RCM [16]. In this method, the elements are renumbered in a cyclic manner. Figure

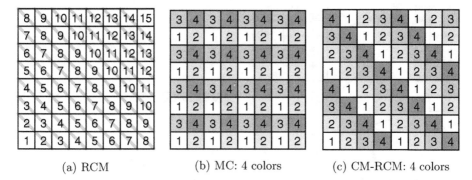

| (a) RCM | (b) MC: 4 colors | (c) CM-RCM: 4 colors |

Fig. 2. Example of hyperplane/RCM, multicoloring and CM-RCM reordering for 2D geometry [10]

2(c) shows an example of CM-RCM reordering. In this case, there are 4 colors; the 1st, 5th, 9th and 13th colors in Fig. 2(a) are classified into the 1st color. There are 16 elements in each color. In CM-RCM, the number of colors should be large enough to ensure that elements in the same color are independent.

3.2   DJDS Reordering

The compressed row storage (CRS) [1] matrix storage format is highly memory-efficient, however the innermost loop is relatively short, order of number of off-diagonal components, for matrix-vector operations as shown below:

```
do i= 1, N
  do j= 1, NU(i)
    k1= indexID(i,j);k2= itemID(k1)
    F(i)= F(i) + A(k1)*X(k2)
  enddo
enddo
```

The following loop exchange is then effective for obtaining a sufficiently long innermost loop for vector operations [5]:

```
do j= 1, NUmax
  do i= 1, N
    k1= indexID(i,j);k2= itemID(k1)
    F(i)= F(i) + A(k1)*X(k2)
  enddo
enddo
```

Descending-order jagged diagonal storage (DJDS) [16] is suitable for this type of operation and involves permuting rows into an order of decreasing number of non-zeros, as in Fig. 3(a). As elements on the same color are independent, performing this permutation inside each color does not affect results. Thus, a 1D array of matrix coefficients with continuous memory access and sufficiently long innermost loops can be obtained, as shown in Fig. 3(b).

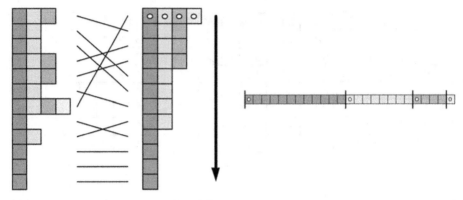

(a) Permutation of rows into order of decreasing number of non-zeros

(b) 1D array of matrix coefficient

Fig. 3. DJDS reordering for efficient vector/parallel processing

3.3 Distribution over SMP Nodes: Parallel DJDS Reordering

The 1D array of matrix coefficients with continuous memory access and sufficiently long innermost loops is suitable for both parallel and vector computing. The loops for this type of array are easily distributed to each PE in an SMP node via loop directives. In order to balance the computational load across PEs in the SMP node, the DJDS array should be reordered again in a cyclic manner. The procedure for this reordering is called parallel DJDS (PDJDS) [10].

3.4 Summary of Reordering Methods

The reordering procedures for increasing parallel/vector performance of the SMP cluster architecture described in this section are summarized as follows:

(1) RCM reordering on the original local matrix for independent sets and CM reordering to obtain innermost loops whose length is sufficiently long and uniform.
(2) DJDS reordering for efficient vector processing, producing 1D arrays of coefficients with continuous memory access and sufficient length of innermost loops.
(3) Cyclic reordering for load-balancing among PEs on an SMP node.

The typical loop structure of the matrix-vector operations for PDJDS/CM-RCM or PDJDS/MC reordered matrices based on the vector and OpenMP directives of the Earth Simulator is described as follows. In this study, IC/ILU factorization and matrix-vector product processes are executed in a same manner using same ordering methods and indexes. In the flat MPI programming model, PEsmpTOT is set to 1 without any option of OpenMP for compiler while PEsmpTOT is set to 8 in the hybrid programming model for the Earth Simulator (Fig. 4).

```
do iv= 1, NCOLORS
  !$omp parallel do private (iv0,j,iS,iE,i,k,kk etc.)
  do ip= 1, PEsmpTOT
    iv0= STACKmc(PEsmpTOT*(iv-1)+ip- 1)
    do j= 1, NLhyp(iv)
      iS= INL(npLX1*(iv-1)+PEsmpTOT*(j-1)+ip-1)
      iE= INL(npLX1*(iv-1)+PEsmpTOT*(j-1)+ip  )
      !CDIR NODEP
      do i= iv0+1, iv0+iE-iS
        k= i+iS - iv0
        kk= IAL(k)
        (Important Computations)
      enddo
    enddo
  enddo
enddo
```

SMP parallel

Vectorized

Fig. 4. Forward/backward substitution procedure for ILU/IC process by PDJDS/CM-RCM reordering.

4   Effect of Reordering

The proposed methods were applied to 3D solid mechanics example cases, as described in Fig. 5, which represent linear elastic problems with homogeneous material properties and boundary conditions. Figure 6 shows the results demonstrating the performance on a single SMP node of the Earth Simulator by hybrid parallel programming model. In this case, the following three cases were compared (Fig. 6):

- PDJDS/CM-RCM reordering
- Parallel descending-order compressed row storage (PDCRS)/CM-RCM reordering
- CRS without reordering

PDCRS/CM-RCM reordering is identical to PDJDS/CM-RCM except that the matrices are stored in a CRS manner [1] after permutation of rows into the order of decreasing number of non-zeros. The length of the innermost loop is shorter than that for PDJDS. The elapsed execution time was measured for

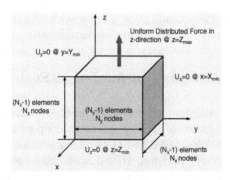

Fig. 5. Problem definition and bopundary conditions for 3D solid mechanics example cases.

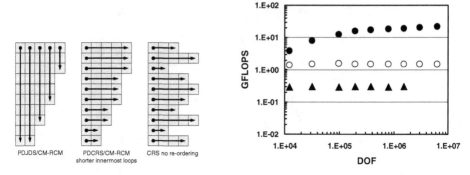

Fig. 6. Effect of coefficient matrix storage method and reordering for the 3D linear elastic problem in Fig. 5 with various problem sizes on the Earth Simulator with a single SMP node. (BLACK Circles: PDJDS/CM-RCM, WHITE Circles: PDCRS/CM-RCM, BLACK Triangles: CRS no reordering).

various problem sizes from 3×16^3 (12,288) DOF to 3×128^3 (6,291,456) DOF on a single SMP node of the Earth Simulator (8 PEs, 64 GFLOPS peak performance, 16 GB memory). The difference between PDCRS and PDJDS for smaller problems is not significant, but PDJDS outperforms PDCRS for larger problems due to larger length of inner-most loops. On the Earth Simulator, the PDCRS performs at a steady 1.5 GLOPS (2.3% of peak performance), while the performance of PDJDS increases from 3.81 GFLOPS to 22.7 GFLOPS with problem size. The loop length provided by PDCRS is order of number of off-diagonal components for each node, which is less than 30 in this case. On the contrast, average loop length provided by PDJDS is more than 2,500 for the case with 3×128^3 (6,291,456) DOF.

The cases without reordering exhibit very poor performance of only 0.30 GFLOPS (0.47% of peak performance). Without reordering, either of parallel and vector computations on a SMP node are impossible for the IC(0) factorization process even in the simple geometry examined in this study. This factorization process represents about 50% of the total computation time in CG solvers with IC(0) preconditioning, as was mentioned before. If this process is not parallelized, the performance decreases significantly.

5   Performance Evaluation for Large-Scale Problems

Figures 7-10 show the results for large-scale problems having simple geometries and boundary conditions as in Fig. 5 implemented up to 176 SMP nodes of the Earth Simulator (1,408 PEs, 11.26 TFLOPS peak performance, 2.8 TB memory). Performance of the hybrid and flat MPI models were evaluated. The problem size for one SMP node was fixed and the number of nodes was varied between 1 and 176. The largest problem size was $176 \times 3 \times 128 \times 128 \times 256$ (2,214,592,512) DOF, for which the performance was about 3.80 TFLOPS, corresponding to

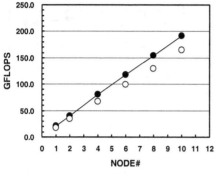
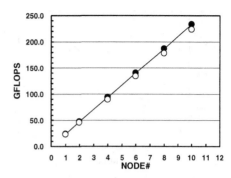

(a) Problem size/SMP = 786,432 DOF (3×64^3). Largest case is 7,864329 DOF on 10 SMP nodes (80PEs). Maximum performance is 192 (Flat MPI) and 165 (Hybrid) GFLOPS.

(b) Problem size/SMP node = 12,582912 DOF ($3 \times 256 \times 128 \times 128$). Largest case is 125,829,120 DOF on 10 SMP nodes (80 PEs). Maximum performance is 233 (Flat MPI) and 224 (Hybrid) GFLOPS.

Fig. 7. Problem size and parallel performance on the Earth Simulator for the 3D linear elastic problem in Fig. 5 using between 1 and 10 SMP nodes. (BLACK Circles: Flat MPI, WHITE Circles: Hybrid). PDJDS/CM-RCM reordering.

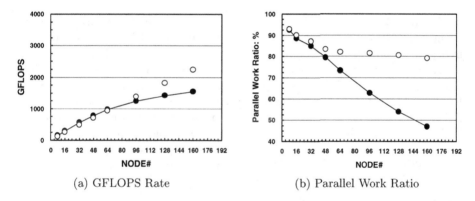

(a) GFLOPS Rate

(b) Parallel Work Ratio

Fig. 8. Problem size and parallel performance on the Earth Simulator for the 3D linear elastic problem in Fig. 5 using between 8 and 160 SMP nodes. (a)GFLOPS rate and (b)Parallel work ratio. Problem size/PE is fixed as 786,432 DOF (3×64^3). Largest case is 125,829,120 DOF on 160 SMP nodes (1280 PEs). Maximum performance is 1.55 (Flat MPI) and 2.23 (Hybrid) TFLOPS (Peak performance= 10.24 TFLOPS). (BLACK Circles: Flat MPI, WHITE Circles: Hybrid). PDJDS/CM-RCM reordering.

33.7% of the total peak performance of the 176 SMP nodes. The parallel work ratio among SMP nodes for MPI is more than 90% if the problem is sufficiently large.

The performance of the hybrid model is competitive with that of the flat MPI model, and both provide robust convergence and good parallel performance for a wide range of problem sizes and SMP node numbers. Iterations for convergence

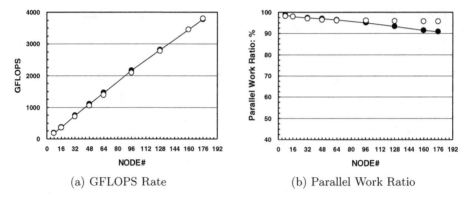

(a) GFLOPS Rate   (b) Parallel Work Ratio

Fig. 9. Problem size and parallel performance on the Earth Simulator for the 3D linear elastic problem in Fig. 5 using between 8 and 176 SMP nodes. (a)GFLOPS rate and (b)Parallel work ratio. Problem size/SMP node is fixed as 12,582,912 DOF $(3 \times 256 \times 128 \times 128)$. Largest case is 2,214,592,512 DOF on 176 SMP nodes (1408 PEs). Maximum performance is 3.78 (Flat MPI) and 3.80 (Hybrid) TFLOPS (Peak performance= 11.26 TFLOPS). (BLACK Circles: Flat MPI, WHITE Circles: Hybrid). PDJDS/CM-RCM reordering.

in the hybrid and flat MPI are almost equal, although the hybrid converges slightly faster as shown in Fig. 10(a). In general, flat MPI performs better the hybrid model for smaller numbers of SMP nodes as shown in Fig. 7, while the hybrid outperforms flat MPI when a large number of SMP nodes are involved (Fig. 8-10), especially if the problem size per node is small as shown in Fig. 8 and Fig. 10. This is mainly because of the latency overhead for MPI communication. According to the performance estimation for finite-volume application code for CFD with local refinement in [8], a greater percentage of time is required by the latency component on larger processor counts, simply due to the available bandwidth being much larger (Fig. 11). Flat MPI requires eight times as many MPI processes as hybrid model. If the node number is large and problem size is small, this effect is significant.

6   Solid Earth Simulations for Complicated Geometries and Boundary Conditions

Figures 12 and 13 show the most recent results in solid earth simulations (ground motion simulation with contact) on the Earth Simulator using parallel iterative methods and parallel programming models developed in this study [11]. Figure 12 describes the model of the Southwest Japan, consisting of 7,767,002 nodes (23,301,006 DOF) and 7,684,072 tri-linear (1st order) hexahedral elements. Special preconditioning method, called selective blocking, has been developed based on the block IC/ILU method in this study for non-linear problems due to fault-zone contact with penalty constraint conditions [11]. Figure 13 provides preliminary results of the simulation on the Earth Simulator. Effects of color num-

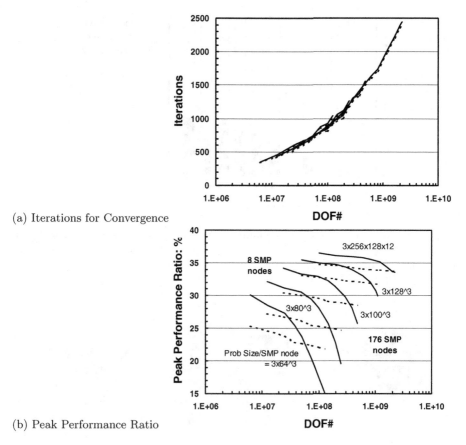

(a) Iterations for Convergence

(b) Peak Performance Ratio

Fig. 10. Problem size and parallel performance on the Earth Simulator for the 3D linear elastic problem in Fig. 5 using between 8 and 176 SMP nodes. (a) Iterations for convergence and (b) Ratio to the peak performance. Problem size/SMP node is fixed (THICK lines: Flat MPI, DASHED lines: Hybrid). PDJDS/CM-RCM reordering.

bers for multicolor reordering [11] in flat MPI and three-level hybrid are shown. Both programming models attained about 30% of the peak performance (640 GFLOPS) even in this complicated geometry with contact boundary conditions, while flat MPI shows better performance.

7   Conclusions and Remarks

This paper described an efficient parallel iterative method for unstructured grids developed for SMP cluster architectures with vector processors, such as the Earth Simulator. The method employs three-level hybrid parallel programming model consisting of the following hierarchy, (1) Inter-SMP node: MPI, (2) Intra-SMP node: OpenMP for parallelization and (3) Individual PE: Compiler directives for vectorization. Multiple reordering methods have been applied

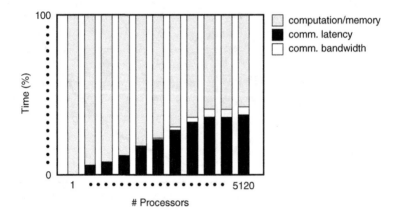

Fig. 11. Performance estimation of a finite-volume application code for CFD with local refinement on the Earth Simulator. Based on the results described in [8]. A greater percentage of time is taken by the latency component on larger processor counts, simply due to its much larger available bandwidth.

Fig. 12. Description of the Southwest Japan model. This model consists of crust (dark gray) and subduction plate (light gray). 7,767,002 nodes (23,301,006 DOF) and 7,684,072 tri-linear (1st order) hexahedral elements are included [11].

in order to attain concurrent local operations, no global dependency, continuous memory access, and sufficiently long innermost loops. PDJDS (parallel descending-order jagged diagonal storage) provides long innermost loops. CM-RCM (cyclic-multicoloring/reverse Cuthill-McKee) provides local and parallel operations.

Simple 3D linear elastic problems with more than 2.2×10^9 DOF were solved by 3×3 block ICCG(0) with additive Schwarz domain decomposition and PDJDS/CM-RCM reordering on 176 SMP nodes of the Earth Simulator, achieving performance of 3.80 TFLOPS (33.7% of peak performance). PDJDS/CM-RCM reordering provides excellent vector and parallel performance on SMP nodes. Without reordering, parallel processing of forward/backward substitution in IC/ILU factorization was impossible due to global data dependencies even in the simple examples in this study. While the three-level hybrid and flat MPI parallel programming models offer similar performance, the hybrid pro-

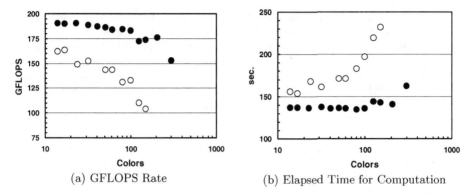

(a) GFLOPS Rate   (b) Elapsed Time for Computation

Fig. 13. Performance on 10 SMP nodes of the Earth Simulator (peak performance = 640 GFLOPS) using CG iterative method with *selective blocking* preconditioning [11] for the 3D elastic contact problem with MPC (linear multiple point constraint) condition ($\lambda = 10^6$) in Fig. 12 (Southwest Japan model, 23,301,006 DOF). Effect of color numbers for multicolor reordering. (a) GFLOPS rate, (b) Elapsed Time for Computation. (BLACK Circles: Flat MPI, WHITE Circles: Hybrid). PDJDS/multicolor (MC) reordering [11].

gramming model outperforms flat MPI in the problems with large numbers of SMP nodes. In the next stage, developed solvers will be applied to realistic applications with complicated geometries and boundary conditions in science and engineering field, as shown in the previous section.

Acknowledgements

This study is a part of the *"Solid Earth Platform for Large-Scale Computation"* project funded by the Ministry of Education, Culture, Sports, Science and Technology, Japan through Special Promoting Funds of Science & Technology.

References

1. Barrett, R., Berry, M., Chan, T.F., Donato, J., Dongarra, J.J., Eijkhout, V., Pozo, R., Romine, C. and van der Vorst, H.: Templates for the Solution of Linear Systems: Building Blocks for Iterative Methods, SIAM, 1994.
2. Cappelo, F. and Etiemble, D.: "MPI versus MPI+OpenMP on the IBM SP for the NAS Benchmarks", SC2000 Technical Paper, Dallas, Texas, 2000.
3. Djomehri, M.J. and Jin, H.H.: "Hybrid MPI+OpenMP Programming of an Overset CFD Solver and Performance Investigations", NASA/NAS Technical Report (NASA Ames Research Center), NAS-02-002, (2002). 16.
4. Doi, S. and Washio: "Using Multicolor Ordering with Many Colors to Strike a Better Balance between Parallelism and Convergence", RIKEN Symposium on Linear Algebra and its Applications, The Institute of Physical and Chemical Research, 1999, pp.19-26.

5. Dongara, J.J., Duff, I.S., Sorensen, D.C. and van der Vorst, H.A.: Solving Linear Systems on Vector and Shared Memory Computers, SIAM, 1990.
6. Falgout, R. and Jones, J.: "Multigrid on Massively Parallel Architectures", Sixth European Multigrid Conference, Ghent, Belgium, September 27-30, 1999.
7. Garatani, K., Nakamura, H., Okuda, H., Yagawa, G.: "GeoFEM: High Performance Parallel FEM for Solid Earth", HPCN Europe 1999, Amsterdam, The Netherlands, Lecture Notes in Computer Science, Vol.1593 (1999) pp.133-140.
8. Kerbyson, D.J., Hoisie, A. and Wasserman, H. "A Comparison Between the Earth Simulator and Alpha Server Systems using Predictive Application Performance Models", LA-UR-02-5222, Los Alamos National Laboratory, USA, 2002.
9. Nakajima, K., Okuda, H.: "Parallel Iterative Solvers with Localized ILU Preconditioning for Unstructured Grids on Workstation Clusters", International Journal for Computational Fluid Dynamics, Vol.12 (1999) pp.315-322.
10. Nakajima, K. and Okuda, H.: "Parallel Iterative Solvers for Unstructured Grids using an OpenMP/MPI Hybrid Programming Model for the GeoFEM Platform on SMP Cluster Architectures", International Workshop on OpenMP: Experiences and Implementations (WOMPEI 2002), Lecture Notes in Computer Science 2327 (2002), pp.437-448.
11. Nakajima, K.: "Parallel Iterative Solvers of GeoFEM with Selective Blocking Preconditioning for Nonlinear Contact Problems on the Earth Simulator", RIST/Tokyo GeoFEM Report 2003-006 (http://geofem.tokyo.rist.or.jp/report_en/2003_006.html/), 2003.
12. Oliker, L, Li X., Husbands, P. and Biswas, R.: "Effects of Ordering Strategies and Programming Paradigms on Sparse Matrix Computations", SIAM Review, Vol.44, No. 3(2002), pp.373-393.
13. Rabenseifner, R.: "Communication Bandwidth of Parallel Programming Models on Hybrid Architectures", International Workshop on OpenMP: Experiences and Implementations (WOMPEI 2002), Lecture Notes in Computer Science 2327 (2002), pp.437-448.
14. Saad, Y.: Iterative Methods for Sparse Linear Systems, PWS Publishing Company, 1996.
15. Smith, B., Bjørstad, P. and Gropp, W.: Domain Decomposition, Parallel Multilevel Methods for Elliptic Partial Differential Equations, Cambridge Press, 1996.
16. Washio, T., Maruyama, K., Osoda, T., Shimizu, F. and Doi, S.: "Efficient implementations of block sparse matrix operations on shared memory vector machines", SNA2000: The Fourth International Conference on Supercomputing in Nuclear Applications, 2000.
17. Accelerated Strategic Computing Initiative (ASCI) Web Site: http://www.llnl.gov/asci/
18. Earth Simulator Center Web Site: http://www.es.jamstec.go.jp/
19. GeoFEM Web Site: http://geofem.tokyo.rist.or.jp/
20. MPI Web Site: http://www.mpi.org/
21. OpenMP Web Site: http//www.openmp.org/

Performance Evaluation of Low Level Multithreaded BLAS Kernels on Intel Processor Based cc-NUMA Systems

Akira Nishida and Yoshio Oyanagi

Department of Computer Science, The University of Tokyo,
7-3-1 Hongo, Bunkyo-ku, Tokyo 113-0033, Japan,
Tel:+81-3-5841-4076, Fax:+81-3-3818-1073,
nishida@is.s.u-tokyo.ac.jp

Abstract. Parallel implementation of the BLAS library for sparse matrix algorithms in computational linear algebra is a critical problem, especially on the shared memory architectures with finite memory bandwidth. In this study, we evaluate the performance of the cc-NUMA systems using low level multithreaded BLAS kernels. The performance of both the compiler and the systems are evaluated on two Intel processor based architectures, NEC TX7/AzusA and IBM xSeries 440.

1 Background

In recent studies, it has been general to use the Lanczos/Arnoldi method as the eigensolver for large-scale sparse matrices [5], in the cases where you have to find only extreme eigenvalues and the associated eigenvectors. Although the QR method, which computes all the eigenpairs of a matrix, can be used for small problems, it cannot deal with large-scale problems because of its $O(n^3)$ complexity for the problem size n. The restarted iterative Lanczos/Arnoldi method is one of the most practical solution for sparse matrices, which is known to be difficult fo parallelize.

The Jacobi-Davidson method [10], which is based on the Davidson method [7] developed for the matrix diagonalization in quantum chemistry, is a method based on the solution of correction equation, and is known to be highly suitable for vector computers.

2 Jacobi-Davidson Method

In the Davidson method, the eigenvalue is computed as follows: Consider an approximate eigenpair of the matrix A,, i.e., Ritz pair (θ_k, u_k) in the partial space $\mathcal{K} = \mathrm{span}\{v_1, ..., v_k\}$ of dimension k, where $v_1, ..., v_k$ is a orthonormal base. To compute u_k, we have to expand the dimension of $\mathcal{K}$, computing the correction equation

$$M_k t = r, \quad M_k = D_A - \theta_k I \tag{1}$$

A. Veidenbaum et al. (Eds.): ISHPC 2003, LNCS 2858, pp. 500–510, 2003.

for the residual $r = Au_k - \theta_k u_k$, where D_A is the diagonal element of A. We get v_{k+1} by orthogonalizing t with $\mathcal{K}$. If we put $V_{k+1} = [v_1....v_{k+1}]$, the new Ritz pair (θ_{k+1}, u_{k+1}) is computed as the eigenpair of the matrix

$$H_{k+1} = V_{k+1}^* A V_{k+1}. \tag{2}$$

In the Jacobi-Davidson method, the element for the update is extracted from the orthogonal complement of u_k, which is assumed to be normalized. We project the eigenproblem $Ax = \lambda x$ onto the orthogonal complement $u_k^\perp$ of u_k. The orthogonal projection of A onto $u_k^\perp$ is defined by

$$A_P = (I - u_k u_k^*)A(I - u_k u_k^*), \tag{3}$$

which can be rewritten as

$$A = A_P + u_k u_k^* A + Au_k u_k^* - \theta_k u_k u_k^*. \tag{4}$$

Correction vector z satisfies

$$A(z + u_k) = \lambda(z + u_k), \quad z \perp u_k, \tag{5}$$

and from the equation (4), we have

$$(A_P - \lambda I)z = -r + (\lambda - \theta_k - u_k^* Az)u_k. \tag{6}$$

From the relations $A_P z \perp u_k$, $z \perp u_k$, and $r \perp u_k$, the coefficient of u_k must be 0, and the problem is reduced to the computation of the correction equation

$$(A_P - \lambda I)z = -r. \tag{7}$$

You do not have to solve the equation (7) exactly, since you cannot know the value of λ. Instead, we solve

$$(I - u_k u_k^*)(A - \theta_k I)(I - u_k u_k^*)z = -r \tag{8}$$

using the approximate eigenvalue θ_k. We orthogonalize the obtained vector against the subspace V_k and define the new basis v_{k+1}. The rightmost eigenvalue of $H_{k+1} = V_{k+1}^* A V_{k+1}$ is obtained as the Ritz value θ_{k+1}.

3   Parallel Implementation on Low Level BLAS Kernels

Although the algorithm can be implemented using BLAS (Basic Linear Algebra Subprograms) [9], which are developed by Dongarra et al., it is hard to fully utilize the cache memories, since the complexity of matrix-vector operations is $\mathcal{O}(n)$ in sparse systems. Therefore, we have to consider the characteristics of the problems [11] if the parallelization at the level of BLAS is an appropriate choice.

To build a parallelized software efficiently, it is desirable to achieve the optimal performance without knowing the hardware details of the architecture. In order to construct such environment, the shared memory architecture is more suitable than the distributed memory environment, which requires explicit data partitioning.

In general, shared memory architecture has lower memory latency, although the frequency of cache misses are higher than distributed memory architectures, and the performance is reduced when the write requests to the same cache line occur. However, it is possible to achieve high performance even in such cases with careful implementation. On shared memory architectures, we can expect enough scalability even in the Level 1 and 2 BLAS routines, which implement vector-vector and matrix-vector operations respectively. We have partitioned the outermost loops of the above subroutines using the OpenMP API.

In the following sections, we discuss the implementation of the above algorithm and the parallelization techniques for shared memory architectures. We have adopted a program called JDQZ, which is a serial Fortran 77 implementation of the Jacobi-Davidson algorithm [8] by Fokkema and van Gijzen et al. The templates [6] are used for description of the iterative method, while the BLAS [9] and LAPACK [3] routines are used for basic linear algebra operations.

For the preliminary evaluation of the algorithm, we have used a 4-way SMP (Dell PowerEdge 6300 with 450NX chipset and 768MB main memory) with Intel Pentium III Xeon processors (550MHz frequency with 512KB L2 cache). The operating system is Solaris 7 and the compiler is PGI Fortran 3.2. for Solaris x86[1].

The eigenvalues of a tridiagonal matrix A_1 of size n with the diagonal elements 2 and the subdiagonal element -1 are given by $2 - 2\cos[k\pi/(n+1)]$ for $k = 1, ..., n$, and the eigenvalues of the pentadiagonal matrix of size $n = N^2$ with

$$
A_2 = \begin{pmatrix} T_N & -I & & O \\ -I & \ddots & \ddots & \\ & \ddots & \ddots & -I \\ O & & -I & T_N \end{pmatrix}, \quad \text{where } T_N = \begin{pmatrix} 4 & -1 & & O \\ -1 & \ddots & \ddots & \\ & \ddots & \ddots & -1 \\ O & & -1 & 4 \end{pmatrix} \tag{9}
$$

are given by $4 - 2(\cos(k\pi/(N+1)) + \cos(j\pi/(N+1)))$ for $j, k = 1, ..., N$, respectively. We compute here the rightmost eigenvalues of A_1 and A_2 with residual error of 10^{-8}. We show the relation between the problem size and the computation time in Table 1.

Table 2 shows the profiling results in the case with A_1 of size 128^2, we can see that the computation time of BiCGSTAB occupies about 85%. Since the condition numbers of A_1 and A_2 can be estimated by

$$
4 \Big/ \left\{ 2 - 2\cos\left(\frac{\pi}{n+1}\right) \right\} \approx \frac{4}{\pi^2} n^2 \tag{10}
$$

---

[1] PGI's numerical library did not supported threads when we started evaluation, but recent libraries such as Intel's Math Kernel Library includes many threaded routines. We are planning to test them in the near future

Table 1. Time for Computation of rightmost eigenvalue by Jacobi-Davidson.

Size	Time(s)	
n	A_1	A_2
32^2	2	1
64^2	50	12
128^2	4328	60
256^2	—	396

Table 2. Result for A_1 with $n = 128^2$.

Function	Calls	Cost(s)	Cost(%)
jdqz	1	4328	100.00
zcgstabl	990	3659	84.55
zgemv	298096	1064	24.59
zdotc	344829	794	18.35
zaxpy	230778	719	16.63
...			

and

$$8 \bigg/ \left\{ 4 - 4 \cos\left(\frac{\pi}{N+1}\right) \right\} \approx \frac{4}{\pi^2} N^2, \tag{11}$$

respectively, the slow convergence of the the iterative method in the case of A_1 shows that the acceleration of the correction equation is inevitable for the Jacobi-Davidson method. The *Cost* includes time spent on behalf of the current function in all children whether profiled or not.

Next, we show the result with A_2 of size 256^2 in Table 3. Five rightmost eigenvalues are computed here. In the table, the *Time* does not include time spent in functions called from this function or line.

Table 3. Result for A_2 with $n = 256^2$.

Function	Calls	Time(s)	Time(%)
zgemv	10246	319	33.40
zdotc	12615	150	15.77
zaxpy	8163	116	12.14
jdqz	1	109	11.50
jdqzmv	3479	54	5.74
zxpay	4193	54	5.68
dznrm2	5402	52	5.50
amul	3540	47	5.00
bmul	3540	31	3.26
zcgstabl	41	12	1.33
zmgs	369	3	0.32
...			

zgemv is the Level 2 BLAS routine of the type

```
y := alpha*A*x + beta*y,
y := alpha*A'*x + beta*y,
```

and

```
y := alpha*conjg( A' )*x + beta*y,
```

and zdotc is the dot product

```
z := conjg(x)*y,
```

while zaxpy and zxpay are the Level 1 BLAS routines of the type

```
y := alpha*x + y
```

and

```
y := x + alpha*y,
```

respectively.

We can see that the Level 1 and 2 BLAS operations occupies dominant part of the computation time in the Jacobi-Davidson method, where the complexity of matrix-vector multiply is estimated to be $\mathcal{O}(n)$ since the matrix is assumed to be sparse. We have discussed some scalability issues on symmetric multiprocessors using Sun UltraEnterprise 10000 in [2].

4   Performance of BLAS Kernels on NUMA Architectures

In the following sections, we target NUMA computing environments as the final goal of parallel implementation, which will enable users to realize high perfor-mance scientific computing without thinking scalability limit. We discuss im-plementation and optimization issues on Linux environment on Intel processor based ccNUMA Server NEC AzusA and IBM x440. The results are compared with the data using SGI's Origin 2000, de facto standard of ccNUMA architec-ture.

NEC AzusA

Figure 1 and 2 [4] show the architecture of AzusA, the 16-way configuration of which has four sets of components plus the external data crossbar. The chip set design is optimized for 16-way or 4-cell configurations and employs a snoop-based coherency mechanism for lower snoop latencies. To fully accommodate Itanium processor system bus bandwidth as well as the I/O traffic, the chip set provides approximately 4.2 Gbytes/s of memory bandwidth per cell (a system total of 16.8 Gbytes/s). The interconnection between the system data controller and the data crossbar chip is also 4.2 Gbytes/s per port (8.4 Gbytes/s bisectional).

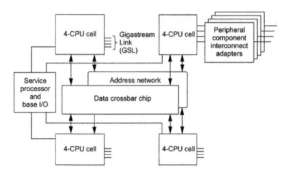

Fig. 1. AzusA system block diagram.

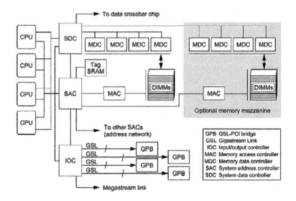

Fig. 2. AzusA chip set components.

The performance in MFLOPS of serial and single threaded parallel codes is shown in Table 4. The low performance on AzusA shown in Figure 4 indicates that the runtime library of Intel compiler had some problem[2].

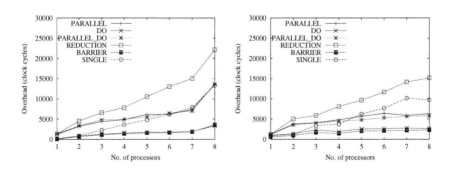

Fig. 3. Synchronization overheads on NEC AzusA and SGI Origin 2000.

Table 4. Performance in MFLOPS of serial and single threaded parallel codes.

Function	AzusA		Origin 2000	
	Serial	Parallel	Serial	Parallel
zaxpy	259.1	138.7	334.0	328.2
zdotc	118.1	21.6	285.2	278.7
dznrm2	95.6	93.4	116.5	150.2
zgemv	113.2	52.2	147.0	206.5

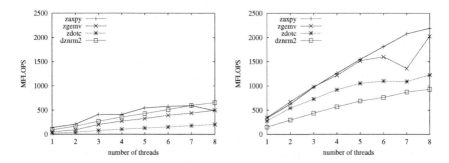

Fig. 4. Parallel BLAS performance on NEC AzusA and SGI Origin 2000.

We used an NEC's AzusA which has eight 733MHz Intel Itanium Processors with 16KB L1 data, 96KB L2, and 2MB L3 caches, and an SGI's Origin 2000 at AIST, which has 32 400MHz MIPS R12000 processors with 32KB L1 data, 8MB L2 caches for reference. Two Itanium processors and 128×4MB PC100 SDRAM are implemented on each cell of AzusA. The operation system running on the AzusA is Red Hat 7.1 based NEC Linux R1.2, and the compiler is `efc` of Intel Compiler for Linux 6.0 and `f90`.

We show the synchronization overheads of EPCC Microbenchmark below. Figure 3 shows overheads on AzusA and Origin 2000[3]. The results shows that the overheads of AzusA are smaller than Origin in general.

To investigate this problem more precisely, we have used STREAM benchmark [1] from Virginia University. STREAM measures the performance of Level 1 BLAS like operations shown in Table 5. Here, we have used OpenMP version `stream_d_omp.c` and the latest version of Intel compiler 6.0. From the results shown in Figure 5 and 6, we can see that AzusA shows comparable performance with Origin, using memory affinity which NEC Linux implements[4].

---

[2] The problem has been solved in the latest version of 6.0.

[3] We used compiler options `-O3 -zero -ip -nodps -w -openmp` for Intel Compiler and `-Ofast -mp` for MIPSpro.

[4] Ahthough it can be specified as a kernel parameter, no APIs to describe memory affinity are provided. The NUMA API is planned to be supported by the kernel 2.6.

Table 5. STREAM benchmark types

Benchmark	Operation	Bytes per iteration
Copy	a[i] = b[i]	16
Scale	a[i] = q * b[i]	16
Add	a[i] = b[i] + c[i]	24
Triad	a[i] = b[i] + q * c[i]	24

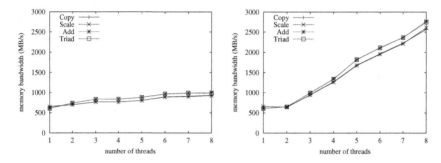

Fig. 5. STREAM benchmark performance in MB/s with array size 20,000,000 on NEC AzusA, without memory affinity (left) and with memory affinity (right)).

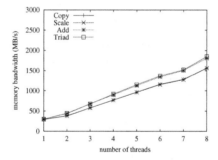

Fig. 6. STREAM benchmark performance in MB/s with array size 80,000,000 on SGI Origin 2000.

IBM x440

IBM x440 is an Intel Xeon processor based commodity ccNUMA server using IBM's Summit chipset, which enables up to 16-way configuration. We have used an IBM x440 which implements two nodes (two 2.4GHz Intel Xeon DP Processors, 32MB PC200 DDR SDRAM node cache and 2GB PC133 SDRAM for each node). The operating system is Red Hat Advanced Server 2.1 with Linux kernel 2.4.21-pre4[5].

---

[5] We are also evaluating the latest development kernels, which are still unstable with multithreaded applications, but shows better scalability.

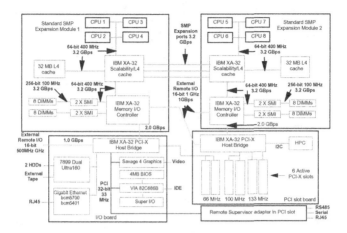

Fig. 7. x440 architecture block diagram.

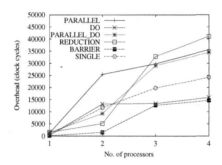

Fig. 8. Synchronization overheads on IBM x440.

The synchronization overheads of the EPCC Microbenchmark are shown in Figure 8. The results show relatively large latency on x440 compared with other architectures, which is partially due to the high frequency of processors and the complexity of multinode configuration.

Figure 9 shows the remarkable effect of node cache architecture observed using the stream benchmark, in which the size of the problem are 4,000,000 and 2,000,000. The latter requires about 44MB memory and it is within the size of the total of two node cache memories.

We have measured the performance of parallel subroutines zgemv, zdotc, zaxpy, and dznrm2. The size of vectors is 256^2 and the size of matrices is $256^2 \times 4$. The performance of BLAS kernels on x440 and AzusA are shown in Figure 10. The results indicates that the node caches compensate for the small onchip cache in the processors and large network latency of x440.

In this study, we have proposed the BLAS-level parallelization of the preconditioned Jacobi-Davidson method, and reported its evaluation on Intel processor based ccNUMA servers NEC AzusA and IBM x440, which have the potential

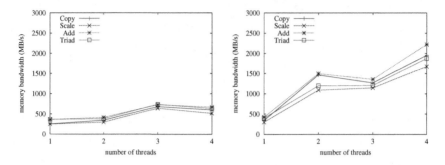

Fig. 9. STREAM benchmark performance in MB/s with array size 4,000,000 and 2,000,000 on IBM x440 (without memory affinity).

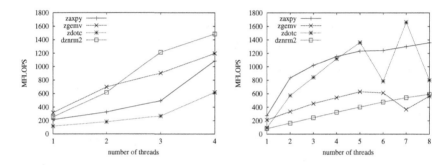

Fig. 10. Parallel BLAS performance on IBM x440 and NEC AzusA.

merit of limitless scalability based on commodity components. The comparative study with SGI Origin 2000 shows that the sustained memory bandwidth will be the most critical factor to be noticed for building more high performance shared memory environments in the near future[6]. Complehensive studies are required for scalable computing on commodity architectures.

Acknowledgment

We are grateful for supports and advices from many people at IBM and NEC, especially from Dr. Martin J. Bligh at IBM, Dr. Kimio Suganuma and Dr. Takahito Kouchi at NEC. This study is partially supported by MEXT Grant-in-aid for Scientific Research 14019030 and 13480080, and a grant from CREST, JST.

---

[6] We are currently working on SGI Altix 3000 for more detailed comparison.

References

1. *STREAM: Sustainable Memory Bandwidth in High Performance Computers.* http://www.cs.virginia.edu/stream/.
2. A. Nishida and Y. Oyanagi. A Parallel Implementation of the Jacobi-Davidson Method using OpenMP and its Evaluation on Shared Memory Architectures. *Proceedings of Joint Symposium on Parallel Processing 2002*, pages 79–86, 2002.
3. E. Anderson, Z. Bai, C. Bischof, S. Blackford, J. Demmel, J. Dongarra, J. Du Croz, A. Greenbaum, S. Hammarling, A. McKenney, and D. Sorensen. *LAPACK Users' Guide*. Society for Industrial and Applied Mathematics, third edition, 1999.
4. F. Aono and M. Kimura. The AzusA 16-Way Itanium Server. *IEEE Micro*, 20(5):54–60, 2000.
5. Z. Bai, J. Demmel, J. Dongarra, A. Ruhe, and H. van der Vorst, editors. *Templates for the Solution of Algebraic Eigenvalue Problems : A Practical Guide*. SIAM, 2000.
6. R. Barrett, M. Berry, T. F. Chan, J. Demmel, J. Donato, J. Dongarra, V. Eijkhout, R. Pozo, C. Romine, and H. van der Vorst. *Templates for the Solution of Linear Systems: Building Blocks for Iterative Methods*. SIAM, 1994.
7. E. R. Davidson. The iterative calculation of a few of the lowest eigenvalues and corresponding eigenvectors of large real symmetric matrices. *J. Comp. Phys.*, 17:87–94, 1975.
8. D. R. Fokkema, G. L. G. Sleijpen, and H. A. van der Vorst. Jacobi-Davidson style QR and QZ algorithms for the partial reduction of matrix pencils. Technical Report 941, Department of Mathematics, Utrecht University, 1996.
9. L. Lawson, R. J. Hanson, D. Kincaid, and F. T. Krogh. Basic Linear Algebra Subprograms for FORTRAN usage. *ACM Trans. Math. Soft.*, 5:308–323.
10. G. L. G. Sleijpen and H. A. van der Vorst. A Jacobi-Davidson iteration method for linear eigenvalue problems. *SIAM J. Matrix Anal. Appl.*, 17(2):401–425, 1996.
11. Sivan. Toledo. Improving Memory-System Performance of Sparse Matrix-Vector Multiplication. *IBM Journal of Research and Development*, 41(6):711–725, 1997.

Support of Multidimensional Parallelism in the OpenMP Programming Model

Haoqiang Jin[1] and Gabriele Jost[1,2]

[1] NAS Division, NASA Ames Research Center, Moffett Field, CA 94035-1000, USA,
{hjin,gjost}@nas.nasa.gov
[2] Computer Sciences Corporation, M/S T27A-1, NASA Ames Research Center

Abstract. OpenMP is the current standard for shared-memory programming. While providing ease of parallel programming, the OpenMP programming model also has limitations which often effect the scalability of applications. Examples for these limitations are work distribution and point-to-point synchronization among threads. We propose extensions to the OpenMP programming model which allow the user to easily distribute the work in multiple dimensions and synchronize the workflow among the threads. The proposed extensions include four new constructs and the associated runtime library. They do not require changes to the source code and can be implemented based on the existing OpenMP standard. We illustrate the concept in a prototype translator and test with benchmark codes and a cloud modeling code.

1   Introduction

OpenMP [11] was introduced as an industrial standard for shared-memory programming with directives. It has gained significant popularity and wide compiler support. The main advantage of the OpenMP programming model is that it is easy to use and allows the incremental parallelization of existing sequential codes. OpenMP provides a fork-and-join execution model in which a program begins execution as a single process or thread. This thread executes sequentially until a PARALLEL construct is found. At this time, the thread creates a team of threads and it becomes its master thread. All threads execute the statements lexically enclosed by the parallel construct. Work-sharing constructs (DO, SECTIONS and SINGLE) are provided to divide the execution of the enclosed code region among the members of a team. All threads are independent and may synchronize at the end of each work-sharing construct or at specific points (specified by the BARRIER directive).

An existing code can be easily parallelized by placing OpenMP directives around time consuming loops which do not contain data dependences. The ease of programming is a big advantage over a parallelization based on data distribution and message passing, as it is required for distributed computer architectures. There are, however, limitations to the programming model which can affect the scalability of OpenMP programs. Comparative studies (i.e. [8]) have shown that high scalability in message passing based codes is often due to user optimized

A. Veidenbaum et al. (Eds.): ISHPC 2003, LNCS 2858, pp. 511–522, 2003.

work distribution and process synchronization. When employing OpenMP the user does not have this level of control anymore. For example, a problem arises if the outer loop in a loop nest does not contain a sufficient number of iterations to keep all of the threads busy. In [1] and [7] it has been shown that directive nesting can be beneficial in these cases.

Another issue is the workflow between threads. Data dependences may require point-to-point synchronization between individual threads. The OpenMP standard requires barrier synchronization at the end of parallel regions which potentially destroy the workflow, especially when nested parallel regions are used. The SPMD programming model is a way to mimic the data distribution and workflow achieved by message passing programs. In this model the programmer expresses manually the data and work distribution among the threads. The data is copied to small working arrays which are then accessed locally by each thread. The work is distributed according to the distribution of the data. The parallelization of the loop nests requires to compute explicitly which threads executes which iteration. The bounds for each loop have to be calculated based on the number of threads and the identifier of each thread, therefore introducing significant changes to the source code. The programming style allows the programmer to carefully manage the workflow, but completely defeats the advantages of the OpenMP programming model.

We are proposing extensions to the OpenMP programming model, which allow the automatic generation of SPMD style code based on user directives. Our goal is to remove some of the performance inhibiting limitations of OpenMP while preserving the ease of programming. In the current work we are addressing the issue of work distribution in multiple dimensions and point-to-point thread synchronization. The rest of the paper is structured as follows. In Section 2 we discuss related work regarding multidimensional parallelization and present our own approach in comparison. In Section 3 we give a description of our prototype implementation. In Section 4 we present two case studies to demonstrate the proposed concept and conclude in Section 5.

2   Multidimensional Parallelism

The OpenMP standard allows the nesting of the OMP PARALLEL directive. But the OMP DO directive cannot be nested without nesting the OMP PARALLEL directive. At this point not many compilers support this type of directive nesting. For example, the SGI compiler does not support nested OMP PARALLEL directives, but provides some support of multidimensional work distribution by means of the NEST clause on the OMP DO directive [10]. The NEST clause requires at least two variables as arguments to identify indices of subsequent DO-loops. The identified loops must be perfectly nested and no code is allowed between the identified DO statements and the corresponding END DO statements. The NEST clause on the OMP DO directive informs the compiler that the entire set of iterations across the identified loops can be executed in parallel. The compiler can then linearize the execution of the loop iteration and divide them among the available single level of threads.

An example of a research platform that supports true nested OpenMP parallelism is the OpenMP NanosCompiler [3]. The OpenMP NanosCompiler accepts Fortran-77 code containing OpenMP directives and generates plain Fortran-77 code with calls to the NthLib thread library [9] currently implemented for the SGI Origin. In contrast to the SGI MP library, NthLib allows for multilevel parallel execution such that inner parallel constructs are not being serialized. The NanosCompiler suuport of multilevel parallelization is based on the concept of thread groups. A group of threads is composed of a subset of the total number of threads available in the team to run a parallel construct. In a parallel construct, the programmer may define the number of groups and the composition of each one. When a thread in the current team encounters a `PARALLEL` construct defining groups, the thread creates a new team and it becomes its master thread. The new team is composed of as many threads as the number of groups. The rest of the threads are used to support the execution of nested parallel constructs. In other words, the definition of groups establishes an allocation strategy for the inner levels of parallelism. To define groups of threads, the NanosCompiler supports the `GROUPS` clause extension to the `PARALLEL` directive. The NanosCompiler also provides the `PRED/SUCC` extensions [4] in order to allow point-to-point synchronization between threads.

We propose the following extensions to OpenMP for support of multidimensional parallelism, for short MOMP directives. We introduce two new directives, `TMAP` and `MDO`, for mapping a team of threads to a grid of multiple dimensions and for distributing work in the multiple dimensions. The concept is illustrated for a two-dimensional grid. It can easily be extended to higher dimensions.

`TMAP (ndim,sfactor1,sfactor2)`

Maps a team of threads to a grid of multiple dimensions. `ndim` is the dimension of the mapped thread grid (currently implemented 1 or 2); `sfactor1` and `sfactor2` are the shape factors for the grid, i.e. the ratio of the two factors is proportional to that of the grid sizes in each dimension. For example, $(2,1,1)$ defines a squared grid; $(2,N,0)$ indicates that the number of threads mapped to the first dimension should not exceed N. See Fig. 1 for an illustration.

`MDO (idim[,gplow,gphigh])`

Binds or distributes a worksharing DO loop to the `idim` dimension of the thread grid. The optional parameters `gplow` and `gphigh` can be used to specify additional ghost iterations to be assigned on the low and high ends of the bound loop. This is to mimic the ghost points concept used in a message-passing program. For the $(2,1,1)$ mapping in Fig. 1, if loops K and J are bound to `idim`=1 and 2, respectively, threads 0,1,2,3 will be bound to the same iterations of loop K, while threads 0,4,8,12 will be bound to the same iterations of loop J. By default, threads are not synchronized at the end of an `MDO` loop, as opposite to the implicit synchronization at the end of an `OMP DO` loop. This selection reflects closer to the SPMD coding style.

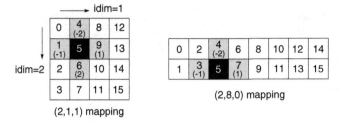

Fig. 1. Examples of thread mapped grids with TMAP for 16 threads. The left (2,1,1) mapping defines a squared shape, while the right (2,8,0) mapping limits the number of threads mapped to the first dimension to 8, which then gives an 8 × 2 topology. The light shaded boxes indicate the neighboring threads of a given thread (5 in the figure) corresponding to idir=-1,1,-2,2

To enforce synchronization, the user needs to use explicit synchronization directives.

To support flexible synchronization among threads in the thread grid, we include two more directives, `TSIGNAL` and `TWAIT`.

`TSIGNAL (idir\[,...\])`
    Sends a signal to the direction 'idir'.
`TWAIT (idir\[,...\])`
    Waits a signal from the direction 'idir'.

For both `TSIGNAL` and `TWAIT`, `idir` takes the following values: -1 for lower-neighbor in the first dimension, 1 for higher-neighbor in the first dimension, -2 for lower-neighbor in the second dimension, and 2 for higher-neighbor in the second dimension (see Fig. 1). Multiple directions can be listed. `TSIGNAL` and `TWAIT` must always be used in a matching pair; else a deadlock will occur. Constructs for signal/wait of a particular thread are not included at this point.

A sample code using MOMP directives is given in Fig. 2. The `TMAP` clause is added to the beginning of a parallel region to define a 2-D thread grid. The first `MDO` distributes the work of the K loop, while the second `MDO` distributes the work of the J loop. The MOMP extensions can work seamlessly with most of the OpenMP directives.

In Fig. 3 we compare code examples using MOMP to the Nanos and SGI OpenMP extensions for multidimensional parallelization. The Nanos extensions support true multilevel OpenMP parallelism. The inner parallel region requires forking and joining of thread teams. The SGI extensions merely support linearization of the loop nest and require tightly coupled loops. They would not be applicable to the code example from Fig. 2.

3   Prototype Implementation

To illustrate the concept of using the MOMP directives described above for multidimensional parallelization, we implemented a prototype translator and

Serial code	Code with MOMP directives
| ```
DO K=1,NZ
   ZETA = K*0.1
   DO J=1,NY
      do more work...
   ENDDO
ENDDO
``` | ```
!$OMP PARALLEL TMAP(2,NZ,0)
!$OMP MDO(1)
   DO K=1,NZ
      ZETA = K*0.1
!$OMP MDO(2)
      DO J=1,NY
         do more work...
      ENDDO
   ENDDO
!$OMP END PARALLEL
``` |

Fig. 2. A sample code using the MOMP directives

Code with MOMP directives	OpenMP with Nanos extensions	OpenMP with SGI extensions
| ```
!$OMP PARALLEL
!$OMP& TMAP(2,NZ,0)
!$OMP MDO(1)
   DO K=1,NZ
!$OMP MDO(2)
      DO J=1,NY
         do more work
      ENDDO
   ENDDO
!$OMP END PARALLEL
``` | ```
!$OMP PARALLEL
!$OMP& GROUPS(NZ)
!$OMP DO
   DO K=1,NZ
!$OMP PARALLEL DO
      DO J=1,NY
         do more work
      ENDDO
!$OMP END PARALLEL DO
   ENDDO
!$OMP END PARALLEL
``` | ```
!$OMP PARALLEL DO
!$SGI+NEST ONTO(NZ,*)
   DO K=1,NZ
      DO J=1,NY
         do more work
      ENDDO
   ENDDO
!$OMP END PARALLEL DO
``` |

Fig. 3. Code examples comparing the MOMP directives with Nanos and SGI OpenMP extensions

the supporting runtime library. The translator simply translates all the MOMP directives into appropriate runtime calls and leaves the other OpenMP directives intact. The resulting code is a standard OpenMP code and can be compiled with any OpenMP compiler and linked with the MOMP runtime library.

In Table 1 we summarize the translation of MOMP directives into the runtime functions for FORTRAN codes; the concept applies to C as well. The translation of TMAP, TSIGNAL and TWAIT is straightforward, simply mapping into the corresponding runtime functions. The MDO directive is translated into a call to momp_get_range that computes the new loop limit after the associated loop is distributed in the defined grid dimension, and the loop is then replaced with the new limit. For simplicity we only consider the block distribution which also simplifies the implementation of momp_tsignal() and momp_twait(). The ghost iterations are translated into the overlap of the iteration space between two neighboring threads. When no ghost iteration is involved, (gplow,gphigh)=(0, 0) is used.
```

Table 1. Translation of MOMP directives into the runtime functions

MOMP directive	Runtime function
`TMAP(ndim,sfactor1,sfactor2)`	`call momp_create_map(ndim,` `&   sfactor1,sfactor2)`

`MDO(idim,gplow,gphigh)`
`  DO lvar=low,high,step`
	`call momp_get_range(idim,`
	`&   gplow,gphigh,low,high,`
	`&   step,new_low,new_high)`
	`DO lvar=new_low,new_high,step`

`MDO(idim)`	`call momp_get_range(idim,` `&   0,0,...)`
`TSIGNAL(idir[,...])`	`call momp_tsignal(idir)` `[call momp_tsignal(...)]`
`TWAIT(idir[,...])`	`call momp_twait(idir)` `[call momp_twait(...)]`

In addition, an environment variable `MOMP_SHAPE` is used to control the grid shape externally. `MOMP_SHAPE` takes a value like "S1xS2," which is equivalent to specifying `sfactor1=S1` and `sfactor2=S2` to `TMAP`. This value overwrites the values given in `TMAP`. It allows a user to freely change the grid shape without recompiling the code. If `S1xS2=N` where `N` is the total number of threads, the shape "S1xS2" defines the current grid topology. An example of the translation of the code given in Fig. 2 is shown in Fig. 4. The translated code is a standard OpenMP code with a proper list of private variables for the new loop limits and can be compiled with any OpenMP compiler.

4   Case Study

In this section we show examples for using our proposed MOMP directives. We will describe the usage of the directives and discuss the performance of the resulting code. All tests were run on an SGI Origin 3000 with 400MHz R12000 CPUs and 2GB local memory per node. We have parallelized two codes (BT and LU) from the NAS Parallel Benchmark suite [2] that are based on the implementation as described in [5]. In addition we have applied our extension to a full-scale cloud modeling code.

4.1   The BT Benchmark

BT is a simulated CFD application. It uses an implicit algorithm to solve the 3D compressible Navier-Stokes equations. The x, y, and z dimensions are decoupled by usage of an Alternating Direction Implicit (ADI) factorization method. In the code, all of the nested parallel loops are at least triple nested. The structure of the loops is such that the two outer most loops can be parallelized and enclose a reasonably large amount of computational work. We applied the `TMAP` and `MDO` directives as shown in Fig. 2 throughout the code. In Fig. 5 we show timings obtained for the BT benchmark, class A problem ($64 \times 64 \times 64$ grid points). We

Code with MOMP directives	Translated OpenMP code

```!$OMP PARALLEL```
```!$OMP& TMAP(2,NZ,0)```
```!$OMP MDO(1)```
```   DO K=1,NZ```
```   ZETA = K*0.1```
```!$OMP MDO(2)```
```   DO J=1,NY```
```   do more work...```
```   ENDDO```
```   ENDDO```
```!$OMP END PARALLEL```
	```!$OMP PARALLEL PRIVATE(K_NLOW,```
	```!$OMP& K_NHIGH,J_NLOW,J_NHIGH)```
	``` CALL MOMP_CREATE_MAP(2,NZ,0)```
	``` CALL MOMP_GET_RANGE(1,0,0,1,```
	```& NZ,1,K_NLOW,K_NHIGH)```
	``` DO K=K_NLOW,K_NHIGH```
	``` ZETA = K*0.1```
	``` CALL MOMP_GET_RANGE(2,0,0,```
	```& 1,NY,1,J_NLOW,J_NHIGH)```
	``` DO J=J_NLOW,J_NHIGH```
	``` do more work...```
	``` ENDDO```
	``` ENDDO```
	```!$OMP END PARALLEL```

Fig. 4. Translation of a sample MOMP code to the standard OpenMP code with MOMP runtime calls

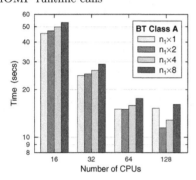

Fig. 5. MOMP timing comparison for various numbers of threads running in different topologies for the BT benchmark. The first dimension n_1 corresponds to the number of CPUs divided by the second topology dimension

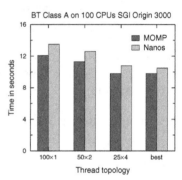

Fig. 6. Time comparisons between Nanos and MOMP for different thread configurations. The columns in category *best* indicate the best time over all tested topologies

compare timings achieved for different numbers of threads running in various topologies. In the figure n_1 denodes the first dimension of the thread topology, which is the total number of threads divided by the second dimension of the thread grid. For example, if we are running on 32 CPUs employing 32 threads, then the topology of $n_1 \times 4$ corresponds to employing 8 threads on the first and 4 threads on the second dimension. The timings show that for more than 64 threads distributing the work in multiple dimensions can improve the performance significantly. The loop length resulting from 64 points does not provide enough iterations in one dimension to keep more than 64 threads busy. Distributing the work in two dimensions allows the exploitation of additional parallelism.

We noticed the ratio of L2 cache misses versus floating instructions per thread increases as more threads are assigned in the second dimension, which causes the increase in running time. The positive impact of better workload balance, however, is stronger than the negative impact of a lack of data locality in the run with 128 threads.

In Fig. 6 we compare timings achieved for different thread topologies for the MOMP version with timings obtained for the nested OpenMP version using the NanosCompiler. For the NanosCompiler we use the GROUPS clause extension as shown in Fig. 3. The timings for the MOMP version are slightly better than for the Nanos version, but not significantly. The extra barrier synchronization points at the end of inner parallel regions do not introduce considerable overhead to the NanosCompiler generated code, which is due to the very efficient NthLib thread library. We also noticed that the thread scheduling applied by the NanosCompiler yields a somewhat better workload balance than our approach. This indicates that additional performance might be gained for MOMP with an optimized runtime library. The SGI NEST clause is not applicable to some of the time consuming loops in the BT benchmark because they are not tightly nested.

4.2   The LU Benchmark

The LU application benchmark is a simulated CFD application that uses the symmetric successive over-relaxation (SSOR) method to solve a seven band block-diagonal system., All of the loops involved carry data dependences that prevent straightforward parallelization. There is, however, the possibility to exploit a certain level of parallelism by using wave-front pipelining which requires explicit synchronization of individual threads. We use the MOMP TMAP and MDO directives to distribute the work in multiple dimensions. Distributing the work in multiple dimensions requires a two-dimensional thread pipeline. We use the MOMP TWAIT and TSIGNAL directives to set up threaded pipeline execution in multiple dimensions. As an example, the source code for the 2D pipeline implemented using the two MOMP directives is given here.

```
!$OMP PARALLEL TMAP(2,NZ,0)
    DO K=2,NZ
!$OMP TWAIT(-1,-2)
!$OMP MDO(1)
      DO J=2,NY
!$OMP MDO(2)
        DO I=2,NX
          V(I,J,K)=V(I,J,K)+A*V(I-1,J,K)+B*V(I,J-1,K)+C*V(I,J,K-1)
            . . .
        ENDDO
      ENDDO
!$OMP TSIGNAL(1,2)
    ENDDO
!$OMP END PARALLEL
```

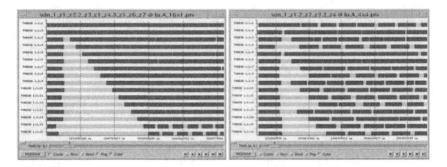

Fig. 7. Timeline views of the thread workflow during the forward substitution phase in LU running on 16 threads. Dark shading indicates time spent in computation, light shading indicates time spent in synchronization. The left images shows a 1D thread pipeline, the right image shows a 2D thread pipeline

At this point we note that a 2D pipeline involving nested OpenMP parallel regions runs into the problem that the inner parallel region imposes an extra barrier synchronization point which inhibits the 2D pipeline.

We used the Paraver [12] performance analysis system to show the effect of the 2D pipeline on the workflow of the threads. In Fig. 7 we show the time line view of the useful thread time during the forward substitution phase of LU for a 16 CPU run. Dark shades indicate time the threads spend in computation, light shades indicate time spent in synchronization or other OpenMP introduced overhead. The left image shows a 16×1 thread topology, corresponding to a 1D pipeline. The right image shows a 4×4 topology corresponding to a 2D pipeline. The use of the 2D pipeline decreases a pipeline startup time for the computations.

Just like in the case of BT, the LU benchmark provides only work for 62 threads on the outer loop level. Even though we were able to achieve a 2D pipelined thread execution, we did not gain any significant speedup by using extra threads and exploiting the inner loop for additional parallelism. Timings for LU benchmark class A problem are shown in Fig. 8. A large increase in L2 cache misses for some of the threads lead to a very imbalanced workflow which decreased the performance. This is clearly indicated in Fig. 9 where the effect of multidimensional parallelism in the LU benchmark on the value of various hardware counters is summarized.

4.3   The Cloud Modeling Code

The Goddard Cumulus Ensemble (GCE) code developed at the NASA Goddard Space Flight Center [13] is used for modeling the evolution of clouds and cloud systems under large-scale thermodynamic forces. The 3-dimensionsal version of this code, GCEM3D, was previously parallelized [6] using the standard OpenMP directives. The parallelization was performed on the outer loops (in many cases the vertical K dimension) to achieve coarse parallel granularity and little OpenMP overhead. However, due to the size limitation of the K dimension,

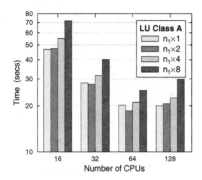

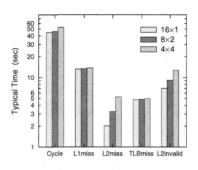

Fig. 8. MOMP timings for various numbers of threads running in various topologies for LU benchmark class A. See Fig. 5 for a note on the topology

Fig. 9. Effect of 2D work distribution on hardware counters for the LU benchmark

the code does not scale beyond 32 CPUs even though the horizontal sizes are much larger (see [6]). Use of the MOMP directives is a natural way to exploit parallelism in the horizontal dimension of the GCEM3D code. We use TMAP to define a 2-D thread mapping and MDO to bind the K loops to the first dimension of the thread grid and the J loops to the second dimension. The code requires the use of the ghost iterations in the MDO directives, as shown here.

```
!$OMP PARALLEL PRIVATE(I,J,K,XY1) TMAP(2,KLES-1,0)
!$OMP MDO(1)
      DO 100 K=2,KLES
!$OMP MDO(2,1,0)
        DO 200 J=1,JLES
          DO 200 I=2,ILES
            XY1(I,J)=...
200     CONTINUE
!$OMP MDO(2)
        DO 300 J=2,JLES
          DO 300 I=2,ILES
            U1(I,J,K)=U1(I,J,K)+(XY1(I,J-1)-XY1(I,J))
300     CONTINUE
        ...
100 CONTINUE
!$OMP END PARALLEL
```

The "DO 200 J" loop computes additional values for the XY1 array in the low end so that XY1(I,J-1) can be used in the next J loop. Since the XY1 array is declared as private, the use of "MDO(2,1,0)" ensures that each thread performs one extra calculation at the low end of the assigned iteration so that the XY1(I,J-1) value is available in the next J loop for each thread.

To test the effect of the MOMP directives, we applied them to the most time consuming routine 'fadvuw' in the GCEM3D code. The parallel code was run

Table 2. Timing obtained for the most time consuming routine 'fadvuw' in GCEM3D

CPUs	32	64	
Topology	32×1	64×1	32×2
Time (secs)	59.3	51.1	35.4
Ratio	-	1.16	1.68

on 32 and 64 CPUs for a problem size of $258 \times 258 \times 34$. The timing results are summarized in Table 2. As one can see, the single dimension parallelization (as indicated by 32×1 and 64×1) only improves the timing slightly from 32 to 64 CPUs; the multidimensional parallelization with the MOMP directives reduces the timing on 64 CPUs from 51.1 seconds to 35.4 seconds.

5   Conclusion

We have proposed extensions to the current OpenMP programming model which allow the user to easily distribute the work in multiple dimensions within nested DO loops and synchronize the work flow between threads. The purpose is to exploit parallelism in multiple dimensions within a nest of loops. We have demonstrated the feasibility of our approach in several case studies. The advantage of the MOMP directives proposed in this work is that they are simple and clean. They can be implemented within the current OpenMP programming model and do not require changes to the OpenMP standard. The proposed work distribution directives support the automatic generation of SPMD style parallelization of loop. They do not impose restrictions on the structure of the loop nest, such as being tightly nested, nor do they require the nesting of parallel regions. Using the MOMP directives it is trivial to create a 2-D pipeline in LU, which would be difficult when using nested parallel regions.

Our case studies also demonstrated that distributing the work in multiple dimensions introduces disadvantages: The data is accessed employing large strides which can lead to severe cache problems. This can be seen in the case of the LU benchmark. The lack of data locality is always an issue with the shared-addressing programming model on cache-based systems. This issue needs to be addressed in future work, for example by automatically restructuring the code for cache optimization. Other possible enhancements to the MOMP directives include "MBARRIER" for barrier synchronization and "MREDUCTION" for reduction on a selected grid dimension.

Acknowledgements

This work was partially supported by NASA contract DTTS59-99-D-00437/ A61812D with Computer Sciences Corporation.

References

1. E. Ayguadé, X. Martorell, J. Labarta, M. Gonzalez, and N. Navarro, "Exploiting Multiple Levels of Parallelism in OpenMP: A Case Study," Proc. Of the 1999 International Conference on Parallel Processing, Ajzu, Japan, September 1999.
2. D. Bailey, T. Harris, W. Saphir, R. Van der Wijngaart, A. Woo, and M. Yarrow, "The NAS Parallel Benchmarks 2.0," RNR-95-020, NASA Ames Research Center, 1995. NPB2.3, http://www.nas.nasa.gov/Software/NPB/.
3. M. Gonzalez, E. Ayguadé, X. Martorell, J. Labarta, N. Navarro and J. Oliver. "NanosCompiler: Supporting Flexible Multilevel Parallelism in OpenMP." Concurrency: Practice and Experience. Special issue on OpenMP. vol. 12, no. 12. pp. 1205-1218. October 2000.
4. M. Gonzalez, E. Ayguadé, X. Martorell, and J. Labarta. "Defining and Supporting Pipelined Executions in OpenMP." 2nd International Workshop on OpenMP Applications and Tools. July 2001.
5. H. Jin, M. Frumkin, and J. Yan, "The OpenMP Implementations of NAS Parallel Benchmarks and Its Performance," NAS Technical Report NAS-99-011, NASA Ames Research Center, 1999.
6. H. Jin, G. Jost, D. Johnson, and W-K. Tao, "Experience on the Parallelization of a Cloud Modeling Code Using Computer-Aided Tools," NAS Technical Report NAS-03-006, NASA Ames Research Center, March 2003.
7. H. Jin, G. Jost, J. Yan, E. Ayguadé, M. Gonzalez, and X. Martorell, "Automatic Multilevel Parallelization Using OpenMP," 3rd European Workshop on OpenMP (EWOMP01), Barcelona, Spain, September 2001.
8. G. Jost, H. Jin, J. Labarta, J. Gimenez, and J. Caubet, "Performance Analysis of Multi-level Parallel Programs on Shared Memory Computer Architectures," Proceedings of the 17th International Parallel and Distributed Processing Symposium (IPDPS03), Nice, France, April 2003.
9. X. Martorell, E. Ayguadé, N. Navarro, J. Corbalan, M. Gonzalez, and J. Labarta. "Thread Fork/join Techniques for Multi-level Parallelism Exploitation in NUMA Multiprocessors." 13th International Conference on Supercomputing (ICS'99), Rhodes, Greece, pp. 294-301, June 1999.
10. MIPSPro 7 Fortran 90 Commands and Directives Reference Manual, 007-3696-03, http://techpubs.sgi.com/.
11. OpenMP Fortran/C Application Program Interface, http://www.openmp.org/.
12. Paraver, http://www.cepba.upc.es/paraver/.
13. W.-K. Tao, "Goddard Cumulus Ensemble (GCE) Model: Application for Understanding Precipitation Processes, AMS Meteorological Monographs," Symposium on Cloud Systems, Hurricanes and TRMM, 2002.

On the Implementation
of OpenMP 2.0 Extensions
in the Fujitsu PRIMEPOWER Compiler

Hidetoshi Iwashita[1], Masanori Kaneko[1], Masaki Aoki[1],
Kohichiro Hotta[1], and Matthijs van Waveren[2]

[1] Software Technology Development Division, Software Group, Fujitsu Ltd., 140
Miyamoto, Numazu-shi, Shizuoka 410-0396, Japan
[2] Fujitsu Systems Europe, 8, rue Maryse Hilsz, Parc de la Plaine, 31500 Toulouse,
France,
waveren@fujitsu.fr

Abstract. The OpenMP Architecture Review Board has released version 2.0 of the OpenMP Fortran language specification in November 2000, and version 2.0 of the OpenMP C/C++ language specification in March 2002. This paper discusses the implementation of the OpenMP Fortran 2.0 WORKSHARE construct, NUM_THREADS clause, COPY-PRIVATE clause, and array REDUCTION clause in the Parallelnavi software package. We focus on the WORKSHARE construct and discuss how we attain parallelization with loop fusion.

1 Introduction

The OpenMP Architecture Review Board has released version 2.0 of the OpenMP Fortran language specification [1] in November 2000, and version 2.0 of the OpenMP C/C++ language specification [2] in March 2002. Currently the OpenMP ARB is working on the development of an integrated OpenMP Fortran/C/C++ language specification.

While OpenMP Fortran 1.1 was aligned with Fortran 77, OpenMP Fortran 2.0 has been updated to be aligned with Fortran 95. The WORKSHARE directive allows parallelization of array expressions in Fortran statements. The new NUM_THREADS clause on parallel regions defines the number of threads to be used to execute that region. The COPYPRIVATE clause has been added on END SINGLE. THREADPRIVATE may now be applied to variables as well as COMMON blocks. REDUCTION is allowed on an array name. COPYIN works on variables as well as COMMON blocks, and reprivatization of variables is now allowed.

While OpenMP C/C++ 1.0 was aligned with OpenMP Fortran 1.0 and C89, OpenMP C/C++ 2.0 has been brought up to the level of C99 and OpenMP Fortran 2.0. This means that some of the extensions mentioned above for OpenMP Fortran 2.0 have been introduced in OpenMP C/C++ 2.0, viz. the num_threads clause, the copyprivate clause, and the extension on the threadprivate directive.

A. Veidenbaum et al. (Eds.): ISHPC 2003, LNCS 2858, pp. 523–528, 2003.

Parallelnavi is a software package for the PRIMEPOWER series server. Parallelnavi Fortran V2.1 supports the OpenMP Fortran API Version 2.0. Parallelnavi C/C++ V2.1 supports the OpenMP C/C++ API Version 2.0. This paper describes the implementation of the OpenMP Fortran 2.0 extensions in the Parallelnavi software package. We focus on the WORKSHARE construct and discuss how we attain parallelization with loop fusion.

The Fujitsu PRIMEPOWER HPC2500 [3] is a parallel computation server supporting up to 128 CPUs and 512 gigabytes of memory. The CPU used is SPARC64GP, which conforms to the SPARC International V9 architecture and loads the Solaris 8 operating system. Each cabinet (node) has 8 system boards and each system board has four CPUs, 16 gigabytes of memory, six PCI cards and a first level (L1) crossbar switch. All system boards, within and between nodes, are connected with the second level (L2) crossbar switch.

A distinguishing feature of the PRIMEPOWER HPC2500 is its behavior as a flat Symmetric Multi-processing (SMP) server. Two-layered crossbar networks allow every CPU to access all memory in the system and guarantee coherency.

2   OpenMP API v2.0 Compiler Support

The Fujitsu Solaris compiler supports the OpenMP API V2.0, which includes the following major features:

- WORKSHARE construct,
- NUM_THREADS clause,
- COPYPRIVATE clause, and
- Array REDUCTION clause.

The NUM_THREADS clause, which specifies the number of threads to execute a parallel region, was implemented in a similar fashion as the OMP_SET_NUM_THREADS library call.

The COPYPRIVATE clause was implemented to make the threads keep the following order: 1) the SINGLE thread specified with the COPYPRIVATE clause writes out the value of the variable to a global temporary, 2) all threads perform barrier synchronization, and 3) every thread reads the global temporary into its own private variable.

The implementation of the array REDUCTION is a simple and natural expansion of the scalar REDUCTION in the OpenMP API V1.1.

The following part of this section introduces the implementation of the WORKSHARE construct, which is the most important new feature of the OpenMP API V2.0.

2.1   Compilation of WORKSHARE Construct

The configuration of the Fujitsu Solaris compiler is shown in Figure 1, focusing on the treatment of OpenMP code. The OMP module, a part of Middle

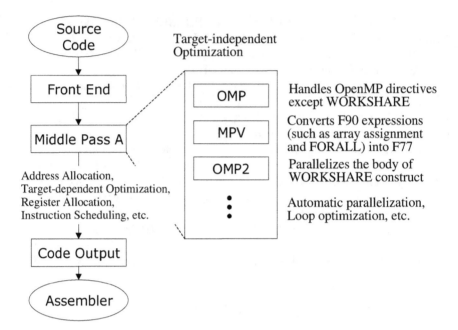

Fig. 1. Configuration of the Fujitsu Solaris compiler that supports the OpenMP API V. 2.0

Pass A, interprets all OpenMP directives except WORKSHARE and generates corresponding execution code and runtime system (RTS) calls.

The OMP2 module was added to support the WORKSHARE construct. It converts the following statements and constructs into code corresponding to DO loops enclosed with OpenMP DO directives.

- Array assignment statement
- FORALL statement/construct
- WHERE statement/construct

The OMP2 module follows the MPV module, which converts array assignment statements, FORALL and WHERE constructs into loops. As a result, the implementation of the WORKSHARE block is reduced to a problem of loop parallelization.

The OMP2 module regards a sequence of scalar assignment statements, as if it were enclosed with the SINGLE construct.

Some Fortran 90 array intrinsic functions are parallelized in the WORKSHARE construct, as described in Sect. 2.3.

2.2   Loop Fusion in WORKSHARE Construct

The Fortran 95 language specification [4] defines an array assignment statement as an execution sequence, which includes several DO loops. For efficiency of

```
      !$omp workshare

1        A(1:N)=0.5*(A(2:N+1)+A(1:N))
            u3           u1       u2
2        C(1:N)=A(1:N)+B(1:N)
                  u4
3        D(1:N)=A(0:N-1)+B(1:N)
                  u5
      !$omp end workshare
```

(a) An example of WORKSHARE construct

```
      do i=1,N
1        A(i)=0.5*(A(i+1)+A(i))
2        C(i)=A(i)+B(i)
3        D(i)=A(i-1)+B(i)
      end do
```

(b) Loop expansion and fusion in Fortran context

```
      !$omp do
         do i=1,N
11          tmp(i)=0.5*(A(i+1)+A(i))
         end do

      !$omp do
         do i=1,N
12          A(i)=tmp(i)
         end do

      !$omp do
         do i=1,N
2           C(i)=A(i)+B(i)
3           D(i)=A(i-1)+B(i)
         end do
```

(c) WORKSHARE conversion with loop fusion

Fig. 2. Example of WORKSHARE conversion with loop fusion

Table 1.

u1 to u3	loop carried (+1)	anti-dependence
u2 to u3	loop independent	anti-dependence
u3 to u4	loop independent	flow-dependence
u3 to u5	loop carried (+1)	flow-dependence

the generated code, the loops must be fused as much as the data dependence allows. In our implementation, the MPV module not only expands the Fortran 90 conventions into loops, but also fuses the loops using data dependence analysis by taking into account the characteristics of the original statements.

An example of the WORKSHARE construct is shown in (a) of Fig. 2. In this program fragment, the following data dependences can be found with data dependency analysis [7].

If the two OpenMP directive lines were absent, the MPV module would perfectly fuse the three array assignment statements into one loop as shown in (b), taking into account avoidance of the problem of data dependence. As the result, the generated loop could not be parallelized however because it would have loop-carried data dependence.

In order to have parallelization instead of fusion, we modified the MPV module and changed the condition of fusion for the body of WORKSHARE as follows:

- If an array assignment statement contains loop-carried data dependence, separated loops connected with temporary arrays will be generated. See lines 11 and 12 of (c) corresponding to line 1 of (a) for example.
- If two array assignment statements have a mutual loop-carried data dependence, fusion of the loops generated from them will be suppressed. As shown

in (c) of Fig. 2, lines 12 and 3 must be separated into different loops because of their loop-carried data dependence.

As the result of the modification, the MPV module still fuses many loops that have no loop-carried data dependence and the OMP2 module can parallelize generated loops, as shown in (c).

2.3   Parallelization of Array Intrinsic Functions

The compiler parallelizes the execution of the Fortran 90 transformational functions MATMUL, DOT_PRODUCT and TRANSPOSE in the body of WORK-SHARE in a manner selected from the following, according to the values and the attributes of the arguments:

- Either the Front End or the MPV modules perform inline-expansion of the function call. Subsequently the OMP2 module receives the result just like DO loops written in the WORKSHARE construct and parallelizes them as much as possible.
- The OMP2 module replaces the function call with a parallel version if it is in the context to be able to be parallelized. The parallel version library assumes that it is invoked in a team of threads concurrently in a redundant region. The OMP2 module also generates barrier synchronization if data dependence with the neighboring statements requires it.

3   SPEC OMPM2001 Results

We have run the SPEC OMPM2001 benchmark suite on the Fujitsu PRIME-POWER HPC2500. This benchmark suite is described in [3]. The metric published for the 128 CPU system is more than 30000, and it constitutes a world record as of 5 August 2003. For the latest results published by SPEC, see http://www.spec.org/hpg/omp2001.

4   Conclusion

This paper discusses the implementation of the OpenMP 2.0 WORKSHARE construct, NUM_THREADS clause, COPYPRIVATE clause and array REDUCTION clause. We focus on the WORKSHARE construct, and show how parallelization with loop fusion can be attained.

References

1. OpenMP Architecture Review Board. OpenMP Fortran Application Program Interface Version 2.0, November 2000. (http://www.openmp.org/specs/mp-documents/fspec20.pdf)

2. OpenMP Architecture Review Board. OpenMP C/C++ Application Program Interface Version 2.0, March 2002. (http://www.openmp.org/specs/mp-documents/cspec20.pdf)
3. Iwashita H., Yamanaka E., Sueyasu N., Waveren M. van, and Miura K: The SPEC OMP2001 Benchmark on the Fujitsu PRIMEPOWER System. In Proc. of EWOMP2001.
4. ISO/IEC 1539-1:1997, Information Technology - Languages - Fortran
5. Iwashita H., Okada S., Nakanishi M., Shindo T., and Nagakura H: VPP Fortran and Parallel Programming on the VPP500 Supercomputer. In Proceedings of the 1994 International Symposium on Parallel Architectures, Algorithms and Networks (poster session papers), pages 165-172, Kanazawa, Japan, December 1994.
6. Iwashita H., Sueyasu N., Kamiya S., and Waveren, M. van. VPP Fortran and the Design of HPF/JA Extensions, Concurrency and Computation: Practice and Experience.14:575-588,2002
7. Chandra R., Dagum L., Kohr D., Maydan D., MacDonald J., and Menon R.. Parallel Programming in OpenMP. Academic Press, San Diego, CA, USA, 2001.
8. Almasi G., Gottlieb A.. Highly Parallel Computing. The Benjamin/Cummings Publ. Company, Inc, Redwood City, CA, USA, 1989.

Improve OpenMP Performance
by Extending BARRIER
and REDUCTION Constructs*

Huang Chun and Yang Xuejun

National Laboratory for Parallel and Distributed Processing, P.R. China,
`compiler@sohu.com`

Abstract. Barrier synchronization and reduction are global operations used frequently in large scale OpenMP programs. To improve OpenMP performance, we present two new directives `BARRIER(0)` and `ALLREDUCTION` to extend `BARRIER` and `REDUCTION` constructs in OpenMP API. The new extensions have been implemented on our portable OpenMP compiler on JIAJIA. Benchmark testing and experiments show that these constructs decrease the system overheads from synchronization, reduction operation and access of reduction variables on SDSM systems significantly. It is predicable that the improvement of performance can be obtained on ccNUMA systems.

1   Introduction

The OpenMP Application Programming Interface(API) [1, 2] provides a flexible and scalable means for programming parallel applications on shared-memory multiprocessors. The OpenMP API uses a directive-based programming paradigm. Data communication and synchronization are managed by OpenMP compiler. The programmer does not need to worry about the subtle details of the underlying architecture and operating system, such as thread management and the implementation of shared memory. It has gained momentum in both industry and academia, and becomes the de-facto standard programming model on shared-memory multiprocessors.

Distributed Shared Memory (DSM) multiprocessors are nowadays an attractive and viable platform for both parallel and mainstream computing. DSM systems are realized either as tightly integrated cache-coherent non-uniform memory access (ccNUMA) systems or as loosely integrated networks of commodity workstations and symmetric multiprocessors (SMPs), equipped with a software DSM (SDSM) layer on top of a message-passing infrastructure [3]. SDSM [4–6] can provide a sufficiently practical emulation of shared memory across physically distributed memories, and are thus a possible candidate for basing an implementation of OpenMP upon. Implementing OpenMP on a SDSM system

---

* This work was supported by National 863 Hi-Tech Programme of China under Grant No. 2002AA1Z2105.

A. Veidenbaum et al. (Eds.): ISHPC 2003, LNCS 2858, pp. 529–539, 2003.

is one possible approach to making OpenMP amenable to distributed memory systems, such as cluster architectures. It turns out to be much more difficult than expected to obtain high performance of OpenMP on these DSM systems. The primary obstacle is the non-uniformity of memory access latencies. For example, accessing remote memory may cost 100 or even 1000 times as much as accessing local memory in the case of SDSM.

Barrier synchronization and reduction are global operations used frequently in large scale OpenMP programs. Synchronization overhead is an intrusive source of bottlenecks in parallel programs for shared memory architectures. Several studies have shown that parallel programs may spend as much as half of their execution time in locks and barriers on moderate and large-scale DSM systems [7]. In large OpenMP programs, the cost of reduction may be another of the crucial factors of the program, such as two-dimensional plasma simulation program with particle clouds in cells methods [8, 9], which contains several reduction of array. Based on the SDSM system JIAJIA, we have fully implemented OpenMP 1.0 and part features of OpenMP 2.0 Fortran API. To improve the performance of OpenMP programs, two new global operations BARRIER(0) and ALLREDUCTION are presented to extend BARRIER and REDUCTION constructs in OpenMP API respectively. BARRIER(0) provides a new economical synchronization mechanism to support the barrier synchronization between the master and slaves, i.e., it allows the slave to continue as soon as the master arrives, while BARRIER requires the slave to wait until all the threads arrive. ALLREDUCTION complements REDUCTION to eliminate redundant system overheads by supporting the reduction operation over private variables. These two constructs are supported by our runtime library nicely and improve the system performance impressively in the benchmark testing and experiments. Both of the new constructs have been defined at the level of the OpenMP directives and are available for programmers to optimize the program performance.

The rest of the paper is organized as follows. Section 2 briefly describes the portable OpenMP compiler based on JIAJIA SDSM system. Section 3 and 4 devote to the presented extensions to barrier and reduction constructs respectively. Section 5 analyzes the effects of the new extensions on ccNUMA systems. Finally, it is concluded that the contributions are applicable to all OpenMP systems on NUMA machines.

2   Portable OpenMP Compiler

We have completely implemented the OpenMP 1.0 Fortran API and part features of OpenMP 2.0 Fortran API on JIAJIA. To achieve portability, we don't translate OpenMP applications into JIAJIA programs directly. Our compiler consists of two parts, a source-to-source translator and a runtime library. The objective of the source-to-source translator is just to translate OpenMP applications into normal Fortran programs with the runtime library calls. The runtime library is the interface to JIAJIA and responsible for thread management and job schedule. When planning to port our compiler to others platforms, one can reuse the existing source-to-source translator and modify the runtime library with the interface provided by the others platforms. Moreover, many techniques of the implementation of OpenMP directives can be adopted in the new platform.

2.1 JIAJIA SDSM System

JIAJIA [4, 5] is the home-based SDSM system. It has two distinguished features compared to other SDSM systems such as TreadMarks [6]. First, it employs a unique (global) address mapping method and utilizes all the memory available on the processors by combining them to form one large global shared space. Second, it provides data coherence with the scope consistency model [10], which is implemented through a locked-based protocol [4].

JIAJIA provides six basic functions, jia_init(), jia_alloc(), jia_lock(), jia_unlock(), jia_barrier() and jia_exit() to the programmer, and two variables jiapid and jiahosts which specify the host identification number and the total number of nodes to execute a parallel program respectively. Besides these functions and variables, JIAJIA also provides several MPI-like message passing primitives to improve performance, such as jia_bcast() and jia_reduce(). Our runtime library is built by using these functions and global variables.

2.2 Source-to-Source Translator

Our implementation is based on the fork-join programming model. First, all the threads are created at the beginning of execution and do some initialization operations, such as setting environment variables, allocating shared spaces, etc. Then only the master thread continues to execute serial region, and all the slave threads will be suspended once initialization is finished. Before entering a parallel region, the master wakes up all the slaves and broadcasts necessary information of the parallel region. The fork-join programming model is compatible to native OpenMP programming model and makes source-to-source translator simpler than the others based on SPMD model in which correctness must be ensured carefully.

The source-to-source translator encapsulates each parallel region into a separate subroutine. The task assignment and schedule are supported by the runtime library based on thread identifier and schedule strategy specified by users.

2.3 Runtime Library

The runtime library is a bridge between OpenMP applications and JIAJIA system. It focuses on three tasks: thread management, job schedule, and implementation of OpenMP library routines and environment variables. The number of execution threads is decided by environment variable OMP_NUM_THREADS at the beginning of execution and always remains fixed over the duration of each parallel region. Nested parallel regions are serialized and dynamic adjustment of the number execution threads is disabled. We have implemented four schedule strategies supported by OpenMP, i.e. static, dynamic, guided and runtime schedule.

3   Extending Barrier Construct

In most SDSM systems [4–6], barrier synchronization plays two roles. The first role is the same as BARRIER directive described in [1, 2], i.e. to synchronize all the threads in a team. When BARRIER directive is encountered, each thread will wait until all the other threads in that team reach the synchronization point. The second role of barrier is to maintain the coherence of shared-memory, because the coherence of SDSM such as JIAJIA is maintained by software cache coherence protocol. Therefore the overhead of barrier synchronization becomes more expensive.

In order to investigate the system overheads incurred by the different OpenMP constructs, we have used the kernel benchmarks, Microbenchmark [11], to obtain a picture of the performance. Fig. 1 shows the performance of our OpenMP compiler before extending BARRIER construct[1]. The vertical axis gives the overheads of the constructs, PARALLEL, PARALLEL DO, BARRIER and REDUCTION respectively, and the horizontal axis gives the number of the used threads during the benchmark testing. OpenMP API does not require any barrier on the entry to a parallel region [1, 2]. However, when implementing OpenMP over SDSM, one must use barrier synchronization to make the values of shared variables updated in the last serial region by the master visible to the slaves. It can be seen that the overheads of barrier synchronization constitute the large parts of the overheads of PARALLEL and PARALLEL DO.

Two implied barriers are needed when encountering a parallel region on SDSM systems. But in fact, it is not necessary for the implied barrier synchronization on the entry to perform all the actions required by OpenMP API.

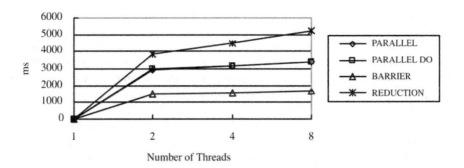

Fig. 1. System Overheads Measured by the OpenMP Microbenchmark

---

[1] In the measurements we use a cluster architecture consisting of 8 alpha nodes. Each node contains one Compaq alpha 21264 processor and 2 Gbyte of memory. All the nodes are running OSF 4.0F and their page size is 8192 bytes.

```
              C$OMP MASTER
                    read(*,*) n
              C$OMP END MASTER
              C$OMP BARRIER
```

Fig. 2. Program Fragment of Using BARRIER

To analyze the necessary actions inside an implementation of the one on the entry to a parallel region, we find that the slaves only need to cooperate with and wait the master to complete the consistency operation while the synchronization among the slaves is not required. Therefore, almost half of barriers in an OpenMP program can be implemented in a simplified way.

In some OpenMP programs, many operations must be done by the master, such as I/O operation. Then MASTER and END MASTER are used. It is noticed that the code enclosed within MASTER and END MASTER directives is executed by the master thread of the team. There is no implied barrier either on entry to or exit from the master section. In a OpenMP program, it is needed to use BARRIER directive after END MASTER explicitly if the slaves want to access the values of shared variables updated by the master. Fig. 2 gives a program fragment of such case. The slaves wait at the BARRIER for the master's arrival which signals that n is safe to use.

The case is similar to the one entering parallel region, i.e., the slaves only need to cooperate with and wait the master to complete the consistency operation and the synchronization among the slaves is not required. So, to decrease the cost, our OpenMP compiler has introduced and implemented a new directive C$OMP BARRIER(0) to extend BARRIER construct. The format of BARRIER(0) is as follows:

    C$OMP BARRIER(0)

BARRIER(0) is implemented by our runtime library function _omp_barrier(0). The function _omp_barrier(0) works as follows. When entering _omp_barrier(0), the master scans the shared variables modified by it and informs the slaves to maintain the memory consistency for these variables if necessary. Each slave waits for the master's arrival and maintains the memory consistency according to the messages from the master. There are two differences between BARRIER(0) and BARRIER. The first is the scope of maintaining the memory consistency. With BARRIER(0), only the shared variables modified by the master need to be made visible to all the threads. The second is that BARRIER(0) allows to the slave to continue as soon as the master arrives, while BARRIER requires the slaves to wait until all the threads arrive. So, in many cases, BARRIER(0) is significantly faster than BARRIER. By using BARRIER(0) instead of BARRIER on entry to a parallel region, our OpenMP compiler reduces the overhead of entering a parallel region significantly.

Figure 3 shows the performance of our OpenMP compiler with BARRIER(0) on the Microbenchmark. After using BARRIER(0), the overheads of PARALLEL and PARALLEL DO are reduced consequently. For example, the overhead of PARALLEL

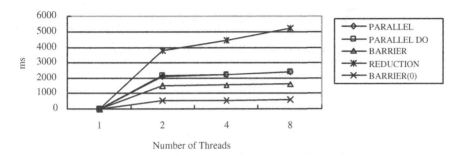

Fig. 3. System Overheads after Introducing `BARRIER(0)`

over 2 CPUs is reduced from 2910.43 microseconds to 2110.43 microseconds, and the one over 4 CPUs is reduced from 3138.24 microseconds to 2218.24 microseconds.

4   Extending Reduction Construct

Reduction is used frequently in parallel programming. For example, in two-dimensional plasma simulation program with particle clouds in cells methods [8, 9], the implementation and performance of reduction influence the performance of the whole program directly. Each particle moves according to Newton-Lorenz equation of relativity and causes the changes of the densities of electric charges of cells independently. So, the computation on particles can be parallelized in coarse grain. Each thread obtains the initial data from the master and computes the densities of electric charges of the cells caused by motivation of particles it manages. Let the size of the Euler cell be $nx \times ny$. The densities of electric charges of the cells computed by the thread m are stored in the shared array `ros1(1:nx,1:ny,m)`, where `ros1` is declared as three-dimensional array, `ros1(nx,ny,nthreads)` and `nthreads` denotes the total number of executing thread. Once each simulation step is completed, the program needs to collect these densities computed by each thread, i.e., to get the final densities stored in `ros(nx,ny)` by sum up `ros1(nx,ny,m)` of all the threads.

OpenMP 1.0 and 1.1 contain REDUCE clause which performs a reduction on the variables that appear in `list`, with the operator `operator` or the intrinsic `intrinsic_procedure_name`. REDUCE clause is of the form [1]:

REDUCTION(operator|intrinsic_procedure_name: list)

where variables in `list` must be named scalar variables of intrinsic type and SHARED in the enclosing context. So, because variable `ros` is a two-dimensional array, a three-level nested loop has to be constructed for computing each element of `ros` as follows.

Comparing to OpenMP 1.0 and 1.1, OpenMP 2.0 allows an array appearing in `list`, and the above nested loop can replaced by a REDUCTION clause as Fig. 5.

```
C$OMP PARALLEL PRIVATE(j,i,k,tmp1,nthreads)
      nthreads=1
C$    nthreads=OMP_GET_MAX_THREADS()
C$OMP DO
      do 100 j=1,ny
      do 100 i=1,nx
        tmp1=0.0d0
        do 101 k=1,ncpus
101       tmp1=tmp1+ros1(i,j,k)
100   ros(i,j)=tmp1
C$OMP END DO NOWAIT
C$OMP END PARALLEL
```

Fig. 4. Program Fragment of a Three-Level Nested Loop

```
C$OMP PARALLEL REDUCTION(+:ros)
      ros =ros+ros1
C$OMP END PARALLEL
```

Fig. 5. Program Fragment of REDUCTION

With REDUCTION, the program code can be simplified and the performance can be improved. Furthermore, REDUCTION makes it possible to store the densities of electric charges in a private two-dimensional array ros1(nx,ny) instead of the original shared three-dimensional array ros1(nx,ny,nthreads), and therefore the cost of maintaining memory consistency will decrease.

JIAJIA provides MPI-like message passing primitive, jia_reduce() to improve reduction operation performance. Based on jia_reduce(), the OpenMP runtime library function _omp_reduce(sendbuf,recvbuf,count,type,op) is implemented, where sendbuf denotes the address of source data, count is the number of elements in source data, type is the data type of elements of source data, and op is the reduction operation. recvbuf is an output parameter to give address of reduction result, and its type must be the same as that of sendbuf. For saving space, recvbuf can share same space with sendbuf. According to JIA-JIA's implementation, the variables in sendbuf and recvbuf must be private.

But in OpenMP, variables that appear in a REDUCTION clause must be SHARED in the enclosing context [1, 2]. Reduction operation of a shared array can be implemented though _omp_reduce and array assignment. The program fragment in Fig. 5 will be implemented as in Fig. 6.

Table 1 shows the execution time of two-dimensional plasma simulation program after 100 simulation steps, where the size of particles is 2×10^6 and the size of Euler cell is 64×64.

```
dimension rc_ros(nx,ny)
rc_ros = rc_ros+ros1
if (omp_get_num_threads().gt.1) then
    call _omp_reduce(rc_ros,rc_ros,nx1*ny1,REAL4,SUM)
    call _omp_barrier()
endif
//assignment from private array rc_ros to shared array ros
if (omp_get_thread_num().eq.0) then
    ros=ros+rc_ros
endif
```

Fig. 6. The Implementation of REDUCTION

Table 1. Execution Time of Two-Dimensional Plasma Simulation Program after 100 Simulation Steps

number of threads	1	2	4	8
wall time	510.6	353.76	262.98	203.47

From the above example, one may find that, the performance of the OpenMP compiler is restricted from at least two aspects by the requirement that the variables appearing in a REDUCTION clause must be SHARED. Firstly, the master has to execute assignment operation to write the results to the related shared variables. Secondly, the cost of accessing shared variables is more expensive that that of accessing private variables when the threads request the value of ros. According to its size, ros may be distributed among several nodes. The assignment and access to ros may refer to the remote memory access, which is costly. So, it is desired to eliminate the system overheads caused by these two aspects. The solution in our OpenMP compiler is to extend reduction construct by introducing a new OpenMP clause ALLREDUCTION, which has the following format:

ALLREDUCTION (operator|intrinsic_procedure_name: list)

ALLREDUCTION combines values of variables in list from all the threads and distributes the result back to all the threads just like MPI_ALLREDUCE function. The rules of parameters are the same as REDUCTION clause except that the variables appearing in ALLREDUCTION clause must be PRIVATED in the enclosing context or appear in THREADPRIVATE common blocks. operator is one of +, *, -, .AND., .OR., .EQV., or .NEQV., and intrinsic_procedure_name refers to one of MAX, MIN, IAND, IOR, or IEOR. The program fragment in Fig. 4 will be replaced by the one in Fig. 7 with the construct ALLREDUCTION.

ALLREDUCTION clause has been implemented in our OpenMP compiler though one reduction operation plus one broadcast operation. In the implementation, the last assignment statement in Fig. 6 is replaced by a broadcast operation.

By using ALLREDUCTION clause, all the threads get the same reduction result of ros, so each thread accesses the private copy of ros simply and the additional

```
C$OMP PARALLEL ALLREDUCTION(+:ros)
        ros =ros+ros1
C$OMP END PARALLEL
```

Fig. 7. Program Fragment of ALLREDUCTION

```
//assignment from private array rc_ros to shared array ros
if (omp_get_num_threads().gt.1) then
   _omp_bcast(ros,nx1*ny1,0)
endif
```

Fig. 8. The Implementation of ALLREDUCTION

overheads introduced by shared array is omitted as desired. Table 2 shows the performance improvement of the example given in Table 1.

Table 2. Execution Time of the Example given in Table 1

number of threads	1	2	4	8
ALLREDUCTION	510.1	286.57	171.61	114.4
REDUCTION	510.6	353.76	262.98	203.47

5   Improving OpenMP Performance on ccNUMA

In the previous sections, we show that the presented extensions to BARRIER and REDUCTION constructs work well to improve OpenMP performance on SDSM. In this section, it will be shown that, on hardware DSM systems such as ccNUMA, BARRIER(0) and ALLREDUCTION will also improve the performance of OpenMP programs. Furthermore, BARRIER(0) and ALLREDUCTION can be implemented with affordable costs. From our experiences, the OpenMP compiler based on JIAJIA can be ported on ccNUMA systems. Benefited from the portable structure, the Source-to-Source translator has been reused almost completely and the main point is to implement the OpenMP runtime library based on Posix thread library provided by ccNUMA systems. Based on implementation of BARRIER and REDUCTION, BARRIER(0) and ALLREDUCTION have been implemented without too many efforts.

Now, we discuss the underlying benefits from the new directives BARRIER(0) and ALLREDUCTION.

Memory consistency is maintained by hardware on ccNUMA systems, while it is maintained by software on SDSM. Although the role of barrier on ccNUMA systems is to synchronize all the threads, BARRIER(0) is significantly faster than BARRIER in many cases. The reason is that BARRIER(0) allows the threads to continue as soon as the master arrives while BARRIER requires the threads to wait until all the threads arrive.

Though the structural organization of ccNUMA systems provides programmers a globally shared memory address space, the actual memory modules are physically distributed among processing nodes. Accessing a remote memory module on a state-of-the-art ccNUMA system costs 3 to 8 times as much as accessing local memory [12–14]. If a thread m needs to access the elements of array A, which stores the result of REDUCTION clause, such access belongs to remote access unless those elements to be accessed are stored on the running node of m exactly. The delay of the remote memory access will be 3 to 8 times as much as that of the local memory access, depending on the distance between the running node of m and the home node of the element to be accessed. However, if ALLREDUCTION is used instead of REDUCTION, one may omit the cost of the remote memory access caused by such access of the shared array. The cost is that ALLREDUCTION needs a broadcast operation. It is predictable that the performance improvement by using ALLREDUCTION becomes larger along with the increases of the number of threads and the size of the array.

6   Conclusions

To improve the performance of our portable OpenMP compiler on JIAJIA SDSM, we present two new constructs, BARRIER(0) and ALLREDUCTION, to extend BARRIER and REDUCTION directives in OpenMP. With BARRIER(0), all the threads maintain the memory consistency only for the shared variables modified by the master, and the slave continues as soon as the master arrives the synchronization point, regardless of the synchronization among the slaves. Therefore BARRIER(0) may decrease the system overheads caused by the usual BARRIER in many cases. ALLREDUCTION combines the values of private variables from all the threads and distributes the result back to all the threads just like MPI_ALLREDUCE function, and eliminates the redundant overheads caused by REDUCTION, which requires the reduction variables to be shared. These two constructs, which have been implemented in our OpenMP compiler, improve the OpenMP performance impressively in the benchmark testing and experiments. It is predicable that the improvement of performance can be obtained on ccNUMA systems. Both of these new constructs have been defined at the level of the OpenMP directives and are available for programmers to optimize the program performance.

References

1. The OpenMP Forum. OpenMP Fortran Application Program Interface, Version 1.0, October 1997 and OpenMP Fortran Application Program Interface, Version 1.1, November. 1999. See http://www.OpenMP.org.
2. The OpenMP Forum. OpenMP Fortran Application Program Interface, Version 2.0, November 2000. See http://www.OpenMP.org.
3. D. Culler, J. P. Singh and A. Gupta. Parallel Computer Architecture, a Hardware/Software Approach. Morgan Kaufmann Publishers, 1998.

4. W. Hu, W. Shi, and Z. Tang. A Lock-Based Cache Coherence Protocol for Scope Consistency. Journal of Computer Science and Technology, 13(2):97-109, Mar. 1998.
5. W.Hu, W.S. Shi, and Z. Tang. JIAJIA: An SVM System Based on A New Cache Coherence Protocol. HPCN'99, LNCS 1593, pp. 463-472, Springer, 1999, Amsterdam, Netherlands.
6. C. Amza, A.L. Cox, S. Dwarkadas, P. Keleher, H. Lu, R. Rajamony, W. Yu, and W. Zwaenepoel. TreadMarks: Shared Memory Computing on Networks of Workstations. IEEE Computer, 29(2):18-28, 1996.
7. S. Kumar, D. Jiang, R. Chandra and J. P. Singh. Evaluating Synchronization on Shared Address Space Multiprocessors: Methodology and Performance. Proc. of the 1999 ACM SIGMETRICS Conference, pp. 23-34, Atlanta(USA), 1999.
8. Birdsall C.K and Longdon A.B, Plasma Physics via Computer Simulation. McGraw Hill Book Company, 1985.
9. Mo Zeyao, Xu Linbao, Zhang Boilin and Shen Longjun. Parallel Computing and Performance Analysis for Two-Dimensional Plasma Simulations with Particle Clounds in Cells Methords. Computational Physics, pp. 496-504, September 1999.
10. L. Iftode, J. P. Singh, and K. Li. Scope Consistency: A Bridge between Release Consistency and Entry Consistency. In Proc. of the 8th ACM Annual Symp. on Parallel Algorithms and Architectures (SPAA'96), pages 277-287, June 1996.
11. J.M. Bull and Darragh O'Neill, A Microbenchmark Suite for OpenMP 2.0. In Proc. of the European Workshop on OpenMP (EWOMP'01), September 1999.
12. J. Laudon and D. Lenoski. The SGI Origin2000: A ccNUMA Highly Scalable Server. Proc. of the 24th Annual International Symposium on Computer Architecture, pp. 241-251, Denver (USA), 1997.
13. Silicon Graphics, Inc.. SGITM OriginTM 3000 Series. Technical Report. 2002.
14. T.Lovett and R.Clapp. STiNG: A CC-NUMA Computer System for the Commercial Marketplace. Proc. of the 23rd Annual International Symposium on Computer Architecture, Philadelphia (USA), 1996.

OpenMP for Adaptive Master-Slave
Message Passing Applications

Panagiotis E. Hadjidoukas, Eleftherios D. Polychronopoulos,
and Theodore S. Papatheodorou

High Performance Information Systems Laboratory (HPCLAB),
Department of Computer Engineering and Informatics,
University of Patras,
Rio 26500, Patras, Greece,
http://www.hpclab.ceid.upatras.gr, {peh,edp,tsp}@hpclab.ceid.upatras.gr

Abstract. This paper presents a prototype runtime environment for
programming and executing adaptive master-slave message passing applications on cluster of multiprocessors. A sophisticated portable runtime
library provides transparent load balancing and exports a convenient application programming interface (API) for multilevel fork-join RPC-like
parallelism on top of the Message Passing Interface. This API can be
used directly or through OpenMP directives. A source-to-source translator converts programs that use an extended version of the OpenMP
workqueuing execution model into equivalent programs with calls to the
runtime library. Experimental results show that our runtime environment
combines the simplicity of OpenMP with the performance of message
passing.

1   Introduction

Master-slave computing is a fundamental approach for parallel and distributed
applications. It has been used successfully on a wide class of applications. On
distributed memory machines, programming a master-slave application is a difficult task that requires the knowledge of the primitives provided by the underlying message passing environment and additional programming effort when the
application exhibits load imbalance. OpenMP [10] is an emerging standard for
programming shared-memory multiprocessors and recently clusters of workstations and SMPs through software distributed shared memory. Since OpenMP
represents a master-slave (fork-join) programming paradigm for shared memory,
it is possible to extend OpenMP for executing the task parallelism of master-
slave message passing applications. OmniRPC [13] is an initial proposal for using
OpenMP on client-server applications through remote procedure calls (RPC).

This paper presents a runtime environment for programming and executing adaptive master-slave message passing applications on cluster of SMPs. A
portable runtime library exports a "shared-memory" API that can be used either directly or through OpenMP directives and a source-to-source translator,
providing thus ease of programming and transparent load balancing without

A. Veidenbaum et al. (Eds.): ISHPC 2003, LNCS 2858, pp. 540–551, 2003.

requiring any knowledge of low-level message passing primitives. The library has been designed to provide application adaptability to the same application code, or even binary, on shared-memory multiprocessors, clusters of SMPs and metacomputing environments.

The primary goal of this work is the easier development of efficient master-slave message passing applications using OpenMP directives rather than an extension of OpenMP. The translator converts programs that use an extended version of the proposed OpenMP workqueuing execution model [15] into equivalent programs with calls to the runtime library. Experimental results on two architectural platforms (shared-memory multiprocessor and cluster of SMPs) and two different operating systems (Windows and Linux) indicate the efficient execution of adaptive OpenMP master-slave message passing programs on our runtime environment.

The rest of this paper is organized as follows: Section 2 describes the general design and architecture of the proposed runtime system. Section 3 presents our OpenMP extensions for master-slave message-passing programming. Experimental evaluation with OpenMP programs is reported in Section 4. Related work is presented in Section 5. We discuss our future research in Section 6.

2   Runtime Library

The proposed runtime library provides an implementation of the Nano-threads Programming Model on top of message passing and specifically the Message Passing Interface (MPI) [7]. This model has been implemented on shared-memory multiprocessors in the context of the NANOS project [9] and recently on clusters of SMPs, through SDSM, within the POP (Performance Portability of OpenMP) project [12]. The NANOS API is targeted by the NanosCompiler [1], which converts programs written in Fortran77 that use OpenMP directives to equivalent programs that use this API. Our runtime library exports a slightly extended version of the NANOS API for master-slave message passing applications.

2.1   Design and Architecture

Figure 1 illustrates the modular design of our runtime system. NanosRTL is the core runtime library that implements the architecture and the functionality of the Nano-threads Programming Model. Its implementation is based on the POSIX Threads API and the interfaces that are exported by the rest components of this runtime environment. Moreover, a POSIX Threads library for Windows and a simple unix2nt software layer allows the compilation of our code, without modifications, on both Unix and Windows platforms.

UTHLIB (Underlying Threads Library) is a portable thread package that provides the primary primitives (creation and context-switching) for managing non-preemptive user-level threads. Its purpose is to be used for implementing two-level thread models (libraries), where virtual processors are system scope POSIX Threads. It is portable since its machine dependent parts are based

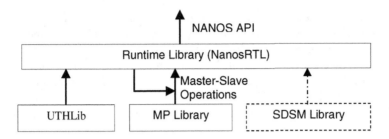

Fig. 1. Modular Design

on either the POSIX `setjmp-longjmp` calls or the `ucontext` operations. The MPI library is used by the runtime system for the management of its internal structures and the explicit, though transparent, movement of data. Since MPI corresponds to an SPMD (Single Program Multiple Data) execution model, the runtime system provides several mechanisms for a fork-join parallel execution model (e.g. memory allocation, broadcasting). Moreover, our two-level thread model requires the thread-safety of the MPI library. An optional component of our runtime environment is an SDSM library, which provides an additional way for implicit movement of data. The integration of the MPI and the SDSM libraries is performed with their concurrent linking and the appropriate setting of the environment variables that each library requires.

In master-slave computing, the master distributes the data to the slaves and waits any processed results to be returned. A task in our runtime environment corresponds to the execution of a function on a set of data that are passed as arguments to this function. Each task is represented with a work descriptor (called nano-thread), a data structure that encapsulates all the necessary information of a task (function pointer, arguments (private data or pointers to them), dependencies, successors) [9]. This descriptor is separated from its execution vehicle, i.e. an underlying user-level thread, which is created on demand, according to the adopted lazy stack allocation policy. Similar to Remote Procedure Call (RPC), a descriptor is associated with the data of its corresponding function. The user defines the quantity (count), MPI datatype, and the passing method for each argument, information that is also stored in the descriptor. This definition is the only major extension in the existing NANOS API.

The runtime library has the same architecture with our OpenMP runtime system on top of SDSM, described thoroughly in [5]. The main difference is that the consistency protocol has been replaced with message passing: there are not shared data or stacks and data movement has to be performed explicitly with MPI calls. Each node of the cluster (process) consists of one or more virtual processors and a special I/O thread, the *Listener*, which is responsible for the dependencies and queue management and supports the transparent and asynchronous movement of data. There are per-virtual processor intra- and inter-node ready queues and per-node global queues. The insertion/stealing of a descriptor in/from a queue that resides in the same node (process) is performed through hardware-shared memory without any data movement. Otherwise, the operations are performed with explicit messages to the Listener of the remote node. This combination of hardware shared-memory with explicit messages is also used to maintain the coherence of the fork-join execution model: each work descriptor (task) is associated with an owner node. If a task is finished on its owner node, its parent is notified directly through (hardware) shared memory. Otherwise, its descriptor returns to the owner node and the notification is performed by the Listener.

2.2   Data Movement

The insertion of the descriptor in a remote queue corresponds to `MPI_Send` calls for the descriptor and any arguments, based on their description. On the other side, the Listener accepts the descriptor, analyzes its information and allocates the necessary space to receive the arguments. The runtime system ensures the coherence of the descriptor and its data by appropriately setting the tag field in each sent MPI message. Specifically, the tag field denotes the local identifier of the virtual processor that sends the message. Once the Listener has received a descriptor, it receives the subsequent data from the source of the message (i.e. the specific virtual processor). A virtual processor executing a descriptor actually executes the function with the locally stored arguments. When it finishes, it sends the descriptor back to its owner node, along with any arguments that represent results. These are received asynchronously by the Listener and copied on their actual memory locations in the address space of the owner. All the aforementioned movement of data is transparent to the user and the only points that remind the underlying MPI programming is the description of the arguments. Currently, the following definitions determine the passing method for an argument:

– `CALL_BY_VAL`: The argument is passed by value. If it is a scalar value, it is stored directly in the descriptor.
– `CALL_BY_PTR`: As above, but the scalar value has to be copied from the specified address. This option can be also used for sending user-defined MPI datatypes.
– `CALL_BY_REF`: The argument represents data that are sent with the descriptor and returned as a result in the owner node's address space.
– `CALL_BY_RES`: No data has to be sent but it will be returned as a result. It is assumed by the target node that receives data initialized to zero.
– `CALL_BY_SVM`: The argument is an address in the shared virtual memory of a software DSM system. Data will be transferred implicitly, through the memory consistency protocol.

For arguments (data) that represent results and return to the master, the runtime system supports some primary reduction operations that can be performed on both single values and arrays. The following definitions can be combined (OR 'ed) with the passing method definitions:

- OP_SUM: The results (returned data) are added to the data that reside at the specific memory address.
- OP_MUL: The results are multiplied to the data that reside at the specific memory address.

The first list can be expanded with other forms of data management/fetching, like MPI one-side communications or SHMEM operations and the second with more reduction operations.

2.3   Load Balancing

According to the Nano-threads Programming Model, the scheduling loop of a virtual processor is invoked when the current user-level thread finishes or blocks. A virtual processor visits in a hierarchical way the ready queues in order to find a new descriptor to execute. The stealing of a descriptor from a remote queue includes the corresponding data movement, given that it does not return to its owner node. The stealing is performed in a synchronous way: the virtual processor that issued the request waits synchronously for an answer from the Listener of the target node. The answer is either a special descriptor denoting that no work was available at that node, or a ready descriptor, which will be executed directly by the virtual processor. Due to the adopted lazy stack allocation policy [4], a descriptor (and its data) can be transferred among the nodes without complications, since an underlying thread is created only at the time of its execution. The two-level thread model of our runtime environment favors the fine decomposition of the applications, which can be exploited for achieving better load balancing. These conditions minimize, or even eliminate, the need for user-level thread migration, which is difficult to support on heterogeneous and metacomputing environments.

2.4   Integration with SDSM

As aforementioned, we have managed to integrate in this runtime environment an SDSM library, which provides a method for implicit movement of data. A distinct classification of data is required, as described in [2], since data that resides in shared virtual memory cannot be sent (MPI_Send) explicitly. On Windows platforms, we have successfully integrated the MPIPro [8] and the SVMLib [11] libraries. With SVMLib, the user can disable the SDSM protocol for a specific memory range, a feature that allows the explicit movement of shared data. On Unix (Linux), we have integrated MPIPro with Mome [6]. According to the Mome's weak consistency model, a process can issue explicit consistency requests for specific views of a shared-memory segment, a feature that provides a receiver-initiated method for data transfer. This integration of MPI and SDSM provides a very flexible and complex programming environment that can be easily exploited with the unification of our OpenMP environment on top of SDSM and the work presented in this paper. However, a thorough study of this integration is beyond the scope of the paper.

3   OpenMP Extensions

In [13], authors propose a parallel programming model for cluster and global computing using OpenMP and a thread-safe remote procedure call facility, OmniRPC. Multiple remote procedure calls can be outstanding simultaneously from multithreaded programs written in OpenMP, using `parallel for` or `sections` constructs. These constructs have the disadvantage that all work units that can be executed are known at the time the construct begins execution and have difficulties handling applications with irregular parallelism. Moreover, in master-slave (client-server) computing, the master (client) usually executes task-specific initializations before distributing the tasks to the slaves (servers). The execution model that addresses these issues and fits better in our case is the proposed OpenMP workqueuing model [15]. This model is a flexible mechanism for specifying units of work that are not pre-computed at the start of the worksharing construct. Conceptually, the `taskq` pragma causes an empty queue to be created by the chosen thread, and then the code inside the `taskq` block is executed single-threaded. The `task` pragma specifies a unit of work, potentially executed by a different thread. The following fragment of code is a typical example that uses the workqueuing constructs in order to execute task parallelism. This pattern is found in most master-slave message passing applications, where the master distributes a number of tasks among the slaves (`do_work`), usually after having performed some task-specific initialization (`j=i*i`):

```
int i, j;
#pragma intel omp parallel taskq
{
    for (i = 0; i < 10; i++) {
        j = i*i;
        #pragma intel omp task
        {
            do_work(j);
        }
    }
}
```

3.1   Workqueuing Constructs

In this section, we present our extensions to the proposed OpenMP workqueuing model for master-slave message passing computing. Since we target a pure distributed memory environment, with private address spaces, a default (private) clause in implied naturally in the OpenMP programs. Our OpenMP compilation environment exploits the two workqueuing constructs presented in the previous example (`intel omp parallel taskq` and `intel omp task`). For convenience, we omit the `intel` keyword and replace `omp` with `domp`. These directives are parsed by a source-to-source translator (`comp2mpi`), which generates an equivalent program with calls to our runtime library. Similar directives have been defined for the Fortran language and a corresponding translator has been developed. None of these two constructs is altered and only the second one is extended with two optional clauses (`schedule`, `callback`):

```
#pragma domp parallel taskq

#pragma domp task [schedule(processor|node))] [callback(<functionname>)]
 <task parallel function>
 [<callback function>]
```

The MPI description for the arguments of a function is provided with the following format:

```
#pragma domp function <functionname> <number of arguments>
#pragma domp function <description of first argument>
 ...
#pragma domp function <description of last argument>
```

where,

```
<description> -> <count>,<datatype>,<passing method>[|<reduction>]
<count> -> number of elements
<datatype> -> valid MPI data type
<passing method> -> CALL_BY_VAL | CALL_BY_PTR | CALL_BY_REF |
                    CALL_BY_RES | CALL_BY_SVM
<reduction> -> OP_SUM | OP_MUL
```

Finally, the schedule clause determines the task distribution scheme that will be followed by the OpenMP translator. Currently, the generated tasks can be distributed on a per-processor (default) or on a per-node basis. The latter can be useful for tasks that generate intra-node (shared-memory) parallelism either by using the NANOS API or through OpenMP directives and the NanosCompiler. Alternatively, there can be one virtual processor per node and the intra-node OpenMP parallelism can be executed using other OpenMP compilation environments.

3.2 Examples

Figure 2 presents two programs that use our OpenMP extensions. The first application is a recursive version of Fibonacci. It exhibits multilevel parallelism and each task generates two new tasks that are distributed across the nodes. If we remove the pragmas that describe the function and the schedule clause, the resulted code is compliant with the proposed workqueuing model. The second example is a typical master-slave application: each task computes the square root of a given number and returns the result to the master. Tasks are distributed across the processors and whenever a task completes, a callback function (cbfunc) is executed asynchronously.

4 Experimental Evaluation

Our runtime environment is portable and allows the execution of the same application binary on both shared-memory multiprocessors and distributed memory

```
/* Fibonacci */                                      /* Master-Slave Demo Application */

void fib(int n, int *res)                            void taskfunc(double in, double *out)
{                                                    {
#pragma domp function fib 2                          #pragma domp function taskfunc 2
#pragma domp function 1,MPI_INT,CALL_BY_VAL          #pragma domp function 1,MPI_DOUBLE,CALL_BY_VAL
#pragma domp function 1,MPI_INT,CALL_BY_REF          #pragma domp function 1,MPI_DOUBLE,CALL_BY_RES
    int res1 = 0, res2 = 0;                              *out = sqrt(in);
    if (n < 2) {                                     }
        *res = n;
    } else if (n < 30) {                             void cbfunc(double in, double *out)
        fib(n-1, &res1);                             {
        fib(n-2, &res2);                                 printf("sqrt(%f)=%f\n", in, *out);
        *res = res1+ res2;                           }
    } else {
        #pragma domp parallel taskq schedule(node)   void main()
        {                                            {
                                                         int cnt = 100, i;
            #pragma domp task                            double *result;
            fib(n-1, &res1);                             double num;
            #pragma domp task
            fib(n-2, &res2);                             result = (double *)malloc(cnt*sizeof(double));
        }
        *res = res1+ res2;                               #pragma domp parallel taskq
    }                                                    for (i=0; i<cnt; i++) {
}                                                            num = (double) i;
                                                             #pragma domp task callback(cbfunc)
void main()                                                  taskfunc(num, &result[i]);
{                                                            cbfunc(num, &result[i]);
    int res, n = 36;                                     }
    fib(n, &res);                                    }
}
```

Fig. 2. OpenMP Examples

environments. These features enable us to present experimental results on both architectural platforms and two different operating systems: Linux and Windows. Our experiments were conducted on a quad-processor 200 MHz Pentium Pro system equipped with 512 MB of main memory, running Windows 2000 Advanced Server and a cluster of 4 Pentium II 266 MHz dual-processor machines, running Linux 2.4.12. Each machine has 128MB of main memory and the cluster is interconnected with Fast Ethernet (100 Mbps). As native compilers, Microsoft Visual C++ 6.0 (Windows) and GCC 3.2.3 (Linux) have been used. The context-switching mechanism of the underlying threads is based on setjmp-longjmp and the thread-safe MPIPro (1.6.3) library [8] has been used.

4.1 Applications

For the experimental evaluation of our runtime system, we have used the following embarrassingly parallel OpenMP applications:

- **Fibonacci (FIB)**: It computes the 36th Fibonacci number using the code illustrated in Figure 2. The application exploits several levels of parallelism, generating threads until the 29th Fibonacci number, which results in a large number of computational intensive threads.
- **EP**: EP is an Embarrassingly Parallel benchmark. It generates pairs of Gaussian random deviates according to a specific scheme. In EP, the number of tasks is equal to the number of available processors.
- **Synthetic Matrix Addition (SMA)**. This application adds two matrices of doubles, introducing a fixed computational cost with each addition. The matrices, of dimension 512x512, are allocated and initialized by the master

thread. SMA creates N (=512) parallel tasks that are distributed cyclically among the available processors. With each task, three rows are sent (3x12 KB total) and one row returns.

These embarrassingly parallel applications can be also executed using OpenMP on top of SDSM. However, as we have shown in [4], they will obtain less performance due to the false-sharing problem of SDSM and the overhead of the paged-based SDSM protocol, which is less efficient compared to the explicit data movement.

4.2 Experimental Results

Figure 3 illustrates the normalized execution times of the three applications (FIB, EP, SMA) on the SMP machine and the Linux cluster. Each column corresponds to a "Processes x Threads" configuration, denoting the number of MPI processes and the per-process virtual processors.

On Windows, the pure context-switching overhead of the underlying threads using `setjmp-longjmp` is less than 171 processor cycles (0.85 μsec). The average overhead to fork (create-enqueue) and join a remotely executed descriptor, with inactive the stealing mechanism, is 8 μsecs for the "1x2" (Processes x Threads) configuration and 325 μsecs for the "2x1" configuration. The low overhead is explained taking into account that MPI/Pro implements inter-process communication on a single machine through shared memory queues and between processes on separate machines through sockets. We observe that all applications manage to scale efficiently. On four processors, FIB performs better on the "2x2" configuration, which balances the overheads caused by the internal shared-memory synchronization ("1x4") and the movement of the descriptors across the processes ("4x1"). On the other hand, SMA scale better when more kernel threads (virtual processors) are used, since many data movements are avoided due to shared memory. The execution times of the applications on a single processor ("1x1") were (in seconds): FIB: 32.05, EP: 309.67, and SMA:40.44.

Except for the last case (8x1), on the cluster of SMPs, the MPI processes always run on different nodes. The overhead for context-switching is 148 pro-

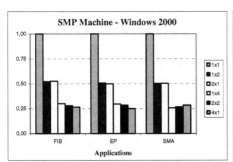 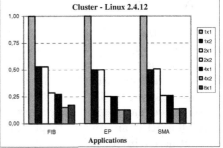

Fig. 3. Normalized Execution Times (Linux - Windows)

Table 1. Execution Times (sec) of Mandelbrot

SMP Machine (Windows)

Processes x Threads	Total CPUS	MAND No stealing	MAND Stealing
1X1	1	31.90	31.90
1X2	2	16.34	15.96
2X1	2	16.36	16.00
1X4	4	10.70	8.56
2X2	4	11.01	8.71
4X1	4	10.72	8.80

Cluster (Linux)

Processes x Threads	Total CPUS	MAND No stealing	MAND Stealing
1X1	1	17.97	17.97
2X1	2	9.22	9.18
4X1	4	6.04	5.05
8X1	8	5.06	2.79

cessor cycles (0.56 μsec). The fork-join overhead is 6 μsecs for a single process with 2 threads and 457 μsecs for two processes. This difference, compared to the SMP machine, is due to the way MPI/Pro implements message-passing between applications that are executed on the same node. The applications exhibit the same behavior with that observed on the shared-memory multiprocessor, with main difference the higher communication latency of the network interface and the overhead of the TCP protocol. The execution times of the applications on the "1x1" configuration (first column) were: FIB: 15.98, EP: 109.54, and SMA: 53.83 (in seconds).

MAND is a master-slave application that computes the Mandelbrot set, a 2-d image (of 512x512 pixels) in which each pixel's value is calculated solely as a function of its coordinates. Because computation is widely varying at each pixel and thus a potential load-balancing problem arises, the application groups pixels into blocks (of 64x64 pixels) and maps them onto processors. For each task, a callback routine stores the calculated values in a two-dimensional matrix. Despite the distribution of blocks, the application fails to scale efficiently without a task stealing mechanism, especially when the number of processors increases. This is observed on both the SMP machine and the cluster (Table 1). On the other hand, the application continues to scale if we activate the stealing mechanism. The performance gain is significantly higher when the application runs on 4 or 8 processors on the cluster.

5   Related Work

OmniRPC [13] is a Grid RPC system for parallel programming in cluster and grids environments, where the client uses OpenMP in order to issue several asynchronous remote procedure calls. Compared to our work, OmniRPC is less flexible, based exclusively on the RPC model, does not target shared memory multiprocessors, neither exploits the OpenMP workqueuing model, which fits better in the master-slave programming paradigm. To the best of our knowledge, only OmniRPC and our work provide OpenMP execution on pure distributed computing environments.

In [2], authors present a number of compiler techniques that can be used to translate OpenMP programs for execution on distributed memory systems.

They use a hybrid communication model, where the data is classified by its usage patterns: distributed data is handled by an explicit message passing mechanism and shared data via an SDSM mechanism. Our runtime system already supports such a hybrid communication model. Its unification with our OpenMP implementation on top of SDSM can provide an OpenMP environment with the appropriate runtime support to most of these compiler techniques.

AMWAT [3] is a library that makes it easier for programmers to develop applications that solve a problem by breaking it into subproblems, distributing the subproblems to multiple processes, and combining the subproblem results into an overall solution. MW [14] is a software framework that allows users to quickly and easily parallelize scientific computations using the master-worker paradigm in an metacomputing environment. MW takes care of systems events like workers joining/leaving allowing users to focus on the implementation of their algorithm.

6   Future Work

We presented a runtime environment for programming and executing adaptive master-slave message passing on clusters of compute nodes. The environment provides a unified dependence-driven and data-flow model of execution, and combines a wide area of parallel programming models, from threads and OpenMP to message passing and RPC. It can be considered as an RPC-like environment with inherent load balancing where every process (or kernel-thread) can operate as both a server and a client. Moreover, it manages to integrate successfully hardware shared-memory, message passing and optionally Software Distributed Shared Memory.

Our future work includes the integration of these OpenMP extensions in the NANOS compiler, the improvement of the load balancing mechanisms, the further study of the integration with an SDSM library, and the exploitation of the application adaptability in a multiprogrammed environment. Advanced features of our runtime environment include the support of thread migration, the execution on heterogeneous platforms (MPI/Globus) and the exploitation of the MPI-2 dynamic process management.

Acknowledgments

We would like to thank the European Center of Parallelism in Barcelona (CEPBA) for providing us access to the Linux cluster, and all our partners in the POP project. This work was supported by the POP IST/FET project (IST-2001-33071).

References

1. E. Ayguadé, J. Labarta, X. Martorell, N. Navarro, and J. Oliver, *NanosCompiler: A Research Platform for OpenMP Extensions*, In Proceedings of the 1st European Workshop on OpenMP, Lund (Sweden), October 1999.

2. R. Eigenmann, J. Hoeflinger, R. H. Kuhn, D. Padua, A. Basumallik, S-J Min and J. Zhu, *Is OpenMP for Grids?*, Workshop on Next-Generation Systems, In Proceedings of the International Parallel and Distributed Processing Symposium (IPDPS'02), Fort Lauderdale, Florida, USA, May 2002.
3. J-P. Goux, S. Kulkarni, J. Linderoth, and M. Yoder, *An Enabling Framework for Master-Worker Applications on the Computational Grid*, In Proceedings of the 9th IEEE International Symposium on High Performance Distributed Computing (HPDC 2000), Pittsburgh, Pennsylvania, USA, August 2000.
4. P. E. Hadjidoukas, E. D. Polychronopoulos, and T. S. Papatheodorou, *Runtime Support for Multigrain and Multiparadigm parallelism*, In Proceedings of the 10th International Conference on High Performance Computing (HIPC' 02), Bangalore, India, December 2002.
5. P. E. Hadjidoukas, E. D. Polychronopoulos, and T. S. Papatheodorou, *OpenMP Runtime Support for Clusters of Multiprocessors*, In Proceedings of the International Workshop on OpenMP Applications and Tools (WOMPAT '03), Toronto, Canada, June 2003.
6. Y. Jegou, *Controlling Distributed Shared Memory Consistency from High Level Programming Languages*, In Proceedings of Parallel and Distributed Processing, IPDPS 2000 Workshops, pages 293-300, May 2000.
7. Message Passing Interface Forum, *MPI: A message-passing interface standard*, International Journal of Supercomputer Applications and High Performance Computing, Volume 8, Number 3/4, 1994.
8. MPI Software Technology, Inc., http://www.mpi-softtech.com.
9. NANOS: Effective Integration of Fine-Grain Parallelism Exploitation and Multiprogramming. ESPRIT IV Framework Project No. 21907. http://www.ac.upc.es/NANOS.
10. OpenMP Architecture Review Board, *OpenMP Specifications*, Available at: http://www.openmp.org.
11. S. M. Paas, M. Dormanns, T. Bemmerl, K. Scholtyssik and S. Lankes, *Computing on a Cluster of PCs: Project Overview and Early Experiences*, In Proceedings of the 1st Workshop on Cluster-Computing, TU Chemnitz-Zwickau, November 1997.
12. POP: Performance Portability of OpenMP. IST/FET project (IST-2001-33071). http://www.cepba.upc.es/pop.
13. M. Sato, T. Boku, and D. Takahashi, *OmniRPC:a Grid RPC ystem for Parallel Programming in Cluster and Grid Environment*, In Proceedings of the 3rd International Symposium on Cluster Computing and the Grid, Tokyo, Japan, May 2003.
14. G. Shao, F. Berman, and R. Wolski, *Master/Slave Computing on the Grid*, In Proceedings of the 9th Heterogeneous Computing Workshop. Cancun, Mexico, May 2000.
15. E. Su, X. Tian, M. Girkar, H. Grant, S. Shah, and P. Peterson, *Compiler Support of the Workqueuing Execution Model for Intel SMP Architectures*, In Proceedings of the 4th European Workshop on OpenMP, Rome, Italy, September 2002.

OpenGR: A Directive-Based Grid Programming Environment

Motonori Hirano[1,2], Mitsuhisa Sato[3], and Yoshio Tanaka[4]

[1] Master's Program in Science and Engineering,
University of Tsukuba,
1-1-1 Teno-dai, Tsukuba, Ibaraki, 305-8573 Japan,
m-hirano@hpcs.is.tsukuba.ac.jp
[2] Software Research Associates, Inc.,
m-hirano@sra.co.jp
[3] Institute of Information Science and Electronics /
Center for Computational Physics,
University of Tsukuba,
msato@is.tsukuba.ac.jp
[4] Grid Technology Research Center,
National Institute of Advanced Industrial Science and Technology,
yoshio.tanaka@aist.go.jp

Abstract. In order to provide a grid programming environment for RPC-based master-worker type task parallelization of existing sequential application programs, we have designed a set of compiler directives called OpenGR, and been implementing this compiler system based on the Omni OpenMP compiler system and a grid-enabled RPC system, Ninf-G, as a parallel execution mechanism. With OpenGR directives, some existing sequential applications can easily be adapted to the grid environment as master-worker type parallel programs using RPC. Furthermore, using both OpenGR directives and the OpenMP directives enables "Hybrid parallelization" of the sequential programs.

Keywords. OpenMP, Grid, GridRPC, Globus, Task parallelism.

1   Introduction

Recently, advances in wide-area networking technology and infrastructure have made it possible to construct large-scale, high-performance distributed computing environments, or computational grids, that provide dependable, consistent and pervasive access to enormous computational resources.

For distributed computing, RPC (Remote Procedure Call) facility has been widely used. For grid computing, several systems also adopt RPC, called grid-enabled RPC, as the basic model of computation, including Ninf-G [1], NetSolve [2], OmniRPC [3], and CORBA [4]. The RPC-style system is particularly useful in that it provides an easy-to-use, intuitive programming interface, allowing users of the grid system to easily make grid-enabled applications. Applications

A. Veidenbaum et al. (Eds.): ISHPC 2003, LNCS 2858, pp. 552–563, 2003.

that require large computing resources such as optimization problems, or parameter searching, are suitable for the grid computing environment. These types of applications often have master-worker parallelism, in which a client application, as the master, makes several calls to the workers in remote computing resources by the grid RPC in parallel.

Although the grid RPC provides a reasonable programming model for a grid environment, it still forces additional work on programmers in order to make the grid application, especially when starting from a serial version of the application program. A programmer will decide which functions should be executed on the server side, make their stubs, and modify their function calls using the grid RPC API. To make the stubs, he may need to define interfaces of their functions. The steps required for using a grid RPC may differ between each RPC system.

Furthermore, especially in the grid-enabled RPC environment, deployment of server side stub programs (server stubs) causes an annoying problem. The server stubs must exist on proper directories (depending on configuration, RPC system, etc.) of the callee (server) system *before* the caller invokes a remote procedure call in most RPC systems. In the grid environment, since it is often heterogeneous and is made up of the resources of many sites, before deployment of the server stubs, the programmer should check and know the hosting environment of each and every server. When the number of servers (or remote hosts) becomes large, it will be an exhaustive task. Some semi-automated shipping mechanisms are required to reduce "manual labor" operations. We have been calling this problem the "server stub shipping problem."

In this paper, we propose a set of directives for C/C++ called *OpenGR* to solve those problems, and have implemented a prototype of an OpenGR compiler using Ninf-G as a grid RPC system. The role of OpenGR directives is very similar to one of OpenMP directives. As by OpenMP, OpenGR provides an incremental programming environment for a grid RPC programming with minimum modification on the serial version the program. The OpenGR directives are translated into a set of runtime library calls by the compiler. Our design objectives for OpenGR are as follows:

1. Provide a grid programming environment for existing application programs with minimal source code modification.
2. With the OpenMP directives, provide a "Hybrid parallelization" for describing parallelism.
3. In order to reduce the complexity and the work of the server stub shipping, provide a semi-automated server stub install facility as a part of the compiler system.

Programmers can generate different executables for both the parallel version and the serial version, from the very same source code only by specifying compile options since these directives are easily ignored during the compiler's parsing phase. Furthermore, it is easy to create server stubs that are parallelized by the thread mechanism only by inserting OpenMP directives into functions that are executed on server side. On the other hand, it is also easy to create client side programs that invoke RPCs in threads created by OpenMP, by using OpenGR

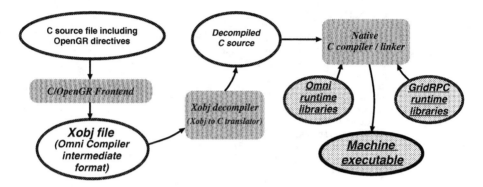

Fig. 1. An overview of OpenGR compiler based on Omni OpenMP compiler system

directives within the OpenMP `parallel` directive closure. These two models are what we mentioned before as "Hybrid parallelization."

Our implementation of the OpenGR compiler system is based on the Omni OpenMP compiler system. Figure 1 shows an overview of that compiler system.

In this paper, we describe the design and implementation of the OpenGR directives and the resulting compiler system. First, in **Section 2**, we illustrate briefly what the grid RPC looks like, using Ninf-G as an example. Next, in **Section 3**, we show sample source code using OpenGR directives. Then, in **Section 4**, we describe an overview of OpenGR directives. And in **Section 5**, we report the results of a preliminary evaluation. In **Section 6**, we present related work. Lastly, in **Section 7** we make concluding remarks and discuss our plans for future work.

2   Background: GridRPC API

GridRPC [5] API is a specification of standard API drafted by the GridRPC working group of the Global Grid Forum (GGF). Ninf-G is one of the implementations of GridRPC API. GridRPC systems should consider security issues, comparing them to the existing LAN-based RPC systems. In order to satisfy this requirement, Ninf-G is written based on the Ninf [6] RPC system, rewriting its communication layer and resource management mechanism with the Globus toolkit [7].

With Ninf-G, programmers write server stubs in IDL, and generate server stub modules using IDL compiler. Figure 2 shows an example of Ninf-G IDL. To build and install(including modules registration to server stub module database managed by GRIS, LDAP based directory service provided by the Globus) server stub, the IDL compiler emits a `Makefile` file, user should issue `make` and `make install` to build and install the server stub. These procedures must be done on each and every server the user wants to use.

To invoke functions in server stub modules existing on arbitrary servers, programmers write source code in the C language using the GridRPC API pro-

```
Module pi;

Define pi_trial(IN int seed, IN long times,
                OUT long * count)
"monte carlo pi computation"
{
  long l, counter = 0;
  double x, y;
  srandom(seed);
  for (l = 0; i < times; l++) {
    x = (double)random() / RAND_MAX;
    y = (double)random() / RAND_MAX;
    if ((x * x + y * y) < 1.0) {
      counter++;
    }
  }
  printf("counter = %d, times = %d\n", counter, times);
  *count = counter;
}
```

Fig. 2. A Ninf-G IDL example

vided by the Ninf-G runtime libraries. Figure 3 shows an example of a program
using the GridRPC API. In this example, functions beginning with "grpc_*"
are GridRPC API functions. These provide mechanisms such as initialization
of runtime libraries, creation of RPC function handles, invocation of RPC, synchronization, etc.

Note that grpc_call_async() (Fig. 3, line 25) is a variable arguments number function (A.K.A. varargs). varargs is convenient especially in this kind of
API (taking the name of the function to be invoked as an RPC and its arguments as function parameters, depending on the function to be invoked), but the
compiler can't check the types of the arguments of the function to be invoked.
This might be a cause of errors that can be hard to find.

3   An Example of OpenGR Programming

Figure 4 shows an OpenGR version of the program shown in Fig. 3.

A define_func directive declares an RPC function stub and related parameter information. A gen_stub directive specifies information needed to generate
a server stub depending on underlying RPC system. A start_func_body and
end_func_body pair specifies a source code region which is generated as a server
stub module. A call_async directive invokes the specified function as an RPC. A
wait_async directive synchronizes the RPCs invoked asynchronously. All other
functions, such as initializing the GridRPC runtime, function handle creation,
etc. are hidden from programmers by the OpenGR compiler system.

In Fig. 4, two OpenMP directives are also included. The first one is in between the start_func_body and end_func_body OpenGR directives (line 15) to
generate a server stub using a thread mechanism on server machines.[1] The sec-

---

[1] If the server has SMP capability

```
 1 #include "grpc.h"
 2 #define NUM_HOSTS 2
 3 char * hosts[] = {"brain.a02.aist.go.jp", "brain.a02.aist.go.jp"};
 4 grpc_function_handle_t handles[NUM_HOSTS];
 5 int port = 4000;
 6
 7 int main(int argc, char *argv[]) {
 8   double pi;
 9   long times, count[NUM_HOSTS], sum;
10   int i;
11   times = atol(argv[2]) / NUM_HOSTS;
12
13   /* Initialize GRPC runtimes. */
14   if (grpc_initialize(argv[1]) != GRPC_OK){
15     exit(2);
16   }
17
18   /* Initialize handles. */
19   for (i = 0; i < NUM_HOSTS; i++) {
20     grpc_function_handle_init(&handles[i], hosts[i], port, "pi/pi_trial");
21   }
22
23   for (i = 0; i < NUM_HOSTS; i++) {
24     /* Parallel non-blocking remote function invocation. */
25     if (grpc_call_async(&handles[i], i, times, &count[i]) == GRPC_ERROR) {
26       grpc_perror("pi_trial");
27       exit(2);
28     }
29   }
30
31   /* Sync. */
32   if (grpc_wait_all() == GRPC_ERROR){
33     grpc_perror("wait_all");
34     exit(2);
35   }
36
37   /* Handler destruction. */
38   for (i = 0; i < NUM_HOSTS; i++) {
39     grpc_function_handle_destruct(&handles[i]);
40   }
41
42   /* Compute and display pi. */
43   for (i = 0, sum = 0; i < NUM_HOSTS; i++) {
44     sum += count[i];
45   }
46   pi = 4.0 * (sum / ((double) times * NUM_HOSTS));
47   printf("PI = %f\n", pi);
48
49   /* Finalize GRPC runtimes. */
50   grpc_finalize();
51
52   return 0;
53 }
```

Fig. 3. A sample program using the GridRPC API

```
 1 #pragma ogr define_func pi_trial(IN int seed, \
 2                                  IN long times, \
 3                                  OUT long *count)
 4 #pragma ogr gen_stub pi_trial(Module pi)
 5
 6 #pragma start_func_body pi_trial
 7 #include <stdio.h>
 8 #include <stdlib.h>
 9
10 static void pi_trial(int seed, long times, long *count) {
11   long l, counter = 0;
12   double x, y;
13
14   srandom(seed);
15 #pragma omp parallel for private(x, y) reduction(+:counter)
16   for (l = 0; i < times; l++) {
17     x = (double)random() / RAND_MAX;
18     y = (double)random() / RAND_MAX;
19     if ((x * x + y * y) < 1.0) {
20       counter++;
21     }
22   }
23
24   printf("counter = %d, times = %d\n", counter, times);
25   *count = counter;
26 }
27 #pragma end_func_body pi_trial
28
29 int main(int argc, char *argv[]) {
30   double pi;
31   long times, sum, numhosts, count;
32   int i;
33
34   times = atol(argv[1]);
35   numhosts = atol(argv[2]);
36
37 #pragma omp parallel for private(count) reduction(+:sum)
38   for (i = 0; i < numhosts; i++) {
39 #pragma ogr call_async pi_trial
40     {
41       pi_trial(i, times, &count);
42       sum += count;
43     }
44   }
45 #pragma ogr wait_async
46
47   pi = 4.0 * (sum / ((double)times * numhosts));
48   printf("PI = %f\n", pi);
49
50   return 0;
51 }
```

Fig. 4. A sample program using OpenGR directives

ond one is just the outer closure of the `call_async` OpenGR directive (line 37), used to invoke RPCs in each thread created by OpenMP.

Compared to the GridRPC API version, the OpenGR version is more intuitive and closer to "what the programmer meant to write in the first place". Actually, the OpenGR version has full compatibility with an ordinary C language program which could be written without any particular knowledge about the RPC system's API. Just as in OpenMP programming, this is the one of the pros of directive-based programming.

4   An Overview of OpenGR Directives

In this section, we describe the design policy for the directives and the compiler system, and the implementation choices we made. Then we describe OpenGR directives for the C/C++ language.

4.1   Design and Implementation

The following points comprise a design policy for the directives and the compiler system, and the implementation choices we made.

1. **A design policy**

   In order to keep the directive specifications compact, directives provided by OpenGR are focused on generic RPC mechanisms. Optional functionalities such as scheduling, fault tolerance, performance measurement, etc., are excluded from the directive set even if underlying RPC systems provide them.

2. **Implementation choices**

   - In order to support various grid-enabled RPC systems other than the GridRPC API, the core RPC invocation and synchronization mechanisms are implemented in an abstraction runtime layer as functions. The complier itself emits the calling sequences of these functions.
   - In order to provide a seamless programming environment between OpenMP and OpenGR, the compiler must understand both of directives correctly. For example, especially in a case OpenGR asynchronous RPC invocations included in an OpenMP parallel region with thread private variables used for reduction (Fig. 4, line 37 - 45), the reduction must be delayed to the point of RPC synchronization, and the variables must be "alive" at least to that point. Due to this, the compiler must change code generation methods depended on usage of both directives. To implement this kind of code generator, we chose the Omni OpenMP compiler system as a base since it is opensource and has structures for which it is relatively easy to modify the code generator.

4.2   Syntax and Directive Categories

An OpenGR directive is used with the following syntax.

#pragma ogr *directive* [*directiveArgs* ...]

OpenGR directives are categorized as follows:

1. Server stub declaration directive: define_func.
2. Server stub generation directives: gen_stub, start_func_body and end_func_body.
3. RPC invocation directives: call and call_async.
4. Synchronization directive: wait_async.

4.3   The define_func Directive

The define_func directive declares the server function stub.

```
#pragma ogr define_func foo(IN double a[N], IN double b[N], \
                            OUT double c[N], IN int N)
```

In this example, declaring function foo as a server function stub that takes double a[N], b[N] and int N as input parameters, and double c[N] as an output parameter. This format is very similar to a C/C++ function prototype declaration. The difference is that the OpenGR declaration has an additional type qualifier IN, OUT, or INOUT to specify the direction of RPC parameter passing. Information specified with this description is used to generate an IDL for the underlying GridRPC system (For Ninf-G IDL, it corresponds to the Define statement) and RPC parameter type checking for the OpenGR compiler.

4.4   The gen_stub Directive

The gen_stub directive generates a server stub module for a function declared by define_func.

```
#pragma ogr gen_stub foo(Module bar, ...)
```

Between "(" and ")", information needed to generate a server stub module depending on the underlying GridRPC system must be specified. This specification is not well-defined, but it is hard to decide syntax and specifiers since the implementation of the underlying GridRPC system might vary. Thus, for compatibility between OpenGR compiler systems, this directive should be used with C pre-processor macros for conditional compilation.

In the current implementation of our OpenGR compiler, Module and the Required sentences of Ninf-G can be specified. Additionally, the external specifier is available, which specifies that the function stub has already been generated from another source code file or by using the underlying GridRPC directly. This enables the use of existing server stub modules generated by the underlying GridRPC system and the external specifier should be mandatory for the gen_stub directive.

4.5   The start_func_body and end_func_body Directives

These two directives are used to specify where the source code should be generated for the server function stub.

```
#pragma ogr start_func_body foo
#include <stdio.h>

int bar(...)
{
    ...
}
void foo(double *a, double *b,
         double *c, int N)
{
    ...
    bar(...);
    ...
}
#pragma ogr end_func_body foo
```

Theoretically, referencing of external symbols within server functions stubs can be resolved in the compilation phase, like, for example, a method introducing some kind of source code level symbol resolver among all source code files comprising the application. Thus, these two directives seem to be unnecessary.

But if the symbols referenced in the functions are defined in the system's include files by C pre-processor macros (such as stdout in stdio.h), the method described above causes problems. As mentioned, since the grid environment consists of various architectures, we have to care about the differences of implementations and the definitions of system headers and symbols, otherwise, it is impossible to generate server stubs suitable for the grid environment with minimal modification of existing source code.

Therefore, we took an easy and practical way out, which is to specify a region of source code that might have C pre-processor statements, to generate a server stub.

4.6   The call and call_async Directives

These two directives are used to invoke a server stub function declared and defined by define_func and [start|end]_func_body directives. The call directive is used for synchronous RPC invocation, and call_async is used for asynchronous RPC invocation.

```
#pragma ogr call foo
{
    if (x == ...) foo(X, Y, Z)
    else foo(X1, Y1, Z)
}
```

The directives replace invocations of specified function name (foo in the example) within the next closure to RPC invocations. If the RPC invocations are done by using the call_async directive, these invocations could be synchronized by the wait_async directive (**Section 4.7**) wherever the programmer desired.

4.7   The wait_async Directive

This directive is used to synchronize RPC invocations issued by the call_async directives.

As a specification, at a place of this directive, all asynchronous RPC invocations that are not synchronized yet are synchronized. If this directive is not used for any asynchronous RPC invocations, the RPC invocations must be synchronized at the end of application execution or at the calling of the exit(3) function by the compiler system's mechanism.

5   A Preliminary Evaluation

Currently, our OpenGR compiler is working, except server stub generation and shipping mechanisms. As a GridRPC system, our implementation is using Ninf-G. The generated execution program can work with existing server stubs generated by Ninf-G. In order to evaluate the effectiveness of OpenGR directives, we applied the directives to an existing sequential parameter search application using the polyhedral homotopy continuation method. Figure 5 is the OpenGR-applied version of the application, and this is the whole modified file. In this case, the main computation function poly_main is generated as a server stub using Ninf-G directly, and it is shipped to the servers in the grid environment manually. The stubs are invoked asynchronously within an OpenMP parallel for closure. As we mentioned, this is a part of the "Hybrid parallelization." This program works using both the grid and SMP environments correctly with current compiler system. Only by inserting four exact OpenGR directives and an OpenMP directive into the sequential program, is it easily converted to a hybrid-parallelized program that works using both the grid and SMP environments. And the source keeps full-compatibility with the original one in the C language.

6   Related Studies

There are several studies enabling distributed parallel programming by using complier directives. Gross*et al.* introduced compiler directives to specify task-parallel regions for HPF, to provide a framework enabling both task parallelism and data parallelism [8, 9]. Foster*et al.* also introduced compiler directives for Fortran 77 using the same approach as Gross [10]. The both studies aim for efficient application execution by using both task parallelism and data parallelism complementally. The target application of the studies is based on pipelining of different kinds of tasks. On the other hand, our approach is mainly targeting applications such as parameter searching, suitable for master-worker parallelism. Sato*et al.* introduce a programming model using OpenMP directives and OmniRPC, a thread-safe RPC system, to execute task parallel programs in the grid environment [3]. Our approach is very similar to this, but in Sato's approach, the programmer have to write and generate the server stubs, and has to write

RPC invocation code suitable for the OmniRPC API. On the other hand, in our approach, programmers only have to use OpenGR directives.

```
#define bufSz 16384

#pragma ogr define_func \
    poly_main(mode_in char file1[size], \
        mode_in char file2[size], \
        mode_in int cell, \
        mode_in int start, \
        mode_in int end, \
        mode_in int size, \
        mode_out char sol[size2], \
        mode_out char stat[size2], \
        mode_in int size2)

#pragma ogr gen_stub poly_main(module poly_main, external)

extern void    poly_main(char *file1, char *file2,
                    int cell, int start, int end, int size,
                    char *sol, char *stat, int size2);
int main(int argc, char* argv[])
{
    int ii, i, startCell, endCell;
    char **SolBuffer, **StatBuffer;
    char fname1[128];
    char fname2[128];
    int nItr = -1;

    startCell = 1;
    endCell = 14;

    nItr = endCell - startCell + 1;
    SolBuffer = (char **)malloc(sizeof(char *) * nItr);
    StatBuffer = (char **)malloc(sizeof(char *) * nItr);
    for (i = 0; i < nItr; i++) {
        SolBuffer[i] = (char *)malloc(sizeof(char) * bufSz);
        StatBuffer[i] = (char *)malloc(sizeof(char) * bufSz);
    }
    strcpy(fname1, argv[1]);
    strcpy(fname2, argv[2]);

#pragma omp parallel for
    for (ii = startCell; ii <= endCell; ii++) {
#pragma ogr call_async poly_main
        {
            poly_main(fname1, fname2, ii, startCell, endCell, 128,
                    SolBuffer[ii - startCell],
                    StatBuffer[ii - startCell], bufSz);
        }
    }
#pragma ogr wait_async
    return 0;
}
```

Fig. 5. OpenGR-applied version of a parameter search application using the polyhedral homotopy continuation method

7   Concluding Remarks and Future Work

In this paper, we described the design and specification of OpenGR directives and the implementation of an OpenGR compiler system. We also reported the results of a preliminary effectiveness evaluation which applies the directives to an existing application. We believe that the OpenGR directives can provide an effective method to parallelize legacy sequential applications easily, especially for parameter searches.

We are planning to implement a server stub generator and shipping mechanism, improvements of the RPC invocation code generation engine, and additional specification design for the barrier set definition for the synchronization mechanism of asynchronous RPC invocations.

Acknowledgement

We thank to Dr. Katsuki Fujisawa (Tokyo Denki University) who provided the application program for this study.

References

1. Tanaka, Y., Nakada, H., Hirano, M., Sato, M., Sekiguchi, S.: Implementation and evaluation of GridRPC based on Globus. IPSJ HPC (in Japanese) **2001** (2001) 165–170
2. Casanova, H., Dongarra, J.: NetSolve: A network server for solving computational science problems. Supercomputer Applications and High Performance Computing **11** (1997) 212–223 http://icl.cs.utk.edu/netsolve.
3. Sato, M., Hirano, M., Tanaka, Y., Sekiguchi, S.: OmniRPC: A Grid RPC facility for Cluster and Global Computing in OpenMP. In: Workshop on OpenMP Application and Tools. (2001)
4. Object Management Group, Inc.: Welcome To The OMG's CORBA Website (1997-2003) http://www.omg.org.
5. Seymour, K., Nakada, H., Matsuoka, S., Dongarra, J., Lee, C., Casanova, H.: GridRPC: A Remote Procedure Call API for Grid Computing. In: Proc. of Grid 2002. (2002)
6. Nakada, H., et.al: Design and implementations of Ninf: towards a global computing infrastructure. *FGCS* **15** (1999) 649–658
7. Foster, I., Kesselman, C.: Globus: A metacomputing infrastructure toolkit. Supercomputer Applications **11** (1997) 115–128 htpp://www.globus.org.
8. Gross, T., O'Hallaron, D.R., Subhlok, J.: Task parallelism in a high performance fortran framework. IEEE Parallel and Distributed Technology **2** (1994) 16–26
9. Subhlok, J., Yang, B.: A new model for integrated nested task and data parallel programming. In: 6th SIGPLAN Symposium on Principles and Practice of Parallel Programming. (1997)
10. Foster, I., David R. Kohr, J., Krishnaiyer, R., Choudhary, A.: Double standards: Bringing Task Parallelism to HPF via the Message Passing Interface. In: Supercomputing 96. (1996)

Author Index

Lecture Notes in Computer Science

For information about Vols. 1–2780
please contact your bookseller or Springer-Verlag